Intelligent Control and Learning Systems

Volume 13

Series Editor

Dong Shen, School of Mathematics, Renmin University of China, Beijing, Beijing, China

The Springer book series Intelligent Control and Learning Systems addresses the emerging advances in intelligent control and learning systems from both mathematical theory and engineering application perspectives. It is a series of monographs and contributed volumes focusing on the in-depth exploration of learning theory in control such as iterative learning, machine learning, deep learning, and others sharing the learning concept, and their corresponding intelligent system frameworks in engineering applications. This series is featured by the comprehensive understanding and practical application of learning mechanisms. This book series involves applications in industrial engineering, control engineering, and material engineering, etc.

The Intelligent Control and Learning System book series promotes the exchange of emerging theory and technology of intelligent control and learning systems between academia and industry. It aims to provide a timely reflection of the advances in intelligent control and learning systems. This book series is distinguished by the combination of the system theory and emerging topics such as machine learning, artificial intelligence, and big data. As a collection, this book series provides valuable resources to a wide audience in academia, the engineering research community, industry and anyone else looking to expand their knowledge in intelligent control and learning systems.

Dong Shen

Variable Gain Design in Stochastic Iterative Learning Control

Springer

Dong Shen
School of Mathematics
Renmin University of China
Beijing, China

ISSN 2662-5458 ISSN 2662-5466 (electronic)
Intelligent Control and Learning Systems
ISBN 978-981-97-8280-2 ISBN 978-981-97-8281-9 (eBook)
https://doi.org/10.1007/978-981-97-8281-9

This Springer imprint is published by the registered company Springer Nature Singapore Pte Ltd.
The registered company address is: 152 Beach Road, #21-01/04 Gateway East, Singapore 189721, Singapore

Preface

Iterative learning control (ILC) is a prominent intelligent control method characterized by its simple control structure and exceptional tracking performance for repetitive systems and tasks. In this context, "repetitive" refers to the system's ability to repeatedly complete a given tracking task within a finite time interval, with each trial referred to as an iteration. Unlike traditional control problems, the duration of each trial or iteration is finite. The essence of the learning process lies in generating the input signal for the current iteration using input signals and tracking errors from previous iterations, aiming to continuously enhance control performance along the iteration axis rather than the time axis employed in traditional control methods. The objective of ILC is to achieve asymptotically perfect tracking performance over the entire operation interval as the number of iterations increases. This objective can be accomplished through ILC's utilization of feedback principles in the domain of iterations. In particular, the conventional P-type update rule in ILC exhibits similar effects to integral control known in classical feedback control theory. Recent research focuses on various aspects including problem scenarios such as varying trial lengths and fading communication environments, algorithm design that emphasizes novel structures for update rules based on specific requirements, and analysis methods beyond commonly used contraction mapping principle and composite energy function theory.

We consider nonlinearity and randomness as two primary challenges in addressing practical application problems, among other potential challenges. Nonlinearity implies limited available information on system dynamics, but this can be effectively addressed by ILC due to its inherent data-driven control feature. For instance, the basic formulation of update rules follows the PID mode, where the learning gain is typically adjusted through trial and error. It should be noted that alternative design methods such as norm optimal ILC algorithms may require certain prior knowledge. Therefore, further efforts are required in this direction.

Randomness is another fundamental characteristic in practical problems, as reflected by practical disturbances, measurement noise, communication noise, and unknown uncertainties, which are often represented by random variables and necessitate the application of probability theory for proper handling. While conventional

ILC update rules with a predefined learning gain may ensure stability in terms of keeping controlled signals such as input and output within a certain range when confronted with unpredictable randomness in system dynamics, they typically fail to achieve steady convergence to a limit due to the presence of random quantities. This poses a significant challenge in stochastic ILC. To tackle this issue, variable gain design emerges as a promising approach.

In this monograph, we explore the incorporation of variable gain design into the update rule to address the stochastic ILC problem. Specifically, we consider additive randomness in the system and replace the fixed learning gain with a variable one that is tailored based on different principles. Notably, this monograph elaborates on four designs of the variable gain: decreasing gain, adaptive gain, event-triggering gain, and optimal gain.

The decreasing gain indicates a variable gain sequence that gradually diminishes to zero as the iteration number increases. The conditions for the decreasing gain have been established to ensure asymptotic convergence in the iteration domain, serving as the fundamental cornerstone and support for subsequent results. The adaptive gain aims to divide the learning process into two stages: an initial stage employing a fixed gain and a later stage transitioning to a randomly diminishing gain sequence. Here, the switching iteration between these two stages is adaptively determined by the learning process itself; furthermore, the diminishing gain is closely tied to practical learning processes rather than being predefined. The event-triggering gain aims to construct a successive sequence of learning stages, with reduction in learning gains occurring only when transitioning from one stage to another. This perspective offers a completely new view of the learning process. The optimal gain seeks an optimal value of the learning gain based on specific performance indices related to the learning process. One approach involves approximating mean-squared input error to generate optimal gain values, while another employs Kalman filtering principles to generate a recursive sequence of gain matrices.

These results constitute the main content of this monograph. The author would like to take this opportunity to express his gratitude to all collaborators and students supervised, as the main results in this monograph originate from the joint articles with them. The authors would like to thank all ILC experts worldwide as they benefited a lot from the previous foundational works when investigating the problems in this book.

These findings constitute the primary content of this monograph. The author would like to take this opportunity to express sincere gratitude to all collaborators and supervised students, as the principal outcomes in this monograph stem from collaborative articles with them. The authors extend their appreciation to ILC experts worldwide for their invaluable contributions, which greatly influenced the investigation of the problems addressed in this book.

Beijing, China
June 2024

Dong Shen

Acknowledgements We are pleased to thank the support of the National Natural Science Foundation of China under grants 62173333, 61673045 and 61304085, and Beijing Natural Science Foundation under grants Z210002 and 4152040. We also thank the National Young Talents Support Program. We sincerely appreciate the publishers IEEE, Wiley, Taylor & Francis and Elsevier for granting us the permission to reuse materials in the copyrighted papers.

Contents

Chapter 1
Introduction

1.1 Iterative Learning Control

In our daily lives, the ability to engage in repetitive tasks fosters continuous improvement. For instance, in basketball set shooting, as the number of attempts increases, the shooter can enhance their shooting accuracy by adjusting variables such as angle and speed on a shot-by-shot basis to minimize deviations. This phenomenon stems from our capacity to learn from experiences and subsequently refine our behaviors.

This fundamental cognitive aspect has spurred research on iterative learning control (ILC), a control methodology that enhances its performance by leveraging insights from previous control iterations. Specifically tailored for systems capable of executing tasks within fixed time intervals and repeating them, ILC utilizes input-output information from past cycles along with the tracking objective to formulate the input signal for subsequent iterations, thereby continuously improving tracking performance as the number of cycles approaches infinity. Consequently, ILC exhibits the following characteristics: (1) timely task completion, (2) resettable system states to identical initial values, and (3) iteration-invariant tracking objectives.

In Fig. 1.1, y_d denotes the reference trajectory. Based on the input of the i-th iteration, u_k, as well as the tracking error $e_k = y_d - y_k$, the input u_{k+1} for the next iteration, i.e., the $(i+1)$-th iteration, is constructed. Meanwhile, the input u_{k+1} is also stored into the memory for the $(i+2)$-th iteration. Thus, a closed-loop feedback is formed along the iteration index. The major advantage of ILC lies in its minimal requirement for plant information, as the design of control law solely relies on tracking references and input/output signals. In other words, even if the knowledge about the plant is limited or completely unknown, this approach remains effective due to its simplicity.

By comparing ILC with our daily lives, we observe that the previous information regarding inputs and outputs of the plant aligns with the experiences encountered in our everyday existence. Individuals typically formulate a strategy for a given task based on prior experiences, while in this context, the strategy corresponds to the

D. Shen, *Variable Gain Design in Stochastic Iterative Learning Control*,
Intelligent Control and Learning Systems 13,
https://doi.org/10.1007/978-981-97-8281-9_1

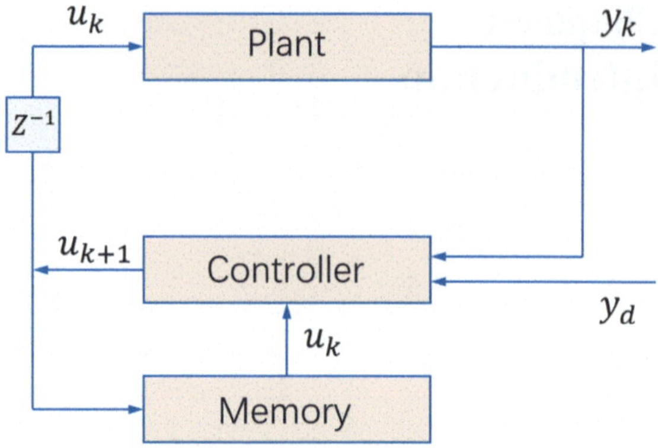

Fig. 1.1 Framework of ILC

input signal of ILC. It is important to note that past experiences aid in enhancing our behavior; thus, it is reasonable to posit that knowledge about previous operations can contribute to improving control performance to some extent.

It is worth noting that ILC adjusts the control based on the iteration index rather than the time index, which distinguishes it from other control methods such as proportional-integral-derivative (PID) control. PID control is a widely employed feedback control technique. However, for iterative systems, PID generates consistent tracking errors in each iteration since it does not utilize previous information, whereas ILC progressively reduces the tracking error with each iteration. Furthermore, ILC differs from adaptive control, which also learns from past operational data. Adaptive control aims to adjust the parameters of a given controller, while ILC focuses on directly constructing the input signal.

The concept of ILC may be traced back to a paper published in 1978 by Uchiyama (1978). However, this paper failed to attract widespread attention as it was written in Japanese. The paper Arimoto et al. (1984) published in 1984 is generally recognized as the initial proposal of ILC concept. Subsequently, large amounts of literature have been published on various related issues Moore (2012); Bien and Xu (2012); Xu and Tan (2003); Ahn et al. (2007); Owens (2016); Wang et al. (2014); Yang et al. (2017); Freeman et al. (2015); Shen (2018); Shen and Li (2019); Shen and Yu (2023). ILC has recently become an important branch of intelligent control, and its use is widespread in many practical applications.

1.2 Formulation of Iterative Learning Control

This section presents fundamental formulations of ILC, followed by traditional convergence results. We consider a discrete-time linear time-invariant system:

$$\begin{aligned} x_k(t+1) &= Ax_k(t) + Bu_k(t), \\ y_k(t) &= Cx_k(t), \end{aligned} \tag{1.1}$$

where $x \in \mathbb{R}^n$, $u \in \mathbb{R}^p$, and $y \in \mathbb{R}^q$ denote the system state, input, and output, respectively. Matrices A, B, and C are system matrices with appropriate dimensions. t denotes the time label, $t = 0, 1, \cdots, N$, where N is the interval length, and $k = 0, 1, 2, \cdots$ denotes the iteration label.

Given the requirement for repeated execution of a specific tracking task, it is necessary to reset the initial state at each iteration. The subsequent reset condition presented below has been widely employed in numerous academic publications.

$$x_k(0) = x_0, \quad \forall k. \tag{1.2}$$

The reference trajectory is denoted as $y_d(t)$, where $t \in [0, N]$. In terms of the resetting condition, it is typically necessary to satisfy $y_d(0) = y_0 \triangleq Cx_0$. The objective of ILC control is to design an appropriate update law for the input $u_k(t)$ in order to enable the corresponding output $y_k(t)$ to closely track $y_d(t)$. To achieve this, we define the tracking error for any given time instant t within the interval $[0, N]$ as follows:

$$e_k(t) = y_d(t) - y_k(t). \tag{1.3}$$

The update law is a function of $u_k(t)$ and $e_k(t)$ to generate $u_{k+1}(t)$ as follows:

$$u_{k+1}(t) = h(u_k(\cdot), \cdots, u_0(\cdot), e_k(\cdot), \cdots, e_0(\cdot)). \tag{1.4}$$

When the aforementioned relationship is solely dependent on the previous iteration, it is referred to as a first-order ILC update law; otherwise, it is termed as a high-order ILC update law. Generally, due to its algorithmic simplicity, most update laws are of first-order nature as follows:

$$u_{k+1}(t) = h(u_k(\cdot), e_k(\cdot)). \tag{1.5}$$

Moreover, the update law typically exhibits linearity. The most elementary form of the update law is as follows:

$$u_{k+1}(t) = u_k(t) + Ke_k(t+1), \tag{1.6}$$

where the learning gain matrix, denoted as K, serves as both a designed parameter and the basis for the update law. In (1.6), $u_k(t)$ represents the input at the current iteration, while $Ke_k(t+1)$ corresponds to the innovation term. This update law, referred to as a P-type ILC update law, can be transformed into a D-type one by replacing the innovation term with $K(e_k(t+1) - e_k(t))$.

The formulation of ILC demonstrates the incorporation of fundamental characteristics inherent to 2D systems, which has been extensively explored by numerous researchers who have adopted a 2D system-based approach as a key technique for ILC design and analysis.

It should be noted that the operation length is constrained by N and subsequently repeated multiple times. Therefore, one could employ the lifting technique, commonly known as supervector lifting, to elevate all input and output vectors:

$$\boldsymbol{u}_k = [u_k^T(0), u_k^T(1), \cdots, u_k^T(N-1)]^T, \tag{1.7}$$

$$\boldsymbol{y}_k = [y_k^T(1), y_k^T(2), \cdots, y_k^T(N)]^T. \tag{1.8}$$

Denote

$$\mathbf{G} = \begin{bmatrix} CB & 0 & 0 & \cdots & 0 \\ CAB & CB & 0 & \cdots & 0 \\ \vdots & \vdots & \vdots & \ddots & \vdots \\ CA^{N-1}B & CA^{N-2}B & \cdots & \cdots & CB \end{bmatrix}, \tag{1.9}$$

then one has

$$\boldsymbol{y}_k = \mathbf{G}\boldsymbol{u}_k + \boldsymbol{d}, \tag{1.10}$$

where

$$\boldsymbol{d} = [(Cx_0)^T, (CAx_0)^T, \cdots, (CA^N x_0)^T]^T. \tag{1.11}$$

Similar to (1.7), (1.8), we define

$$\boldsymbol{y}_d = [y_d^T(1), y_d^T(2), \cdots, y_d^T(N)]^T,$$

$$\boldsymbol{e}_k = [e_k^T(1), e_k^T(2), \cdots, e_k^T(N)]^T.$$

Then, it leads to

$$\boldsymbol{u}_{k+1} = \boldsymbol{u}_k + \mathbf{K}\boldsymbol{e}_k, \tag{1.12}$$

where $\mathbf{K} = \text{diag}\{K, K, \cdots, K\}$. We obtain

$$\begin{aligned} \boldsymbol{e}_{k+1} =& \boldsymbol{y}_d - \boldsymbol{y}_{k+1} \\ =& \boldsymbol{y}_d - \mathbf{G}\boldsymbol{u}_k - \mathbf{G}\mathbf{K}\boldsymbol{e}_k - \boldsymbol{d} \\ =& \boldsymbol{e}_k - \mathbf{G}\mathbf{K}\boldsymbol{e}_k \\ =& (\mathbf{I} - \mathbf{G}\mathbf{K})\boldsymbol{e}_k. \end{aligned}$$

Therefore, we derive a condition that is sufficient to ensure the convergence of ILC, namely, $\|\mathbf{I} - \mathbf{G}\mathbf{K}\| < 1$. The lifting technique not only aids in obtaining the convergence condition but also provides an intrinsic understanding of ILC. In the lifted model (1.10), the evolutionary process of each iteration is integrated into $\mathbf{G}$, emphasizing the relationship between adjacent iterations. Consequently, the lifted model (1.10) solely varies along the iteration axis, while having no influence on the time axis.

It is worth noting that the primary focus of ILC lies in enhancing tracking performance iteratively with respect to the iteration index, as evident from the design of the update law and lifted model (1.10). Consequently, extending the system from a linear time-invariant case to a linear time-varying case does not introduce any additional complexities. This is attributed to the fact that for any fixed time instance, updating along the iteration index remains invariant.

The assumption is typically made that the reference trajectory $y_d(t)$ can be realized, implying the existence of an appropriate initial state x_0 and input $u_d(t)$ such that expression (1.1) still holds with k replaced by d. In other words, we have $\boldsymbol{y}_d = \mathbf{G}\boldsymbol{u}_d + \boldsymbol{d}$, where $\boldsymbol{u}_d$ is defined in a similar manner as (1.7). Consequently, the discussion regarding convergence of the system output to the reference trajectory $\lim_{k\to\infty} \boldsymbol{y}_k = \boldsymbol{y}_d$ becomes one focused on convergence of the system input to the objective input $\lim_{k\to\infty} \boldsymbol{u}_k = \boldsymbol{u}_d$. This approach proves more convenient when dealing with systems subject to stochastic noises.

One may be interested in the scenario where the reference trajectory is unattainable, meaning that there exists no control input capable of producing the desired trajectory; thus, achieving perfect tracking becomes impossible. In this case, the objective of the ILC algorithm shifts from guaranteeing asymptotically accurate tracking to converging towards the closest achievable trajectory to the given reference. Consequently, the tracking problem transforms into an optimization problem. However, it should be noted that in practical applications, it is usually assumed that the reference trajectory can indeed be realized; hence, this assumption lacks strict rigor.

1.3 Stochastic Iterative Learning Control

The definition of stochastic iterative learning control (SILC) will be provided in this section. Throughout the monograph, SILC is employed to refer to ILC for systems incorporating stochastic signals, which are characterized by random variables in probability theory. These signals encompass various sources of uncertainty such as system noises during operation, measurement errors, and random packet losses. The salient attributes of a stochastic signal include its inherent uncertainty, randomness, and absence of an upper bound. In the literature, the current research mainly focuses on the following scenarios:

- linear systems with system disturbances and/or measurement noises;
- nonlinear systems with system disturbances and/or measurement noises;
- systems with other kind of stochastic signals, such as random packet losses and random asynchronism.

Currently, control theorists and engineers have made significant contributions to providing comprehensive explanations of ILC for deterministic systems. Numerous noteworthy results have been published and successfully applied to real-world systems. However, practical operations inevitably encounter interferences and noises, making disturbance rejection a crucial research focus in ILC. Most studies assume that disturbances are bounded by a specific limit, which leads to the convergence of algorithms within a small range determined by this limit. Specifically, we denote the disturbance as $\boldsymbol{w}_k$ and assume $\|\boldsymbol{w}_k\| \leq \epsilon$, where ϵ is an appropriate constant. It has been generally demonstrated that ILC algorithms guarantee tracking error $\boldsymbol{e}_k$ satisfying $\|\boldsymbol{e}_k\| \leq l(\epsilon)$, with $l(\epsilon) \xrightarrow[\epsilon\to 0]{} 0$. However, it is evident that this approach fails in the case of SILC.

Therefore, greater emphasis should be placed on addressing stochastic variables when considering SILC. In pursuit of this objective, several innovative methodologies have been proposed and has been comprehensively reviewed Shen and Wang (2014). However, it is important to note that these efforts represent only the initial stride towards understanding SILC, necessitating further research in this evolving field.

1.4 Variable Gain Design

To demonstrate the necessity of variable gain design, we first revisit the lifted linear systems with additional noise:

$$\boldsymbol{y}_k = \mathbf{G}\boldsymbol{u}_k + \boldsymbol{d} + \boldsymbol{w}_k, \tag{1.13}$$

where $\boldsymbol{w}_k$ denotes the random noise satisfying $\mathbb{E}\boldsymbol{w}_k = 0$ and $W \triangleq \sup_k \mathbb{E}\|\boldsymbol{w}_k\|^2 < \infty$. That is, the random noise is of zero mean and bounded variance. Clearly, the common Gaussian white noise satisfies this condition.

Accordingly, the desired tracking reference $\boldsymbol{y}_d$ satisfies the following equation:

$$\boldsymbol{y}_d = \mathbf{G}\boldsymbol{u}_d + \boldsymbol{d}. \tag{1.14}$$

Then, we obtain the tracking error as follows

$$\boldsymbol{e}_k = \boldsymbol{y}_d - \boldsymbol{y}_k = \mathbf{G}(\boldsymbol{u}_d - \boldsymbol{u}_k) - \boldsymbol{w}_k. \tag{1.15}$$

We revisit the traditional P-type update rule:

$$\boldsymbol{u}_{k+1} = \boldsymbol{u}_k + \rho\mathbf{K}\boldsymbol{e}_k, \tag{1.16}$$

where the additional ρ is a constant learning gain and $\mathbf{K}$ serves as the direction regulation matrix such as all eigenvalues of $\mathbf{KG}$ are positive constants.

Subtracting both sides of (1.16) from $\boldsymbol{u}_d$ and substituting (1.15) yields

$$\Delta\boldsymbol{u}_{k+1} = \Delta\boldsymbol{u}_k - \rho\mathbf{KG}\Delta\boldsymbol{u}_k + \rho\mathbf{K}\boldsymbol{w}_k, \tag{1.17}$$

where $\Delta\boldsymbol{u}_k \triangleq \boldsymbol{u}_d - \boldsymbol{u}_k$ denotes the input error.

We argue that the constant learning gain ρ cannot guarantee zero-error convergence of the input error. Specifically, it is impossible to ensure $\Delta\boldsymbol{u}_k \to 0$ as $k \to \infty$ due to the presence of random noise $\boldsymbol{w}_k$. To illustrate this point, let us assume that $\Delta\boldsymbol{u}_k \to 0$. Taking limits on both sides of (6.5) implies $\lim_{k\to\infty} \boldsymbol{w}_k = 0$, which contradicts the inherent nature of random noise. Therefore, in order to achieve asymptotic convergence of $\Delta \boldsymbol{u}_k$ towards zero, a novel design for the learning gain ρ needs to be considered.

In particular, we consider the derivation of $\mathbb{E}\|\Delta \boldsymbol{u}_k\|^2$ as follows:

$$\begin{aligned}\mathbb{E}\|\Delta \boldsymbol{u}_{k+1}\|^2 &= \mathbb{E}\|(\mathbf{I}-\rho\mathbf{KG})\Delta \boldsymbol{u}_k\|^2 + \mathbb{E}\|\rho\mathbf{K}\boldsymbol{w}_k\|^2 \\ &\le \lambda_{\max}((\mathbf{I}-\rho\mathbf{KG})^T(\mathbf{I}-\rho\mathbf{KG}))\mathbb{E}\|\Delta \boldsymbol{u}_k\|^2 + \mathbb{E}\|\rho\mathbf{K}\boldsymbol{w}_k\|^2 \\ &\le \lambda_{\max}((\mathbf{I}-\rho\mathbf{KG})^T(\mathbf{I}-\rho\mathbf{KG}))\mathbb{E}\|\Delta \boldsymbol{u}_k\|^2 + \rho^2\lambda_{\max}(\mathbf{K}^T\mathbf{K})W, \end{aligned} \quad (1.18)$$

which further leads to

$$\limsup_{k\to\infty}\mathbb{E}\|\Delta \boldsymbol{u}_k\|^2 \le \frac{\rho^2\lambda_{\max}(\mathbf{K}^T\mathbf{K})W}{1-\lambda_{\max}((\mathbf{I}-\rho\mathbf{KG})^T(\mathbf{I}-\rho\mathbf{KG}))}. \quad (1.19)$$

The upper bound of the convergence limit is observed to be proportional to ρ^2, indicating that the final tracking performance of the conventional P-type update law can be controlled by adjusting the learning gain. Specifically, a smaller learning gain ρ leads to improved tracking performance. However, it should be noted that a zero-error tracking performance cannot be achieved with a learning gain of zero, as evident from the denominator in (1.19). Therefore, an effective approach is to gradually decrease the learning gain ρ during iterations, which motivates further investigation into variable gain design.

In summary, the variable gain design entails two primary considerations:

- Gradual reduction of the variable learning gain with increasing iterations is necessary to mitigate the impact of random noise on tracking performance.
- It is crucial to avoid excessively rapid decrease in the variable learning gain in order to ensure adequate learning and steady convergence of input error.

These observations can be considered as the guiding principles, which will be further expounded upon in this monograph.

1.5 Outline of This Book

In Chap. 2, we present preliminary results of probability theory and iterative sequences to facilitate the readership and subsequent analysis. Subsequently, we elaborate on four variable gain designs: decreasing gain, adaptive gain, event-triggering gain, and optimal gain. These designs apply different principles to enhance the learning process. Specifically, the decreasing gain is a comprehensive design that ensures neither too fast nor too slow decrease in learning gains based on a series of conditions. These conditions serve as fundamental requirements for the following design. The adaptive gain divides the learning process into two stages; during the first stage, the gain remains unchanged to ensure rapid convergence while in the latter stage it randomly decreases to asymptotically eliminate noise effects. The event-triggering gain further divides the learning process into consecutive stages where only when switching between stages does the gain decrease while remaining constant within each stage. Lastly, with optimal gain design, we aim to optimize learning gains based on specific performance indices.

In Chap. 3, we investigate the general decreasing gain sequence. We represent the learning gain sequence as $\{\rho_k\}$. The conditions to guarantee asymptotic convergence of input error in a mean square sense are that $\rho_k > 0$, $\rho_k \to 0$, and the sum of all terms in the series from $k = 1$ to infinity diverges, namely, $\sum_{k=1}^{\infty} \rho_k = \infty$. Additionally, to ensure almost sure convergence, it is necessary that the sum of squares of all terms in this series remains finite, namely, $\sum_{k=1}^{\infty} \rho_k^2 < \infty$. This design has been validated through four specific learning control problems: point-to-point tracking; learning control for saturated sensor systems affected by observation noise; learning consensus for multi-agent systems influenced by communication noise; and finally, learning control for stochastic multi-sensor systems.

In Chap. 4, the adaptive gain design is investigated by dividing the learning process into two stages. During the initial stage comprising several iterations, the tracking error may be dominated by input deviation from the desired input, which can significantly exceed random noise. Therefore, to ensure rapid learning, the learning gain remains unchanged. In contrast, during the subsequent stage where random noise may dominate the tracking error, a randomized decreasing mode is adopted for the learning gain to mitigate its impact. This design has been applied to three specific learning control problems: point-to-point tracking problem, learning control under Markov nonuniform trial lengths, and learning temperature control for large-scale buildings.

In Chap. 5, we introduce a novel multi-stage adaptive gain strategy for application in stochastic systems, with the objective of achieving a balance between a constant gain, which can expedite convergence speed, and a vanishing sequence, which can enhance tracking performance. The design principles and procedures are meticulously elucidated. In this proposed strategy, the iteration domain is partitioned into multiple stages consisting of successive iterations. Within each stage, the learning gain remains constant but decreases during transitions between stages. Furthermore, three progressive schemes for generating the gain sequence are presented based on an assessment of the descending speed of mean-squared input error and its estimated upper bound.

In Chap. 6, two optimal gain designs are proposed. The first design aims to optimize the scalar gain ρ_k based on a performance evaluation index, while also providing a careful estimation of the convergence speed for the optimal gain sequence. This approach is then applied to address learning control problems in probabilistic quantization communication environments and large-scale systems. The second design considers the scalar gain ρ_k and direction regular matrix $\mathbf{K}$ as a unified matrix denoted by $\mathbf{K}_k$. Saab proposes an optimal calculation method for this matrix Saab (2001a). The optimal gain matrix design has been successfully employed to solve learning control problem under fading communication environments.

Chapter 2
Preliminary Results

In this chapter, we present fundamental concepts and findings of probability theory and iterative sequences. Readers with a solid foundation in probability theory may choose to bypass this chapter.

2.1 Probability Theory

2.1.1 Probability Space

The fundamental space is denoted as Ω. An elementary event or sample, represented by the point $\omega \in \Omega$, is considered. The set of points within Ω is denoted as A, where A is a subset of Ω.

The family $\mathcal{F}$ consists of sets in Ω that satisfy the following conditions:

- $\Omega \in \mathcal{F}$;
- If $A \in \mathcal{F}$, then A^c, the complement of A, also belongs to $\mathcal{F}$, $A^c \in \mathcal{F}$;
- If $A_i \in \mathcal{F}$, $i = 1, 2, \ldots$, then $\bigcup_{i=1}^{\infty} A_i \in \mathcal{F}$.

Then, $\mathcal{F}$ is called the σ-algebra. The element A of $\mathcal{F}$ is called the measurable set, or random event.

It is evident that $\bigcap_{i=1}^{\infty} A_i \in \mathcal{F}$, if $A_i \in \mathcal{F}$, $i = 1, 2, \ldots$.

A set function ϕ defined on $\mathcal{F}$ is termed as σ-additive if it satisfies the condition that $\phi(\bigcup_{i=1}^{\infty} A_i) = \sum_{i=1}^{\infty} \phi(A_i)$ for any sequence of disjoint events $A_i \in \mathcal{F}$. It should be noted that, by definition, the values $+\infty$ and $-\infty$ are excluded from being taken by ϕ. A nonnegative set function satisfying the property of being σ-additive is referred to as a measure.

D. Shen, *Variable Gain Design in Stochastic Iterative Learning Control*,
Intelligent Control and Learning Systems 13,
https://doi.org/10.1007/978-981-97-8281-9_2

Let $\mathbb{P}$ be a set function defined on $\mathcal{F}$ with the following properties:

- $\mathbb{P}(A) \geq 0, \forall A \in \mathcal{F}$;
- $\mathbb{P}(\Omega) = 1$;
- $\mathbb{P}(\bigcup_{i=1}^{\infty} A_i) = \sum_{i=1}^{\infty} \mathbb{P}(A_i)$, if A_i, $i = 1, 2, \ldots$, are disjoint.

Then, $\mathbb{P}$ is referred to as a probability measure on the σ-algebra $\mathcal{F}$. The triple $(\Omega, \mathcal{F}, \mathbb{P})$ constitutes a probability space. The value $\mathbb{P}(A)$ denotes the probability associated with the random event A.

The relationship between random variables is said to hold almost surely or with probability one if it holds for all ω, except possibly on a set of measure zero.

2.1.2 Random Variable and Distribution Function

The Borel σ-algebra, denoted by $\mathcal{B}$, is the smallest σ-algebra in $\mathbf{R}$ that contains all intervals. By "smallest," we mean that if there exists a σ-algebra $\mathcal{B}_1$ containing all intervals, then it must hold that $\mathcal{B} \subset \mathcal{B}_1$, indicating that $B \in \mathcal{B}_1$ for any $B \in \mathcal{B}$. Similarly, the Borel σ-algebra in $\mathbf{R}^r$, denoted by $\mathcal{B}^r$, can be defined. Any set belonging to either $\mathcal{B}$ or $\mathcal{B}^r$ in higher dimensions is referred to as a Borel set.

A real function $\xi = \xi(\omega)$ defined on $(\Omega, \mathcal{F}, \mathbb{P})$ is called measurable if $\{\omega : \xi(\omega) < x\} \in \mathcal{F}$ for all $x \in \mathbf{R}$. If $\xi(\cdot)$ is a real measurable function defined on $(\Omega, \mathcal{F}, \mathbb{P})$ and $\mathbb{P}(\omega : |\xi(\omega)| < \infty) = 1$, then ξ is referred to as a random variable. If $f(\cdot)$ is a measurable function, then $f(\xi)$ is also a random variable provided that $|f(\xi)| < \infty$ almost surely.

Let ξ be a random variable, and its distribution is defined as $F_\xi(x) = \mathbb{P}(\omega : \xi(\omega) < x)$ for all $x \in \mathbf{R}$. A random vector $\boldsymbol{\xi} = [\xi_1, \ldots, \xi_n]^T$ consists of multiple components ξ_i, where each component ξ_i, $i = 1, 2, \ldots$, is a random variable. The distribution function of the random vector $\boldsymbol{\xi}$ is defined as $F_{\boldsymbol{\xi}}(x_1, \ldots, x_n) = \mathbb{P}(\omega : \xi_1 < x_1, \ldots, \xi_n < x_n)$. If the distribution function $F_\xi(x)$ is differentiable, then its derivative $f_\xi(x) = \frac{\mathrm{d}F_\xi(x)}{\mathrm{d}x}$ is referred to as the density function of ξ.

2.1.3 Expectation

Let ξ be a random variable and $\xi \geq 0$ almost surely. Let $A_{n_i} = \{\omega : i2^{-n} < \xi \leq (i+1)2^{-n}\}$ and define

$$\mathbb{E}\xi = \int_\Omega \xi \mathrm{d}\mathbb{P} = \lim_{n\to\infty} \left[\sum_{i=0}^{n2^n-1} i2^{-n}\mathbb{P}(A_{n_i} + n\mathbb{P}(\xi > n)) \right].$$

$\mathbb{E}\xi$ is called the expectation of ξ.

For any random variable ξ, we define its positive part as $\xi^+ = \max \xi, 0$ and its negative part as $\xi^- = \max -\xi, 0$. The expectation of ξ can then be expressed as the difference between the expectations of these two parts: $\mathbb{E}\xi = \mathbb{E}\xi^+ - \mathbb{E}\xi^-$. This definition holds when at least one of the expectations, either $\mathbb{E}\xi^+$ or $\mathbb{E}\xi^-$, is finite. Furthermore, if the absolute value of the random variable ξ satisfies $\mathbb{E}|\xi| = \mathbb{E}\xi^+ + \mathbb{E}\xi^- < \infty$, it is referred to as an integrable random variable. Finally, for a measurable function $f(\cdot)$, we can define the expectation of $f(\cdot)$ applied to a random variable by $\mathbb{E}f(\xi) = \int_{-\infty}^{\infty} f(x)\mathrm{d}F_\xi(x)$.

2.1.4 Convergence Theorems and Inequalities

Let $\{\xi_k\}$ be a sequence of random variables and let ξ be another random variable. If $\mathbb{P}(\xi_k \to \xi) = 1$, then we say that ξ_k converges to ξ almost surely. If for any given value $\epsilon > 0$, there exists $\mathbb{P}(|\xi_k - \xi| > \epsilon) \to 0$ as $k \to \infty$, then we say that ξ_k converges to ξ in probability. If the distribution function $F_{\xi_k}(x)$ of the random variable ξ_k in the sequence converges pointwise to the distribution function $F_\xi(x)$ at every point where F_ξ is continuous, then we say that ξ_k converges to ξ in distribution. Finally, if $\mathbb{E}|\xi_k - \xi|^2 \to 0$, then we say that ξ_k converges to ξ in the mean square sense.

Theorem 2.1 (Monotone Convergence Theorem) *If random variables ξ_k nondecreasingly (nonincreasingly) converges to ξ a.s., $\xi_k \geq \eta$ ($\xi_k \leq \eta$) a.s. and $\mathbb{E}\eta^- < \infty$ ($\mathbb{E}\eta^+ < \infty$), then $\mathbb{E}\xi_k \uparrow \mathbb{E}\xi$ ($\mathbb{E}\xi_k \downarrow \mathbb{E}\xi$).*

Theorem 2.2 (Dominated Convergence Theorem) *If $\xi_k \to \xi$ in distribution and there exists an integrable random variable η such that $|\xi_k| \leq \eta$, then $\mathbb{E}|\xi| < \infty$, $\mathbb{E}\xi_k \to \mathbb{E}\xi$, and $\mathbb{E}|\xi_k - \xi| \to 0$ as $k \to \infty$.*

Theorem 2.3 (Fatou Lemma) *If $\xi_k \geq \eta$ ($\xi_k \leq \eta$), $k = 1, 2, \ldots$, for some random variable η with $\mathbb{E}\eta^- < \infty$ ($\mathbb{E}\eta^+ < \infty$), then*

$$\mathbb{E} \liminf_{k\to\infty} \xi_k \leq \liminf_{k\to\infty} \mathbb{E}\xi_k, \quad (\limsup_{k\to\infty} \mathbb{E}\xi_k \leq \mathbb{E} \limsup_{k\to\infty} \xi_k).$$

Chebyshev Inequality:

$$\mathbb{P}(|\xi| > \epsilon) \leq \frac{\mathbb{E}|\xi|}{\epsilon}, \quad \text{for any } \epsilon > 0.$$

Lyapunov Inequality:

$$(\mathbb{E}|\xi|^s)^{1/s} \leq (\mathbb{E}|\xi|^t)^{1/t}, \quad \forall 0 < s \leq t.$$

Hölder Inequality: Let $1 < p < \infty$, $1 < q < \infty$, and $\frac{1}{p} + \frac{1}{q} = 1$. Then,

$$\mathbb{E}|\xi\eta| \leq (\mathbb{E}|\xi|^p)^{1/p} (\mathbb{E}|\xi|^q)^{1/q}.$$

2.1.5 Independence

Let $A_i, i = 1, 2, \ldots$, be a sequence of events. If for any set of indices $\{i_1, i_2, \ldots, i_k\}$, $\mathbb{P}(\bigcap_{j=1}^k A_{i_j}) = \prod_{j=1}^k \mathbb{P}(A_{i_j})$, then $\{A_i\}$ is called mutually independent.

Let $\mathcal{F}_i, i = 1, 2, \ldots$, be a sequence of σ-algebras. If events $A_1, \ldots, A_k$ are mutually independent whenever $A_i \in \mathcal{F}_i$, $i = 1, 2, \ldots, k$, $\forall k \geq 1$, then the family $\{\mathcal{F}_i\}$ of σ-algebras is called mutually independent.

Let $\{\xi_i\}$ be a sequence of random variables and let $\mathcal{F}^{\xi_i}$ be the σ-algebra generated by ξ. If $\{\mathcal{F}^{\xi_i}\}$ is mutually independent, then the sequence $\{\xi_i\}$ of random variables is called mutually independent.

2.1.6 Conditional Expectation

Let $(\Omega, \mathcal{F}, \mathbb{P})$ be a probability space. $\mathcal{F}_1$ is called a sub-σ-algebra of $\mathcal{F}$ if $\mathcal{F}_1$ is a σ-algebra and $\mathcal{F}_1 \subset \mathcal{F}$, by which it is meant that any $A \in \mathcal{F}_1$ implies $A \in \mathcal{F}$. Let $\mathcal{F}^\eta$ be the smallest σ-algebra containing all sets $\{\omega : \eta(\omega) < x\}$, $\forall x \in \mathbf{R}$. $\mathcal{F}^\eta$ is called the σ-algebra generated by η.

The conditional expectation $\mathbb{E}(\xi|\eta)$ of ξ given η is defined as $\mathbb{E}(\xi|\eta) = \mathbb{E}(\xi|\mathcal{F}^\eta)$ almost surely. Let A be an event. Conditional probability $\mathbb{P}(A|\mathcal{F}_1)$ of A given $\mathcal{F}_1$ is defined by $\mathbb{P}(A|\mathcal{F}_1) = \mathbb{E}(I_A|\mathcal{F}_1)$ almost surely, where I_A is an indicator function of the event A. The followings are properties of the conditional expectation:

- $\mathbb{E}(a\xi + b\eta|\mathcal{F}_1) = a\mathbb{E}(\xi|\mathcal{F}_1) + b\mathbb{E}(\eta|\mathcal{F}_1)$ for constants a and b;
- $\mathbb{E}(\mathbb{E}(\xi|\mathcal{F}_1)) = \mathbb{E}\xi$;
- $\mathbb{E}(\eta\xi|\mathcal{F}_1) = \eta\mathbb{E}(\xi|\mathcal{F}_1)$, if η is $\mathcal{F}_1$-measurable and $\mathbb{E}(|\xi||\mathcal{F}_1) < \infty$ a.s.;
- $\mathbb{E}(\mathbb{E}(\xi|\mathcal{F}_1)|\mathcal{F}_2) = \mathbb{E}(\xi|\mathcal{F}_2)$, if $\mathcal{F}_2 \subset \mathcal{F}_1 \subset \mathcal{F}$;
- $\mathbb{E}(\xi|\mathcal{F}_1) = \mathbb{E}\xi$, if ξ is independent of $\mathcal{F}_1$.

2.2 Convergence Lemmas of Iterative Sequence

Lemma 2.1 *((Polyak, 1987), p. 45, Lemma 3) Let* $\{a_k\}$, $\{b_k\}$, *and* $\{c_k\}$ *be real number sequences satisfying* $0 < b_k \leq 1$, $c_k \geq 0$, $b_k \to 0$, $\sum_{k=1}^\infty b_k = \infty$, $c_k/b_k \to 0$ *and* $a_{k+1} \leq (1 - b_k)a_k + c_k$. *If* $a_k \geq 0$, *then* $\lim_{k\to\infty} a_k = 0$.

Lemma 2.2 *((Polyak, 1987), p. 50, Robbins-Siegmund Lemma) Let r_k, s_k, κ_k, and τ_k be nonnegative random variables. Assume*

$$\mathbb{E}[r_{k+1}|\mathcal{G}_k] \leq (1+\kappa_k)r_k - s_k + \tau_k, \tag{2.1}$$

$$\sum_{k=1}^{\infty} \kappa_k < \infty, \sum_{k=1}^{\infty} \tau_k < \infty, \tag{2.2}$$

where $\mathcal{G}_k$ is a σ-algebra generated by r_i, s_i, κ_i, and τ_i, $1 \leq i \leq k$. Then, r_k converges almost surely and $\sum_{k=1}^{\infty} s_k < \infty$.

Chapter 3
Decreasing Gain Design

3.1 Decreasing Gain Sequence in SILC

As elucidated in Sect. 1.4, the inclusion of a variable gain sequence in stochastic ILC is imperative for ensuring consistent convergence of the input sequence amidst random noise. To substantiate this assertion, we revisit the lifted linear systems with additional noise

$$\boldsymbol{y}_k = \mathbf{G}\boldsymbol{u}_k + \boldsymbol{d} + \boldsymbol{w}_k, \tag{3.1}$$

where $\boldsymbol{w}_k$ denotes the random noise satisfying $\mathbb{E}\boldsymbol{w}_k = 0$ and $W \triangleq \sup_k \mathbb{E}\|\boldsymbol{w}_k\|^2 < \infty$.

Accordingly, the desired tracking reference $\boldsymbol{y}_d$ satisfies the following equation:

$$\boldsymbol{y}_d = \mathbf{G}\boldsymbol{u}_d + \boldsymbol{d}. \tag{3.2}$$

Then, we obtain the tracking error as follows:

$$\boldsymbol{e}_k = \boldsymbol{y}_d - \boldsymbol{y}_k = \mathbf{G}(\boldsymbol{u}_d - \boldsymbol{u}_k) - \boldsymbol{w}_k. \tag{3.3}$$

The traditional P-type update rule is revised by replacing the constant learning gain with a decreasing gain as follows:

$$\boldsymbol{u}_{k+1} = \boldsymbol{u}_k + \rho_k \mathbf{K}\boldsymbol{e}_k, \tag{3.4}$$

where ρ_k is a variable learning gain and $\mathbf{K}$ serves as the direction regulation matrix such as all eigenvalues of $\mathbf{KG}$ are positive constants.

In general, the decreasing gain sequence $\{\rho_k\}$ is usually required to satisfy the following condition:

D. Shen, *Variable Gain Design in Stochastic Iterative Learning Control*,
Intelligent Control and Learning Systems 13,
https://doi.org/10.1007/978-981-97-8281-9_3

$$\rho_k > 0, \quad \rho_k \to 0, \quad \sum_{k=1}^{\infty} \rho_k = \infty. \tag{3.5}$$

This condition ensures the asymptotic convergence of the input sequence $\boldsymbol{u}_k$ generated by (3.4) in terms of mean-square sense. Moreover, to achieve the asymptotic convergence in the almost sure sense, an additional condition is typically required as follows:

$$\sum_{k=1}^{\infty} \rho_k^2 < \infty. \tag{3.6}$$

Stochastic approximation theory is a fundamental theory that underpins the aforementioned findings (Chen, 2003; Borkar, 2009). In fact, many control problems can be reformulated as a root-finding problem of unknown functions affected by random noise, and stochastic approximation theory provides effective solutions for such problems. Herein, we present a classic convergence theorem of stochastic approximation.

Theorem 3.1 *(Chen, 2003) Let $\{H_k\}$ and H be matrices with dimension of $l \times l$. Assume H is stable and $H_k \to H$ as $k \to \infty$. Let a_k satisfy the conditions*

$$a_k > 0, a_k \to 0, \sum_{k=1}^{\infty} a_k = \infty, \sum_{k=1}^{\infty} a_k^2 < \infty, \tag{3.7}$$

and let l-dimensional vectors $\{\mu_k\}$, $\{\nu_k\}$ satisfy the following conditions:

$$\sum_{k=1}^{\infty} a_k \mu_k < \infty, \quad \nu_k \xrightarrow[k\to\infty]{} 0. \tag{3.8}$$

Then, $\{\alpha_k\}$ generated by the following recursion with an arbitrary initial value α_0 converges to zero with probability one:

$$\alpha_{k+1} = \alpha_k + a_k H_k \alpha_k + a_k(\mu_k + \nu_k). \tag{3.9}$$

Here, by stability of a matrix we mean that all its eigenvalues are with negative real parts.

3.2 Decreasing Gain Design in Point-to-Point Tracking Problem

In this section, we employ the decreasing gain design in ILC for stochastic point-to-point tracking systems. We present the general formulation of the point-to-point control problem and introduce an associated performance index for stochastic linear systems. Rigorous proofs are provided to establish the almost sure convergence and asymptotic properties of the proposed recursive algorithm.

3.2.1 Point-to-Point Tracking Problem

Typically, conventional ILC methods require the system output to accurately track a desired objective throughout the entire time interval. However, in practical applications, it may be sufficient to only focus on accurately tracking specific desired points while disregarding others. For instance, consider a basketball player repeatedly shooting from a fixed position; their primary concern is whether the basketball hits the target rather than following a predetermined trajectory. This scenario gives rise to terminal iterative learning control (TILC), where only the final point is considered and can be viewed as a special case of point-to-point ILC. In various practical applications such as 'pick and place' robotic tasks, satellite positioning, and production line automation, multiple key points may need to be tracked for effective point-to-point tracking.

The general point-to-point ILC can be categorized into two types based on the required input and output positions. One type involves dividing the entire operation range into two adjacent intervals, namely the actuation interval and observation interval. The other type involves interspersing input positions with output positions. Residual vibration suppression is a typical case of the former type of point-to-point tracking problem (Park et al., 2006; Van de Wijdeven and Bosgra, 2008; Ding and Wu, 2007). In Park et al. (2006), an iterative learning input shaping technique was developed for the first three axes of a six-degrees-of-freedom industrial robot, where ILC was used to update control parameters. A similar problem was also addressed in Van de Wijdeven and Bosgra (2008), where an additional signal was added to compensate for residual vibrations in the command input. In Ding and Wu (2007), ILC was employed to minimize new performance indices related to position error instead of tracking error for a high-acceleration positioning table. It should be noted that these publications utilized standard ILC techniques and may not have fully exploited the potential benefits offered by unrequested pass points for improved performance. In contrast, Freeman et al. (2011) proposed a novel approach to solve multiple point-to-point tracking problems by iteratively updating reference values between trials rather than modifying input profiles, thereby effectively utilizing trajectory freedom. Son et al. (2013) also proposed an ILC algorithm based on reference trajectory updates, along with convergence analysis. The distinction between Freeman et al. (2011); Son et al. (2013) is that the former employs discrete Fourier transform techniques in the frequency domain, while the latter utilizes interpolation techniques in the time domain. Another promising approach for addressing point-to-point problems involves directly updating the control signal based on specified tracking data Son and Ahn (2011); Son et al. (2013), as well as a performance index. The note presented in Son and Ahn (2011) introduces ILC algorithms for multiple pass points tracking problems, where the performance index represents a quadratic form solely considering pass points and inputs rather than the entire trajectory. In contrast, Son et al. (2013) linearly parameterizes the control input using basis functions constructed from system information matrices. The performance index employed therein encompasses error costs, control efforts, and variations in control amounts. Additionally, reference Owens et al. (2013) presents a norm-optimal ILC solution to continuous-time point-to-point problems and provides a comparison between experimental performance and theoretical results.

The literature emphasizes that, in the case of an MIMO system, the pass point required is the entire output vector for any given time. However, in practical applications, it may be sufficient to satisfy constraints only for certain components of the output vector. For instance, if we consider a three-dimensional vector representing spatial location as the output, we can impose a constraint on altitude while leaving the other two dimensions unconstrained at a given time. This specific point-to-point tracking problem has been addressed separately for linear systems in Freeman and Tan (2013) and nonlinear systems in Freeman and Tan (2011). The detailed formulation of such point-to-point control was initially presented in Freeman and Tan (2013), which also extensively analyzed gradient descent-based ILC and Newton method-based ILC with various mixed constraints. For concise results and experimental verification, readers can refer to Freeman (2012).

However, the above publications do not consider any stochastic noise, which further motivates the work presented here. The objective of this section is to address ILC for a stochastic point-to-point tracking system with a general form of tracking reference as described in Freeman and Tan (2011, 2013). In our formulation, we take into account both system noise and measurement noise, which are characterized by random variables. Therefore, it is necessary to introduce probability-based construction and convergence analysis techniques for ILC. Specifically, we first formulate the stochastic point-to-point tracking problem using a variant expression based on Freeman and Tan (2013). The output vectors are arranged from left to right according to time instances, resulting in a matrix dimension equaling the product of the output vector dimension and the time interval length. The difference between our formulations in this study and those in Freeman and Tan (2013) lies in Freeman's selection direction of required output points: row by row for selecting time instances with fixed one dimension of the output; whereas our selection direction is column by column for selecting dimensions with fixed time instance. This formulation presented in this section provides clearer understanding and construction.

3.2.2 Problem Formulation

Consider the following LTI system:

$$\begin{aligned} x_k(t+1) &= Ax_k(t) + Bu_k(t) + w_k(t+1), \\ y_k(t) &= Cx_k(t) + v_k(t), \end{aligned} \tag{3.10}$$

where subscripts k denotes different cycles, $k = 1, 2, \cdots$ and t denotes an arbitrary time in an operation cycle, $0 \le t \le N$, where N is the length of the operation cycle. $x_k(t) \in \mathbf{R}^n$, $u_k(t) \in \mathbf{R}^p$, and $y_k(t) \in \mathbf{R}^q$ are system state vector, input vector and output vector, respectively. System matrices A, B and C are with appropriate dimensions. Additionally, noise signals include system noise $w_k(t)$ and measurement noise $v_k(t)$, both assumed to be zero-mean Gaussian white noise with normal distribution in this section. Besides, for different cycles and different time instances, the noise signals are uncorrelated. For any t, both $\{w_k(t), k = 1, 2, \cdots\}$ and $\{v_k(t), k = 1, 2, \cdots\}$ are independent and identically distributed (i.i.d.) sequences.

Here, it is assumed that the system relative degree is one, i.e. $CB \neq 0$. Then, the input and output sequences can be rewritten into super-vector form given by

$$\begin{aligned} \boldsymbol{u}_k &= [u_k^T(0), u_k^T(1), \cdots, u_k^T(N-1)]^T \in \mathbf{R}^{pN}, \\ \boldsymbol{y}_k &= [y_k^T(1), y_k^T(2), \cdots, y_k^T(N)]^T \in \mathbf{R}^{qN}. \end{aligned}$$

In addition, let

$$\begin{aligned} \mathcal{G} &= \begin{bmatrix} CB & 0 & \cdots & 0 \\ CAB & CB & \cdots & 0 \\ \vdots & \vdots & \ddots & \vdots \\ CA^{N-1}B & CA^{N-2}B & \cdots & CB \end{bmatrix} \in \mathbf{R}^{qN \times pN}, \\ \mathcal{M} &= [(CA)^T \ (CA^2)^T \ \cdots \ (CA^N)^T]^T \in \mathbf{R}^{qN \times n}. \end{aligned}$$

Then, one has the following relationship between input and output:

$$\boldsymbol{y}_k = \mathcal{G}\boldsymbol{u}_k + \mathcal{M}x_k(0) + \epsilon_k, \tag{3.11}$$

where the stochastic noise term ϵ_k is expressed by

$$\epsilon_k = \begin{bmatrix} v_k(1) + Cw_k(1) \\ v_k(2) + Cw_k(2) + CAw_k(1) \\ \vdots \\ v_k(N) + Cw_k(N) + \cdots + CA^{N-1}w_k(1) \end{bmatrix}.$$

Thus ϵ_k is considered as zero-mean Gaussian process noise with covariance Q, i.e. $\epsilon_k \sim N(0, Q)$.

In this section, we consider two scenarios for the initialization condition: the identical and asymptotic conditions. Without loss of generality, we assume $x_k(0) = 0$ for the former scenario and $x_k(0) \to 0$ for the latter.

For the standard ILC framework, the tracking objective is

$$\boldsymbol{y}_d = [y_d^T(1), y_d^T(2), \cdots, y_d^T(N)]^T \in \mathbf{R}^{qN}. \tag{3.12}$$

Denote standard tracking error as $\boldsymbol{e}_k = \boldsymbol{y}_d - \boldsymbol{y}_k$.

In many practical applications, it is unnecessary to track the entire trajectory $\boldsymbol{y}_d$, but rather only a subset of it. However, most point-to-point control methods consider tracking an l-size subset of the time interval $\{1, 2, \cdots, N\}, l \leq N$. Specifically, the tracking points are $\{y_d(n_j)\}$ with $j = 1, 2, \cdots, l$, where $1 \leq n_1 < n_2 < \cdots < n_l \leq N$. In this section, we present a more general formulation that allows for partial output tracking at any given time instant l instead of requiring full output tracking.

Here we propose a modified modeling approach, which is equivalent to the one presented in Freeman and Tan (2013). Let us consider that only l_j components of the output at time j need to be tracked, where $0 \leq l_j \leq q$, and $j = 1, 2, \cdots, N$. If $l_j = 0$, it indicates that the output at time instant j is completely disregarded. On the other hand, if $l_j \neq 0$, we denote the tracking components as $1 \leq n_{j,1} < n_{j,2} < \cdots < n_{j,l_j} \leq q$. By eliminating all unnecessary points from the original objective $\boldsymbol{y}_d$, we obtain a new reference trajectory $\boldsymbol{y}_r$ with dimensionality of l, where $l = \sum_{j=1}^{N} l_j$. In other words, $\boldsymbol{y}_r$ is a condensed reference trajectory of $\boldsymbol{y}_d$ satisfying the following relationship:

$$\boldsymbol{y}_r = \Phi \boldsymbol{y}_d, \tag{3.13}$$

where $\Phi \in \mathbf{R}^{l \times qN}$ is defined in (3.15).

Remark 3.1 The distinction between the model presented here and that proposed in Freeman and Tan (2013) lies in the selection order of tracking points. In Freeman and Tan (2013), reference points are chosen based on the dimensionality of the output vector, meaning they first determine a specific dimension of the output vector and then identify which time instances need to be tracked. Conversely, our formulation selects reference points based on time, whereby we first fix a particular time and then ascertain which dimensions of the output require tracking. For our objectives and subsequent analysis, this formulation offers greater intuitiveness and convenience. However, it is important to note that these two modes ultimately yield equivalent relationships.

The row vector $\theta \in \mathbf{R}^{qN}$ is initially defined, which has the same dimension as $\boldsymbol{y}_d$. If the i-th component of $\boldsymbol{y}_d$ needs to be tracked, then the i-th component of θ is set to one, i.e., $\theta_i = 1$. Otherwise, it is set to zero. This can be expressed as follows:

$$\theta_i = \begin{cases} 1, & \text{if the } i\text{-th component is tracked, } 1 \leq i \leq qN \\ 0, & \text{otherwise} \end{cases} \tag{3.14}$$

Then Φ could be constructed as follows:

$$\Phi_{i,j} = \begin{cases} 1, & \text{if } \theta_j = 1 \text{ and } \sum_{k=1}^{j} \theta_k = i \\ 0, & \text{otherwise} \end{cases} \tag{3.15}$$

Specifically, there are l components of θ that equal one while the rest are zero. The l rows of matrix Φ correspond to these non-zero elements. For the i-th non-zero element of θ, locate its position in the entire vector and denote it as j. Then set $\Phi_{i,j} = 1$ and all other elements to zero. This constitutes the main concept behind the aforementioned definition. The construction procedure for matrix Φ is presented in Algorithm 3.1.

Algorithm 3.1

1. Let all elements of Φ equals zero, $\Phi \in \mathbf{R}^{l\times qN}$;
2. Let the first row of Φ be replaced by θ;
3. Set $k = 1$;
4. For the k-th row of Φ, find the first non-zero element, then move all the subsequent part of the row down to the next row and replace the left elements of the k-th row with zero;
5. Let $k := k + 1$, back to step 4) until $k = l$.

According to the definition and construction process of Φ, it is evident that Φ has full row rank, denoted as $rank(\Phi) = l$. For further analysis, we observe the following property regarding the matrix $\Phi\mathcal{G}$.

Lemma 3.1 *Assume that CB is of full row rank, then $\Phi\mathcal{G}$ is of full row rank. That is, $rank(\Phi\mathcal{G}) = l$.*

Proof. Since CB is of full row rank, we have $rank(CB) = q$. From the definition of $\mathcal{G}$, it is evident $rank(\mathcal{G}) = qN$. By Sylvester's rank inequality, $rank(\Phi) + rank(\mathcal{G}) - qN \leq rank(\Phi\mathcal{G})$. This further yields that $l \leq rank(\Phi\mathcal{G})$. On the other hand, $rank(\Phi\mathcal{G}) \leq \min\{rank(\Phi), rank\mathcal{G}\} = l$. Thus, we have $rank(\Phi\mathcal{G}) = l$.

Remark 3.2 The matrix Φ can be expressed in a simplified form when the point-to-point control is settled to the whole output, i.e., when all elements of $y_d(j)$ need to be tracked for arbitrary fixed time j. Denote the tracking time instances as $1 \leq m_1 < m_2 < \cdots < m_l \leq N$. The block-wise definition of matrix Φ is given by

$$\Phi_{i,j} = \begin{cases} I_q, & \text{if } j = m_i, i = 1, 2, \cdots, l \\ 0, & \text{otherwise} \end{cases}$$

Readers may see the following example for a quick understanding.

Example 3.1 Let $N = 5$ and dimension of output $q = 3$. We first show standard point-to-point control formulation. Suppose $y_d(1)$, $y_d(3)$, and $y_d(4)$ are required pass points, then Φ is constructed as

$$\Phi = \begin{bmatrix} I_q & 0 & 0 & 0 & 0 \\ 0 & 0 & I_q & 0 & 0 \\ 0 & 0 & 0 & I_q & 0 \end{bmatrix} \in \mathbf{R}^{9\times 15}.$$

While for the general formulation in this note, suppose $y_d^{(1)}(1)$, $y_d^{(2)}(3)$, $y_d^{(3)}(3)$, $y_d^{(1)}(4)$ and $y_d^{(3)}(4)$ are reference points, then Φ is as follows:

$$\Phi = \begin{bmatrix} 1&0&0&0&0&0&0&0&0&0&0&0&0&0&0 \\ 0&0&0&0&0&0&0&1&0&0&0&0&0&0&0 \\ 0&0&0&0&0&0&0&0&1&0&0&0&0&0&0 \\ 0&0&0&0&0&0&0&0&0&1&0&0&0&0&0 \\ 0&0&0&0&0&0&0&0&0&0&0&1&0&0&0 \end{bmatrix}.$$

The objective of standard ILC is to determine an input sequence such that $\|\boldsymbol{e}_k\| \to 0$ as $k \to \infty$. However, this control objective is not suitable for stochastic point-to-point tracking systems due to two reasons. Firstly, our focus lies on the required reference points, which means that not all components of $\boldsymbol{e}_k$ may be available in practical applications. Secondly, the presence of stochastic noises prevents the tracking errors from converging to zero. Therefore, a new performance index is necessary.

Note that $\boldsymbol{e}_k = \boldsymbol{y}_d - \boldsymbol{y}_k$ and $\boldsymbol{y}_r = \Phi \boldsymbol{y}_d$. Thus, we obtain that $\Phi \boldsymbol{e}_k = \Phi(\boldsymbol{y}_d - \boldsymbol{y}_k) = \boldsymbol{y}_r - \Phi \boldsymbol{y}_k$ is the practical tracking information for update.

There are stochastic noises involved in $\boldsymbol{y}_k$, thus we could not make $\|\Phi \boldsymbol{e}_k\| \to 0$. To evaluate the tracking performance, we consider the following averaged tracking error index:

$$\min \lim_{n\to\infty} \frac{1}{n} \sum_{k=1}^{n} \|\Phi \boldsymbol{e}_k\|^2. \tag{3.16}$$

In order to make more insights, we first define nondecreasing σ-algebras as $\{\mathcal{F}_k\}$, where $\mathcal{F}_k \triangleq \sigma(\mathbf{x}_i(0), w_i(t), v_i(t), 0 \le i \le k, 0 \le t \le N)$. Since the input signal for the $(k+1)$-th is designed based on the data from initial cycle up to the k-th cycle, we know $\boldsymbol{u}_{k+1} \in \mathcal{F}_k$. Remind that the noises are uncorrelated and normally distributed, we have $\mathbb{E}\{\epsilon_k|\mathcal{F}_{k-1}\} = 0$ and $\mathbb{E}\{\|\epsilon_k\|^2|\mathcal{F}_{k-1}\} < \infty$.

Lemma 3.2 *(Chen and Guo 1991, Theorem 2.8) Let $\{X(t), \mathcal{F}_t\}$ be a martingale difference sequence and $\{M(t), \mathcal{F}_t\}$ an adapted sequence, $\|M(t)\| < \infty$, $\forall t \ge 0$. If $\sup_{t\ge0} \mathbb{E}[\|X(t)\|^2|\mathcal{F}_{t-1}] < \infty$, a.s., then as $n \to \infty$*

$$\sum_{t=0}^{n} M(t)X(t+1) = O\left(\left(\sum_{t=0}^{n} \|M(t)\|^2\right)^{\frac{1}{2}+\eta}\right) \quad a.s.,\ \forall \eta > 0.$$

By Lemma 3.2 we have

$$\begin{aligned}\frac{1}{n}\sum_{k=1}^{n}\|\Phi \boldsymbol{e}_k\|^2 &= \frac{1}{n}\sum_{k=1}^{n}\|\boldsymbol{y}_r - \Phi\mathcal{G}\boldsymbol{u}_k - \Phi\epsilon_k\|^2 \\ &= \frac{1}{n}\sum_{k=1}^{n}\|\boldsymbol{y}_r - \Phi\mathcal{G}\boldsymbol{u}_k\|^2(1+o(1)) + \frac{1}{n}\sum_{k=1}^{n}\|\Phi\epsilon_k\|^2,\end{aligned}$$

which further implies

$$\sum_{k=1}^{n}\|\boldsymbol{y}_r - \Phi\mathcal{G}\boldsymbol{u}_k\|^2 = o(n) \ \Leftrightarrow \ \lim_{n\to\infty}\frac{1}{n}\sum_{k=1}^{n}\|\Phi\boldsymbol{e}_k\|^2 = Q',$$

where Q' is the covariance matrix of $\Phi\epsilon_k$. That is, the index (3.16) is minimized if the left side of the last relationship is satisfied.

Now define $\Phi\widehat{\boldsymbol{e}}_k \triangleq \boldsymbol{y}_r - \Phi G\boldsymbol{u}_k$. In order to ensure the above performance index, it suffices to prove that

$$\Phi\widehat{\boldsymbol{e}}_k \to 0 \quad \text{as } k\to\infty. \tag{3.17}$$

This is the control objective of this section.

3.2.3 ILC Update Law and Its Convergence

Let $\{a_k\}$ be a sequence satisfying the following conditions,

$$\begin{aligned} &a_k > 0,\ a_k \xrightarrow[k\to\infty]{} 0,\ \sum_{k=1}^{\infty} a_k = \infty, \\ &\frac{a_k - a_{k+1}}{a_k a_{k+1}} \xrightarrow[k\to\infty]{} \alpha \ge 0. \end{aligned} \tag{3.18}$$

The selection of $a_k = \frac{\alpha}{k}$ with $\alpha > 0$ satisfied all the above requirements.

The ILC update law is now defined as follows:

$$\boldsymbol{u}_{k+1} = \boldsymbol{u}_k + a_k\mathcal{L}(\boldsymbol{y}_r - \Phi\boldsymbol{y}_k), \tag{3.19}$$

where $\mathcal{L}$ is the direction regulation matrix.

Now, we present the main convergence results. We first consider the identical initialization condition, i.e., $x_k(0) = 0, \forall k$.

Theorem 3.2 *For system (3.11) with initial value $x_k(0) = 0$, we design $\mathcal{L}$ such that $-\Phi\mathcal{G}\mathcal{L}$ is stable, then the ILC update law (3.19) with arbitrary initial input $\boldsymbol{u}_0$ would* *guarantee that $\Phi\widehat{\boldsymbol{e}}_k \to 0$ as $k \to \infty$. In other words, the reference $\mathbf{y}_r$ is best tracked according to performance index (3.16).*

Proof. It is obvious that the ILC update law (3.19) leads to

$$\Phi\widehat{\boldsymbol{e}}_{k+1} = (I_l - a_k \Phi \mathcal{G}\mathcal{L})\Phi\widehat{\boldsymbol{e}}_k + a_k \Phi \mathcal{G}\mathcal{L}\Phi \epsilon_k. \tag{3.20}$$

Denote $J = \Phi\mathcal{G}\mathcal{L}$. We define

$$\Psi_{k,j} \triangleq \prod_{i=j}^{k}(I - a_i J), \forall k \geq j, \ \Psi_{j,j+1} \triangleq I. \tag{3.21}$$

We first give an estimation of the growth of Ψ. In particular, we show that there exist constants $c_0 > 0$ and $c > 0$ such that

$$\|\Psi_{k,j}\| \leq c_0 \exp[-c \sum_{i=j}^{k} a_i], \ \forall k \geq j, \forall j \geq 0. \tag{3.22}$$

Consider $P = 2\int_0^\infty e^{-J^T t} e^{-Jt} \mathrm{d}t$. Since $-J$ is stable, the positive definite matrix P is well defined. Integrating by parts, it yields

$$\begin{aligned}
P &= -2\int_0^\infty e^{-J^T t} \mathrm{d} e^{-Jt} J^{-1} \\
&= -2e^{-J^T t} e^{-Jt} J^{-1} |_0^\infty - 2J^T \int_0^\infty e^{-J^T t} e^{-Jt} \mathrm{d}t J^{-1} \\
&= 2J^{-1} - J^T P J^{-1},
\end{aligned}$$

which implies $J^T P + P J = 2I$. Consequently,

$$\begin{aligned}
\Psi_{k,j}^T P \Psi_{k,j} &= \Psi_{k-1,j}^T (I - a_k J)^T P (I - a_k J) P \Psi_{k-1,j} \\
&= \Psi_{k-1,j}^T (P + a_k^2 J^T P J - a_k J^T P - a_k P J)\Psi_{k-1,j} \\
&= \Psi_{k-1,j}^T (P + a_k^2 J^T P J - 2a_k I)\Psi_{k-1,j} \\
&= \Psi_{k-1,j}^T P^{\frac{1}{2}} (I - 2a_k P^{-1} + a_k^2 P^{-\frac{1}{2}} J^T P J P^{-\frac{1}{2}}) P^{\frac{1}{2}} \Psi_{k-1,j}.
\end{aligned}$$

Thus, there exists a sufficiently large constant k_0 such that $\forall k \geq k_0$,

$$\|I - 2a_k P^{-1} + a_k^2 P^{-\frac{1}{2}} J^T P J P^{-\frac{1}{2}}\| \leq 1 - 2ca_k < e^{-2ca_k} \tag{3.23}$$

for some constant $c > 0$, where the first inequality is because $a_k \to 0$ as $k \to \infty$ and $P^{-1} > 0$. This further leads to $\left\|\Psi_{k,j}^T P \Psi_{k,j}\right\| \le \exp(-2c\sum_{i=j}^{k} a_i)$ for $j \ge k_0$, and thus,

$$\|\Psi_{k,j}\| \le \lambda_{\min}^{-\frac{1}{2}}(P)\exp(-c\sum_{i=j}^{k} a_i), \tag{3.24}$$

where $\lambda_{\min}(P)$ denotes the minimum eigenvalue of P.

If $j < k_0$, notice that

$$\|\Psi_{k_0-1,j}\| \le \prod_{i=j}^{k_0-1}(1 + a_i\|J\|) \le \prod_{i=1}^{k_0-1}(1 + a_i\|J\|).$$

By combing the above inequality and (3.24), we obtain

$$\begin{aligned}\|\Psi_{k,j}\| \le& \|\Psi_{k,k_0}\| \cdot \|\Psi_{k_0-1,j}\| \\ \le& \lambda_{\min}^{-\frac{1}{2}}(P)\left(\exp(-c\sum_{i=k_0}^{k} a_i)\right)\prod_{i=1}^{k_0-1}(1 + a_i\|J\|) \\ \le& \lambda_{\min}^{-\frac{1}{2}}(P)\left(\exp(c\sum_{i=j}^{k_0-1} a_i)\right)\prod_{i=1}^{k_0-1}(1 + a_i\|J\|) \\ &\times \left(\exp(-c\sum_{i=j}^{k} a_i)\right),\end{aligned}$$

which means the estimation (3.22) holds.

From (3.20), it follows that

$$\Phi\widehat{e}_{k+1} = \Psi_{k,0}\Phi\widehat{e}_0 + \sum_{j=0}^{k}\Psi_{k,j+1}J\Phi a_j\epsilon_{j+1}. \tag{3.25}$$

To show the zero convergence of $\Phi\widehat{e}_{k+1}$, it is sufficient to show that the right-hand side of (3.25) tends to zero as $k \to \infty$.

For any fixed j, $\|\Psi_{k,j}\| \to 0$ as $k \to \infty$ because of (3.18) and (3.22). Thus, $\Psi_{k,0}\Phi\widehat{e}_0 \to 0$ as $k \to \infty$ for any initial value.

We estimate the second term $\sum_{j=0}^{k}\Psi_{k,j+1}J\Phi a_j\epsilon_{j+1}$. Define $S_k = J\Phi\sum_{j=0}^{k} a_j \epsilon_{j+1}$, $S_{-1} = 0$. Using convergence theorem of martingale difference sequence (Chow and Teicher, 2003), it leads to $S_n \to S < \infty$. Hence, for any $\eta > 0$, there exists k_1 such that $\|S_j - S\| \le \eta$, $\forall j \ge k_1$. In addition, k_1 is set to be large enough such that $0 < ca_j < 1$. By a partial summation, we have

$$\sum_{j=0}^{k} \Psi_{k,j+1} J \Phi a_j \epsilon_{j+1} = \sum_{j=0}^{k} \Psi_{k,j+1}(S_j - S_{j-1}) = S_k - \sum_{j=0}^{k} (\Psi_{k,j+1} - \Psi_{k,j}) S_{j-1}$$

$$= S_k - \sum_{j=0}^{k} (\Psi_{k,j+1} - \Psi_{k,j}) S - \sum_{j=0}^{k} (\Psi_{k,j+1} - \Psi_{k,j})(S_{j-1} - S)$$

$$= S_k - S + \Psi_{k,0} S - \sum_{j=0}^{k_1} (\Psi_{k,j+1} - \Psi_{k,j})(S_{j-1} - S) + \sum_{j=k_1+1}^{k} \Psi_{k,j+1} a_j J (S_{j-1} - S),$$

where the sum of other terms tends to zero as $k \to \infty$ except the last term.

Since $\|S_j - S\| \le \eta, \forall j \ge k_1$, from (3.22), we have

$$\left\| \sum_{j=k_1+1}^{k} \Psi_{k,j+1} a_j J (S_{j-1} - S) \right\|$$

$$\le \eta \|J\| \sum_{k_1+1}^{k} c_0 \left(\exp(-c \sum_{i=j+1}^{k} a_i) \right) a_j,$$

which tends to zero as $k \to \infty$ and $\eta \to 0$ because of the following fact:

$$\sum_{k_1+1}^{k} c_0 \left(\exp(-c \sum_{i=j+1}^{k} a_i) \right) a_j$$

$$\le 2c_0 \sum_{k_1+1}^{k} \left(\exp(-c \sum_{i=j+1}^{k} a_i) \right) \left(a_j - \frac{c a_j^2}{2} \right)$$

$$\le \frac{2c_0}{c} \sum_{k_1+1}^{k} \left(\exp(-c \sum_{i=j+1}^{k} a_i) \right) (1 - e^{-c a_j})$$

$$= \frac{2c_0}{c} \sum_{k_1+1}^{k} \left[\exp(-c \sum_{i=j+1}^{k} a_i) - \exp(-c \sum_{i=j}^{k} a_i) \right]$$

$$\le \frac{2c_0}{c}.$$

As mentioned above, $0 < c a_j < 1$ holds for $j > k_1$. Therefore, the first and second inequalities above hold. The proof is thus completed.

Next, we consider the asymptotic initialization condition, i.e., $x_k(0) \to 0$ as $k \to \infty$. Under this circumstance, the convergence properties are still valid.

Theorem 3.3 *For system (3.11) with initial value $x_k(0) \to 0$, we design $\mathcal{L}$ such that $-\Phi \mathcal{G} \mathcal{L}$ is stable, then the ILC update law (3.19) with arbitrary initial input $\boldsymbol{u}_0$ would* *guarantee that $\Phi\widehat{\boldsymbol{e}}_k \to 0$ as $k \to \infty$. In other words, the reference $\boldsymbol{y}_r$ is best tracked according to the performance index (3.16).*

Proof. As the initial value $x_k(0)$ is no longer equal to zero, the modification is adding one extra term to (3.20), which becomes

$$\Phi\widehat{\boldsymbol{e}}_{k+1} = (I_l - a_k\Phi\mathcal{G}\mathcal{L})\Phi\widehat{\boldsymbol{e}}_k + a_k\Phi\mathcal{G}\mathcal{L}\Phi\epsilon_k + a_k\Phi\mathcal{G}\mathcal{L}\Phi\mathcal{M}x_k(0). \tag{3.26}$$

Consequently, (3.25) is modified as

$$\Phi\widehat{\boldsymbol{e}}_{k+1} = \Psi_{k,0}\Phi\widehat{\boldsymbol{e}}_0 + \sum_{j=0}^{k}\Psi_{k,j+1}J\Phi a_j\epsilon_{j+1} + \sum_{j=0}^{k}\Psi_{k,j+1}J\Phi a_j\mathcal{M}x_{j+1}(0). \tag{3.27}$$

Thus, to accomplish this proof, it suffices to show that the last term on the right hand-side side of the last equation tends to zero. Because $x_k(0) \to 0$ as $k \to \infty$, for any small $\eta > 0$, there is k_2 such that $\|J\Phi\mathcal{M}x_k(0)\| < \eta, \forall k \geq k_2$. From (3.22), we obtain

$$\begin{aligned}
&\left\|\sum_{j=0}^{k}\Psi_{k,j+1}J\Phi a_j\mathcal{M}x_{j+1}(0)\right\| \\
&\leq c_0\sum_{j=0}^{k_2-1}\left[\exp\left(-c\sum_{i=j+1}^{k}a_i\right)\right]a_j\|J\Phi\mathcal{M}x_{j+1}(0)\| \\
&+ \eta c_0\sum_{j=k_2}^{k}\left(\exp\left(-c\sum_{i=j+1}^{k}a_i\right)\right)a_j.
\end{aligned}$$

The first term at the right hand-side of the last inequality tends to zero by the definition of $\{a_k\}$, while the second term has been proved in Theorem 3.2. The proof is completed.

The design condition on the direction regulation matrix $\mathcal{L}$ in the aforementioned theorems is to ensure stability of $-\Phi\mathcal{G}\mathcal{L}$. Specifically, it is necessary to guarantee that all eigenvalues of $\Phi\mathcal{G}\mathcal{L}$ have positive real parts. Notably, there exists a wide range of admissible choices for $\mathcal{L}$ due to the full row rank property of $\Phi\mathcal{G}$.

The first and apparent choice is $\mathcal{L} = (\Phi\mathcal{G})^T$, which ensures that $\Phi\mathcal{G}\mathcal{L}$ is a positive definite matrix. This selection corresponds to the gradient algorithm, as previously employed in Freeman and Tan (2013); Butcher et al. (2008). However, there are two concerns associated with this approach. Firstly, it necessitates complete knowledge of the system matrices derived from the formulation of $\mathcal{G}$. Secondly, one might be concerned about the computational burden when dealing with large iteration lengths N.

To address the aforementioned drawbacks, an alternative choice for $\mathcal{L}$ is proposed as follows. It should be noted that $\mathcal{G}$ represents a block lower triangular matrix with

CB serving as its diagonal blocks. Consequently, these diagonal blocks assume a crucial role in control. Therefore, we can construct another block diagonal matrix denoted as $\mathcal{G}_1 = \text{diag}\{CB, \cdots, CB\}$. Subsequently, an alternative selection for $\mathcal{L}$ can be made by taking $\mathcal{L} = (\Phi\mathcal{G}_1)^T$.

In this case, $\mathcal{G}\mathcal{G}_1$ remains a block lower triangular matrix with the diagonal block being $(CB)(CB)^T$. Since CB has full row rank, the diagonal block $(CB)(CB)^T$ is positive definite. Consequently, all eigenvalues of $\mathcal{G}G_1$, despite its lack of symmetry, possess positive real parts. Similarly, the eigenvalues of $\Phi\mathcal{G}\mathcal{L}$ also exhibit positive real parts. Notably, only information regarding the input-output coupling matrix CB is required for these observations. Furthermore, the coupling matrix CB represents crucial control direction information essential for controller design. Additionally, selecting $\mathcal{L} = (\Phi\mathcal{G}_1)^T$ effectively avoids non-causality in controller design.

Furthermore, the computational burden can be further reduced by decomposing the lifted update law (3.19) into non-lifted forms. In other words, we employ the lifted update law (3.19) for analytical convenience. To illustrate this point, let us consider a specific example based on a SISO LTI system with an overall output dimension of N. Denoting the tracking error as $\hat{e}_{1,k}, \cdots, \hat{e}_{N,k}$, we assume that there are $\kappa \leq N$ outputs that need to be tracked at locations $j_1, \cdots, j_\kappa$ (with $j_0 = 0$). Consequently, the update law along the time axis for the case of $\mathcal{L} = (\Phi\mathcal{G})^T$ is given as follows:

$$u_{k+1}(t) = \begin{cases} u_k(t) + \sum_{m=i}^{\kappa} CA^{j_m - t - 1} B\hat{e}_{j_m,k} & j_{i-1} \leq t < j_i, 1 \leq i \leq \kappa \\ u_k(t) & j_\kappa \leq t \leq N-1 \end{cases} \tag{3.28}$$

while for $\mathcal{L} = (\Phi\mathcal{G}_1)^T$ case the update law is given as

$$u_{k+1}(t) = \begin{cases} u_k(t) + CB\widehat{e}_{j_i,k}, & t = j_i, \ -1, \ 1 \leq i \leq \kappa \\ u_k(t), & otherwise \end{cases} \tag{3.29}$$

3.2.4 Asymptotic Property of Input Limits

This subsection presents a comprehensive analysis of the asymptotic properties of ILC algorithms with a decreasing gain sequence. It has been rigorously proven that as the iterations approach infinity, $\Phi\widehat{e}_k$ tends to zero, providing a path-wise result. Considering the presence of stochastic noise in our system, we demonstrate that $\frac{1}{\sqrt{a_k}}\Phi\widehat{e}_k$ converges to a normal distribution as $k \to \infty$, establishing an important statistical finding.

We first introduce a central limit theorem on double-indexed random variables (Hall and Heyde 1980, Chapter 3) and formulate it as a lemma.

Lemma 3.3 *(Hall and Heyde 1980, Chapter 3) Let ξ_{ki}, $1 \leq i \leq k$ be an array of l-dimensional random vectors. Denote*

$$S_{ki} \triangleq \mathbb{E}\xi_{ki}\xi_{ki}^T, \quad R_{ki} = \mathbb{E}\{\xi_{ki}\xi_{ki}^T|\xi_{k1}, \cdots, \xi_{k,i-1}\}$$

and $S_k = \sum_{i=1}^k S_{ki}$. *Assume*

$$\mathbb{E}\{\xi_{ki}|\xi_{k1}, \cdots, \xi_{k,i-1}\} = 0, \tag{3.30}$$

$$\sup_{k>1}\sum_{i=1}^k \mathbb{E}\|\xi_{ki}\|^2 < \infty, \quad \lim_{k\to\infty} S_k = S, \tag{3.31}$$

$$\lim_{k\to\infty}\sum_{i=1}^k \mathbb{E}\|S_{ki} - R_{ki}\| = 0, \tag{3.32}$$

$$\lim_{k\to\infty}\sum_{i=1}^k \mathbb{E}\|\xi_{ki}\|^2 I_{[\|\xi_{ki}\|>\varepsilon]} = 0, \forall\varepsilon > 0. \tag{3.33}$$

Then, we obtain that

$$\sum_{i=1}^k \xi_{ki} \xrightarrow[k\to\infty]{d} N(0, S). \tag{3.34}$$

Before presenting the main theorems, we give several auxiliary lemmas.

Lemma 3.4 *Redefine* $\Psi_{k,i}$ *as*

$$\Psi_{k,j} \triangleq \prod_{i=j}^k (I - a_i J_i), \forall k \geq j, \ \Psi_{j,j+1} \triangleq I, \tag{3.35}$$

where $J_k \xrightarrow[k\to\infty]{} J$ *and* J *is stable. Then, for any* $r > 0$, *we obtain*

$$\sup_k \sum_{i=1}^k a_i\|\Psi_{k,i+1}\|^r < \infty. \tag{3.36}$$

Proof. Through slight modification of the proof of (3.22), it is easy to find that the estimation is still valid. Then, it follows that

$$\sum_{i=1}^k a_i\|\Psi_{k,i+1}\|^r \leq c_0^r \exp\left(-cr\sum_{i=j}^k a_i\right). \tag{3.37}$$

Function $1 - e^{-x} + \frac{x^2}{2} - x$ is equal to zero at $x = 0$ and its derivative satisfies $e^{-x} - 1 + x \geq 0$. Thus, we obtain

$$1 - e^{-x} + \frac{x^2}{2} \geq x, \quad \forall x \geq 0.$$

Using this inequality, we obtain that

$$\begin{aligned}
&\sum_{i=1}^{k} a_i \exp\left(-cr \sum_{j=i+1}^{k} a_j\right) \\
&\leq \frac{1}{cr} \sum_{i=1}^{k} \left(1 - e^{-cra_i} + \frac{(cra_i)^2}{2}\right) \exp\left(-cr \sum_{j=i+1}^{k} a_j\right) \\
&= \frac{1}{cr} \sum_{i=1}^{k} \left[\exp\left(-cr \sum_{j=i+1}^{k} a_j\right) - \exp\left(-cr \sum_{j=i}^{k} a_j\right)\right] \\
&+ \frac{cr}{2} \sum_{i=1}^{k} a_i^2 \exp\left(-cr \sum_{j=i+1}^{k} a_j\right),
\end{aligned}$$

which further implies that

$$\begin{aligned}
&\sum_{i=1}^{k} \left(1 - \frac{cr}{2} a_i\right) a_i \exp\left(-cr \sum_{j=i+1}^{k} a_j\right) \\
&\quad \leq \frac{1}{cr} \left[1 - \exp\left(-cr \sum_{j=1}^{k} a_j\right)\right] < \frac{1}{cr}.
\end{aligned} \tag{3.38}$$

Consider that k_3 is a sufficiently large number such that $\frac{cr}{2} a_i < \frac{1}{2}, \forall i \geq k_3$. We obtain that

$$\begin{aligned}
&\sum_{i=1}^{k} a_i \exp\left(-cr \sum_{j=i+1}^{k} a_j\right) \\
&< \sum_{i=1}^{k_3} a_i + \sum_{i=k_3+1}^{k} a_i \exp\left(-cr \sum_{j=i+1}^{k} a_j\right) \\
&< \sum_{i=1}^{k_3} a_i + 2 \sum_{i=k_3+1}^{k} \left(1 - \frac{cr}{2} a_i\right) a_i \exp\left(-cr \sum_{j=i+1}^{k} a_j\right) \\
&< \sum_{i=1}^{k_3} a_i + \frac{2}{cr},
\end{aligned} \tag{3.39}$$

where for the last inequality (3.38) is invoked.

Combining (3.37) and (3.39) yields (3.36), which completes the proof.

Lemma 3.5 *Set*

$$\vartheta_{k,i} = \exp(-J\sum_{j=i}^{k} a_j) - \Psi_{k,i}, \tag{3.40}$$

where $\Psi_{k,i}$ is given by (3.35). Then, we obtain that

$$\sup_{k,i} \|\vartheta_{k,i}\| < \infty, \tag{3.41}$$

$$\|\vartheta_{k,i}\| \to 0 \text{ as } i \to \infty \text{ uniformly with respect to } k : k \geq i, \tag{3.42}$$

$$\|\vartheta_{k,i}\| \to 0 \text{ as } k \to \infty \text{ uniformly with respect to } i : i \leq k. \tag{3.43}$$

Proof. Expanding $e_{-a_k J}$ to the series $e^{-a_k J} = I - a_k J + M_k$ with $M_k = O(a_k^2)$, we obtain that

$$\begin{aligned}\vartheta_{k,i} &= (I - a_k J + M_k)\exp(-J\sum_{j=i}^{k-1} a_j) - (I - a_k J_k)\Psi_{k-1,i} \\ &= (I - a_k J_k)\vartheta_{k-1,i} + (a_k(J_k - J) + M_k)\exp(-J\sum_{j=i}^{k-1} a_j) \\ &= \sum_{s=i}^{k} \Psi_{k,s+1}(a_s(J_s - J) + M_s)\exp(-J\sum_{j=i}^{s-1} a_j),\end{aligned} \tag{3.44}$$

where by definition $\sum_{j=i}^{i-1} a_j \triangleq 0$.

Because $-J$ is a stable matrix, there exist constants $c_1 > 0$ and $\delta > 0$ such that

$$\left\|\exp(-J\sum_{j=i}^{k-1} a_j)\right\| \leq c_1 \exp(-\delta\sum_{j=i}^{k-1} a_j). \tag{3.45}$$

Considering (3.44) and (3.45), we derive that for any $\sigma > 0$

$$\begin{aligned}\|\vartheta_{k,i}\| &\leq \sum_{s=i}^{k} \|\Psi_{k,s+1}\|(o(1) + O(a_s))c_1\exp(-\delta\sum_{j=i}^{k-1} a_j) \\ &\leq c_1 \sum_{s=i}^{k} a_s \|\Psi_{k,s+1}\|(o(1) + O(a_s)) \\ &\leq \frac{\sigma}{2} c_1 \sum_{s=i_1}^{k} a_s \|\Psi_{k,s+1}\| \\ &\quad + c_0 c_1 \sum_{s=i}^{i_1-1} a_s[\exp(-c\sum_{j=s+1}^{k} a_j)](o(1) + O(a_s)),\end{aligned} \tag{3.46}$$

where for the last inequality, i_1 is assumed to be sufficiently large such that $(o(1) + O(a_s)) \leq \frac{\sigma}{2}, \forall s \geq i_1$, and (3.22) with modifications in Lemma 3.4 is used.

Notice that $\sum_{j=1}^{\infty} a_j = \infty$ and $\sigma > 0$ can be arbitrarily small, the conclusion follows from (3.46) by Lemma 3.4.

Lemma 3.6 *Let D, E and F be $l \times l$ matrices and let D and E be stable. Then, we obtain that*

$$\lim_{k\to\infty} \sum_{i=1}^{k} a_i \left(\exp(D \sum_{j=i+1}^{k} a_j)\right) F \left(\exp(E \sum_{j=i+1}^{k} a_j)\right) = \int_0^{\infty} e^{Dz} F e^{Ez} \mathrm{d}z. \tag{3.47}$$

Proof. For any $T > 0$, we define

$$s(k, T) = \min\{s : \sum_{i=s}^{k} a_i \leq T\}.$$

Considering $a_i \to 0$, we obtain $\sum_{i=s(k,T)}^{k} a_i \to T$ for a fixed T. Denote $\sum_{j=s(k,T)}^{i} a_j$ by z. We obtain $\sum_{j=i+1}^{k} a_j \xrightarrow[k\to\infty]{} T - z$. Consequently, the term

$$\sum_{i=s(k,T)}^{k} a_i \left(\exp(D \sum_{j=i+1}^{k} a_j)\right) F \left(\exp(E \sum_{j=i+1}^{k} a_j)\right)$$

serves as an integral sum for $\int_0^T e^{D(T-z)} F e^{E(T-z)} \mathrm{d}z$, or equivalently $\int_0^T e^{Dz} F e^{Ez} \mathrm{d}z$. Thus, we obtain that

$$\lim_{k\to\infty} \sum_{i=s(k,T)}^{k} a_i \exp(D \sum_{j=i+1}^{k} a_j) F \exp(E \sum_{j=i+1}^{k} a_j) = \int_0^{T} e^{Dz} F e^{Ez} \mathrm{d}z.$$

Therefore, to prove (3.47), it is sufficient to show that

$$\lim_{T\to\infty} \lim_{k\to\infty} \sum_{i=1}^{s(k,T)-1} a_i \exp(D \sum_{j=i+1}^{k} a_j) F \exp(E \sum_{j=i+1}^{k} a_j) = 0. \tag{3.48}$$

Similar to (3.39), by stability of D, there is a constant $\eta_0 < \infty$ such that

$$\sup_{k,T} \sum_{i=1}^{s(k,T)-1} a_i \left\| \exp(D \sum_{j=i+1}^{s(k,T)-1} a_j) \right\| \leq \eta_0.$$

By stability of D and E, constants $\eta_1 > 0$, $\eta_2 > 0$ and $\eta >$ exist such that

$$\left\| F \exp\left(\sum_{j=i+1}^{k} a_j \right) \right\| \leq \eta_1, \quad \forall k \geq i+1,$$
$$\|e^{Dz}\| \leq \eta_2 e^{-\eta z}, \quad \forall z \geq 0.$$

Therefore, we obtain that

$$\begin{aligned}
&\limsup_{k\to\infty} \sum_{i=1}^{s(k,T)-1} a_i \left(\exp(D \sum_{j=i+1}^{k} a_j) \right) F \left(\exp(E \sum_{j=i+1}^{k} a_j) \right) \\
&\leq \eta_1 \limsup_{k\to\infty} \left[\left\| \exp\left(D \sum_{j=s(k,T)}^{k} a_j \right) \right\| \right. \\
&\times \left. \sum_{i=1}^{s(k,T)-1} a_i \left\| \exp\left(D \sum_{j=i+1}^{s(k,T)-1} a_j \right) \right\| \right] \\
&\leq \eta_0 \eta_1 \eta_2 \limsup_{k\to\infty} e^{-\eta \sum_{j=s(k,T)}^{k} a_j} = \eta_0 \eta_1 \eta_2 e^{-\eta T} \xrightarrow[T\to\infty]{} 0,
\end{aligned}$$

which verifies (3.48) and completes the proof.

The asymptotic normal property of the input sequence generated by (3.19) is given as follows.

Theorem 3.4 *For system (3.11) with initial value $x_k(0) = 0$, we design L such that $-\Phi \mathcal{G} \mathcal{L} + \frac{\alpha}{2} I$ is stable, where α is defined by (3.18). Then, the ILC update law* (3.19) *with arbitrary initial input $\mathbf{u}_0$ would guarantee that $\frac{1}{\sqrt{a_k}} \Phi \widehat{\boldsymbol{e}}_k$ is asymptotically normal, i.e.*

$$\frac{1}{\sqrt{a_k}} \Phi \widehat{\boldsymbol{e}}_k \xrightarrow[k\to\infty]{d} \mathbf{N}(0, S), \tag{3.49}$$

where $\mathbf{N}(0, S)$ *denotes the normal distribution and*

$$S = \int_0^{\infty} e^{(-J+\frac{\alpha}{2}I)z} J Q' J^T e^{(-J^T+\frac{\alpha}{2}I)z} \mathrm{d}t. \tag{3.50}$$

Proof. Using the condition of a_k in (3.18), we obtain that

$$\begin{aligned}\left(\frac{a_k}{a_{k+1}}\right)^{\frac{1}{2}} &= \left(\frac{a_k - a_{k+1}}{a_{k+1}} + 1\right)^{\frac{1}{2}} \\ &= 1 + \frac{1}{2}a_k\frac{a_k - a_{k+1}}{a_k a_{k+1}} + O\left(\left(\frac{a_k - a_{k+1}}{a_{k+1}}\right)^2\right) \\ &= 1 + \frac{1}{2}\alpha a_k + o(a_k).\end{aligned}$$

Then, we obtain

$$\begin{aligned}\frac{1}{\sqrt{a_{k+1}}}\Phi\widehat{\boldsymbol{e}}_{k+1} &= \left(\frac{a_k}{a_{k+1}}\right)^{\frac{1}{2}}\frac{1}{\sqrt{a_k}}\Phi\widehat{\boldsymbol{e}}_{k+1} \\ &= (1 + \frac{1}{2}\alpha a_k + o(a_k))\left(\frac{I - a_k J}{\sqrt{a_k}}\Phi\widehat{\boldsymbol{e}}_k + \sqrt{a_k}J\Phi\epsilon_k\right) \\ &= (I + a_k W_k)\frac{\Phi\widehat{\boldsymbol{e}}_k}{\sqrt{a_k}} + \sqrt{a_k}\epsilon_k',\end{aligned}$$

where

$$\begin{aligned}W_k &= -J + \frac{\alpha}{2}I - \frac{1}{2}\alpha a_k J + o(1)(I - a_k J) \\ &\xrightarrow[k\to\infty]{} W = -J + \frac{\alpha}{2}I,\end{aligned} \tag{3.51}$$

and

$$\epsilon_k' \triangleq (1 + \frac{1}{2}\alpha a_k + o(a_k))J\Phi\epsilon_k.$$

This further yields that

$$\begin{aligned}\frac{1}{\sqrt{a_{k+1}}}\Phi\widehat{\boldsymbol{e}}_{k+1} &= \sum_{i=1}^{k}\Gamma_{k,i+1}\sqrt{a_i}J\Phi\epsilon_i \\ &\quad + \sum_{i=1}^{k}\Gamma_{k,i+1}\sqrt{a_i}J\Phi\epsilon_i\left(\frac{1}{2}\alpha a_i + o(a_i)\right),\end{aligned} \tag{3.52}$$

where $\Gamma_{k,i}$ is defined as $\Gamma_{k,i} \triangleq (I + a_k W_k)\cdots(I + a_i W_i)$.

Based on (3.51) and stability of W, using similar steps to the analysis of Ψ in Theorem 3.2 and Lemma 3.4, we find that there exist constants $b_0 > 0$ and $b > 0$ such that

$$\|\Gamma_{k,i}\| \leq b_0 \exp\left(-b\sum_{j=i}^{k} a_j\right). \tag{3.53}$$

By Lemma 3.4 and similar steps of (3.46), we obtain

$$\begin{aligned}
&\mathbb{E}\|\sum_{i=1}^{k}\Gamma_{k,i+1}\sqrt{a_i}J\Phi\epsilon_i(\frac{1}{2}\alpha a_i + o(a_i))\| \\
&\leq \sum_{i=1}^{k}\|\Gamma_{k,i+1}\|a_i\|J\|\|Q'\|^{\frac{1}{2}}(\frac{\alpha}{2}\sqrt{a_i} + o(\sqrt{a_i})) \to 0.
\end{aligned}$$

Therefore, the last term of (3.52) tends to zero in probability. Therefore, to prove (3.49), it suffices to show that

$$\sum_{i=1}^{k}\Gamma_{k,i+1}\sqrt{a_i}J\Phi\epsilon_i \xrightarrow[k\to\infty]{d} \mathbf{N}(0, S). \tag{3.54}$$

Lemma 3.3 is applied, where $\xi_{k,i}$ corresponds to $\Gamma_{k,i+1}\sqrt{a_i}J\Phi\epsilon_i$. Thus, we need to check conditions of this lemma.

Because $\{\epsilon_k\}$ are i.i.d. normally distributed random variable sequence with zero mean, conditions (3.30) and (3.32) are satisfied.

Next, we verify the condition (3.31). Notice that

$$\begin{aligned}
S_k &= \sum_{i=1}^{k}\mathbb{E}\left[a_i\Gamma_{k,i+1}J\Phi\epsilon(\Phi\epsilon)^T J^T\Gamma_{k,i+1}^T\right] \\
&= \sum_{i=1}^{k}a_i\Gamma_{k,i+1}JQ'J^T\Gamma_{k,i+1}^T.
\end{aligned} \tag{3.55}$$

We show that the right hand-side of (3.55) tends to (3.50). Taking W and $\Gamma_{k,i}$ as J and $\Psi_{k,i}$ in Lemma 3.5, respectively, and using Lemmas 3.4 and 3.5, we obtain that

$$\sum_{i=1}^{k}a_i\left\|\exp\left(W\sum_{j=i+1}^{k}a_j\right) - \Gamma_{k,i+1}\right\|\|Q'\|\|J\|^2\|\Gamma_{k,i+1}\| \xrightarrow[k\to\infty]{} 0.$$

Considering (3.55), we find that

$$\left\|S_k - \sum_{i=1}^{k}a_i\left[\exp\left(W\sum_{j=i+1}^{k}a_j\right)\right]JQ'J^T\left[\exp\left(W^T\sum_{j=i+1}^{k}a_j\right)\right]\right\| \xrightarrow[k\to\infty]{} 0.$$

Using Lemma 3.6, we obtain

$$\|S_k - S\| \xrightarrow[k\to\infty]{} 0.$$

Last, we verify the condition (3.33). From (3.53), we obtain

$$\begin{aligned}&\left[\|\Gamma_{k,i+1}\sqrt{a_i}J\Phi\epsilon_i\| > \varepsilon\right]\\ &\subset \left[\|\Phi\epsilon_i\| > b_0^{-1}\|J\|^{-1}\frac{\varepsilon}{\sqrt{a_i}}\exp\left(b\sum_{j=i+1}^{k} a_j\right)\right].\end{aligned}$$

Note that $\sqrt{a_i}\exp(-b\sum_{j=i+1}^{k} a_j) \xrightarrow[k\to\infty]{} 0$ uniformly with respect to $i : i \geq 1$ because of $a_i \xrightarrow[i\to\infty]{} 0$, or equivalently, $g_{k,i} \triangleq \frac{1}{\sqrt{a_i}}\exp(b\sum_{j=i+1}^{k} a_j) \xrightarrow[k\to\infty]{} \infty$ uniformly with respect to $i : i \geq 1$. Because $\{\Phi\epsilon_k\}$ is an i.i.d. normally distributed random variable sequence, we obtain that

$$\lim_{N\to\infty} \mathbb{E}\|\Phi\epsilon_k\|^2 I_{[\|\Phi\epsilon_k\|>N]} = 0, \quad \forall k \geq 1.$$

Then, we have

$$\sup_i \mathbb{E}\left[\|\Phi\epsilon_i\|^2 I_{[b_0\|J\|\|\Phi\epsilon_i\|>\varepsilon g(k,i)]}\right] \xrightarrow[k\to\infty]{} 0.$$

Therefore, for any $\varepsilon > 0$, by Lemma 3.4, we obtain

$$\begin{aligned}&\sum_{i=1}^{k}\mathbb{E}\left[\|\Gamma_{k,i+1}\sqrt{a_i}J\Phi\epsilon_i\|^2 I_{[\|\Gamma_{k,i+1}\sqrt{a_i}J\Phi\epsilon_i\|>\varepsilon]}\right]\\ &\quad\leq \sum_{i=1}^{k} a_i\|\Gamma_{k,i+1}\|^2\|J\|^2\mathbb{E}\left[\|\Phi\epsilon_i\|^2 I_{[\|\Gamma_{k,i+1}\sqrt{a_i}J\Phi\epsilon_i\|>\varepsilon]}\right]\\ &\quad\xrightarrow[k\to\infty]{} 0.\end{aligned}$$

Thus, all conditions of Lemma 3.3 hold, implying that (3.54) holds. The proof is completed.

The aforementioned asymptotic normal property remains valid even when the initial state cannot be precisely reset but can be asymptotically reset with high accuracy, provided that the convergence speed of the initial value to zero satisfies certain requirements. Specifically, we present the following theorem.

Theorem 3.5 *For system (3.11), where the initial value $x_k(0) \to 0$ with a speed $\|x_k(0)\| = o(\sqrt{a_k})$, we design $\mathcal{L}$ such that $-\Phi\mathcal{G}\mathcal{L} + \frac{\alpha}{2}I$ is stable, where α is defined by (3.18). Then, the results of Theorem 3.4 still hold.*

Proof. When the initial value $x_k(0)$ is not zero, the recursive behavior of tracking error is (3.26). Define $\{\varphi_k\}$ by the following recursion:

$$\varphi_{k+1} = (I_l - a_k \Phi \mathcal{G}\mathcal{L})\varphi_k + a_k \Phi \mathcal{G}\mathcal{L}\Phi \epsilon_k, \tag{3.56}$$

which actually is (3.20). By Theorem 3.4, we find that $\frac{1}{\sqrt{a_k}}\varphi_k$ is asymptotically normal.

Subtracting (3.56) from (3.26) leads to

$$\Phi \widehat{\boldsymbol{e}}_{k+1} - \varphi_{k+1} = (I_l - a_k \Phi \mathcal{G}\mathcal{L})(\Phi \widehat{\boldsymbol{e}}_k - \varphi_k) + a_k \Phi \mathcal{G}\mathcal{L}\Phi \mathcal{M} x_k(0). \tag{3.57}$$

Then, we obtain

$$\begin{aligned}
&\frac{\Phi \widehat{\boldsymbol{e}}_{k+1} - \varphi_{k+1}}{\sqrt{a_{k+1}}} \\
&= (1 + \frac{1}{2}\alpha a_k + o(a_k)) \left[(I_l - a_k J)\frac{\Phi \widehat{\boldsymbol{e}}_k - \varphi_k}{\sqrt{a_k}} + \sqrt{a_k}\mathcal{M} x_k(0) \right] \\
&= (I_l + a_k W_k)\frac{\Phi \widehat{\boldsymbol{e}}_k - \varphi_k}{\sqrt{a_k}} + \sqrt{a_k}\mathcal{M}\widehat{x}_k(0),
\end{aligned} \tag{3.58}$$

where W_k is defined in (3.51) and $\widehat{x}_k(0) \triangleq (1 + \frac{1}{2}\alpha a_k + o(a_k))x_k(0) = o(a_{k+1})$. Consequently, by the notation of $\Gamma_{k,i}$, we obtain

$$\frac{\Phi \widehat{\boldsymbol{e}}_{k+1} - \varphi_{k+1}}{\sqrt{a_{k+1}}} = \Gamma_{k,1}\left(\frac{\Phi \widehat{\boldsymbol{e}}_1 - \varphi_1}{\sqrt{a_1}} \right) + \sum_{i=1}^{k} \Gamma_{k,i}\sqrt{a_i}\mathcal{M}\widehat{x}_i(0). \tag{3.59}$$

The first term on the right hand-side of (3.59) tends to zero as $k \to \infty$ according to (3.53). The second term is estimated as follows:

$$\begin{aligned}
\lim_{k\to\infty} \|\sum_{i=1}^{k} \Gamma_{k,i}\sqrt{a_i}\mathcal{M}\widehat{x}_i(0)\| &\le \lim_{k\to\infty} \sum_{i=1}^{k} a_i \|\Gamma_{k,i}\| o(1) \\
&= \lim_{k\to\infty} \sum_{i=1}^{k} a_i \|\Gamma_{k,i}\|^{\frac{1}{2}} \|\Gamma_{k,i}\|^{\frac{1}{2}} o(1) = 0,
\end{aligned}$$

which is based on Lemma 3.4 and (3.53). This indicates that $\frac{\Phi \widehat{e}_k}{\sqrt{a_k}}$ and $\frac{\varphi_k}{\sqrt{a_k}}$ have the same limit distribution if they exist. Consequently, the proof of this theorem has been transformed to the case of Theorem 3.4. The proof is completed.

3.2.5 Numerical Simulation

Consider an LTI system with the following system matrices

$$A = \begin{bmatrix} 0.5 & -0.07 & -0.26 \\ -0.03 & 0.38 & -0.3 \\ -0.1 & -0.13 & 0.4 \end{bmatrix}, \quad B = \begin{bmatrix} -1 & 0.1 & 0 \\ 0 & -1 & 0.4 \\ 0.2 & 0 & -1 \end{bmatrix}, \quad C = \begin{bmatrix} 1.8 & 0.5 & 0.3 \\ 0.4 & 1.6 & 1.2 \end{bmatrix}.$$

For a simple illustration, we let $N = 6$; then, we obtain that $\boldsymbol{y}_k \in \mathbf{R}^{12}$ and $\boldsymbol{u}_k \in \mathbf{R}^{18}$. The noise ϵ_k is subject to a zero-mean Gaussian distribution with the covariance $Q = 0.06^2 I_{12}$.

Consider the reference points to be $y_d^{(1)}(1)$, $y_d^{(2)}(3)$, $y_d^{(1)}(4)$, and $y_d^{(1)}(6)$. In other words, we obtain the matrix

$$\Phi = \begin{bmatrix} 1&0&0&0&0&0&0&0&0&0&0&0 \\ 0&0&0&0&0&1&0&0&0&0&0&0 \\ 0&0&0&0&0&0&1&0&0&0&0&0 \\ 0&0&0&0&0&0&0&0&0&0&1&0 \end{bmatrix}.$$

It is easy to verify that $rank(\Phi \mathcal{G}) = 4$. The selected reference trajectory is $\boldsymbol{y}_r = [1 \ \ 2 \ \ 1.5 \ \ 1]^T$.

The iteration-varying gain is selected as $a_k = \frac{1}{k+1}$, and we define $L = (\Phi \mathcal{G})^T$. The algorithm is executed for 200 iterations. Figure 3.1 illustrates the Euclidean norm of the modified tracking error, denoted as $\|\Phi \widehat{\boldsymbol{e}}_k\|$. It can be observed that the error profile rapidly converges to zero. This indicates that, after a sufficient number of learning iterations, the tracking error will solely arise from system and measurement noises which cannot be eliminated by any learning control algorithm.

The complete output $\boldsymbol{y}_{200}$ and the reference points are depicted in Fig. 3.2, where the dashed line with diamonds represents the 12-dimensional output, while four circles indicate the reference points $y_d^{(1)}(1)$, $y_d^{(2)}(3)$, $y_d^{(1)}(4)$, and $y_d^{(1)}(6)$. The x-label corresponds to the dimension of the output vector. It can be observed from the figure that even in a noisy environment, the system output effectively tracks the reference points.

In our algorithm, we exclusively focus on the requested outputs, thereby providing ample opportunities to explore additional optimization objectives. For instance, we can optimize control efforts and minimize energy consumption associated with unrequested outputs. This aspect presents an intriguing yet unresolved research problem.

To examine the asymptotic normality, we conduct 1000 simulations and gather all tracking errors from the final iteration (i.e., the 200th cycle) for each operation. Figure 3.3 illustrates the empirical cumulative distribution function (CDF) of the tracking error data at each reference point. Additionally, Fig. 3.4 presents probability density functions derived from the accumulated tracking error data.

It is worth noting that the convergence is established in relation to the modified tracking error. In previous research, convergence has typically been proven based on

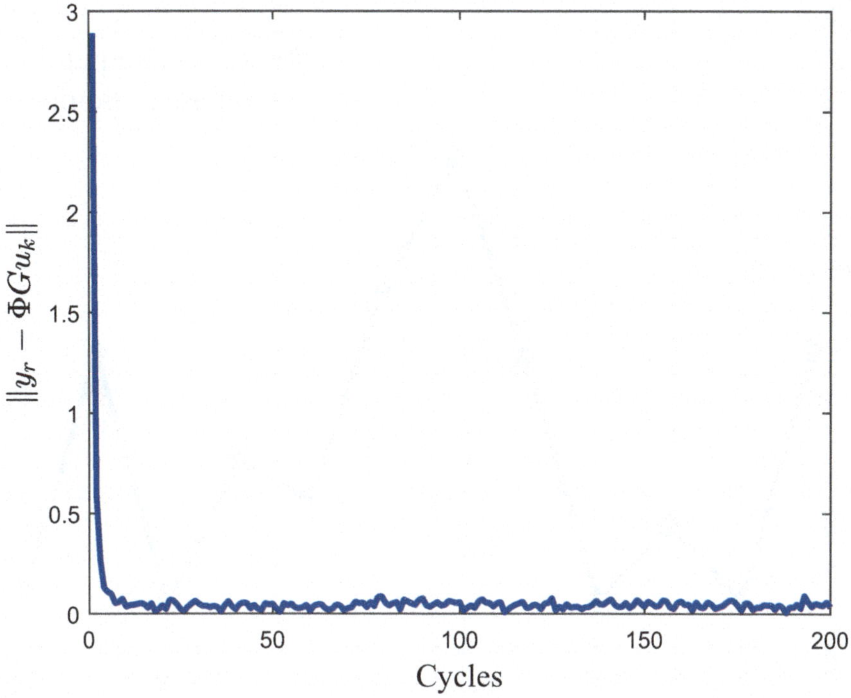

Fig. 3.1 Norm profile of the tracking error $\|\boldsymbol{y}_r - \Phi \mathcal{G} \boldsymbol{u}_k\|$

the generated input sequence. However, this approach requires a realistic assumption: there exists only one control input that can precisely generate the reference for the nominal plant. Intuitively, for stochastic point-to-point tracking systems, it can be expected that the input sequence will converge along the iteration axis. This phenomenon can be verified through simulations. Figure 3.5 illustrates the iteration-wise profiles of each dimension of the input sequence at time instants 1, 3, and 5.

3.3 Decreasing Gain Design in Saturated Sensor Systems

The focus of this section is the study of ILC for a class of nonlinear systems, where the output is measured by sensors that have saturation characteristics. Additionally, the system outputs are subject to stochastic noises arising from uncertain environments. Consequently, the noise-induced output may occasionally exceed the sensor range randomly, resulting in measurement signals that contain uncertainties originating from both saturation and noise effects. To address this issue, we propose an ILC algorithm with a decreasing gain sequence and provide rigorous proof that the generated input sequence converges almost surely to the desired trajectory.

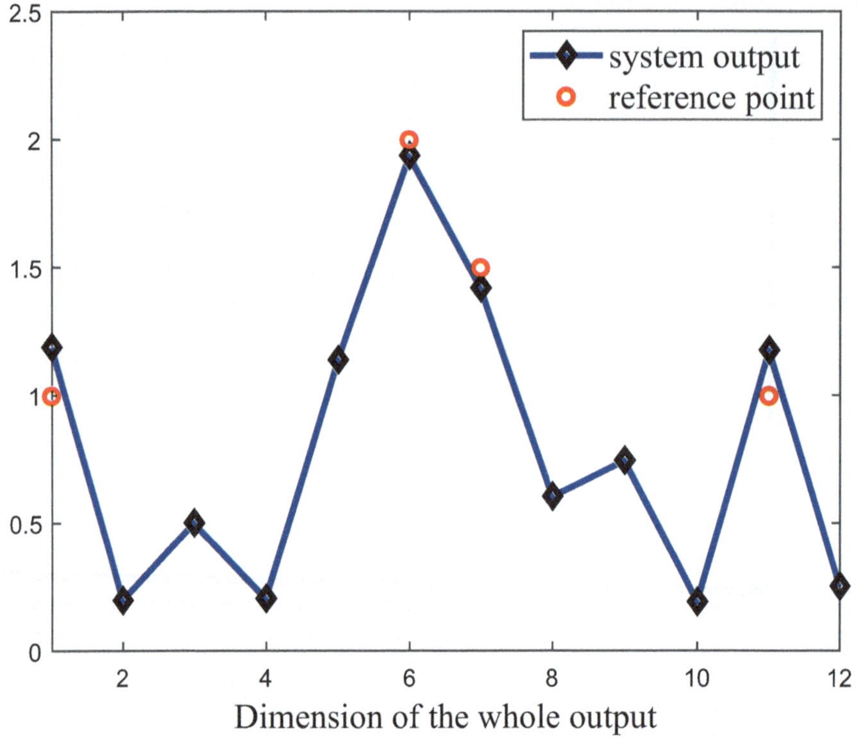

Fig. 3.2 Output y_{200} and the reference y_r

3.3.1 Saturation

Saturation is a prevalent nonlinearity in mechanical systems due to various range limitations in devices, encompassing both input and output saturation. For instance, inexpensive devices with inadequate range and amplifier saturation of electronic circuits can result in saturation. While input saturation has been extensively studied in the literature (Bernstein and Michel, 1995), the exploration of output saturation, which is common in numerous industrial applications, deserves thorough investigation. Kong et al. (2010) demonstrated the presence of saturation characteristics in electric motor systems where speed or angular velocity was limited by maximum voltage. Additionally, Chau et al. (2010) illustrated how the battery recovery effect observed in wireless sensor networks was subject to a saturation threshold determined by random sensing activities. Another example provided in Rousseau et al. (2003) highlighted linear behavior for small inputs and saturated response for large inputs from sensor devices. These industrial illustrations underscore the existence of output saturation across many mechanical systems, warranting significant efforts towards control and optimization analysis.

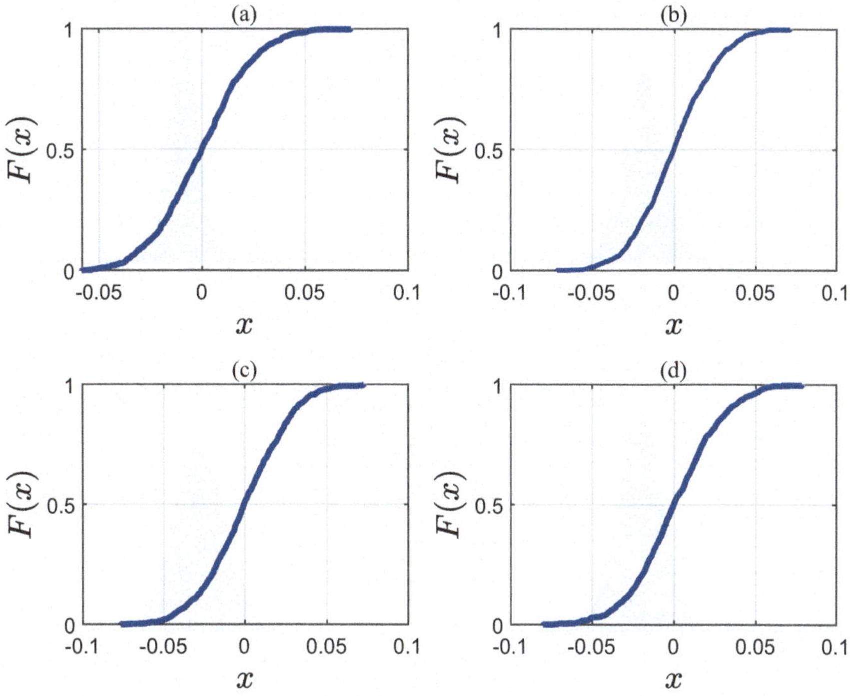

Fig. 3.3 Empirical CDF for the tracking error data at the reference points: **a** $y_d^{(1)}(1)$, **b** $y_d^{(2)}(3)$, **c** $y_d^{(1)}(4)$, and **d** $y_d^{(1)}(6)$

However, the literature on analysis and design for the output saturation problem is significantly less extensive compared to the input saturation case. Currently, only local results have been obtained in this area of research. Stabilization plays a crucial role in control system design when dealing with saturation cases. Early studies Kreisselmeier (1996), Lin and Hu (2001) addressed this issue for SISO systems, establishing semi-globally asymptotic stabilization conditions for linear feedback control laws in the context of output saturation problems. The extension to MIMO systems was later explored by authors in Grip et al. (2010), who demonstrated the possibility of stabilization without additional restrictions. In Cao et al. (2003), researchers further tackled the output feedback $\mathcal{H}_\infty$ control problem for linear systems subject to sensor saturation using linear matrix inequality techniques, which were validated through a flight control system case study. The work Wang et al. (2024) discussed $\mathcal{H}_\infty$ filtering problems for a class of nonlinear systems with randomly occurring sensor saturation, while Yang and Li (2009) introduced set-membership filtering specifically designed for discrete time-varying systems subject to sensor saturation constraints. Nevertheless, it should be noted that existing results on output saturation problems are scattered and mostly provide constructive approaches tailored to specific issues at hand; thus indicating that there is still room for improvement in research related to controlling saturated systems.

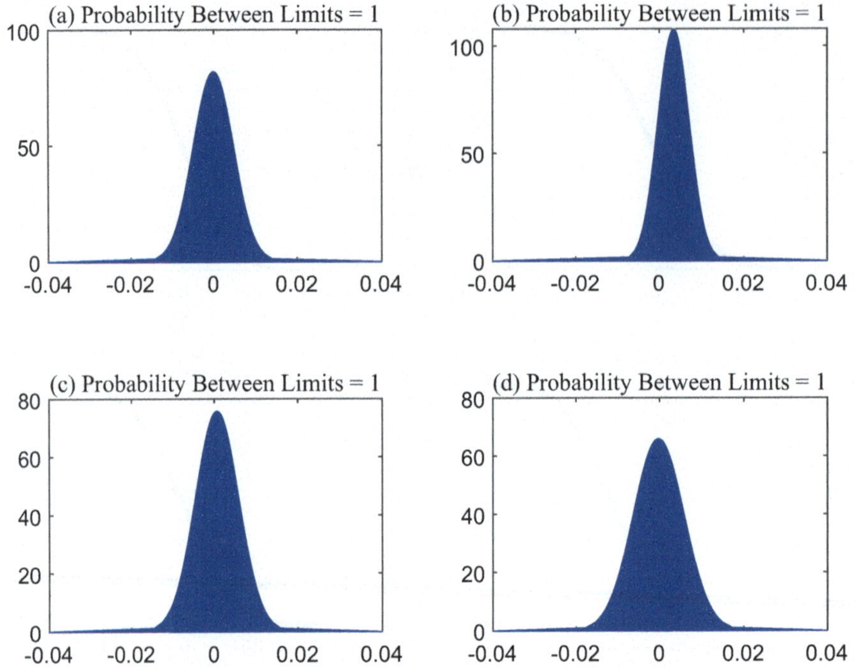

Fig. 3.4 Probability density function of the tracking error data at the reference points: **a** $y_d^{(1)}(1)$, **b** $y_d^{(2)}(3)$, **c** $y_d^{(1)}(4)$, and **d** $y_d^{(1)}(6)$

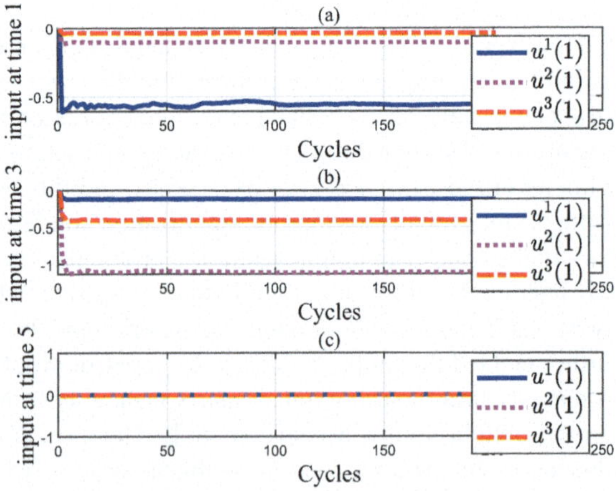

Fig. 3.5 Trajectory behaviors of input at time instants 1, 3, and 5

In the ILC field, few papers are found on the algorithm design and performance analysis for systems with sensor saturation. Most papers related to saturation in the ILC filed pay attention to the input saturation case. Xu et al. (2004); Hou et al. (2008) considered the input saturation problem, where the composite energy function method, as well as an inequality property of saturation function, was introduced to give the convergence analysis. Moreover, a reference governor method was proposed in Tan et al. (2011) to resolve the ILC control problem for dynamic systems with input saturation. Furthermore, for parameterized systems with input saturation, Zhang et al. (2015, 2016) provided an adaptive ILC scheme to update the associated parameters and deal with the saturation problem. This method was also applied to high-speed trains control in Ji et al. (2016). However, there is a scarcity of literature reporting progress on the ILC for systems subject to output saturation. A primary attempt was made in a paper (Zhang and Fang, 2011) that focused on ILC and repetitive control for linear systems, demonstrating convergence through the conventional contraction mapping method in the λ-norm sense. It should be noted that all the aforementioned papers solely consider deterministic systems, and when considering systems with stochastic noises, these proposed approaches fail to provide a rigorous analysis of convergence. This remains an open problem. Furthermore, numerous papers have addressed ILC for nonlinear systems such as Xu (1997); however, they assume full accessibility of the output and do not account for output signal saturation in Xu (1997) and other related works.

3.3.2 Problem Formulation

Consider the following SISO nonlinear system with sensor saturation

$$\begin{aligned} x_k(t+1) &= \boldsymbol{f}(t, x_k(t)) + \boldsymbol{b}(t, x_k(t))u_k(t), \\ y_k(t) &= \boldsymbol{c}^T(t)x_k(t), \\ z_k(t) &= \mathbf{Sat}(y_k(t) + w_k(t)), \end{aligned} \tag{3.60}$$

where $k = 1, 2, \cdots$ labels the iteration number, $t = 0, 1, \cdots, N$ denotes different time instant of one iteration, and N is the length of each iteration. $u_k(t) \in \mathbf{R}$, $x_k(t) \in \mathbf{R}^n$, $y_k(t) \in \mathbf{R}$ are the input, state, and output, respectively, whereas $z_k(t)$ is the measured output by a saturated sensor and $w_k(t)$ is the measurement noise. It should be emphasized that the actual system output $y_k(t)$ is not available for the design of updating laws because of the existence of saturation mechanism. In other words, only the measured output $z_k(t)$ can be used in the control law design. $\boldsymbol{f}(\cdot, \cdot) : \mathbf{R} \times \mathbf{R}^n \to \mathbf{R}^n$ and $\boldsymbol{b}(\cdot, \cdot) : \mathbf{R} \times \mathbf{R}^n \to \mathbf{R}^n$ are nonlinear continuous functions. In addition, these functions, $\boldsymbol{f}(t, x_k(t))$ and $\boldsymbol{b}(t, x_k(t))$, together with $\boldsymbol{c}(t) \in \mathbf{R}^n$ are unknown, thus they are not available for the design of the learning law. The sensor saturation is defined as

$$\mathbf{Sat}(v) = \begin{cases} m, & v \geq m \\ v, & -m < v < m \\ -m, & v \leq -m \end{cases} \tag{3.61}$$

for some unknown positive constant m. In other words, when the output combined with random noise surpasses the measurement range, the boundary value represents the actual measurement. The presence of saturation nonlinearity in this context is typically attributed to practical constraints inherent in various measurement devices.

The control objective of this section is to determine an input sequence $\{u_k(t), t = 0, 1, \cdots, N-1\}$ based on the measured information $z_k(t)$ rather than the original output $y_k(t)$. The aim is for the output $y_k(t)$ to accurately track the desired reference $y_d(t)$ as the iteration number tends towards infinity.

The following assumptions are needed for system (3.60).

Assumption 3.1 The desired reference $y_d(t)$ is realizable in the sense that there exist unique $x_d(0)$ and $u_d(t)$ such that

$$\begin{aligned} x_d(t+1) &= \boldsymbol{f}(t, x_d(t)) + \boldsymbol{b}(t, x_d(t))u_d(t), \\ y_d(t) &= \boldsymbol{c}^T(t)x_d(t). \end{aligned} \tag{3.62}$$

In addition, the desired reference lies in the measurable range, i.e., $|y_d(t)| < m, \forall t$.

Assumption 3.2 The initial states can be precisely reset in the sense that $x_d(0) - x_k(0) \to 0$ as $k \to \infty$ asymptotically.

Remark 3.3 In the field of ILC, it is typically necessary to precisely reset the initial state, i.e., $x_k(0) = x_d(0), \forall k$. This requirement is commonly referred to as identical initialization condition (i.i.c.). Assumption 3.2 relaxes this condition, which has been extensively studied in literature, for instance in Chen et al. (1999). Additionally, for discrete-time systems, the well-established technique of initial rectification can be employed when there are variations in the initial state. However, since this issue falls outside the scope of our current section, we adopt Assumption 3.2.

Assumption 3.3 The functions $\boldsymbol{f}(t, \cdot)$ and $\boldsymbol{b}(t, \cdot)$ are continuous with respect to the second argument, $\forall t$.

Assumption 3.4 The real number $\boldsymbol{c}^T(t+1)\boldsymbol{b}(t, x)$ coupling the input and output is an unknown nonzero constant, but its sign, characterizing the control direction, is assumed known. Without loss of any generality, we assume that $\boldsymbol{c}^T(t+1)\boldsymbol{b}(t, x) > 0$.

Assumption 3.5 For each time instant t, the measurement noise $\{w_k(t)\}$ is a sequence of independent and identical distributed random variables such that $\mathbb{E}w_k(t) = 0$, $\sup_k w_k^2(t) < \infty$, and $\limsup_{n\to\infty} \frac{1}{n}\sum_{k=1}^{n} w_k^2(t) = R_w^t$ a.s., $\forall t$, where R_w^t is unknown.

Remark 3.4 Let us provide further remarks on the assumptions. Assumption 3.1 assumes the realizability of the desired references, encompassing two key points: the existence of an optimal control signal and limitations on the range of the desired reference. If the desired reference exceeds its saturation range, achieving perfect tracking becomes challenging due to inherent constraints imposed by the saturated output $z_k(t)$. Assumption 3.4 asserts that knowledge about control direction is a fundamental requirement for controller design. In cases where control direction is unknown, introducing an adaptive detection mechanism similar to Shen and Wang (2015) would significantly increase algorithm complexity. Assumption 3.5 imposes conditions on stochastic noises. Considering practical environments where processes are repeatable from iteration to iteration, it is reasonable to assume that these noises are typically bounded and random.

For simplicity of writing, let us set $\boldsymbol{f}_k(t) \triangleq \boldsymbol{f}(t, x_k(t))$, $\boldsymbol{f}_d(t) \triangleq \boldsymbol{f}(t, x_d(t))$, $\boldsymbol{b}_k(t) \triangleq \boldsymbol{b}(t, x_k(t))$, $\boldsymbol{b}_d(t) \triangleq \boldsymbol{b}_d(t, x_d(t))$, $\delta x_k(t) \triangleq x_d(t) - x_k(t)$, $\delta u_k(t) \triangleq u_d(t) - u_k(t)$, $\delta \boldsymbol{f}_k(t) \triangleq \boldsymbol{f}_d(t) - \boldsymbol{f}_k(t)$, $\delta \boldsymbol{b}_k(t) \triangleq \boldsymbol{b}_d(t) - \boldsymbol{b}_k(t)$, $\boldsymbol{c}^+ \boldsymbol{f}_k(t) \triangleq \boldsymbol{c}^T(t+1)\boldsymbol{f}_k(t)$, $\boldsymbol{c}^+ \boldsymbol{b}_k(t) \triangleq \boldsymbol{c}^T(t+1)\boldsymbol{b}_k(t)$.

The subsequent lemma provides a foundation for the convergence discussed in the following subsection.

Lemma 3.7 *Assume Assumptions 3.1-3.3 hold for system* (3.60). *If* $\lim_{k\to\infty} \delta u_k(s) = 0$, $s = 0, 1, \cdots, t-1$, *then at the time instant* t, $\|\delta x_k(t)\| \to 0$, $\|\delta \boldsymbol{f}_k(t)\| \to 0$, $\|\delta \boldsymbol{b}_k(t)\| \to 0$ *as* $k \to \infty$.

Proof. We prove the lemma by mathematical induction. From (3.60) and (3.62) it follows that

$$\begin{aligned}\delta x_k(t+1) &= \boldsymbol{f}_d(t) - \boldsymbol{f}_k(t) + \boldsymbol{b}_d(t)u_d(t) - \boldsymbol{b}_k(t)u_k(t) \\ &= \delta \boldsymbol{f}_k(t) + \delta \boldsymbol{b}_k(t)u_d + \boldsymbol{b}_k(t)\delta u_k(t).\end{aligned} \tag{3.63}$$

Initial Step. For $t = 0$, from Assumptions 3.2 and 3.3 we have that $\delta\|\boldsymbol{f}_k(0)\| \to 0$, $\delta\|\boldsymbol{b}_k(0)\| \to 0$ as $k \to \infty$, which further implies that the first two terms on the right-hand side of (3.63) tend to zero as $k \to \infty$. Noticing $\|\boldsymbol{b}_k(0)\| \le \|\boldsymbol{b}_d(0)\| + \|\delta \boldsymbol{b}_k(0)\|$, we find that $\boldsymbol{b}_k(0)$ is bounded. Thus, the fact $\lim_{k\to\infty} \delta u_k(0) = 0$ leads that the third term on the right-hand side of (3.63) also goes to zero. Hence, it follows that $\|\delta x_k(1)\| \to 0$ as $k \to \infty$. Further, by noting Assumption 3.3 we have that $\|\delta \boldsymbol{f}_k(1)\| \to 0$ and $\|\delta \boldsymbol{b}_k(1)\| \to 0$ as $k \to \infty$.

Inductive Step. Assume the conclusions of the lemma are true for $s = 0, 1, \cdots, t-1$, then we have that $\|\delta x_k(t)\| \to 0, k \to \infty$. Now we proceed to show the conclusion for time instant t under the condition $\lim_{k\to\infty} \delta u_k(t) = 0$. To this end, noticing (3.63), by completely the same steps as that used above, we find that $\|\delta x_k(t+1)\| \to 0$ as $k \to \infty$. That is, the conclusions are also valid for t. This completes the proof.

Lemma 3.7 reveals the essential effect of the input at previous time instants (e.g., $0, 1, \cdots, t-1$) on the state at current time instant (e.g., t). Specifically, the state error at current time instant would approach zero as long as the input errors at all previous time instants approach to zero. In other words, the effect of the historical tracking error is asymptotically negligible if the precise convergence of the input sequence can be achieved asymptotically. Therefore, based on Lemma 3.7, the analysis objective in the following is to show the zero-error convergence of the input sequence to the desired input defined in Assumption 3.1. The details of the analysis will be specified in the next subsection.

3.3.3 ILC Algorithm and Its Convergence

In this subsection, we will design the ILC law using a decreasing gain sequence, establish the relationship between the measured information and the original output signal, and present the main theorem regarding asymptotic convergence. Recall the fact that the actual output $y_k(t)$ is not available for algorithm design. That is, only the measured output $z_k(t)$ can be used. Thus the available error is defined as $e_k(t) \triangleq y_d(t) - z_k(t)$, which will be employed in the updating law. We now define the learning algorithm as follows:

$$u_{k+1}(t) = u_k(t) + a_k e_k(t+1), \tag{3.64}$$

where a_k is a prior defined decreasing gain such that $a_k > 0$, $a_k \to 0$ as $k \to \infty$, $\sum_{k=1}^{\infty} a_k = \infty$, and $\sum_{k=1}^{\infty} a_k^2 < \infty$. It is obvious that $a_k = a/k$ meets all these requirements with $a > 0$ being a suitable constant. This decreasing sequence is introduced from the stochastic approximation algorithm, which mainly aims to ensure the almost sure convergence and suppresses the influence of stochastic noises.

The following convergence analysis demonstrates that the classic P-type learning algorithm (3.64) effectively addresses the output saturation problem. When utilizing saturated output information to update the input signal, tracking performance is improved, albeit not as significantly as when using original output information for updating. In other words, while the convergence speed may be slightly slowed down by the output saturation problem, the convergence property remains guaranteed.

For any fixed t we have

$$\begin{aligned} e_k(t+1) &= y_d(t+1) - z_k(t+1) \\ &= y_d(t+1) - \mathbf{Sat}(y_k(t+1) + w_k(t+1)) \\ &= \gamma_k(t)(y_d(t+1) - y_k(t+1) - w_k(t+1)), \end{aligned} \tag{3.65}$$

where $\gamma_k(t)$ is defined as follows:

$$\gamma_k(t) = \begin{cases} \frac{y_d(t+1)-m}{y_d(t+1)-y_k(t+1)-w_k(t+1)}, & \text{if } y_k(t+1)+w_k(t+1) \geq m \\ 1, & \text{if } -m < y_k(t+1)+w_k(t+1) < m \\ \frac{y_d(t+1)+m}{y_d(t+1)-y_k(t+1)-w_k(t+1)}, & \text{if } y_k(t+1)+w_k(t+1) \leq -m \end{cases} \tag{3.66}$$

It is noticed that $0 < \gamma_k(t) \leq 1, \forall t, k$ and $\gamma_k(t)$ is a random variable depending on $w_k(t+1)$.

Remark 3.5 From (3.65) we find that

$$\begin{aligned} & |y_d(t+1) - \textbf{Sat}(y_k(t+1) + w_k(t+1))| \\ = & |\gamma_k(t)(y_d(t+1) - y_k(t+1) - w_k(t+1))| \\ \leq & |y_d(t+1) - y_k(t+1) - w_k(t+1)|. \end{aligned}$$

This aligns with Property 1 presented in Xu et al. (2004). However, due to the presence of stochastic noise, the conventional contraction mapping method is not applicable. Consequently, we introduce the random variable $\gamma_k(t)$ as a measure of the contraction coefficient and demonstrate convergence from a probabilistic perspective using stochastic approximation techniques. This represents a technical distinction between this section and existing findings. Furthermore, it is important to emphasize that the coefficient $\gamma_k(t)$ is purely virtual and derived from the original learning update law (3.64). In other words, this coefficient is solely employed for convergence analysis purposes and does not need to be known for updating.

Next, we substitute the expression (3.65) of available error $e_k(t+1)$ into the learning algorithm (3.64) to derive a specified regression of the input sequence. From (3.60), (3.62), (3.63) and (3.65), we can rewrite (3.64) as

$$\begin{aligned} u_{k+1}(t) = & \ u_k(t) + a_k\gamma_k(t)\boldsymbol{c}^+\boldsymbol{b}_k(t)(u_d(t) - u_k(t)) \\ & + a_k\varphi_k(t) - a_k\gamma_k(t)w_k(t+1), \end{aligned} \tag{3.67}$$

where

$$\varphi_k(t) = \gamma_k(t)(\boldsymbol{c}^+\delta\boldsymbol{f}_k(t) + \boldsymbol{c}^+\delta\boldsymbol{b}_k(t)u_d(t)) \tag{3.68}$$

denotes the structure noise. The regression function is

$$h_{t,k}(u) = \gamma_k(t)\boldsymbol{c}^+\boldsymbol{b}_k(t)(u_d(t) - u). \tag{3.69}$$

It should be pointed out that the above regression function depends on k for any fixed t. However, the root of the function is independent of k.

We now give the first theorem of this section.

Theorem 3.6 *Assume Assumptions 3.1-3.5 hold for system* (3.60). *Then the input sequence* $\{u_k(t)\}$ *generated by the learning control algorithm* (3.64) *converges to the* *desired input $u_d(t)$ as $k \to \infty$ almost surely, $t = 0, 1, \cdots, N-1$. In addition, the output $y_k(t)$ will converge to the desired reference $y_d(t)$ as $k \to \infty$ almost surely, $\forall t$.*

Proof. Noticing (3.60) and Assumption 3.1, we find that it is sufficient to prove that $u_k(t) \to u_d(t)$ a.s., $t = 0, 1, \cdots, N-1$. To this end, the proof is carried out by mathematical induction method along the time axis t with the help of Lemma 3.7 established in the last subsection.

Initial Step: We first verify the validity of the theorem for $t = 0$. In this case, (3.67) is rewritten as

$$\begin{aligned} u_{k+1}(0) = &\ u_k(0) + a_k\gamma_k(0)\boldsymbol{c}^+\boldsymbol{b}_k(0)(u_d(0) - u_k(0)) \\ &+ a_k\varphi_k(0) - a_k\gamma_k(0)w_k(1). \end{aligned} \tag{3.70}$$

Subtracting both sides of last equation from $u_d(0)$ leads to

$$\delta u_{k+1}(0) = (1 - a_k\gamma_k(0)\boldsymbol{c}^+\boldsymbol{b}_k(0))\delta u_k(0) - a_k\varphi_k(0) + a_k\gamma_k(0)w_k(1). \tag{3.71}$$

Since $\boldsymbol{b}_k(0)$ is continuous in its second argument, i.e., the initial state, it follows from Assumption 3.2 that $\boldsymbol{b}_k(0) \to \boldsymbol{b}_d(0)$. Moreover, Assumption 3.4 implies convergence of the coupling number $\boldsymbol{c}^+\boldsymbol{b}_k(0)$ to a positive constant. By utilizing Property 1.6 of Benaim (1996), which states that a sequence generated by a stochastic approximation algorithm is naturally bounded if the regression function is globally Lipschitz and satisfies certain stability conditions, we can easily deduce that the input error sequence generated from (3.71) remains bounded due to linearity of the regression function in (3.71). Consequently, based on system (3.60), it can be inferred that the output $y_k(1)$ is also bounded. Combining this with Assumption 3.5, we can conclude that there exists a nonzero lower bound for $\gamma_k(0)$; specifically, there exists a suitable $0 < \nu < 1$ such that $\gamma_k(0) > \nu$, for all k.

Set

$$\phi_{i,j} \triangleq (1 - a_i\gamma_i(0)\boldsymbol{c}^+\boldsymbol{b}_i(0)) \cdots (1 - a_j\gamma_j(0)\boldsymbol{c}^+\boldsymbol{b}_j(0)),\ i \ge j,\ \phi_{i,i+1} \triangleq 1. \tag{3.72}$$

Note that the notation $\phi_{i,j}$ actually is a successive production of the contraction $1 - a_k\gamma_k(0)\boldsymbol{c}^+\boldsymbol{b}_k(0)$ with the subscript k increasing from j up to i. Therefore, the notation is well defined in (3.72) when $i \ge j$. It is apparent that $\phi_{i,i} = 1 - a_i\gamma_i(0)\boldsymbol{c}^+\boldsymbol{b}_i(0)$ is a special case. While for the case $i < j$, the increase from j up to i is with no sense, and thus, we define $\phi_{i,j} = 1$ with $i < j$ for the completeness of the notation. It is clear that $1 - a_j\gamma_j(0)\boldsymbol{c}^+\boldsymbol{b}_j(0) > 0$ and $\boldsymbol{c}^+\boldsymbol{b}_j(0) \ge c$ for all sufficient large j, say $j \ge j_0$, and a suitable constant $c > 0$. Then for any $i > j$, $j \ge j_0$, we have that

$$\begin{aligned} \phi_{i,j} &= (1 - a_i\gamma_i(0)\boldsymbol{c}^+\boldsymbol{b}_i(0))\phi_{i-1,j} \\ &\le (1 - a_ic\nu)\phi_{i-1,j} \\ &\le \exp(-a_ic_1)\phi_{i-1,j}, \end{aligned}$$

where the basic inequality $1 - x \le e^{-x}$, $\forall x > 0$, is used for the last inequality and $c_1 = c\nu$. This further yields that $\phi_{i,j} \le c_2 \exp\left(-c_1 \sum_{k=j}^{i} a_k\right)$, $\forall j \ge j_0$, $i > j$, for some suitable $c_2 > 0$. Consequently,

$$|\phi_{i,j}| \le c_3 \exp\left(-c_1 \sum_{k=j}^{i} a_k\right), \quad \forall i > j, \ j \ge j_0. \tag{3.73}$$

To complete the estimation, we need to supplement the case $i > j_0$, $j < j_0$. For this case, we have

$$|\phi_{i,j}| \le |\phi_{i,j_0}||\phi_{j_0-1,j}| \le c_4 \exp\left(-c_1 \sum_{k=j}^{i} a_k\right) \tag{3.74}$$

for some suitable $c_4 > 0$ and $c_4 \ge c_3$. Combining (3.73) and (3.74) leads to the following estimation for the product $\phi_{i,j}$,

$$|\phi_{i,j}| \le c_4 \exp\left(-c_1 \sum_{k=j}^{i} a_k\right), \quad \forall i \ge j_0, j > 0. \tag{3.75}$$

When $i < j_0$, the above estimation is still valid due to the finiteness of j_0, as long as we choose a sufficiently large c_4.

Now from (3.71) we have

$$\delta u_{k+1}(0) = \phi_{k,1}\delta u_1(0) - \sum_{j=1}^{k} \phi_{k,j+1} a_j \varphi_j(0) + \sum_{j=1}^{k} \phi_{k,j+1} a_j \gamma_j(0) w_j(1), \tag{3.76}$$

where the first term on the right-hand side of last equation tends to zero as $k \to \infty$ because of (3.74).

Now it comes to the last two terms of (3.76). From Assumption 3.2, Assumption 3.3, and (3.68), it is evident that $\varphi_k(0) \to 0$ as $k \to \infty$. Therefore, for any $\epsilon > 0$, there is a sufficient large integer k_1 such that $|\varphi_k(0)| < \epsilon$, $\forall k \ge k_1$. Then for $k > k_1$, we have

$$\begin{aligned}\left|\sum_{j=1}^{k} \phi_{k,j+1} a_j \varphi_j(0)\right| \le{}& c_4 \sum_{j=1}^{k_1-1} \exp\left(-c_1 \sum_{i=j+1}^{k} a_i\right) a_j |\varphi_j(0)| \\ &+ \epsilon c_4 \sum_{j=k_1}^{k} \exp\left(-c_1 \sum_{i=j+1}^{k} a_i\right) a_j,\end{aligned} \tag{3.77}$$

where the first term on the right-hand side is a finite summation due to the finiteness of k_1, in which each term tends to zero as $k \to \infty$ because the starting number is bounded by k_1. Consequently, the finite summation would tend to zero as $k \to \infty$. For the last term on the right-hand side of (3.77), the following estimation can be obtained

$$\begin{aligned}
&\epsilon c_4 \sum_{j=k_1}^{k} \exp\left(-c_1 \sum_{i=j+1}^{k} a_i\right) a_j \\
\leq & 2\epsilon c_4 \sum_{j=k_1}^{k} \left(a_j - \frac{c_1 a_j^2}{2}\right) \exp\left(-c_1 \sum_{i=j+1}^{k} a_i\right) \\
\leq & \frac{2\epsilon c_4}{c_1} \sum_{j=k_1}^{k} (1 - e^{-c_1 a_j}) \exp\left(-c_1 \sum_{i=j+1}^{k} a_i\right) \\
= & \frac{2\epsilon c_4}{c_1} \sum_{j=k_1}^{k} \left[\exp\left(-c_1 \sum_{i=j+1}^{k} a_i\right) - \exp\left(-c_1 \sum_{i=j}^{k} a_i\right)\right] \\
\leq & \frac{2\epsilon c_4}{c_1}.
\end{aligned} \tag{3.78}$$

Hence the second term on the right-hand side of (3.77) tends to zero as $k \to \infty$ and $\epsilon \to 0$.

Now only the last term of (3.76) is left to verify. To this end, we first show that the summation $\sum_{j=1}^{k} a_j \gamma_j(0) w_j(1)$ converges to some unknown constant. From Assumption 3.5, it is noticed that $\{w_j(1)\}$ is a sequence of independent and identical distributed random variables along iteration axis with finite second moments. Moreover,

$$\sum_{k=1}^{\infty} \mathbb{E}(a_k \gamma_k(t) w_k(1))^2 \leq \sum_{k=1}^{\infty} \mathbb{E}(a_k w_k(1))^2 \leq R_w^1 \sum_{k=1}^{\infty} a_k^2 < \infty.$$

This further implies that $\sum_{k=1}^{\infty} a_k \gamma_k(t) w_k(1) < \infty$ a.s. by Khintchine-Kolmogorov convergence theorem Chow and Teicher (2003).

Let $\lambda_k = \sum_{j=1}^{k} a_j \gamma_j(0) w_j(1)$, $\lambda_0 = 0$. Then $\lambda_k \to \lambda < \infty$ and therefore, for any $\epsilon > 0$, there is a sufficient large integer $k_2 > k_1$ such that $|\lambda_k - \lambda| \leq \epsilon$, $\forall k \geq k_2$. By a partial summation, we have

$$\begin{aligned}\sum_{j=1}^{k}\phi_{k,j+1}a_j\gamma_j(0)w_j(1) &= \sum_{j=1}^{k}\phi_{k,j+1}(\lambda_j-\lambda_{j-1})\\ &= \lambda_k-\sum_{j=1}^{k}(\phi_{k,j+1}-\phi_{k,j})\lambda_{j-1}\\ &= \lambda_k-\sum_{j=1}^{k}(\phi_{k,j+1}-\phi_{k,j})\lambda-\sum_{j=1}^{k}(\phi_{k,j+1}-\phi_{k,j})(\lambda_{j-1}-\lambda)\\ &= \lambda_k-\lambda+\phi_{k,1}\lambda-\sum_{j=1}^{k}(\phi_{k,j+1}-\phi_{k,j})(\lambda_{j-1}-\lambda), \end{aligned}\tag{3.79}$$

where it is obvious that the terms on the right-hand side all tend to zero as $k \to \infty$ except the last one. While for the last term, we further have

$$\begin{aligned}&\sum_{j=1}^{k}(\phi_{k,j+1}-\phi_{k,j})(\lambda_{j-1}-\lambda)\\ &=\left(\sum_{j=1}^{k_2}+\sum_{j=k_2+1}^{k}\right)(\phi_{k,j+1}-\phi_{k,j})(\lambda_{j-1}-\lambda).\end{aligned}$$

It is evident that $\lambda_k \to \lambda$ and $\sum_{j=1}^{k_2}(\phi_{k,j+1}-\phi_{k,j})(\lambda_{j-1}-\lambda) \to 0$ as $k \to \infty$. The latter result holds because that it is a finite summation and $\phi_{k,j} \to 0$ as $k \to \infty$ as j is bounded by k_2. Moreover,

$$\sum_{j=k_2+1}^{k}(\phi_{k,j+1}-\phi_{k,j})(\lambda_{j-1}-\lambda)=\sum_{j=k_2+1}^{k}\phi_{k,j+1}a_j\gamma_j(0)\boldsymbol{c}^{+}\boldsymbol{b}_j(0)(\lambda_{j-1}-\lambda).\tag{3.80}$$

Note that $|\lambda_j - \lambda| \le \epsilon$, $\forall j \ge k_2$. In addition, $\boldsymbol{c}^{+}\boldsymbol{b}_j(0)$ will converge to a constant. Then it follows that

$$\begin{aligned}&\left|\sum_{j=k_2+1}^{k}\phi_{k,j+1}a_j\gamma_j(0)\boldsymbol{c}^{+}\boldsymbol{b}_j(0)(\lambda_{j-1}-\lambda)\right|\\ &\le\epsilon\sup_{1\le j<\infty}|\boldsymbol{c}^{+}\boldsymbol{b}_j(0)|\sum_{j=k_2+1}^{k}c_4\exp\left(-c_1\sum_{i=j+1}^{k}\right)a_j,\end{aligned}\tag{3.81}$$

which tends to zero as $k \to \infty$ and $\epsilon \to 0$. Consequently, $\delta u_{k+1}(0) \to 0$ as $k \to 0$. In other words, the conclusion of this theorem is true for $t = 0$.

Inductive Step: Assume that the theorem is valid for $s = 0, 1, \cdots, t-1$, and we proceed to show that the theorem is also true for t. From (3.67), we can get the counterpart of (3.71) for time instant t,

$$\delta u_{k+1}(t) =(1 - a_k \gamma_k(t)\boldsymbol{c}^+\boldsymbol{b}_k(t))\delta u_k(t) - a_k\varphi_k(t) + a_k\gamma_k(t)w_k(t+1), \quad (3.82)$$

where $\varphi_k(t) = \gamma_k(t)(\boldsymbol{c}^+\delta \boldsymbol{f}_k(t) + \boldsymbol{c}^+\delta\boldsymbol{b}_k(t)u_d(t))$.

Comparing (3.82) with (3.71), we find that the terms $\delta u_k(t)$, $1 - a_k\gamma_k(t)\boldsymbol{c}^+\boldsymbol{b}_k(t)$, $\varphi_k(t)$, and $w_k(t+1)$ correspond to $\delta u_k(0)$, $1 - a_k\gamma_k(0)\boldsymbol{c}^+\boldsymbol{b}_k(0)$, $\varphi_k(0)$, and $w_k(1)$, respectively. Thus, the proof can be completed following the same steps as long as we ensure that $\varphi_k(t) \to 0$ as $k \to \infty$, since all the other terms are with the same properties. By the induction condition, we have that $\lim_{k\to\infty} \delta u_k(s) = 0$ for $s = 0, 1, \cdots, t-1$, then applying the results in Lemma 3.7 leads to a direct conclusion that $\|\delta \boldsymbol{f}_k(t)\| \to 0$ and $\|\delta\boldsymbol{b}_k(t)\| \to 0$ as $k \to 0$. Therefore, we have $\varphi_k(t) \to 0$ by the boundedness of all the left terms in the expression of $\varphi_k(t)$. As a consequence, using completely the same steps to the case $t = 0$, we can prove that $u_k(t) \to u_d(t)$ as $k \to \infty$. The proof of the theorem is thus completed by mathematical induction principle.

Theorem 3.6 establishes the asymptotic convergence of the input sequence in an almost sure sense. This theorem demonstrates that employing a P-type learning law with a decreasing gain ensures stable convergence of the input sequence to the desired input. In essence, the P-type learning law exhibits robustness against output saturation and stochastic noises.

Remark 3.6 The measurement output $z_k(t)$ in this section is subject to stochastic noises, as indicated by the last equation of (3.60). In certain applications, the introduction of stochastic noise $w_k(t)$ during transmission or its generation by other factors results in an additive term to the saturated output. In other words, the measurement output equation of (3.60) can be formulated as $z_k(t) = \textbf{Sat}(y_k(t)) + w_k(t)$. For such a formulation, the proposed ILC algorithm (3.64) remains effective. The convergence proof can be conducted following similar steps with minor modifications in the derivations.

3.3.4 Extension to MIMO Case

The proposed ILC scheme is extended to MIMO nonlinear systems in this subsection by introducing a direction regulation matrix. Firstly, the control direction Assumption 3.4 will be modified according to the multi-dimensional system formulation. Then, we will provide the modifications on the learning algorithm and its convergence analysis in sequence. Specifically, we consider the following nonlinear system:

$$\begin{aligned} x_k(t+1) &= \boldsymbol{f}(t, x_k(t)) + B(t, x_k(t))u_k(t), \\ y_k(t) &= C(t)x_k(t), \\ z_k(t) &= \mathbf{Sat}(y_k(t) + w_k(t)), \end{aligned} \tag{3.83}$$

where $u_k(t) \in \mathbf{R}^p$, $y_k(t) \in \mathbf{R}^q$ and the associated matrix $B(t,x) \in \mathbf{R}^{n\times p}$, $C(t) \in \mathbf{R}^{q\times n}$. For a vector $\boldsymbol{v} = [v_1, \cdots, v_q]^T \in \mathbf{R}^q$, the saturation $\mathbf{Sat}(\boldsymbol{v}) \triangleq [\mathbf{Sat}(v_1), \cdots, \mathbf{Sat}(v_q)]^T$ and $\mathbf{Sat}(v_i)$, $\forall i$, is defined in (3.61).

Assumptions 3.1–3.3 are still valid with the vectors $\boldsymbol{b}(t,x)$ and $\boldsymbol{c}(t)$ being replaced by matrix $B(t,x)$ and $C(t)$. Assumption 3.5 is slightly modified in the expression of R_w^t, which should be $R_w^t = \limsup_{n\to\infty} \frac{1}{n}\sum_{k=1}^n w_k(t)w_k^T(t)$. In the rest of this section, we will directly refer to these assumptions without detailing the differences again to avoid boring reduplication. Moreover, in Assumption 3.4, the control direction is required to be known prior as it is important for controller design. For the SISO system, the control direction is determined by the sign of the coupling value; however, for the MIMO system, determining the control direction becomes more complex. Therefore, Assumption 3.4 has been modified accordingly.

Assumption 3.6 The matrix $C(t+1)B(t,x)$ coupling the input and output is of full-column rank.

Remark 3.7 A direct consequence of Assumption 3.6 is that $q \geq p$. Moreover, Assumption 3.6 also implies that the system possesses a relative degree of one, which characterizes the fundamental relationship between the input and output. In case the system's relative degree exceeds one, denoted as η, the results presented here remain valid if we substitute the tracking error $e(t+1)$ used in our algorithm with $e(t+\eta)$. For brevity, we restrict our analysis to scenarios where the relative degree is equal to one.

Now the learning law (3.64) is modified as follows:

$$u_{k+1}(t) = u_k(t) + a_k L_t e_k(t+1), \tag{3.84}$$

where $L_t \in \mathbf{R}^{p\times q}$ is the direction regulation matrix to be defined later.

Remark 3.8 The role of L_t in the algorithm (3.84) is the updating direction term that is incorporated with the coupling matrix $C^+B(t,x) \triangleq C(t+1)B(t,x)$ to ensure an asymptotical convergence as the iteration number goes to infinity. To be specific, the design condition of L_t is that all the eigenvalues of $L_tC(t+1)B(t,x_d(t))$ are with positive real parts, $\forall t$. This condition is specified in the following theorem.

Theorem 3.7 *Assume Assumptions 3.1-3.3, 3.5, and 3.6 hold for system* (3.83). *Then the input sequence* $\{u_k(t)\}$ *generated by the learning control algorithm* (3.84) *converges to the desired input* $u_d(t)$ *as* $k\to\infty$ *almost surely,* $t = 0, 1, \cdots, N-1$, *if the direction regulation matrix* L_t *satisfies that all eigenvalues of* $L_tC(t+1)B(t,x_d(t))$ *are with positive real parts for all time instances. In addition, the output* $y_k(t)$ *will converge to the desired reference* $y_d(t)$ *as* $k\to\infty$ *almost surely,* $\forall t$.

Proof. The proof can be performed similarly to that of Theorem 3.6. Here, we mainly give the major differences caused by the multi-dimensional formulation. The mathematical induction method is also used here, where the inductive step is simply routine and thus we main check the validity of the initial step. To this end, consider the case $t = 0$ and the algorithm (3.84) results in

$$\begin{aligned} u_{k+1}(0) &= u_k(0) + a_k L_0[y_d(1) - \mathbf{Sat}(y_k(1) + w_k(1))] \\ &= u_k(0) + a_k L_0 \Gamma_k(0)[y_d(1) - y_k(1) - w_k(1)] \\ &= u_k(0) + a_k L_0 \Gamma_k(0) C^+ B_k(0)(u_d(0) - u_k(0)) \\ &+ a_k L_0 \Gamma_k(0)\psi_k(0) - a_k L_0 \Gamma_k(0) w_k(1), \end{aligned} \tag{3.85}$$

where $\Gamma_k(0) = \text{diag}\{\gamma_k^{(1)}(0), \cdots, \gamma_k^{(q)}(0)\}$ is a diagonal matrix, $C^+ B_k(0) \triangleq C(1)B(0, x_k(0))$, $\delta B_k(0) = B(0, x_d(0)) - B(0, x_k(0))$, and $\psi_k(0) = C(1)\delta \boldsymbol{f}_k(0) + C(1)\delta B_k(0)u_d(0)$, which converges to zero as $k \to \infty$.

Then, subtracting both sides of last equation leads to

$$\delta u_{k+1}(0) = (I - a_k L_0 \Gamma_k(0) C^+ B_k(0))\delta u_k(0) - a_k L_0 \Gamma_k(0)\psi_k(0) + a_k L_0 \Gamma_k(0) w_k(1), \tag{3.86}$$

which is similar to (3.71).

Set

$$\Phi_{i,j} \triangleq (I - a_i L_0 \Gamma_i(0) C^+ B_i(0)) \times \cdots \times (I - a_j L_0 \Gamma_j(0) C^+ B_j(0)), \ i \geq j,$$

and $\Phi_{i,i+1} = I$. Compared with the proof of Theorem 3.6, to show the almost sure convergence for the MIMO case, it is sufficient to verify that there exist suitable constants $c_5 > 0$ and $c_6 > 0$ such that

$$\|\Phi_{i,j}\| \leq c_5 \exp\left(-c_6 \sum_{k=j}^{i} a_k\right), \quad \forall i \geq j, j \geq 1. \tag{3.87}$$

Then by applying the same steps as in the proof of Theorem 3.6, the estimations of all terms are still valid.

First, recall the property that there is a lower bound of $\gamma_k^{(i)}(0)$, $\forall 1 \leq i \leq q$, say ν for simplicity. Then we have $\Gamma_k(0) \geq \nu I$. Note that $C^+ B_k(0) \to C^+ B_d(0) = C(t+1)B(0, x_d(0))$ and L_0 is designed such that all eigenvalues of $L_0 C^+ B_d(0)$ are with positive real parts. Thus, according to the Lyapunov stability for linear system (see, for example, pp. 37 of Chen et al. (2012)), for any given positive definite matrix Q, there exists a positive definite matrix P such that

$$P(-L_0 C^+ B_d(0)) + (-L_0 C^+ B_d(0))^T P = -Q, \tag{3.88}$$

or equivalently,

$$P(L_0 C^+ B_d(0)) + (L_0 C^+ B_d(0))^T P = Q. \tag{3.89}$$

Consequently, we can choose $Q = 2I$ and compute the corresponding positive definite matrix P.

Since $C^+B_k(0) \to C^+B_d(0)$ as $k \to \infty$ and $\Gamma_k(0)$ is a diagonal matrix, there is a suitable $k_0 > 0$ such that for $k \geq k_0$,

$$P(-L_0\Gamma_k(0)C^+B_k(0)) + (-L_0\Gamma_k(0)C^+B_k(0))^T P \leq -\nu I. \tag{3.90}$$

For simplicity, let us denote $H_k \triangleq -L_0\Gamma_k(0)C^+B_k(0)$. As a consequence, $\forall i \geq k_0$,

$$\begin{aligned}
\Phi_{i,j}^T P\Phi_{i,j} &= \Phi_{i-1,j}^T(I + a_i H_i)^T P(I + a_i H_i)\Phi_{i-1,j} \\
&= \Phi_{i-1,j}^T(P + a_i^2 H_i^T P H_i + a_i H_i^T P + a_i P H_i)\Phi_{i-1,j} \\
&\leq \Phi_{i-1,j}^T(P + a_i^2 H_i^T P H_i - \nu a_i I)\Phi_{i-1,j} \\
&= \Phi_{i-1,j}^T P^{\frac{1}{2}}(I - \nu a_i P^{-1} + a_i^2 P^{-\frac{1}{2}} H_i^T P H_i P^{-\frac{1}{2}})P^{\frac{1}{2}}\Phi_{i-1,j}.
\end{aligned} \tag{3.91}$$

Without loss of generality, when k_0 is sufficiently large, it is guaranteed that, $\forall i \geq k_0$,

$$\|I - \nu a_i P^{-1} + a_i^2 P^{-\frac{1}{2}} H_i^T P H_i P^{-\frac{1}{2}}\| \leq 1 - 2c_6 a_i \leq \exp(-2c_6 a_i), \tag{3.92}$$

by noticing the boundedness of $P^{-\frac{1}{2}} H_i^T P H_i P^{-\frac{1}{2}}$. Combining (3.91) and (3.92) leads to

$$\Phi_{i,j}^T P\Phi_{i,j} \leq c_7 \exp\left(-2c_6 \sum_{k=j}^{i} a_k\right)$$

with a suitable constant $c_7 > 0$. Hence,

$$\|\Phi_{i,j}\| \leq \lambda_{\min}^{-\frac{1}{2}}(P)\sqrt{c_7} \exp\left(-c_6 \sum_{k=j}^{i} a_k\right). \tag{3.93}$$

As a result, we define $c_5 \triangleq \lambda_{\min}^{-\frac{1}{2}}(P)\sqrt{c_7}$, then the estimation of (3.87) is guaranteed.

The following steps of this proof are completely the same as in the proof of Theorem 3.6 and thus are omitted for space concise. The proof is completed.

Theorem 3.7 extends the asymptotic convergence results from SISO systems to MIMO systems. Furthermore, the convergence condition is primarily imposed on the input/output coupling matrix $C^+B(t, x)$. On one hand, feasible solutions for the learning matrix L_t always exist as long as $C^+B(t, x)$ has full-column rank. On the other hand, to ensure precise tracking (cf. (Huang et al., 2002)), a certain degree of full-column rank is necessary. Thus, Theorem 3.7 employs a more relaxed condition for convergence under output saturation and stochastic noises.

3.3.5 *Illustrative Simulations*

In this subsection, the proposed algorithms are applied to two examples. The first is a single-link manipulator system, which in essence is an SISO system. The second is an artificial system adopted from the existing paper, which is a two-input-two-output system.

3.3.5.1 Example 1: Single-Link Manipulator

A single-link manipulator can be described by the following model (Canudas de Wit et al., 1991),

$$(mr^2 + I)\ddot{q} + mgr\cos(q) + f(q, \dot{q}) = \tau, \tag{3.94}$$

where q, I, and τ are the link displacement, the moment of inertia, and the applied motor torque, respectively. The punctual mass m is payload, which is located at distance r from the center of the coordinate. The friction force f is difficult to model exactly. Letting h be a sampling period, one can discretize the model by using the Euler method as follows,

$$\begin{aligned}\tau(ih) =&(mr^2 + I)\frac{q(ih + 2h) - 2q(ih + h) + q(ih)}{h^2}\\ &+ mgr\cos(q(ih)) + f\left(q, \frac{q(ih + h) - q(ih)}{h}\right).\end{aligned}$$

Let $x^{(1)}(i) = q(ih), x^{(2)}(i) = q(ih + h)$, and $u(i) = \tau(ih)$. Then the above equation can be rewritten as the following state-space form,

$$\begin{aligned}x^{(1)}(i + 1) =&x^{(2)}(i)\\ x^{(2)}(i + 1) =&2x^{(2)}(i) - x^{(1)}(i) - \frac{mgrh^2}{mr^2 + h}\cos(x^{(1)}(i))\\ &- \frac{h^2}{mr^2 + I}f\left(x^{(1)}(i), \frac{x^{(2)}(i) - x^{(1)}(i)}{h}\right) + \frac{h^2}{mr^2 + I}u(i).\end{aligned} \tag{3.95}$$

The parameter used in simulation are set to be $r = 1.0$m, $m = 2.0$g, $I = 1.5$kg·m^2. The sampling period is $h = 0.001$s, and the tracking period or iteration length is $T = 2.632$s. Moreover, the friction function is $f = (2 - 2\sqrt{|\dot{q}|} + 10|\dot{q}|)\mathrm{sgn}(\dot{q})$, which denotes a simplified model of frictions including Coulomb stiction, asymmetries, and downwards bends (Canudas de Wit et al., 1991), and sgn(·) is a sign function. The system output is given as $y(i) = x^{(2)}(i)$. Here we assume the output is bounded with the upper bound being 1.4. That is, if the output is larger than 1.4, the measured output is actually 1.4. The involved stochastic noises satisfy $w_k(t) \sim N(0, 0.1^2)$. The reference trajectory is $y_d(t) = \sin(t) - \frac{1}{5}\cos(5t) - 0.2t + 0.2$.

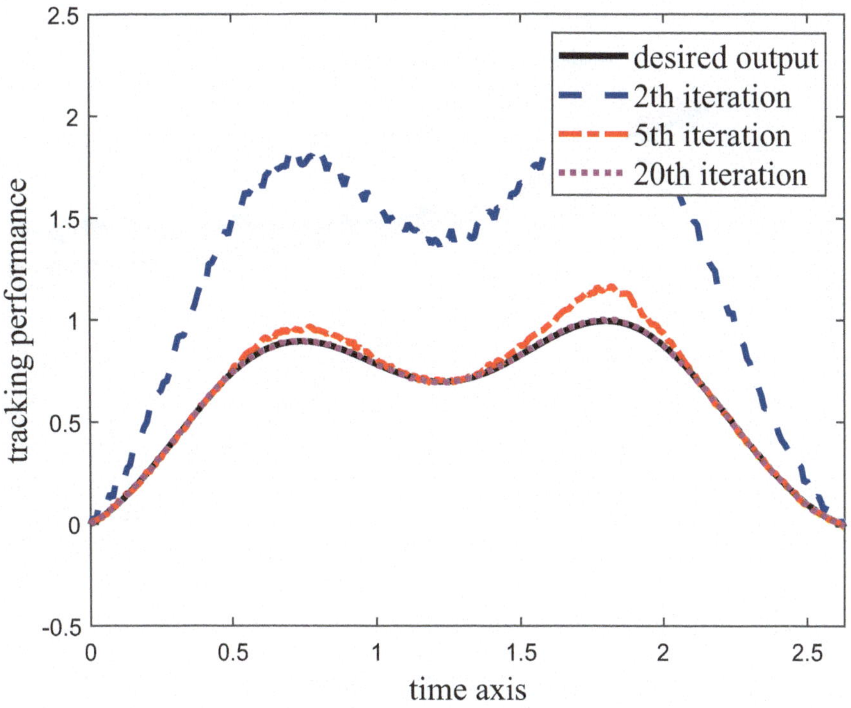

Fig. 3.6 Output tracking performance at the 1st, 5th, and last iteration

The initial input is simply given as $u_0(t) = 0$, $\forall t$. The learning gain is given as $a_k = 2/k$, $\forall k \geq 1$. It is apparent that the requirements of a_k are satisfied. The algorithm runs for 20 iterations.

The tracking performance is illustrated in Fig. 3.6, where the solid line represents the reference. It can be observed that the output at the 1st iteration deviates significantly from the reference, whereas significant improvement in performance is achieved by the 5th iteration. Remarkably, at the 20th iteration, the output almost perfectly aligns with the reference. These results demonstrate a commendable tracking performance of our proposed algorithm.

To demonstrate the diminishing property of errors, we define the actual tracking error as $e_k(t) = y_d(t) - y_k(t)$, where subscript k denotes the iteration number. Additionally, we define the measured tracking error as $\hat{e}_k(t) = y_d(t) - z_k(t)$, where $z_k(t) = \mathbf{Sat}(y_k(t) + w_k(t))$. Figure 3.7 illustrates the convergence behavior of our proposed algorithm, with the maximum tracking error defined as $\max_t\{e_k(t)\}$ and $\max_t\{\hat{e}_k(t)\}$ for actual and measured tracking errors, respectively. The maximum tracking error profiles are represented by circles and squares for actual and measured cases respectively. It is important to note that stochastic noise is present in our

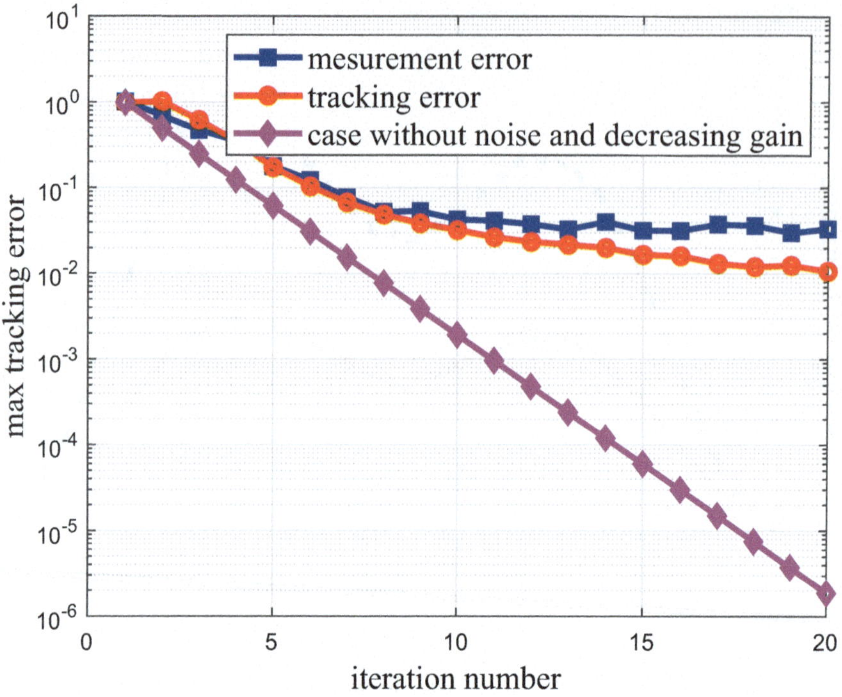

Fig. 3.7 Maximal error profiles

system; therefore, we introduce a decreasing learning gain a_k to mitigate its influence. Consequently, complete reduction of maximum error profiles to zero is not achievable; however, our results already exhibit satisfactory tracking performance.

Moreover, the decreasing gain a_k is solely introduced to mitigate the impact of stochastic noises. However, a drawback of such a gain is its relatively slower convergence speed. Nevertheless, in cases where the system is deterministic or when stochastic noises are negligible, it is possible to substitute the decreasing gain with a constant one. In Fig. 3.7, for comparison purposes, we also simulate the scenario of a noise-free system using a constant gain denoted by the solid line marked with diamonds. The chosen constant learning gain value is set at 0.5 and results in exponential convergence.

To demonstrate sensor saturation, the system output with measurement noise $y_k(t) + w_k(t)$ and the measured output $z_k(t)$ at the 1st iteration are depicted in Fig. 3.8 by the dotted line and dashed line, respectively. It is important to note that this illustration of the 1st iteration serves solely for explanatory purposes without any specific objective. As evident from the figure, when the system output exceeds the measurement range, only an upper bound can be obtained.

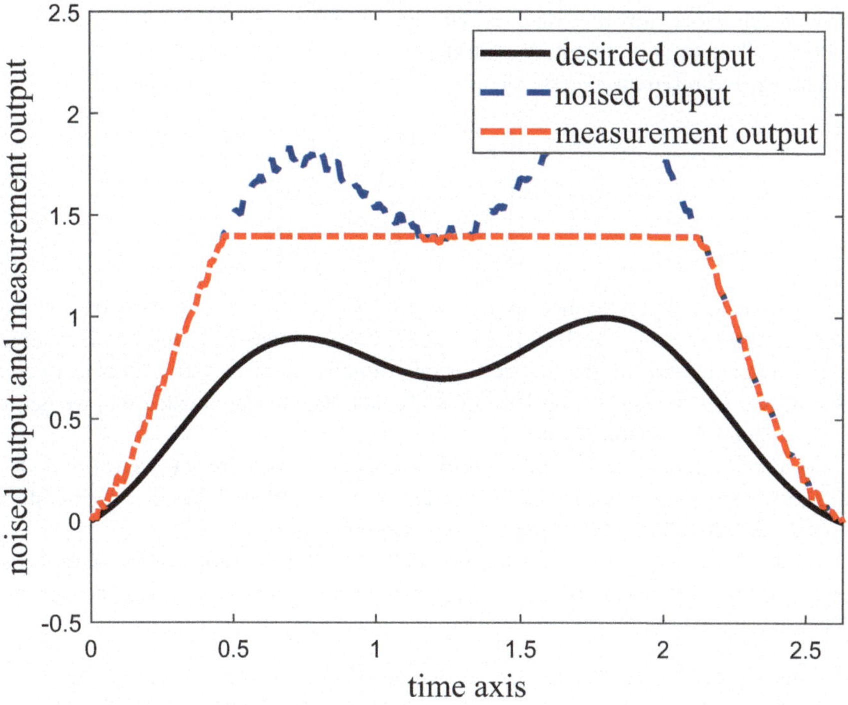

Fig. 3.8 System output with noises and measurement output at the 1st iteration

3.3.5.2 Example 2: MIMO System Case

Consider the following MIMO system (Liu et al., 2014)

$$\begin{aligned} x^{(1)}(t+1) &= x^{(2)}(t), \\ x^{(2)}(t+1) &= f(x(t)) + g(x(t))u(t), \\ y(t) &= x^{(1)}(t), \end{aligned} \tag{3.96}$$

where $u(t) = [u^{(1)}(t), u^{(2)}(t)]^T \in \mathbf{R}^2$ and $y(t) = [y^{(1)}(t), y^{(2)}(t)]^T \in \mathbf{R}^2$ are the input and output, respectively. The state is $x(t) = [(x^{(1)}(t))^T, (x^{(2)}(t))^T]^T \in \mathbf{R}^4$ with $x^{(i)}(t) = [x^{(i1)}(t), x^{(i2)}(t)]^T \in \mathbf{R}^2$, $i = 1, 2$. The unknown functions are defined as

$$\begin{aligned} f(x(t)) &= \begin{bmatrix} 0.4x_{11}(t)/(1+x_{21}^2(t)) \\ x_{22}(t) \times (0.1 + 0.005\cos(x_{12}(t))) \end{bmatrix}, \\ g(x(t)) &= \begin{bmatrix} 2 & 0 \\ 0 & 2 \end{bmatrix}. \end{aligned}$$

The initial value for the state is chosen as $x(0) = [3.5355\ 1.4142,\ 3.6448\ 1.4579]^T$. The iteration length $N = 400$.

The desired references are given as

$$y_d(t) = \begin{bmatrix} 5\sin\left(\frac{0.5\pi t}{50} + \frac{\pi}{4}\right) \\ 2\cos\left(\frac{0.5\pi t}{50} - \frac{\pi}{4}\right) \end{bmatrix}.$$

The stochastic noise is termed as $w(t) = [w^{(1)}(t), w^{(2)}(t)]^T$ with each dimension satisfying that $w^{(i)}(t) \sim N(0, 0.1^2)$, $i = 1, 2$. For the measured output, we assume that the upper bounds for the first and second dimensions are 6 and 2.5, respectively. That is, if $y^{(1)}(t)$ plus $w^{(1)}(t)$ is larger than 6, then the measured value that we get is 6. The second dimension is similar.

For the simulation, the initial input is simply set to be zero, i.e., $u_0(t) = 0$, $\forall t$. The decreasing gain in this example chooses $a_k = 1/k$, $\forall k \geq 1$. The learning matrix is set $L_t = 0.8I_2$. The algorithm runs for 50 iterations.

The tracking performance for the first and second dimensions of the output are shown in Figs. 3.9 and 3.10, respectively, where the solid line denotes the desired

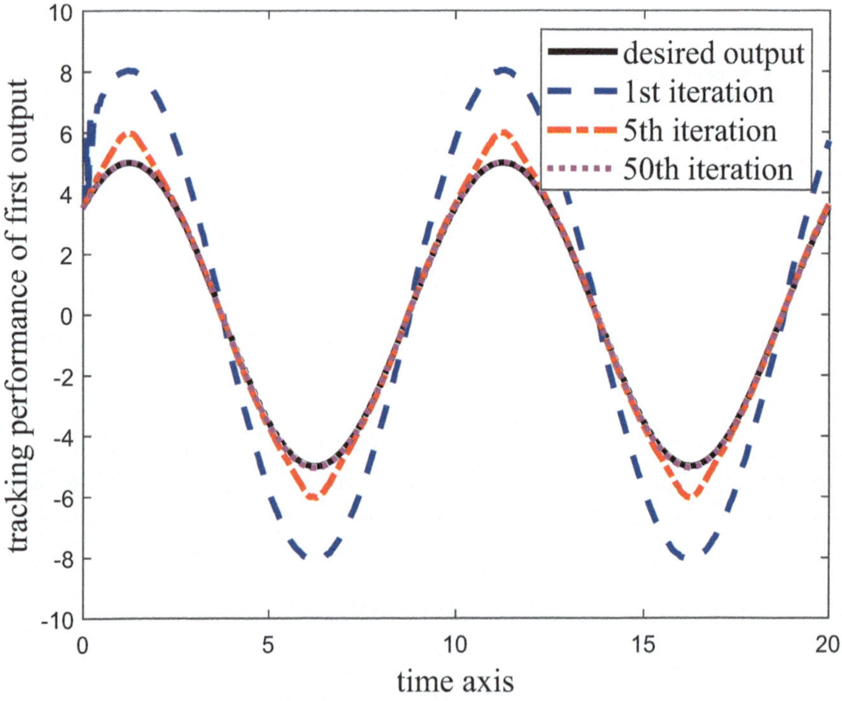

Fig. 3.9 First output tracking performance at the 1st, 5th, and 50th iteration

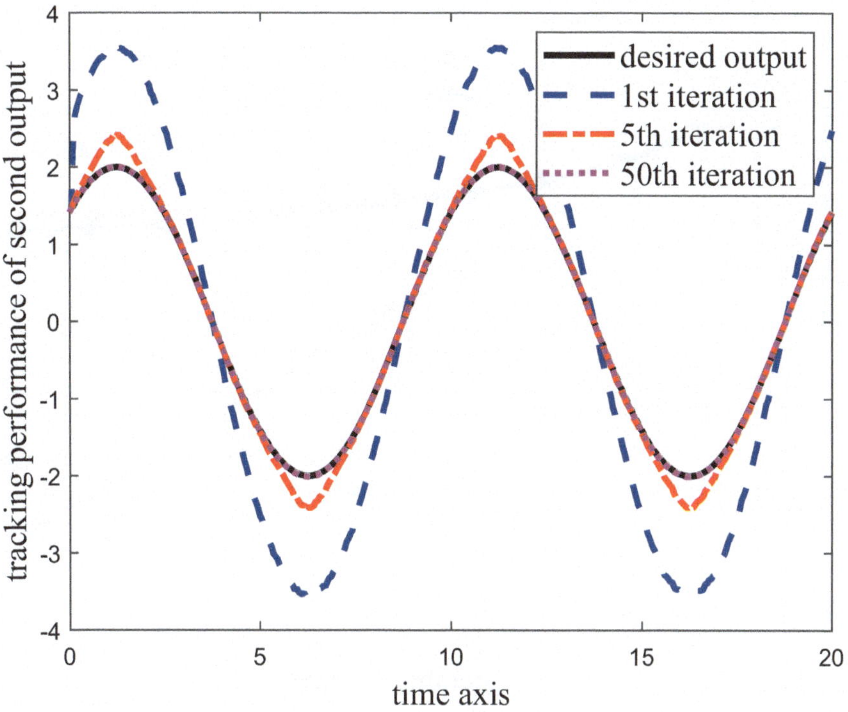

Fig. 3.10 Second output tracking performance at the 1st, 5th, and 50th iteration

reference while the dotted, dash-dot, dashed lines denote the actual outputs at the 1st, 5th, and 50th iterations, respectively. One can find that the tracking performance at the 5th iteration has been much improved, where the deviations mainly occur at peaks and valleys. The output at the 50th iteration almost coincides with the desired reference.

Similar to Example 1, Figs. 3.11 and 3.12 show the maximum tracking error profiles for actual and measured tracking errors, denoted by solid lines marked with circles and squares in two dimensions. The decreasing trend of these lines indicates the convergence of the proposed algorithm. However, due to stochastic noise, complete reduction of maximum tracking error is not achievable. To address this issue, a decreasing gain a_k is introduced which slightly slows down the convergence speed.

The simulation also includes a noise-free scenario, where the decreasing gain a_k is replaced by a constant gain of 0.15. The line marked with diamonds in both Figs. 3.11 and 3.12 represents the maximum tracking error profiles.

To demonstrate the impact of sensor saturation, we present a graphical representation of the first dimension of the output as an illustrative example. The actual output with measurement noise $y^{(1)}(t) + w^{(1)}(t)$, the measured output, and the desired reference are depicted in Fig. 3.13, denoted by dotted, dashed, and solid lines.

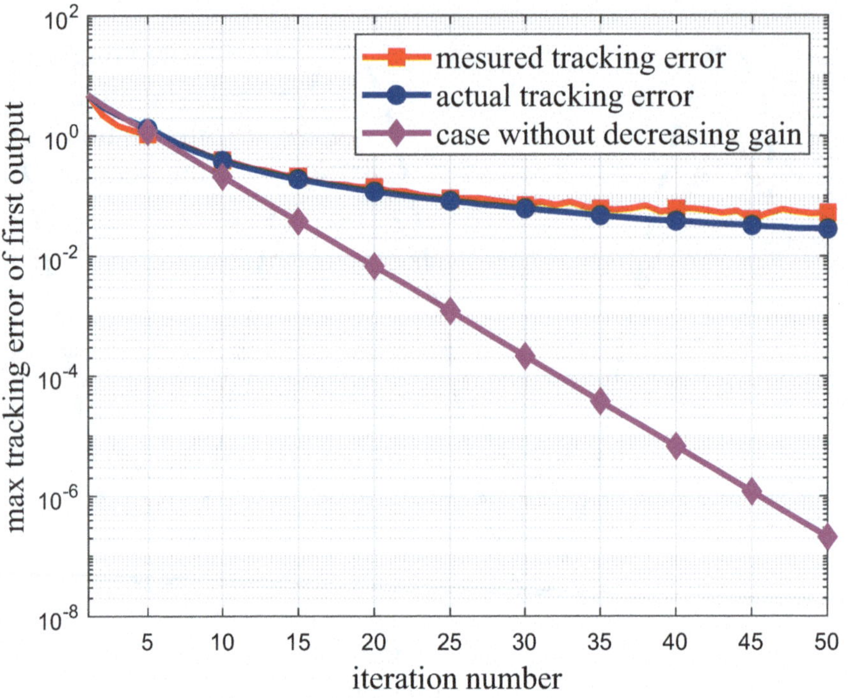

Fig. 3.11 Maximal error profiles of first output

3.4 Decreasing Gain Design in Multi-agent Systems

The ILC method is introduced in this section to address the consensus tracking problem of a multi-agent system (MAS) with random noises and measurement range limitation. A distributed learning control algorithm is proposed for all agents by leveraging their nearest neighbor's measured information from previous iterations.

3.4.1 Learning Control of Multi-agent Systems

A MAS consists of a group of agents, which is required to complete group-level cooperative tasks. The consensus problem for the MAS has attracted considerable research interest due to its wide applications in many practical areas. In the literature, there are many results devoted to studying the distributed consensus problem, see Chen et al. (2011); Han et al. (2013) for cluster consensus, Liu et al. (2012); Ni and Cheng (2010); Cheng et al. (2010) for leader-following consensus, and Li and Zhang (2010); Liu and Chen (2008) for robust consensus. These mentioned results mainly contribute to the consensus problem of the MAS in a traditional control framework;

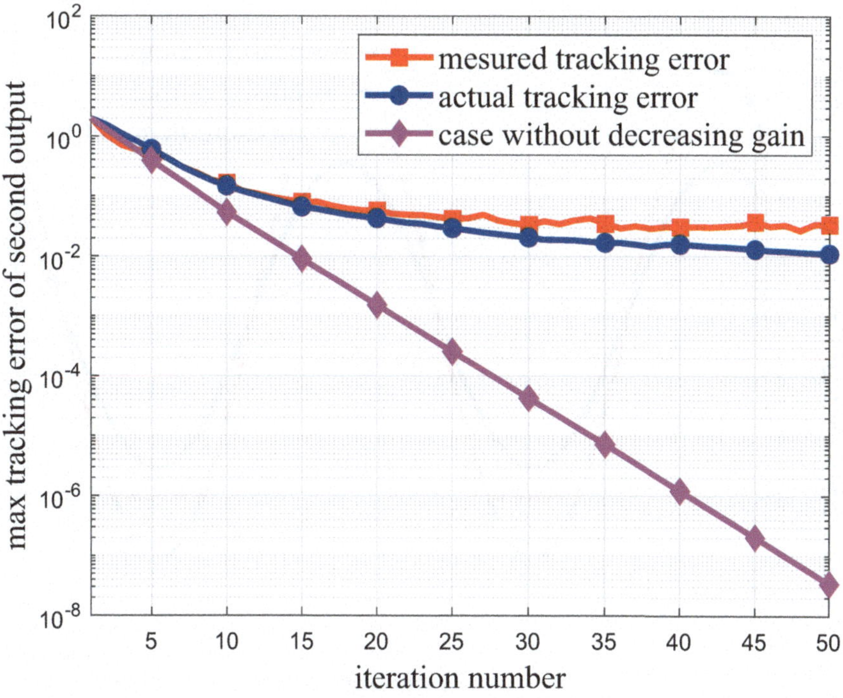

Fig. 3.12 Maximal error profiles of second output

that is, the involved systems mainly evolve along the time axis and generate time-domain-based asymptotical property.

ILC was early employed for the MAS in Ahn and Chen (2009), where the communication graph is of fixed-type. Later, the learning consensus has attracted rapid developments such as Meng et al. (2013); Yang and Xu (2012); Shen and Xu (2018) due to its distinct difference from the conventional framework that the MAS is dynamically dependent on two independent variables: time instant and iteration number. In particular, when using ILC, both time and iteration axes should be taken into account simultaneously for deriving the asymptotical property. Furthermore, in Shen and Xu (2018), the output constraint was considered, requiring each agent's output to be within a specified bound while converging towards the leader's trajectory. However, it is worth noting from Meng et al. (2013); Yang and Xu (2012); Ahn and Chen (2009); Shen and Xu (2018) that all of these findings rely on precise information exchange; specifically, the received neighbor information is assumed to be free from any noise or quality degradation. This observation motivates us to address the problem of learning consensus in the presence of communication noise and measurement range limitations. In particular, communication noise is inevitable in wireless networks. We assume that each agent measures its neighbor information using its own devices which introduces certain limitations due to measurement range.

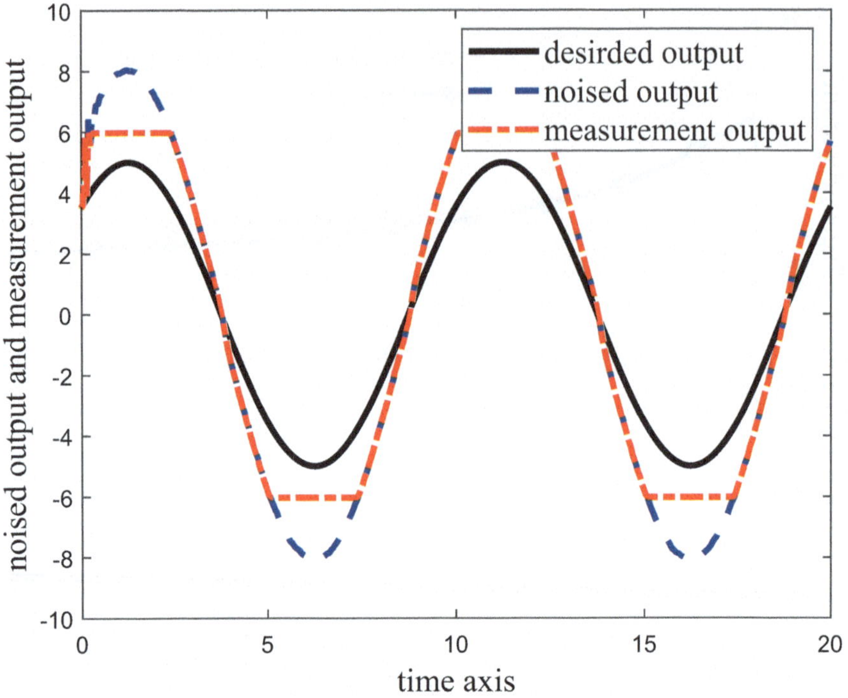

Fig. 3.13 The first system output with noises and measurement output at the 1st iteration

The studies on ILC for MAS typically assume a fixed communication topology, which represents the structure of information exchange among agents. However, in practical applications, the communication topologies can dynamically switch within a given set of possibilities. Therefore, it is crucial to develop methods that consider switching topologies. In Yang et al. (2017), an iteration-varying topology was observed to be strongly connected along the iteration axis and at least one follower agent could obtain the leader's information. A D-type distributed learning rule was proposed to ensure consensus tracking performance. Nevertheless, there remains a lack of comprehensive results regarding ILC for MAS with switching topologies; hence, we will investigate both fixed and time-switching topology cases and provide an analysis technique based on Lyapunov functions in the presence of random noises.

In this section, we investigate the ILC for a class of MAS with both fixed and time-switching topologies using lower-quality data. Specifically, each agent's output is subject to range limitation and random noise; if the output affected by random noises exceeds the measurement range, the received information becomes saturated. To address the asymptotical consensus problem in the presence of randomness and nonlinearity described above, we introduce a decreasing gain sequence into the distributed update algorithms to effectively suppress the influence of randomness along the iteration axis. We propose distributed learning algorithms and rigorously demonstrate their effectiveness in ensuring almost sure convergence of the generated input sequence to its desired value.

Graph Theory Nations: Let $\mathcal{G} = \{\mathcal{V}(\mathcal{G}), \mathcal{E}(\mathcal{G})\}$ be an nth order directed graph. The vertex set and the edge set are denoted by $\mathcal{V}(\mathcal{G})$ and $\mathcal{E}(\mathcal{G})$, respectively. In addition, $\mathcal{V}(\mathcal{G}) = \{v_i : i \in g_n\}$ and $\mathcal{E}(\mathcal{G}) = \{(v_i, v_j) : v_i, v_j \in \mathcal{V}(\mathcal{G})\}$. If (v_i, v_j) is an edge, it indicates that the vertex v_i is able to receive the information from vertex v_j; that is, v_j is a neighbor of v_i or v_i and v_j are described to be adjacent. Moreover, the index set of all neighbors of v_i is denoted by $\mathcal{N}_i = \{j : (v_i, v_j) \in \mathcal{E}(\mathcal{G})\}$. The graph $\mathcal{G}$ is described to have a spanning tree as if there is a special vertex which has paths to all other vertices, and the special vertex is said the root vertex. Moreover, $\mathcal{A} = [a_{i,j}]$ is defined as the adjacency matrix of graph $\mathcal{G}$. $a_{i,j} = 1$ if $(v_i, v_j) \in \mathcal{E}(\mathcal{G})$, otherwise $a_{i,j} = 0$. The degree matrix and the Laplacian matrix are denoted by Δ and $\mathfrak{L}$, respectively. The degree matrix Δ is defined as $\Delta = \mathrm{diag}\left\{\sum_{j\in\mathcal{N}_1} a_{1,j}, \sum_{j\in\mathcal{N}_2} a_{2,j}, \ldots, \sum_{j\in\mathcal{N}_n} a_{n,j}\right\}$, and the Laplacian matrix $\mathfrak{L}$ is defined as $\mathfrak{L} = \Delta - \mathcal{A}$.

3.4.2 Problem Formulation

In this section, we consider a MAS consisting of n MIMO discrete-time agents, with the jth agent governed by the following linear time-variant system

$$\begin{aligned} x_{k,j}(t+1) &= A(t)x_{k,j}(t) + B(t)u_{k,j}(t), \\ y_{k,j}(t) &= C(t)x_{k,j}(t), \end{aligned} \tag{3.97}$$

where the iteration number and the agent number are denoted by $k \in \mathbf{N}_+$ and $j = 1, 2, \ldots, n$, respectively. $t \in \mathbf{Z}_N$ denotes different time instant of each iteration and N is the trial length. The input vector, output vector and state vector are denoted by $u_{k,j}(t) \in \mathbf{R}^{s_i}$, $y_{k,j}(t) \in \mathbf{R}^{s_o}$ and $x_{k,j}(t) \in \mathbf{R}^{s_s}$, respectively, and $A(t) \in \mathbf{R}^{s_s \times s_s}$, $B(t) \in \mathbf{R}^{s_s \times s_i}$, and $C(t) \in \mathbf{R}^{s_o \times s_s}$ are system matrices with appropriate dimensions.

In this section, we discuss three cases of the communication mechanism in MAS.

Case 1 (Noisy output and precise transmission): Assume the ith agent is in the neighborhood of the jth agent. For the ith agent, it measures its output, which is subject to random measurement noise $v_{k,i}(t)$, using a range-limited sensor. Subsequently, it precisely broadcasts the measured output to its neighboring agents. The communication mechanism between agents i and j is illustrated in Fig. 3.14.

The measured output for the ith agent is as follows:

$$z_{k,i}(t) = \mathbf{Sat}\big(y_{k,i}(t) + v_{k,i}(t)\big),$$

where $\mathbf{Sat}(\cdot)$ is a saturation function denoting the measurement range limitation. For a vector $v = [v_1, v_2, \ldots, v_{s_o}]^T \in \mathbf{R}^{s_o}$, the saturation function is given as $\mathbf{Sat}(v) \triangleq [\mathbf{Sat}(v_1), \mathbf{Sat}(v_2), \ldots, \mathbf{Sat}(v_{s_o})]^T$, where $\mathbf{Sat}(v_i)$, $\forall i$, is defined as follows:

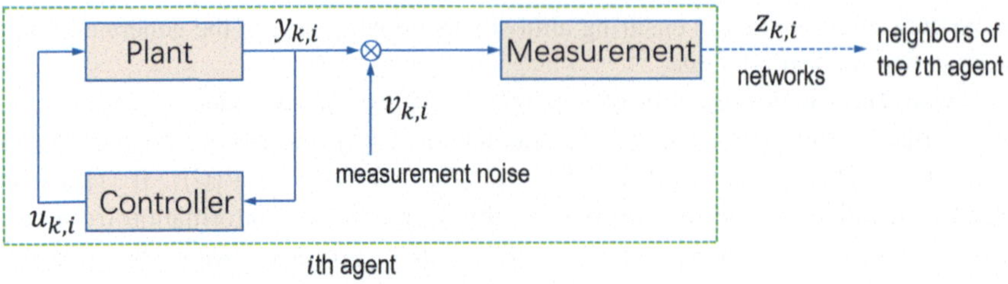

Fig. 3.14 Communication mechanism between agents in Case 1

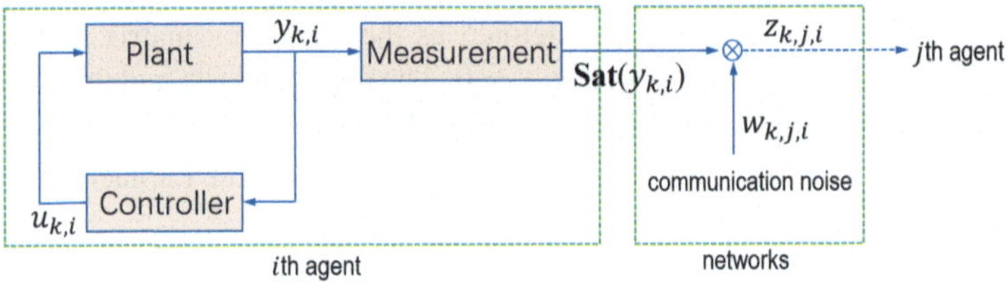

Fig. 3.15 Communication mechanism between agents in Case 2

$$\textbf{Sat}(v_i) = \begin{cases} r_0, & v_i \geqslant r_0 \\ v_i, & -r_0 < v_i < r_0 \\ -r_0, & v_i \leqslant -r_0 \end{cases} \tag{3.98}$$

for some unknown positive constant $r_0 > 0$. If the actual value exceeds the specified range, the measured information is constrained by sensor saturation; in other words, the output combined with measurement noise is guaranteed not to surpass the measurement range.

Case 2 (Precise output and noisy transmission): Each agent measures its actual output using a range-limited sensor and subsequently transmits it to other agents through unreliable networks. During the transmission from the ith agent to the jth agent, additive communication noise $\omega_{k,j,i}(t)$ is present. The communication mechanism between agents i and j is illustrated in Fig. 3.15.

The received output of the ith agent by the jth agent is described as follows:

$$z_{k,j,i}(t) = \textbf{Sat}\big(y_{k,i}(t)\big) + \omega_{k,j,i}(t).$$

Case 3 (Noisy output and noisy transmission): The present case encompasses both Cases 1 and 2, incorporating the consideration of measurement noise, $v_{k,i}(t)$, as well as communication noise, $\omega_{k,j,i}(t)$. The schematic representation of the communication mechanism between agents i and j is depicted in Fig. 3.16.

The measured output by other agents of the ith agent is described by

$$z_{k,j,i}(t) = \textbf{Sat}\big(y_{k,i}(t) + v_{k,i}(t)\big) + \omega_{k,j,i}(t).$$

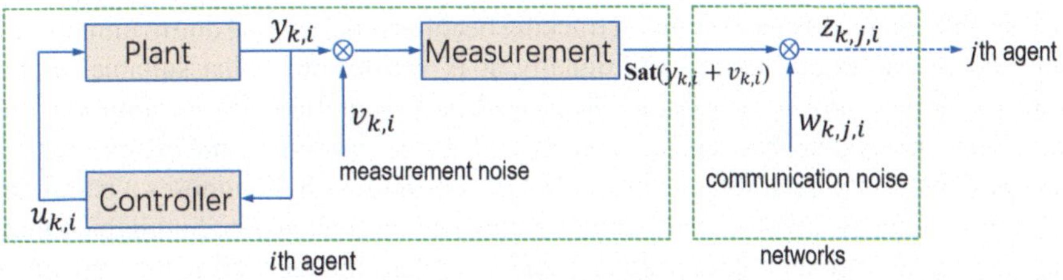

Fig. 3.16 Communication mechanism between agents in Case 3

In this section, we consider the scenario where only a subset of agents possess knowledge of the desired trajectory and treat it as a virtual leader, denoted by vertex 0 in the graph. Consequently, we can represent the complete information flow by $\overline{\mathcal{G}} = \{\mathcal{V}(\mathcal{G}) \bigcup 0, \overline{\mathcal{E}}(\mathcal{G})\}$.

The control objective for the jth agent is to design a distributed ILC scheme such that the input sequence $u_{k,j}(t)$, where $t \in \mathbf{Z}_{N-1}$, precisely drives the jth agent to track the desired trajectory $y_d(t)$ as the iteration number approaches infinity. The input sequence is generated using available neighbors' information $z_{k,i}(t)$ (in Case 1) or $z_{k,j,i}(t)$ (in Cases 2 and 3), $\forall i \in \mathcal{N}_j$.

The following assumptions are needed for MAS (3.97).

Assumption 3.7 For the jth agent, the initial states can be asymptotically reset in the sense that $x_d(0) - x_{k,j}(0) \to 0$ as the iteration number increases.

Assumption 3.8 The desired trajectory $y_d(t)$ is realisable in the sense that there exist $u_d(t)$ and $x_d(0)$ such that

$$\begin{aligned} x_d(t+1) &= A(t)x_d(t) + B(t)u_d(t), \\ y_d(t) &= C(t)x_d(t), \end{aligned} \tag{3.99}$$

Moreover, the desired trajectory $y_d(t)$ should lie in the measurable range, i.e., $|y_d^i(t)| \leqslant r_0, \forall t, i = 1, 2, \ldots, s_o$, where

$$y_d(t) = [y_d^1(t), y_d^2(t), \ldots, y_d^{s_o}(t)]^T.$$

Assumption 3.9 The matrix $C(t+1)B(t)$ is assumed to be of full-column rank for all t.

Assumption 3.10 The measurement noise $\{\nu_{k,j}(t)\}$ is a sequence of independent and identically distributed random variables such that $\mathbb{E}\nu_{k,j}(t) = 0$, $\sup_k \mathbb{E}\|\nu_{k,j}(t)\|^2 < \infty$, and $\lim_{n\to\infty} \frac{1}{n} \sum_{k=1}^{n} \nu_{k,j}(t)\nu_{k,j}^T(t) = R_{\nu,j}^t$, $\forall t$, where $R_{\nu,j}^t$ is an unknown matrix. In addition, the communication noise $\{\omega_{k,i,j}(t)\}$ has the same property as $\{\nu_{k,j}(t)\}$.

Remark 3.9 Assumption 3.8 presupposes the realizability of the desired reference signal. It is important to emphasize that the desired reference should fall within the measurable range; otherwise, perfect tracking becomes challenging due to limitations imposed by saturated output. Additionally, it is worth noting that suitable initial state $x_d(0)$ and input $u_d(t)$ exist as mentioned in Assumption 3.8. Assumption 3.9 states that the system has a relative degree of 1 and an output dimension greater than or equal to its input dimension ($s_o \geqslant s_i$). Assumption 3.10 imposes a common condition on unknown stochastic noises. Furthermore, it should be highlighted that this noise assumption is made with respect to the iteration axis rather than the time axis. In practical scenarios, processes are repeatable from one iteration to another, thus making this assumption non-restrictive.

Assumption 3.11 The directed graph $\overline{\mathcal{G}}$ has a spanning tree with the virtual leader being the root.

Remark 3.10 The virtual leader v_0 in Assumption 3.11 generates the desired trajectory $y_d(t)$ independently, unaffected by other agents within the MAS. Furthermore, Assumption 3.11 is a necessary condition for achieving solvability of the consensus tracking problem. Specifically, if there exists an isolated agent in graph $\overline{\mathcal{G}}$, it becomes impossible for that agent to track the desired trajectory.

In order to simplify the writing, for the jth agent we set $\delta u_{k,j}(t) \triangleq u_d(t) - u_{k,j}(t)$, $\delta x_{k,j}(t) \triangleq x_d(t) - x_{k,j}(t)$. The proof of convergence necessitates the inclusion of the subsequent lemmas.

Lemma 3.8 *(Hu and Hong, 2007) When the virtual agent has a path to any follower agent, the eigenvalues of $\mathcal{L} + \mathcal{D}$ have positive real parts, where $\mathcal{L}$ is the Laplacian matrix of graph $\overline{\mathcal{G}}$ and $\mathcal{D} \triangleq diag(d_1, d_2, \ldots, d_n)$.*

The jth agent is observed to have a value of $d_j = 1$ if the desired trajectory $y_d(t)$ can be accessed by the agent v_j, and $d_j = 0$ otherwise. Consequently, the pinning gain matrix $\mathcal{D}$ is associated with the accessibility of the desired trajectory for follower agents in $\overline{\mathcal{G}}$. Furthermore, when the graph $\overline{\mathcal{G}}$ is undirected, the eigenvalues of $\mathcal{L} + \mathcal{D}$ are positive real numbers.

Lemma 3.9 *Assume that Assumptions 3.7 and 3.8 hold for* (3.97). *If* $\lim_{k\to\infty} \delta \boldsymbol{u}_k (s) = \mathbf{0}$, $s = 0, 1, \ldots, t-1$, *then at the time instant* t, $\|\delta \boldsymbol{x}_k(t)\| \to 0$, *where*

$$\begin{aligned}
\delta \boldsymbol{u}_k(t) &= [\delta u_{k,1}^T(t), \delta u_{k,2}^T(t), \ldots, \delta u_{k,n}^T(t)]^T, \\
\delta \boldsymbol{x}_k(t) &= [\delta x_{k,1}^T(t), \delta x_{k,2}^T(t), \ldots, \delta x_{k,n}^T(t)]^T.
\end{aligned}$$

Proof. We prove this lemma by mathematical induction. By using (3.97) and (3.99), we can obtain the following equation:

$$\begin{aligned}
\delta \boldsymbol{x}_k(t+1) &= \big(1_n \otimes A(t)\boldsymbol{x}_d(t)\big) - \big(I_n \otimes A(t)\big)\boldsymbol{x}_k(t) \\
&\quad + \big(1_n \otimes B(t)\boldsymbol{u}_d(t)\big) - \big(I_n \otimes B(t)\big)\boldsymbol{u}_k(t) \\
&= \big(I_n \otimes A(t)\big)\delta \boldsymbol{x}_k(t) + \big(I_n \otimes B(t)\big)\delta \boldsymbol{u}_k(t), \qquad (3.100)
\end{aligned}$$

where $\boldsymbol{x}_k(t) = [x_{k,1}^T(t), x_{k,2}^T(t), \ldots, x_{k,n}^T(t)]^T$, $\boldsymbol{u}_k(t) = [u_{k,1}^T(t), u_{k,2}^T(t), \ldots, u_{k,n}^T(t)]^T$.

Initial step. For $t = 0$, Assumption 3.7 implies that the first term on the right-hand side of last equation tends to zero. In addition, $\lim_{k\to\infty} \delta\boldsymbol{u}_k(0) = \boldsymbol{0}$ leads that the second term on the right-hand side of last equation also tends to zero. Therefore, we have $\|\delta\boldsymbol{x}_k(1)\| \to 0$ as $k \to \infty$.

Inductive step. Assume the conclusion of the lemma is true for $s = 0, 1, \ldots, t-1$, then we have $\|\delta\boldsymbol{x}_k(t)\| \to 0$ as $k \to \infty$. For the time instant t, under the condition $\lim_{k\to\infty} \delta\boldsymbol{u}_k(t) = \boldsymbol{0}$, by using the same steps as that used above, we find that $\|\delta\boldsymbol{x}_k(t+1)\| \to 0$ as $k \to \infty$. That is, the conclusion is valid for t. The proof is thus completed.

Considering Lemma 3.9, if the input errors at all previous time instants $\delta\boldsymbol{u}_k(s)$, $s = 0, 1, \ldots, t-1$ approach zero, then the state error at the current time instant $\delta\boldsymbol{x}_k(t)$ will also approach zero. Specifically, as long as the convergence of the input sequence can be achieved asymptotically, it guarantees the asymptotic convergence of tracking error. Therefore, the objective of analysis in the next subsection is to demonstrate precise convergence of the input sequence towards the desired input defined in Assumption 3.8. The detailed analysis will be presented in subsequent subsections.

3.4.3 *Consensus Convergence Under Fixed Topology Case*

The distributed ILC algorithm will be designed in this subsection with an appropriate gain matrix, establishing the connection between the actual output and the measured output during the design process. Furthermore, the main theorem on asymptotic convergence will be provided. For smooth readability, a detailed proof is included in the Appendix as mentioned earlier. It should be noted that only measured information, rather than the actual output, is utilized in the algorithm design. In the subsequent part of this section, we elaborate on three separate cases.

3.4.3.1 Case 1: Noisy Output and Precise Transmission

Let the stacked vector of $y_{k,j}$ and $z_{k,j}$ for all agents be as follows:

$$\boldsymbol{y}_k(t) = [y_{k,1}^T(t), y_{k,2}^T(t), \ldots, y_{k,n}^T(t)]^T,$$
$$\boldsymbol{z}_k(t) = [z_{k,1}^T(t), z_{k,2}^T(t), \ldots, z_{k,n}^T(t)]^T.$$

For the jth agent, the measured error at the kth iteration is defined as

$$e_{k,j}(t) = y_d(t) - z_{k,j}(t). \tag{3.101}$$

For the jth agent, $\xi_{k,j}(t)$ denotes the distributed information at the kth iteration:

$$\xi_{k,j}(t) = \sum_{i \in \mathcal{N}_j} a_{j,i}(z_{k,i}(t) - z_{k,j}(t)) + d_j(y_d(t) - z_{k,j}(t)). \tag{3.102}$$

The first term on the right-hand side of (3.102) represents the inter-agent information exchange between agent j and its neighboring agents. Here, $a_{j,i} = 1$ if $(j, i) \in \bar{\mathcal{E}}(\mathcal{G})$, indicating that agent i is a neighbor of agent j. The set of neighbors for agent j is denoted as $\mathcal{N}_j$. On the other hand, the last term on the right-hand side of (3.102) signifies the communication of information between a virtual leader and agent j.

Using (3.102), we can define the distributed P-type ILC update law for the jth agent as follows:

$$u_{k+1,j}(t) = u_{k,j}(t) + \lambda_k \Gamma(t)\xi_{k,j}(t+1), \quad j = 1, 2, \ldots, n, \tag{3.103}$$

where λ_k is a decreasing gain such that $\lambda_k > 0$, $\lim_{k\to\infty} \lambda_k \to 0$, $\sum_{k=1}^{\infty} \lambda_k = \infty$ and $\sum_{k=1}^{\infty} \lambda_k^2 < \infty$. In addition, the gain matrix is denoted by $\Gamma(t) \in \mathbf{R}^{s_i \times s_o}$ for all agents and $\Gamma(t)$ is bounded, i.e., $\|\Gamma(t)\| \le u_\Gamma$ with u_Γ being a suitable positive constant for $k \in \mathbb{N}_+$ and $t \in \mathbb{Z}_{N-1}$.

By using (3.101), the distributed information measured by the jth agent in (3.102) can be rewritten as follows:

$$\xi_{k,j}(t) = \sum_{i \in \mathcal{N}_j} a_{j,i}(e_{k,j}(t) - e_{k,i}(t)) + d_j(e_{k,j}(t)). \tag{3.104}$$

The equation (3.104) can be reformulated in a concise manner as follows:

$$\boldsymbol{\xi}_k(t) = [(\mathcal{L} + \mathcal{D}) \otimes I_{s_o}]\boldsymbol{e}_k(t), \tag{3.105}$$

where

$$\boldsymbol{e}_k(t) = [e_{k,1}^T(t), e_{k,2}^T(t), \ldots, e_{k,n}^T(t)]^T,$$
$$\boldsymbol{\xi}_k(t) = [\xi_{k,1}^T(t), \xi_{k,2}^T(t), \ldots, \xi_{k,n}^T(t)]^T.$$

For the sake of simplicity in our expressions, we consider homogeneous agents without any loss of generality; that is to say, each agent possesses identical dynamics. Furthermore, owing to the uniform learning structure among all agents, we can rewrite (3.103) concisely:

$$\boldsymbol{u}_{k+1}(t) = \boldsymbol{u}_k(t) + \lambda_k[(\mathcal{L} + \mathcal{D}) \otimes \Gamma(t)]\boldsymbol{e}_k(t+1). \tag{3.106}$$

Remark 3.11 For the jth agent, the initial control input $\boldsymbol{u}_{0,j}(t)$ can be arbitrarily chosen. Without loss of generality, we can set the initial control input $\boldsymbol{u}_{0,j}(t)$ to zero for simplicity, i.e., $\boldsymbol{u}_{0,j}(t) \equiv 0$ for all t. Moreover, feedback can be employed in the MAS to design $\boldsymbol{u}_{0,j}(t)$ and enhance transient behavior during the learning process.

Remark 3.12 We focus on the homogeneous MAS in this section, ensuring that the direction regulation matrix remains identical for all agents. However, it is worth noting that the direction regulation matrix can vary across agents within a heterogeneous MAS, as long as it satisfies the condition stated in Theorem 3.8. If we extend our findings to a heterogeneous MAS, it becomes necessary for each agent to have its own distinct direction regulation matrix. By making slight adjustments in notation and definitions, we can apply the following derivations to accommodate a heterogeneous MAS.

From (3.101) we have

$$\begin{aligned} e_{k,j}(t+1) =& y_d(t+1) - \mathbf{Sat}\big(y_{k,j}(t+1) + \nu_{k,j}(t+1)\big) \\ =& \Upsilon_{k,j}(t)\big(y_d(t+1) - y_{k,j}(t+1) - \nu_{k,j}(t+1)\big), \end{aligned} \tag{3.107}$$

where $\Upsilon_{k,j}(t) = \text{diag}\{\gamma_{k,j}^{(1)}(t), \cdots, \gamma_{k,j}^{(s_o)}(t)\}$.

Let $y_{k,j}(t) + \nu_{k,j}(t) = [m_{k,j}^{(1)}(t), \cdots, m_{k,j}^{(s_o)}(t)]^T$. The transform factor $\gamma_{k,j}^{(i)}(t)$ is described as

$$\gamma_{k,j}^{(i)}(t) = \begin{cases} \dfrac{y_d^{(i)}(t+1) - r_0}{y_d^{(i)}(t+1) - m_{k,j}^{(i)}(t+1)}, & \text{if } m_{k,j}^{(i)}(t+1) \geqslant r_0 \\ 1, & \text{if } -r_0 < m_{k,j}^{(i)}(t+1) < r_0 \\ \dfrac{y_d^{(i)}(t+1) + r_0}{y_d^{(i)}(t+1) - m_{k,j}^{(i)}(t+1)}, & \text{if } m_{k,j}^{(i)}(t+1) \leqslant -r_0. \end{cases} \tag{3.108}$$

The desired reference signal, denoted as $y_d(t)$, should be within the measurable range. Hence, it is evident that $0 < \gamma_{k,j}^{(i)}(t) \leqslant 1$, for all $j = 1, 2, \ldots, n$, $\forall t, k$ and $i = 1, \ldots, s_o$. Consequently, the transform factor matrix $\Upsilon_{k,j}(t)$ becomes a special diagonal matrix with positive real numbers not exceeding 1 as its diagonal elements; in other words, $\|\Upsilon_{k,j}(t)\| \leqslant 1$.

Remark 3.13 The decreasing gain sequence is introduced to mitigate the influence of stochastic noises and ensure the asymptotic convergence of the actual output. Furthermore, although utilizing the actual output information $\boldsymbol{y}_k(t)$ instead of the measured output information $z_k(t)$ in learning algorithm (3.103) yields better results, it is not available in MAS. Despite potentially slowing down convergence speed, consensus tracking problem in MAS (3.97) is indeed achieved.

Remark 3.14 From (3.107) we can easily obtain

$$\begin{aligned}
&\|y_d(t+1) - \mathbf{Sat}\big(y_{k,j}(t+1) + \nu_{k,j}(t+1)\big)\| \\
=& \|\Upsilon_{k,j}(t)\big(y_d(t+1) - y_{k,j}(t+1) - \nu_{k,j}(t+1)\big)\| \\
\leq& \|\Upsilon_{k,j}(t)\|\|y_d(t+1) - y_{k,j}(t+1) - \nu_{k,j}(t+1)\| \\
\leq& \|y_d(t+1) - y_{k,j}(t+1) - \nu_{k,j}(t+1)\|.
\end{aligned}$$

The stochastic noise $\nu_{k,j}(t)$ plays a crucial role in the MAS, rendering the conventional contraction mapping method unsuitable. To address this issue, we introduce a transform factor $\gamma_{k,j}^{(i)}(t)$ to characterize the contraction coefficient for the ith dimension of the associated vectors. Furthermore, it is important to emphasize that this contraction coefficient $\gamma_{k,j}^{(i)}(t)$ is virtual; in other words, its sole purpose lies in facilitating convergence analysis.

We define

$$e'_{k,j}(t) = y_d(t) - y_{k,j}(t) - \nu_{k,j}(t).$$

In a concise format, we can rephrase the aforementioned equation as follows:

$$\begin{aligned}
\boldsymbol{e}'_k(t) &= 1_n \otimes y_d(t) - \boldsymbol{y}_k(t) - \boldsymbol{\nu}_k(t), \\
\boldsymbol{e}'_k(t) &= [e'^{\,T}_{k,1}(t), e'^{\,T}_{k,2}(t), \ldots, e'^{\,T}_{k,n}(t)]^T,
\end{aligned} \tag{3.109}$$

where $\boldsymbol{\nu}_k(t) = [\nu_{k,1}^T(t), \nu_{k,2}^T(t), \ldots, \nu_{k,n}^T(t)]^T$.

Using (3.107) and (3.109), we can rewrite (3.106) as follows:

$$\boldsymbol{u}_{k+1}(t) = \boldsymbol{u}_k(t) + \lambda_k[(\mathfrak{L} + \mathfrak{D}) \otimes \Gamma(t)]\Upsilon_k(t)\boldsymbol{e}'_k(t+1), \tag{3.110}$$

where

$$\Upsilon_k(t) = \mathrm{diag}\{\Upsilon_{k,1}(t), \Upsilon_{k,2}(t), \ldots, \Upsilon_{k,n}(t)\}.$$

From (3.97), (3.99), (3.100) and (3.107), we can reformulate the learning algorithm (3.110) as follows:

$$\begin{aligned}
\boldsymbol{u}_{k+1}(t) = \boldsymbol{u}_k(t) &+ \lambda_k\big[(\mathfrak{L} + \mathfrak{D}) \otimes \Gamma(t)\big]\Upsilon_k(t)\Big[I_n \otimes \big(C(t+1)B(t)\big)\Big]\delta\boldsymbol{u}_k(t) \\
&+ \lambda_k\big[(\mathfrak{L} + \mathfrak{D}) \otimes \Gamma(t)\big]\Upsilon_k(t)\Big[I_n \otimes \big(C(t+1)A(t)\big)\Big]\delta\boldsymbol{x}_k(t) \\
&- \lambda_k\big[(\mathfrak{L} + \mathfrak{D}) \otimes \Gamma(t)\big]\Upsilon_k(t)\boldsymbol{\nu}_k(t+1).
\end{aligned} \tag{3.111}$$

The first theorem is presented below.

Theorem 3.8 *Assume that Assumptions 3.7-3.11 hold for MAS* (3.97) *and the learning consensus algorithm* (3.106). *The communication mechanism between two neighbor agents is modeled in Case 1. If the direction regulation matrix $\Gamma(t)$ satisfies that* *all eigenvalues of* $(\mathcal{L}+\mathcal{D})\otimes\big(\Gamma(t)C(t+1)B(t)\big)$ *are with positive real parts for all time instances, then the actual output* $\boldsymbol{y}_k(t)$ *will converge to the desired reference vector* $1_n\otimes y_d(t)$, $\forall t$, *and the input vector* $\boldsymbol{u}_k(t)$ *which is generated by the learning algorithm* (3.106) *converges to the desired input vector* $1_n\otimes u_d(t)$ *as* $k\to\infty$, $t=0,1,\dots,N-1$. *That is, the consensus tracking problem for the MAS is solved as* $k\to\infty$.

Proof. It is sufficient to prove that $\boldsymbol{u}_k(t)\to 1_n\otimes u_d(t)$ a.s., $t=0,1,\dots,N-1$ based on (3.97) and Lemma 3.9. To this end, with the help of Lemma 3.8 and 3.9, we use the mathematical induction method to obtain this proof along the time axis t.

Initial step: First, we prove the validity of the Theorem 3.8 for $t=0$. In this case, (3.111) can be reformulated as

$$\begin{aligned}\boldsymbol{u}_{k+1}(0)=\boldsymbol{u}_k(0)&+\lambda_k\big[(\mathcal{L}+\mathcal{D})\otimes\Gamma(0)\big]\Upsilon_k(0)\Big[I_n\otimes\big(C(1)B(0)\big)\Big]\delta\boldsymbol{u}_k(0)\\&+\lambda_k\big[(\mathcal{L}+\mathcal{D})\otimes\Gamma(0)\big]\Upsilon_k(0)\Big[I_n\otimes\big(C(1)A(0)\big)\Big]\delta\boldsymbol{x}_k(0)\\&-\lambda_k\big[(\mathcal{L}+\mathcal{D})\otimes\Gamma(0)\big]\Upsilon_k(0)\boldsymbol{v}_k(1).\end{aligned}\tag{3.112}$$

Subtracting both sides of last equation from $1_n\otimes\boldsymbol{u}_d(0)$, then we have

$$\begin{aligned}\delta\boldsymbol{u}_{k+1}(0)=&\Big\{I-\lambda_k\big[(\mathcal{L}+\mathcal{D})\otimes\Gamma(0)\big]\Upsilon_k(0)\Big[I_n\otimes\big(C(1)B(0)\big)\Big]\Big\}\delta\boldsymbol{u}_k(0)\\&-\lambda_k\big[(\mathcal{L}+\mathcal{D})\otimes\Gamma(0)\big]\Upsilon_k(0)\Big[I_n\otimes\big(C(1)A(0)\big)\Big]\delta\boldsymbol{x}_k(0)\\&+\lambda_k\big[(\mathcal{L}+\mathcal{D})\otimes\Gamma(0)\big]\Upsilon_k(0)\boldsymbol{v}_k(1).\end{aligned}\tag{3.113}$$

It is noted that in system (3.97) the output $y_{k,j}(1)$ $j=1,2,3,\dots,n$ is bounded. Furthermore, combing with the Assumption 3.10, we can conclude that there exists a nonzero lower bound of $\gamma_{k,j}^{(i)}(0)$; that is, there is a suitable $0<\nu<1$ that $\gamma_{k,j}^{(i)}(0)>\nu$, $\forall k,j$. In addition, we can derive that $\|\Upsilon_k(0)\|>\nu$.

Set

$$\begin{aligned}\Phi_{l,m}\triangleq&\Big\{I-\lambda_l\big[(\mathcal{L}+\mathcal{D})\otimes\Gamma(0)\big]\Upsilon_l(0)\Big[I_n\otimes\big(C(1)B(0)\big)\Big]\Big\}\cdots\\&\Big\{I-\lambda_m\big[(\mathcal{L}+\mathcal{D})\otimes\Gamma(0)\big]\Upsilon_m(0)\Big[I_n\otimes\big(C(1)B(0)\big)\Big]\Big\},\\&l\geq m,\ \Phi_{l,l+1}\triangleq I.\end{aligned}\tag{3.114}$$

It is noted that $\Phi_{l,m}$ is a successive production of the contraction $\Big\{I-\lambda_k\big[(\mathcal{L}+\mathcal{D})\otimes\Gamma(0)\big]\Upsilon_k(0)\Big[I_n\otimes\big(C(1)B(0)\big)\Big]\Big\}$ with the subscript k increasing from m up to l. It is obvious that $\Phi_{l,l}=\Big\{I-\lambda_l\big[(\mathcal{L}+\mathcal{D})\otimes\Gamma(0)\big]\Upsilon_l(0)\Big[I_n\otimes\big(C(1)B(0)\big)\Big]\Big\}$ is a special case. In addition, When $l < m$, the notation $\Phi_{l,m}$ has no sense, and thus, we define $\Phi_{l,m} = I$ for the completeness of the notation. It is clear that $\|I - \lambda_m \big[(\mathfrak{L} + \mathfrak{D}) \otimes \Gamma(0)\big] \Upsilon_m(0) \Big[I_n \otimes \big(C(1)B(0)\big)\Big] \| > 0$ for all sufficient large m.

Then for any $l \geq m$, $m > 0$, it is sufficient to verify that

$$\|\Phi_{l,m}\| \leq c_2 \exp(-c_1 \sum_{k=m}^{l} \lambda_k) \tag{3.115}$$

for some suitable $c_1, c_2 > 0$.

Note that $\Upsilon_l(0) \geq \nu I$ and all eigenvalues of $[(\mathfrak{L} + \mathfrak{D}) \otimes (\Gamma(t)C(t+1)B(t))]$ are with positive real parts. Therefore, by the Lyapunov stability for linear system, for any given positive definite matrix Q, there exists a positive definite matrix P such that

$$\begin{aligned} -\Big\{ P\Big[(\mathfrak{L} + \mathfrak{D}) \otimes \big(\Gamma(t)C(t+1)B(t)\big)\Big] \\ + \Big[(\mathfrak{L} + \mathfrak{D}) \otimes \big(\Gamma(t)C(t+1)B(t)\big)\Big]^T P \Big\} = -Q. \end{aligned}$$

Here, we let $Q = 2I_{ns_i}$ and compute the corresponding matrix P. Moreover, $\Upsilon_k(0)$ is a diagonal matrix. For a sufficient large number k, we have

$$\begin{aligned} -\Big\{ P\big[(\mathfrak{L} + \mathfrak{D}) \otimes \Gamma(0)\big] \Upsilon_k(0) \Big[I_n \otimes \big(C(1)B(0)\big)\Big] \\ + \big[(\mathfrak{L} + \mathfrak{D}) \otimes \Gamma(0)\big] \Upsilon_k(0) \Big[I_n \otimes \big(C(1)B(0)\big)\Big]^T P \Big\} \leq -\nu I. \end{aligned}$$

For simplicity, we denote $W_k \triangleq -\big[(\mathfrak{L} + \mathfrak{D}) \otimes \Gamma(0)\big] \Upsilon_k(0) \Big[I_n \otimes \big(C(1)B(0)\big)\Big]$. Thus, for a sufficiently large number l,

$$\begin{aligned} \Phi_{l,m}^T P \Phi_{l,m} &= \Phi_{l-1,m}^T (I + \lambda_l W_l)^T P (I + \lambda_l W_l) \Phi_{l-1,m} \\ &= \Phi_{l-1,m}^T (P + \lambda_l^2 W_l^T P W_l + \lambda_l W_l^T P + \lambda_l P W_l) \Phi_{l-1,m} \\ &\leq \Phi_{l-1,m}^T (P + \lambda_l^2 W_l^T P W_l - \nu \lambda_l I) \Phi_{l-1,m} \\ &= \Phi_{l-1,m}^T P^{\frac{1}{2}} (I - \nu \lambda_l P^{-1} + \lambda_l^2 P^{-\frac{1}{2}} W_l^T P W_l P^{-\frac{1}{2}}) P^{\frac{1}{2}} \Phi_{l-1,m}. \end{aligned} \tag{3.116}$$

Without loss of generality, when the iteration number l is sufficiently large, we obtain

$$\|I - \nu \lambda_l P^{-1} + \lambda_l^2 P^{-\frac{1}{2}} W_l^T P W_l P^{-\frac{1}{2}}\| \leq 1 - 2c_1 \lambda_l \leq \exp(-2c_1 \lambda_l), \tag{3.117}$$

where $c_1 > 0$ is a suitable constant and the basic inequality $1 - x \leq e^{-x}$ is applied.

Noticing the boundedness of $P^{-\frac{1}{2}}W_l^T P W_l P^{-\frac{1}{2}}$ and combining (3.116) and (3.117) leads to

$$\Phi_{l,m}^T P \Phi_{l,m} \leq c_3 \exp(-2c_1 \sum_{k=m}^{l} \lambda_k) I$$

for some suitable $c_3 > 0$. Thus,

$$\|\Phi_{l,m}\| \leq \lambda_{min}^{-\frac{1}{2}}(P)\sqrt{c_3} \exp(-c_1 \sum_{k=m}^{l} \lambda_k). \tag{3.118}$$

Consequently, if we define $c_2 \triangleq \lambda_{min}^{-\frac{1}{2}}(P)\sqrt{c_3}$, then (3.115) is guaranteed.

It follows from (3.113) that

$$\begin{aligned}
&\delta \boldsymbol{u}_{k+1}(0)\\
&= \Phi_{k,1}\delta \boldsymbol{u}_1(0) - \sum_{m=1}^{k} \Phi_{k,m+1}\lambda_m \left[(\mathcal{L} + \mathcal{D}) \otimes \Gamma(0)\right] \Upsilon_m(0) \left[I_n \otimes \left(C(1)A(0)\right)\right] \delta \boldsymbol{x}_m(0)\\
&+ \sum_{m=1}^{k} \Phi_{k,m+1}\lambda_m \left[(\mathcal{L} + \mathcal{D}) \otimes \Gamma(0)\right] \Upsilon_m(0) \boldsymbol{v}_m(1),
\end{aligned} \tag{3.119}$$

where the first term on the right-hand side of (3.119) tends to 0 as $k \to \infty$ because of (3.118).

From Assumptions 3.7, it is obvious that $\|\delta \boldsymbol{x}_k(0)\| \to 0$ as iteration number increases. Hence, for any $\varepsilon > 0$, there exists a sufficient large integer k_1 such that $\|\delta \boldsymbol{x}_k(0)\| < \varepsilon$, $\forall k \geqslant k_1$. Then for the second term of (3.119), we have

$$\begin{aligned}
&\sum_{m=1}^{k} \|\Phi_{k,m+1}\lambda_m \left[(\mathcal{L} + \mathcal{D}) \otimes \Gamma(0)\right] \Upsilon_m(0) \left[I_n \otimes \left(C(1)A(0)\right)\right] \delta \boldsymbol{x}_m(0)\|\\
&\leq \sum_{m=1}^{k_1-1} \|\Phi_{k,m+1}\lambda_m \left[(\mathcal{L} + \mathcal{D}) \otimes \Gamma(0)\right] \Upsilon_m(0) \left[I_n \otimes \left(C(1)A(0)\right)\right] \delta \boldsymbol{x}_m(0)\|\\
&+ \sum_{m=k_1}^{k} \|\Phi_{k,m+1}\lambda_m \left[(\mathcal{L} + \mathcal{D}) \otimes \Gamma(0)\right] \Upsilon_m(0) \left[I_n \otimes \left(C(1)A(0)\right)\right] \delta \boldsymbol{x}_m(0)\|\\
&\leq \sum_{m=1}^{k_1-1} \|\Phi_{k,m+1}\| \|\lambda_m \left[(\mathcal{L} + \mathcal{D}) \otimes \Gamma(0)\right] \Upsilon_m(0) \left[I_n \otimes \left(C(1)A(0)\right)\right] \| \|\delta \boldsymbol{x}_m(0)\|\\
&+ \sum_{m=k_1}^{k} \|\Phi_{k,m+1}\| \|\lambda_m \left[(\mathcal{L} + \mathcal{D}) \otimes \Gamma(0)\right] \Upsilon_m(0) \left[I_n \otimes \left(C(1)A(0)\right)\right] \| \|\delta \boldsymbol{x}_m(0)\|.
\end{aligned} \tag{3.120}$$

It is noted that the first term on the right-hand side of (3.120) is a finite summation, and each term tends to zero as $k \to \infty$ because the starting number is bounded by k_1. Therefore, the finite summation tends to 0 as iteration number increases. Note that $\lim_{k\to\infty} \lambda_k \to 0$ and $\|\Gamma\| \leqslant u_\Gamma$, then for the second term of (3.120), we have the following estimation:

$$
\begin{aligned}
&\sum_{m=k_1}^{k} \|\Phi_{k,m+1}\|\|\lambda_m \left[(\mathcal{L}+\mathcal{D})\otimes\Gamma(0)\right]\Upsilon_m(0)\Big[I_n\otimes\big(C(1)A(0)\big)\Big]\|\|\delta\boldsymbol{x}_m(0)\| \\
&\le \varepsilon c_2\|(\mathcal{L}+\mathcal{D})\otimes\Gamma(0)\|\|I_n\otimes\big(C(1)A(0)\big)\|\sum_{m=k_1}^{k}\lambda_m\exp(-c_1\sum_{l=m+1}^{k}\lambda_l) \\
&\le 2\varepsilon c_2\|(\mathcal{L}+\mathcal{D})\otimes\Gamma(0)\|\|I_n\otimes\big(C(1)A(0)\big)\|\sum_{m=k_1}^{k}(\lambda_m-\frac{c_1\lambda_m^2}{2})\exp(-c_1\sum_{l=m+1}^{k}\lambda_l) \\
&\le \frac{2\varepsilon c_2}{c_1}\|(\mathcal{L}+\mathcal{D})\otimes\Gamma(0)\|\|I_n\otimes\big(C(1)A(0)\big)\|\sum_{m=k_1}^{k}(1-e^{-c_1\lambda_m})\exp(-c_1\sum_{l=m+1}^{k}\lambda_l) \\
&= \frac{2\varepsilon c_2}{c_1}\|(\mathcal{L}+\mathcal{D})\otimes\Gamma(0)\|\|I_n\otimes(C(1)A(0))\|\sum_{m=k_1}^{k}\Big[\exp(-c_1\sum_{l=m+1}^{k}\lambda_l)-\exp(-c_1\sum_{l=m}^{k}\lambda_l)\Big] \\
&\le \frac{2\varepsilon c_2}{c_1}\|(\mathcal{L}+\mathcal{D})\otimes\Gamma(0)\|\|I_n\otimes\big(C(1)A(0)\big)\|.
\end{aligned}
$$

Hence, the second term on the right-hand side of (3.120) tends to 0 as $k \to \infty$. In summary, (3.120) tends to 0 as $k \to \infty$.

For the last term of (3.119), based on Assumption 3.10 and the fact that the gain matrix is bounded, it is clear that $\{\boldsymbol{\nu}_k(1)\}$ is a sequence of independent and identical distributed random variables. Furthermore,

$$
\begin{aligned}
&\sum_{k=1}^{\infty}\mathbb{E}\|\lambda_k\Upsilon_k(0)\boldsymbol{\nu}_k(1)\|^2 \\
\le &\sum_{k=1}^{\infty}\mathbb{E}\|\lambda_k\boldsymbol{\nu}_k(1)\|^2 \le \|\boldsymbol{R}_\nu^1\|\sum_{k=1}^{\infty}(\lambda_k)^2 < \infty,
\end{aligned}
$$

where

$$\boldsymbol{R}_\nu^1 = [{R_{\nu,1}^1}^T, {R_{\nu,2}^1}^T, \ldots, {R_{\nu,n}^1}^T]^T.$$

This fact implies $\sum_{k=1}^{\infty}\lambda_k\Upsilon_k(0)\boldsymbol{\nu}_k(1) < \infty$ a.s. by the Khintchine-Kolmogorov convergence theorem (Chow and Teicher, 2003).

Now we set $\boldsymbol{\theta}_k = \sum_{m=1}^{k}\lambda_m\Upsilon_m(0)\boldsymbol{\nu}_m(1)$, $\boldsymbol{\theta}_0 = \boldsymbol{0}$. Then $\boldsymbol{\theta}_k \to \boldsymbol{\theta}$ as $k \to \infty$. For any $\varepsilon > 0$, there is a sufficient large k_2 such that $\|\boldsymbol{\theta}_k - \boldsymbol{\theta}\| \le \varepsilon, \forall k \ge k_2$.

$$
\begin{aligned}
&\sum_{m=1}^{k} \Phi_{k,m+1}\lambda_m \big[(\mathcal{L}+\mathcal{D})\otimes\Gamma(0)\big]\Upsilon_m(0)\boldsymbol{v}_m(1)\\
&=\sum_{m=1}^{k} \Phi_{k,m+1}\big[(\mathcal{L}+\mathcal{D})\otimes\Gamma(0)\big](\boldsymbol{\theta}_m-\boldsymbol{\theta}_{m-1})\\
&=\big[(\mathcal{L}+\mathcal{D})\otimes\Gamma(0)\big]\boldsymbol{\theta}_k-\sum_{m=1}^{k}(\Phi_{k,m+1}-\Phi_{k,m})\big[(\mathcal{L}+\mathcal{D})\otimes\Gamma(0)\big]\boldsymbol{\theta}_{m-1}\\
&=\big[(\mathcal{L}+\mathcal{D})\otimes\Gamma(0)\big]\boldsymbol{\theta}_k-\sum_{m=1}^{k}(\Phi_{k,m+1}-\Phi_{k,m})\big[(\mathcal{L}+\mathcal{D})\otimes\Gamma(0)\big]\boldsymbol{\theta}\\
&-\sum_{m=1}^{k}(\Phi_{k,m+1}-\Phi_{k,m})\big[(\mathcal{L}+\mathcal{D})\otimes\Gamma(0)\big](\boldsymbol{\theta}_{m-1}-\boldsymbol{\theta})\\
&=\big[(\mathcal{L}+\mathcal{D})\otimes\Gamma(0)\big]\boldsymbol{\theta}_k-\big[(\mathcal{L}+\mathcal{D})\otimes\Gamma(0)\big]\boldsymbol{\theta}+\Phi_{k,1}\big[(\mathcal{L}+\mathcal{D})\otimes\Gamma(0)\big]\boldsymbol{\theta}\\
&-\sum_{m=1}^{k}(\Phi_{k,m+1}-\Phi_{k,m})\big[(\mathcal{L}+\mathcal{D})\otimes\Gamma(0)\big](\boldsymbol{\theta}_{m-1}-\boldsymbol{\theta}),
\end{aligned}
$$

where it is obvious that $\|\big[(\mathcal{L}+\mathcal{D})\otimes\Gamma(0)\big]\boldsymbol{\theta}_k-\big[(\mathcal{L}+\mathcal{D})\otimes\Gamma(0)\big]\boldsymbol{\theta}+\Phi_{k,1}\big[(\mathcal{L}+\mathcal{D})\otimes\Gamma(0)\big]\boldsymbol{\theta}\|\to 0$ as $k\to\infty$. For the last term, we have

$$
\begin{aligned}
&\sum_{m=1}^{k}(\Phi_{k,m+1}-\Phi_{k,m})\big[(\mathcal{L}+\mathcal{D})\otimes\Gamma(0)\big](\boldsymbol{\theta}_{m-1}-\boldsymbol{\theta})\\
&=(\sum_{m=1}^{k_2}+\sum_{m=k_2+1}^{k})(\Phi_{k,m+1}-\Phi_{k,m})\big[(\mathcal{L}+\mathcal{D})\otimes\Gamma(0)\big](\boldsymbol{\theta}_{m-1}-\boldsymbol{\theta}).
\end{aligned}
$$

It is clear that $\sum_{m=1}^{k_2}\|(\Phi_{k,m+1}-\Phi_{k,m})\big[(\mathcal{L}+\mathcal{D})\otimes\Gamma(0)\big](\boldsymbol{\theta}_{m-1}-\boldsymbol{\theta})\|\to 0$ as $k\to\infty$. Due to the fact that m is bounded by k_2, we have $\|\Phi_{k,m}\|\to 0$ as $k\to\infty$. On the other hand,

$$
\begin{aligned}
&\sum_{m=k_2+1}^{k}(\Phi_{k,m+1}-\Phi_{k,m})\big[(\mathcal{L}+\mathcal{D})\otimes\Gamma(0)\big](\boldsymbol{\theta}_{m-1}-\boldsymbol{\theta})\\
&=\sum_{m=k_2+1}^{k}\Phi_{k,m+1}\Big\{\lambda_m\big[(\mathcal{L}+\mathcal{D})\otimes\Gamma(0)\big]\Upsilon_m(0)\Big[I_n\otimes\big(C(1)B(0)\big)\Big]\Big\}\\
&\quad\times\big[(\mathcal{L}+\mathcal{D})\otimes\Gamma(0)\big](\boldsymbol{\theta}_{m-1}-\boldsymbol{\theta}).
\end{aligned}
$$

By noting that $\|\boldsymbol{\theta}_m-\boldsymbol{\theta}\|\le\varepsilon,\forall m\ge k_2,\|\Gamma\|\le u_\Gamma$ and $\|\Upsilon_k(0)\|>\nu,0<\nu<1$, the above equation can be rewritten as

$$\sum_{m=k_2+1}^{k} \|\Phi_{k,m+1}\Big\{\lambda_m\big[(\mathcal{L}+\mathcal{D})\otimes\Gamma(0)\big]\Upsilon_m(0)\Big[I_n\otimes\big(C(1)B(0)\big)\Big]\Big\}$$
$$\times\big[(\mathcal{L}+\mathcal{D})\otimes\Gamma(0)\big](\boldsymbol{\theta}_{m-1}-\boldsymbol{\theta})\|$$
$$\leq \varepsilon c_2\|(\mathcal{L}+\mathcal{D})\otimes\Gamma(0)\|\|I_n\otimes\big(C(1)B(0)\big)\|\|(\mathcal{L}+\mathcal{D})\otimes\Gamma(0)\|$$
$$\times\sum_{m=k_2+1}^{k}\lambda_m\exp(-c_1\sum_{l=m+1}^{k}\lambda_l).$$

It is clear that the above equation tends to zero as $k\to\infty$ and $\varepsilon\to 0$. Above all, $\delta\boldsymbol{u}_{k+1}(0)\to 0$ as $k\to\infty$, the conclusion of this theorem is true for $t=0$.

Inductive step: Assume that for $s=0,1,\ldots,t-1$ the theorem is true, then we continue to prove that the theorem is true for t. From (3.111), we can obtain the counterpart of (3.113) for time instant t as

$$\begin{aligned}\delta\boldsymbol{u}_{k+1}(t)=&\Big\{I-\lambda_k\big[(\mathcal{L}+\mathcal{D})\otimes\Gamma(t)\big]\Upsilon_k(t)\Big[I_n\otimes\big(C(t+1)B(t)\big)\Big]\Big\}\delta\boldsymbol{u}_k(t)\\&-\lambda_k\big[(\mathcal{L}+\mathcal{D})\otimes\Gamma(t)\big]\Upsilon_k(t)\Big[I_n\otimes\big(C(t+1)A(t)\big)\Big]\delta\boldsymbol{x}_k(t)\\&+\lambda_k\big[(\mathcal{L}+\mathcal{D})\otimes\Gamma(t)\big]\Upsilon_k(t)\boldsymbol{v}_k(t+1).\end{aligned}\tag{3.121}$$

It is easy to observe that $\delta\boldsymbol{u}_k(t)$ corresponds to $\delta\boldsymbol{u}_k(0)$, $I-\lambda_k\big[(\mathcal{L}+\mathcal{D})\otimes\Gamma(t)\big]\Upsilon_k(t)\Big[I_n\otimes\big(C(t+1)B(t)\big)\Big]$ corresponds to $I-\lambda_k\big[(\mathcal{L}+\mathcal{D})\otimes\Gamma(0)\big]\Upsilon_k(0)\Big[I_n\otimes\big(C(1)B(0)\big)\Big]$, $\big[(\mathcal{L}+\mathcal{D})\otimes\Gamma(t)\big]\Upsilon_k(t)\Big[I_n\otimes\big(C(t+1)A(t)\big)\Big]$ corresponds to $\big[(\mathcal{L}+\mathcal{D})\otimes\Gamma(0)\big]\Upsilon_k(0)\Big[I_n\otimes\big(C(1)A(0)\big)\Big]$, $\delta\boldsymbol{x}_k(t)$ corresponds to $\delta\boldsymbol{x}_k(0)$ and $\boldsymbol{v}_k(t+1)$ corresponds to $\boldsymbol{v}_k(1)$. Hence, the proof can be completed following the same steps as far as we guarantee $\|\delta\boldsymbol{x}_k(t)\|\to 0$ as $k\to\infty$. By the induction assumption we have $\|\delta\boldsymbol{u}_k(s)\|\to 0$ as $k\to\infty$ for $s=0,1,\ldots,t-1$. Based on Lemma 3.9, we can conclude that $\|\delta\boldsymbol{x}_k(t)\|\to 0$ as $k\to\infty$. As a result, we can easily prove that $\|\delta\boldsymbol{u}_k(t)\|\to 0$ as $k\to\infty$. The proof of Theorem 3.8 is completed.

3.4.3.2 Case 2: Precise Output and Noisy Transmission

The measured tracking error at the kth iteration for the jth agent is defined as follows:

$$e_{k,j}(t)=y_d(t)-\mathbf{Sat}\big(y_{k,j}(t)\big).\tag{3.122}$$

This error is not applicable to neighboring agents and its definition is solely intended for convergence analysis.

The distributed information measured at the kth iteration is described as

$$\begin{aligned}\xi_{k,j}(t) &= \sum_{i\in\mathcal{N}_j} a_{j,i}\Big(z_{k,j,i} - \mathbf{Sat}\big(y_{k,j}(t)\big)\Big) + d_j\Big(y_d(t) - \mathbf{Sat}\big(y_{k,j}(t)\big)\Big)\\ &= \sum_{i\in\mathcal{N}_j} a_{j,i}\Big(\mathbf{Sat}\big(y_{k,i}(t)\big) - \mathbf{Sat}\big(y_{k,j}(t)\big)\Big)\\ &\quad + d_j\Big(y_d(t) - \mathbf{Sat}\big(y_{k,j}(t)\big)\Big) + \sum_{i\in\mathcal{N}_j} a_{j,i}\omega_{k,j,i}(t),\end{aligned} \tag{3.123}$$

In a compact form, we can reformulate (3.123) as follows:

$$\boldsymbol{\xi}_k(t) = \big[(\mathcal{L}+\mathcal{D})\otimes I_{s_o}\big]\boldsymbol{e}_k(t) + \big[\tilde{\mathcal{A}}\otimes I_{s_o}\big]\boldsymbol{\omega}_k(t), \tag{3.124}$$

where

$$\begin{aligned}\boldsymbol{\omega}_{k,j}(t) &= [\omega_{k,j,1}^T(t), \omega_{k,j,2}^T(t), \ldots, \omega_{k,j,n}^T(t)]^T,\\ \boldsymbol{\omega}_k(t) &= [\boldsymbol{\omega}_{k,1}^T(t), \boldsymbol{\omega}_{k,2}^T(t), \ldots, \boldsymbol{\omega}_{k,n}^T(t)]^T,\\ \tilde{\mathcal{A}} &= \begin{bmatrix} \mathcal{A}_1 & 0 & \cdots & 0\\ 0 & \mathcal{A}_2 & \cdots & 0\\ \vdots & \vdots & \ddots & \vdots\\ 0 & 0 & \cdots & \mathcal{A}_n\end{bmatrix},\end{aligned}$$

$\mathcal{A}_i$ denotes the ith row of the adjacency matrix $\mathcal{A}$.

Following similar steps as in Case 1, we present the learning consensus algorithm as follows:

$$\boldsymbol{u}_{k+1}(t) = \boldsymbol{u}_k(t) + \lambda_k\big[(\mathcal{L}+\mathcal{D})\otimes\Gamma(t)\big]\boldsymbol{e}_k(t+1) + \lambda_k\big[\tilde{\mathcal{A}}\otimes\Gamma(t)\big]\boldsymbol{\omega}_k(t+1). \tag{3.125}$$

For Case 2, no measurement noise exists in the output. To connect the measured tracking error $e_{k,j}(t)$ and the actual tracking error $y_d(t) - y_{k,j}(t)$, we modify the transform factor $\gamma_{k,j}^{(i)}(t)$ as follows:

$$\gamma_{k,j}^{(i)}(t) = \begin{cases} \dfrac{y_d^{(i)}(t+1) - r_0}{y_d^{(i)}(t+1) - y_{k,j}^{(i)}(t+1)}, & \text{if } y_{k,j}^{(i)}(t+1) \geqslant r_0\\ 1, & \text{if } -r_0 < y_{k,j}^{(i)}(t+1) < r_0\\ \dfrac{y_d^{(i)}(t+1) + r_0}{y_d^{(i)}(t+1) - y_{k,j}^{(i)}(t+1)}, & \text{if } y_{k,j}^{(i)}(t+1) \leqslant -r_0\end{cases}$$

Then, we obtain

$$\begin{aligned}&\|y_d(t+1)-\mathbf{Sat}\big(y_{k,j}(t+1)\big)\|\\=&\|\Upsilon_{k,j}(t)\big(y_d(t+1)-y_{k,j}(t+1)\big)\|\\\le&\|\Upsilon_{k,j}(t)\|\|y_d(t+1)-y_{k,j}(t+1)\|\\\le&\|y_d(t+1)-y_{k,j}(t+1)\|.\end{aligned}$$

Consequently, we can rewrite (3.125) as follows:

$$\begin{aligned}\boldsymbol{u}_{k+1}(t)=\boldsymbol{u}_k(t)&+\lambda_k\big[(\mathfrak{L}+\mathfrak{D})\otimes\Gamma(t)\big]\Upsilon_k(t)\Big[I_n\otimes\big(C(t+1)B(t)\big)\Big]\delta\boldsymbol{u}_k(t)\\&+\lambda_k\big[(\mathfrak{L}+\mathfrak{D})\otimes\Gamma(t)\big]\Upsilon_k(t)\Big[I_n\otimes\big(C(t+1)A(t)\big)\Big]\delta\boldsymbol{x}_k(t)\\&+\lambda_k\big[\tilde{A}\otimes\Gamma(t)\big]\boldsymbol{\omega}_k(t+1).\end{aligned}\tag{3.126}$$

The theorem for Case 2 is presented below.

Theorem 3.9 *Assume that Assumptions 3.7-3.11 hold for MAS* (3.97) *and learning consensus algorithm* (3.125). *The communication mechanism between two neighbor agents is modeled in Case 2. If the direction regulation matrix $\Gamma(t)$ satisfies that all eigenvalues of $(\mathfrak{L}+\mathfrak{D})\otimes\big(\Gamma(t)C(t+1)B(t)\big)$ are with positive real parts for all time instances, then the actual output $\boldsymbol{y}_k(t)$ will converge to the desired reference vector $1_n\otimes y_d(t)$, $\forall t$, and the input vector $\boldsymbol{u}_k(t)$ converges to the desired input vector $1_n\otimes u_d(t)$ as $k\to\infty$, $t=0,1,\ldots,N-1$.*

Proof. We copy the learning consensus algorithm (3.126) as follows:

$$\begin{aligned}\boldsymbol{u}_{k+1}(t)=&\boldsymbol{u}_k(t)+\lambda_k\big[(\mathfrak{L}+\mathfrak{D})\otimes\Gamma(t)\big]\Upsilon_k(t)\Big[I_n\otimes\big(C(t+1)B(t)\big)\Big]\delta\boldsymbol{u}_k(t)\\&+\lambda_k\big[(\mathfrak{L}+\mathfrak{D})\otimes\Gamma(t)\big]\Upsilon_k(t)\Big[I_n\otimes\big(C(t+1)A(t)\big)\Big]\delta\boldsymbol{x}_k(t)\\&+\lambda_k\big[\tilde{A}\otimes\Gamma(t)\big]\boldsymbol{\omega}_k(t+1).\end{aligned}$$

It is noticed that the recursion is same to (3.111) except the last term on the form of additive noises. Thus, the proof can be completed following similar steps as in Theorem 3.8.

For the initial step, $t=0$. Subtracting both sides of the above equation from the desired input $\boldsymbol{u}_d(t)$ yields

$$\begin{aligned}&\delta\boldsymbol{u}_{k+1}(0)\\&=\Phi_{k,1}\delta\boldsymbol{u}_1(0)-\sum_{m=1}^{k}\Phi_{k,m+1}\lambda_m\big[(\mathfrak{L}+\mathfrak{D})\otimes\Gamma(0)\big]\Upsilon_m(0)\Big[I_n\otimes\big(C(1)A(0)\big)\Big]\delta\boldsymbol{x}_m(0)\\&-\sum_{m=1}^{k}\Phi_{k,m+1}\lambda_m[\tilde{A}\otimes\Gamma(t)\big]\boldsymbol{\omega}_m(1).\end{aligned}\tag{3.127}$$

It is evident that the expression for $\Phi_{l,m}$ as defined in (3.114) remains applicable to Case 2. Consequently, the estimation (3.118) can be employed. By utilizing this estimation of $\Phi_{l,m}$, we can deduce that the first two terms on the right-hand side of (3.127) tend to zero following analogous steps as outlined in Theorem 3.8. Regarding the last term of (3.127), it can be verified that $\sum_{k=1}^{\infty} \lambda_k \boldsymbol{\omega}_k(1) < \infty$ by observing that $\sum_{k=1}^{\infty} \mathbb{E}\|\lambda_k \boldsymbol{\omega}_k(1)\|^2 < \infty$.

We let $\boldsymbol{\theta}_k = \sum_{m=1}^{k} \lambda_m \boldsymbol{\omega}_m(1)$, $\boldsymbol{\theta}_0 = \mathbf{0}$. Then $\boldsymbol{\theta}_k \to \boldsymbol{\theta}$ as $k \to \infty$ where $\boldsymbol{\theta} = \sum_{m=1}^{\infty} \lambda_m \boldsymbol{\omega}_m(1)$. For any $\varepsilon > 0$, there exists a sufficiently large k_3 such that $\|\boldsymbol{\theta}_k - \boldsymbol{\theta}\| \le \varepsilon, \forall k \ge k_3$. Then, we have

$$\begin{aligned}
&\sum_{m=1}^{k} \Phi_{k,m+1}\lambda_m \left[\tilde{\mathcal{A}} \otimes \Gamma(0)\right] \boldsymbol{\omega}_m(1) \\
&= \sum_{m=1}^{k} \Phi_{k,m+1} \left[\tilde{\mathcal{A}} \otimes \Gamma(0)\right] (\boldsymbol{\theta}_m - \boldsymbol{\theta}_{m-1}) \\
&= \left[\tilde{\mathcal{A}} \otimes \Gamma(0)\right] \boldsymbol{\theta}_k - \sum_{m=1}^{k} (\Phi_{k,m+1} - \Phi_{k,m}) \left[\tilde{\mathcal{A}} \otimes \Gamma(0)\right] \boldsymbol{\theta}_{m-1} \\
&= \left[\tilde{\mathcal{A}} \otimes \Gamma(0)\right] \boldsymbol{\theta}_k - \sum_{m=1}^{k} (\Phi_{k,m+1} - \Phi_{k,m}) \left[\tilde{\mathcal{A}} \otimes \Gamma(0)\right] \boldsymbol{\theta} \\
&\quad - \sum_{m=1}^{k} (\Phi_{k,m+1} - \Phi_{k,m}) \left[\tilde{\mathcal{A}} \otimes \Gamma(0)\right] (\boldsymbol{\theta}_{m-1} - \boldsymbol{\theta}) \\
&= \left[\tilde{\mathcal{A}} \otimes \Gamma(0)\right] \boldsymbol{\theta}_k - \left[\tilde{\mathcal{A}} \otimes \Gamma(0)\right] \boldsymbol{\theta} + \Phi_{k,1} \left[\tilde{\mathcal{A}} \otimes \Gamma(0)\right] \boldsymbol{\theta} \\
&\quad - \sum_{m=1}^{k} (\Phi_{k,m+1} - \Phi_{k,m}) \left[\tilde{\mathcal{A}} \otimes \Gamma(0)\right] (\boldsymbol{\theta}_{m-1} - \boldsymbol{\theta}).
\end{aligned}$$

It is obvious that $\| \left[\tilde{\mathcal{A}} \otimes \Gamma(0)\right] \boldsymbol{\theta}_k - \left[\tilde{\mathcal{A}} \otimes \Gamma(0)\right] \boldsymbol{\theta} + \Phi_{k,1} \left[\tilde{\mathcal{A}} \otimes \Gamma(0)\right] \boldsymbol{\theta} \| \to 0$ as $k \to \infty$. For the last term at the right-hand side of the above equation, we have

$$\begin{aligned}
&\sum_{m=1}^{k} (\Phi_{k,m+1} - \Phi_{k,m}) \left[\tilde{\mathcal{A}} \otimes \Gamma(0)\right] (\boldsymbol{\theta}_{m-1} - \boldsymbol{\theta}) \\
&= \left(\sum_{m=1}^{k_3} + \sum_{m=k_3+1}^{k} \right) (\Phi_{k,m+1} - \Phi_{k,m}) \left[\tilde{\mathcal{A}} \otimes \Gamma(0)\right] (\boldsymbol{\theta}_{m-1} - \boldsymbol{\theta}).
\end{aligned}$$

It is clear that $\sum_{m=1}^{k_3} \|(\Phi_{k,m+1} - \Phi_{k,m}) \left[\tilde{\mathcal{A}} \otimes \Gamma(0)\right] (\boldsymbol{\theta}_{m-1} - \boldsymbol{\theta})\| \to 0$ as $k \to \infty$. Due to the fact that m is bounded by k_3, we have $\|\Phi_{k,m}\| \to 0$ as $k \to \infty$. The rest part can be proved converging to 0 by using the same steps as in Theorem 3.8. In short, the convergence of $\delta \boldsymbol{u}_k(0) \to 0$ is proved. The validation of the initial step is thus completed. The Inductive step can be completed using the same steps as outlined in Theorem 3.8, thereby concluding the proof.

3.4.3.3 Case 3: Noisy Output and Noisy Transmission

Clearly, Case 3 represents a synthesis of the aforementioned cases; thus, the design and analysis of the learning consensus algorithm can be accomplished by integrating the previous two subsections. Herein, we present the fundamental framework for this particular case. The measured error at iteration k for agent j is defined as follows:

$$e_{k,j}(t) = y_d(t) - \mathbf{Sat}\big(y_{k,j}(t) + v_{k,j}(t)\big). \tag{3.128}$$

The distributed information for the jth agent at the kth iteration is described as follows:

$$\begin{aligned}
\xi_{k,j}(t) &= \sum_{i\in\mathcal{N}_j} a_{j,i}\Big(z_{k,j,i}(t) - \mathbf{Sat}\big(y_{k,j}(t) + v_{k,j}(t)\big)\Big)\\
&\quad + d_j\Big(y_d(t) - \mathbf{Sat}\big(y_{k,j}(t) + v_{k,j}(t)\big)\Big)\\
&= \sum_{i\in\mathcal{N}_j} a_{j,i}\Big(\mathbf{Sat}\big(y_{k,i}(t) + v_{k,i}(t)\big) - \mathbf{Sat}\big(y_{k,j}(t) + v_{k,j}(t)\big)\Big)\\
&\quad + d_j\Big(y_d(t) - \mathbf{Sat}\big(y_{k,j}(t) + v_{k,j}(t)\big)\Big) + \sum_{i\in\mathcal{N}_j} a_{j,i}\omega_{k,j,i}(t),
\end{aligned} \tag{3.129}$$

where $z_{k,j,i}(t) = \mathbf{Sat}\big(y_{k,i}(t) + v_{k,i}(t)\big) + \omega_{k,j,i}(t)$ is the available information of the ith agent.

In compact form, we can reformulate (3.129) as

$$\boldsymbol{\xi}_k(t) = \big[(\mathcal{L} + \mathcal{D}) \otimes I_{s_o}\big]\boldsymbol{e}_k(t) + \big[\tilde{\mathcal{A}} \otimes I_{s_o}\big]\boldsymbol{\omega}_k(t). \tag{3.130}$$

Then, the compact form of the learning consensus algorithm is given as follows:

$$\boldsymbol{u}_{k+1}(t) = \boldsymbol{u}_k(t) + \lambda_k\big[(\mathcal{L} + \mathcal{D}) \otimes \Gamma(t)\big]\boldsymbol{e}_k(t+1) + \lambda_k\big[\tilde{\mathcal{A}} \otimes \Gamma(t)\big]\boldsymbol{\omega}_k(t+1). \tag{3.131}$$

Applying the transform factor $\gamma_{k,j}^{(i)}(t)$ in (3.108) yields

$$\begin{aligned}
\boldsymbol{u}_{k+1}(t) &= \boldsymbol{u}_k(t) + \lambda_k\big[(\mathcal{L} + \mathcal{D}) \otimes \Gamma(t)\big]\Upsilon_k(t)\Big[I_n \otimes \big(C(t+1)B(t)\big)\Big]\delta\boldsymbol{u}_k(t)\\
&\quad + \lambda_k\big[(\mathcal{L} + \mathcal{D}) \otimes \Gamma(t)\big]\Upsilon_k(t)\Big[I_n \otimes \big(C(t+1)A(t)\big)\Big]\delta\boldsymbol{x}_k(t)\\
&\quad - \lambda_k\big[(\mathcal{L} + \mathcal{D}) \otimes \Gamma(t)\big]\Upsilon_k(t)\boldsymbol{v}_k(t+1)\\
&\quad + \lambda_k\big[\tilde{\mathcal{A}} \otimes \Gamma(t)\big]\boldsymbol{\omega}_k(t+1).
\end{aligned} \tag{3.132}$$

It is clear that Case 3 introduces a combination of the noise in Cases 1 and 2.

The third theorem is presented below.

Theorem 3.10 *Assume that Assumptions 3.7-3.11 hold for MAS* (3.97) *and learning consensus algorithm* (3.131). *The communication mechanism between two neighbor agents is modeled in Case 3. If the direction regulation matrix $\Gamma(t)$ satisfies that all eigenvalues of $(\mathcal{L}+\mathcal{D})\otimes\big(\Gamma(t)C(t+1)B(t)\big)$ are with positive real parts for all time instances, then the actual output $\boldsymbol{y}_k(t)$ will converge to the desired reference vector $1_n\otimes y_d(t)$, $\forall t$, and the input vector $\boldsymbol{u}_k(t)$ converges to the desired input vector $1_n\otimes u_d(t)$ as $k\to\infty$, $t=0,1,\ldots,N-1$.*

The proof can be completed by combining the analysis for Theorems 3.8 and 3.9. We omit it for the sake of brevity.

Theorems 3.8, 3.9, and 3.10 establish the relationship between the asymptotic convergence of input information and actual output information. Moreover, by employing a decreasing gain sequence λ_k and an appropriate gain matrix $\Gamma(t)$, we can achieve asymptotic convergence of the input $\boldsymbol{u}_k(t)$ to the desired value defined in (3.99). Additionally, considering Lemma 3.9, we obtain almost sure asymptotic convergence of the actual output.

The key idea of the proof is to identify a continuous contraction along the iteration axis, denoted as $\Phi_{l,m}$ in (3.118). It is worth noting that for MIMO MAS, the function $\Phi_{l,m}$ can be expressed as a product of successive matrices. Furthermore, by leveraging the fundamental property of Kronecker product, we can ensure convergence by selecting $\Gamma(t)$ such that all eigenvalues of $\Gamma(t)C(t+1)B(t)$ are positive real numbers. In the subsequent section, we will delve into the case of switching topologies where the communication graph among agents dynamically changes.

3.4.4 Extensions to Time-Switching Topologies Case

The proposed ILC learning algorithm is extended to accommodate time-switching topologies in this subsection. Assumption 3.11 will be slightly modified to account for the switching topologies. It should be noted that the system equation remains unchanged, as stated in (3.97).

Every agent corresponds to a vertex in the dynamically changing directed graph $\mathcal{G}(t)$. Each edge $(v_i, v_j)\in\mathcal{E}(t)$ denotes the information flow between agent v_i and v_j at the time step t. Moreover, the adjacency matrix of dynamically changing directed graph $\mathcal{G}(t)$ is denoted by $\mathcal{A}(t)=[a_{i,j}(t)]$. Therefore, the index set of all neighbors of v_i and the Laplacian matrix associated with $\mathcal{G}(t)$ are denoted by $\mathcal{N}_i(t)$ and $\mathcal{L}(t)$, respectively. In addition, the set of all directed graphs for the agents are denoted by $\mathcal{G}_\alpha=\{\mathcal{G}_{\alpha_1},\mathcal{G}_{\alpha_2},\ldots,\mathcal{G}_{\alpha_q}\}$; that is, $\mathcal{G}(t)\in\mathcal{G}_\alpha$ for all $t\in\mathbb{Z}_N$.

As mentioned in the previous subsections, the desired trajectory is only available to a portion of follower agents. It should be emphasized that $d_j(t)=1$ indicates that the agent v_j can get the desired trajectory $y_d(t)$ at the time step t, otherwise $d_j(t)=0$. Then we can easily modify $\mathcal{D}$ as $\mathcal{D}(t)=diag\{d_1(t),d_2(t),\ldots,d_n(t)\}$.

Note that $\mathcal{G}(t) \in \mathcal{G}_\alpha$, and $\mathcal{G}_\alpha$ consists of q directed graphs. In addition, we can find that $\mathcal{D}(t) \in \mathcal{D}_\alpha$, where $\mathcal{D}_\alpha = \{\mathcal{D}_{\alpha_1}, \mathcal{D}_{\alpha_2}, \ldots, \mathcal{D}_{\alpha_q}\}$ and $\mathcal{D}_{\alpha_i}$ indicates the accessibility of the desired reference for the follower agents in $\mathcal{G}_{\alpha_i}$, $i = 1, 2, \ldots, q$.

The concept of the virtual leader has been discussed in previous subsections. The directed graph $\overline{\mathcal{G}}(t) \in \overline{\mathcal{G}}_\alpha = \{\overline{\mathcal{G}}_{\alpha_1}, \overline{\mathcal{G}}_{\alpha_2}, \ldots, \overline{\mathcal{G}}_{\alpha_q}\}$ comprises of $n+1$ agents denoted as $v_0, v_1, \ldots, v_n$. It is important to note that the switching between graphs can occur arbitrarily at each time instant t, without any restrictions on the transition between any two graphs.

Based on the above supplement, Assumption 3.11 is modified as follows:

Assumption 3.12 The directed graph $\overline{\mathcal{G}}_{\alpha_i}$ has a spanning tree with the virtual leader being the root for $i = 1, 2, \ldots, q$.

If Assumption 3.12 holds, v_0 is the root vertex of the directed graph $\overline{\mathcal{G}}(t)$ for all $t \in \mathbb{Z}_N$, where $\overline{\mathcal{G}}(t) \in \overline{\mathcal{G}}_\alpha$. The virtual leader v_0 generates the desired trajectory and will not receive any information from any other agents.

We elaborate Case 3 in this subsection because the other two are special cases of Case 3. Based on the above analysis, for the jth agent we can rewrite (3.103) in compact forms as

$$\begin{aligned} \boldsymbol{u}_{k+1}(t) = \boldsymbol{u}_k(t) &+ \lambda_k \Big[\big(\mathcal{L}(t) + \mathcal{D}(t)\big) \otimes \Gamma(t) \Big] \boldsymbol{e}_k(t+1) \\ &+ \lambda_k \big[\tilde{\mathcal{A}}(t) \otimes \Gamma(t) \big] \boldsymbol{\omega}_k(t+1), \end{aligned} \tag{3.133}$$

where λ_k is a decreasing gain, $\mathcal{L}(t)$ and $\mathcal{D}(t)$ are time-varying matrices, and the gain matrix is denoted by $\Gamma(t) \in \mathbf{R}^{s_i \times s_o}$ for all agents. Moreover, $\Gamma(t)$ is bounded, i.e., $\|\Gamma(t)\| \leqslant u_\Gamma$ for all $t \in \mathbb{Z}_N$.

The theorem of this subsection is presented below.

Theorem 3.11 *Assume that Assumptions 3.7–3.10 and 3.12 hold for MAS* (3.97). *The learning control algorithm* (3.133) *is applied for MAS* (3.97) *with dynamic changing directed graphs $\overline{\mathcal{G}}(t) \in \overline{\mathcal{G}}_\alpha$. If the direction regulation matrix $\Gamma(t)$ satisfies that all eigenvalues of $\Big[\big(\mathcal{L}(t) + \mathcal{D}(t)\big) \otimes \big(\Gamma(t)C(t+1)B(t)\big)\Big]$ are with positive real parts then the actual output $\boldsymbol{y}_k(t)$ will converge to the desired reference vector $1_n \otimes y_d(t)$, $\forall t$, and the input $\boldsymbol{u}_k(t)$ which is generated by the learning algorithm* (3.133) *converges to the desired input $1_n \otimes u_d(t)$ as $k \to \infty$, $t = 0, 1, \ldots, N-1$. That is, the MAS solves the consensus tracking problem as $k \to \infty$.*

Proof. The proof can be performed similarly to the proof of Theorem 3.8, if we can establish the same estimation (3.115) for the time-switching topologies case. In fact, the recursion for the switching topology case is as follows,

$$
\begin{aligned}
u_{k+1}(t) = & u_k(t) + \lambda_k\Big[\big(\mathcal{L}(t)+\mathcal{D}(t)\big)\otimes\Gamma(t)\Big]\Upsilon_k(t)\Big[I_n\otimes\big(C(t+1)B(t)\big)\Big]\delta u_k(t)\\
& + \lambda_k\Big[\big(\mathcal{L}(t)+\mathcal{D}(t)\big)\otimes\Gamma(t)\Big]\Upsilon_k(t)\Big[I_n\otimes\big(C(t+1)A(t)\big)\Big]\delta x_k(t)\\
& - \lambda_k\Big[\big(\mathcal{L}(t)+\mathcal{D}(t)\big)\otimes\Gamma(t)\Big]\Upsilon_k(t)\omega_k(t+1) + \lambda_k\big[\tilde{A}(t)\otimes\Gamma(t)\big]\omega_k(t+1).
\end{aligned} \tag{3.134}
$$

Thus, similar to $\Phi_{l,m}$, we define $\Psi_{l,m}$ for the time-switching topologies case, taking the initial step $t=0$ as example,

$$
\begin{aligned}
\Psi_{l,m} \triangleq & \Big\{I - \lambda_l\Big[\big(\mathcal{L}(0)+\mathcal{D}(0)\big)\otimes\Gamma(0)\Big]\Upsilon_l(0)\Big[I_n\otimes\big(C(1)B(0)\big)\Big]\Big\}\cdots\\
& \Big\{I - \lambda_m\Big[\big(\mathcal{L}(0)+\mathcal{D}(0)\big)\otimes\Gamma(0)\Big]\Upsilon_m(0)\Big[I_n\otimes\big(C(1)B(0)\big)\Big]\Big\},
\end{aligned}
$$

when $l \geq m$ and $\Psi_{l,l+1} \triangleq I$.

Note that all eigenvalues of $\Big[\big(\mathcal{L}(t)+\mathcal{D}(t)\big)\otimes\big(\Gamma(t)C(t+1)B(t)\big)\Big]$ are with positive real parts. Therefore, there exists a positive-definite matrix $P(0)$ such that

$$
\begin{aligned}
& -\Big[\big(\mathcal{L}(0)+\mathcal{D}(0)\big)\otimes\Gamma(0)\Big]\Upsilon_k(0)\Big[I_n\otimes\big(C(1)B(0)\big)\Big]^T P(0)\\
& - P(0)\Big[\big(\mathcal{L}(0)+\mathcal{D}(0)\big)\otimes\Gamma(0)\Big]\Upsilon_k(0)\Big[I_n\otimes\big(C(1)B(0)\big)\Big] \leq -\nu' I
\end{aligned}
$$

for some suitable constant $\nu' > 0$.

We denote $W_k \triangleq -\big[\big(\mathcal{L}(0)+\mathcal{D}(0)\big)\otimes\Gamma(0)\big]\Upsilon_k(0)\Big[I_n\otimes\big(C(1)B(0)\big)\Big]$. Thus, for positive integers l and m with $l \geq m$, we have

$$
\begin{aligned}
\Psi_{l,m}^T P(0)\Psi_{l,m} &= \Psi_{l-1,m}^T (I+\lambda_l W_l)^T P(0)(I+\lambda_l W_l)\Psi_{l-1,m}\\
&= \Psi_{l-1,m}^T\big(P(0)+\lambda_l^2 W_l^T P(0)W_l + \lambda_l W_l^T P(0) + \lambda_l P(0)W_l\big)\Psi_{l-1,m}\\
&\leq \Psi_{l-1,m}^T\big(P(0)+\lambda_l^2 W_l^T P(0)W_l - \nu'\lambda_l I\big)\Psi_{l-1,m}\\
&= \Psi_{l-1,m}^T P^{\frac{1}{2}}(0)\big(I-\nu'\lambda_l P^{-1}(0)\\
&\quad + \lambda_l^2 P^{-\frac{1}{2}}(0)W_l^T P(0)W_l P^{-\frac{1}{2}}(0)\big)P^{\frac{1}{2}}(0)\Psi_{l-1,m}.
\end{aligned} \tag{3.135}
$$

Without loss of generality, when the iteration number m is sufficiently large, we have

$$
\begin{aligned}
& \|I-\nu'\lambda_l P^{-1}(0) + \lambda_l^2 P^{-\frac{1}{2}}(0)W_l^T P(0)W_l P^{-\frac{1}{2}}(0)\|\\
& \leq 1-2c_4\lambda_{l+1} \leq \exp(-2c_4\lambda_{l+1}),
\end{aligned} \tag{3.136}
$$

where $c_4 > 0$ is a suitable constant and the basic inequality $1-x \leq e^{-x}$ is applied.

Noticing the boundedness of $P^{-\frac{1}{2}}(0)W_l^T P(0)W_l P^{-\frac{1}{2}}(0)$ and combining (3.135) and (3.136) leads to

$$\Psi_{l,m}^T P(0)\Psi_{l,m} \le c_5 \exp(-2c_4 \sum_{k=m}^{l} \lambda_k) I$$

for some suitable $c_5 > 0$. Thus,

$$\|\Psi_{l,m}\| \le \lambda_{min}^{-\frac{1}{2}}\big(P(0)\big)\sqrt{c_5}\exp(-c_4 \sum_{k=m}^{l} \lambda_k). \tag{3.137}$$

The rest of the proof can be completed similarly to Theorems 3.8 and 3.9. This completes the proof.

We have successfully demonstrated the convergence of ILC for MAS with limited measurement range and noise, considering both fixed and time-switching topologies. Theorem 3.11 extends the learning consensus results from the fixed topology case to scenarios involving dynamic changing topologies.

Remark 3.15 In scenarios involving switching topologies where graphs dynamically change over time and iterations, convergence towards the desired output can be achieved if all eigenvalues of $M_{\alpha_i} \triangleq \Big[\big(\mathcal{L}_{\alpha_i}(t) + \mathcal{D}_{\alpha_i}(t)\big) \otimes \big(\Gamma(t)C(t+1)B(t)\big)\Big]$ have positive real parts and there exists a shared positive-definite matrix P satisfying $-\{M_{\alpha_i}^T P + PM_{\alpha_i}\}$ being negative-definite matrices for all $i = 1, \cdots, q$. However, this condition imposes significant requirements. Essentially, finding such a matrix P becomes necessary for establishing the existence of a common Lyapunov function that ensures successive contraction along the iteration axis.

3.4.5 Illustrative Simulations

In this section, the leaning algorithms are simulated for both fixed and time-switching topologies case.

3.4.5.1 MIMO System with Fixed Topology

The MAS with a fixed topology $\overline{\mathcal{G}}$ shown in Fig. 3.17 is considered. Additionally, we consider $t \in \mathbb{Z}_{30}$. The virtual leader has directed edges to agents 2 and 3, while the communication among followers is also directed. Furthermore, the Laplacian matrix and pinning gain matrix are described as follows:

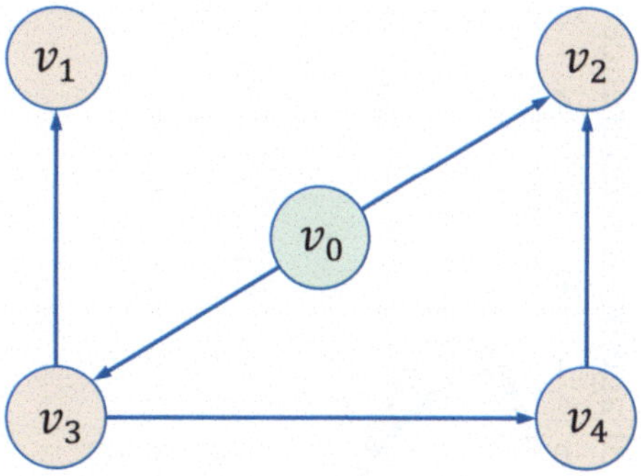

Fig. 3.17 Topology of the communication graph

$$\mathfrak{L} = \begin{bmatrix} 1 & 0 & -1 & 0 \\ 0 & 1 & 0 & -1 \\ 0 & 0 & 0 & 0 \\ 0 & 0 & -1 & 1 \end{bmatrix}, \mathfrak{D} = \begin{bmatrix} 0 & 0 & 0 & 0 \\ 0 & 1 & 0 & 0 \\ 0 & 0 & 1 & 0 \\ 0 & 0 & 0 & 0 \end{bmatrix}.$$

In this example, the system matrices for all agents are describe as follows:

$$A(t) = \begin{bmatrix} -0.05\sin(0.2t) & 0 & 0.01t \\ 0 & -0.01t & -0.02\cos(0.5t) \\ 0.2 & 0.4 & 0.3 + 0.1\cos(0.2t) \end{bmatrix},$$

$$B(t) = \begin{bmatrix} 1 - 0.1\sin^2(0.5t\pi) & 0 \\ 0.01t & 0.01t \\ 0.1 & 1 + 0.1\sin(0.5t\pi) \end{bmatrix},$$

$$C(t) = \begin{bmatrix} 1 + 0.1\sin^2(0.5t\pi) & 0 & 0 \\ 0 & 0 & 1 - 0.2\sin^2(0.5t\pi) \end{bmatrix}.$$

The initial state is chosen as $x_{1,j}(0) = [0,\ 0,\ 0]^T$ for $j = 1, 2, 3, 4$. Moreover, the zero initial input will be used for all agents, i.e., $u_{0,j} \equiv 0$ for all agents. The measurement range is set to be $(-2.5, 2.5)$. That is, the upper bound for the first and second dimensions is 2.5 and the lower bound for the first and second dimensions is -2.5. In addition, for the jth agent the measurement noise is termed as $\nu_{k,j}(t) = [\omega_{k,j_1}(t), \omega_{k,j_2}(t)]^T$ with each dimension satisfying that $\omega_{k,j_\sigma}(t) \sim N(0, 0.05^2)$, $\sigma = 1, 2$. Similarly, the communication noise $\omega_{k,i,j_\sigma}(t) \sim N(0, 0.05^2)$, $\sigma = 1, 2$. The algorithm runs for 40 iterations.

The desired trajectory is described as

$$y_d(t) = \begin{bmatrix} 2\sin(0.1t\pi) \\ 2\cos(0.1t\pi) - \frac{2}{t+1} \end{bmatrix} \quad \text{for } t \in \mathbb{Z}_{30}. \tag{3.138}$$

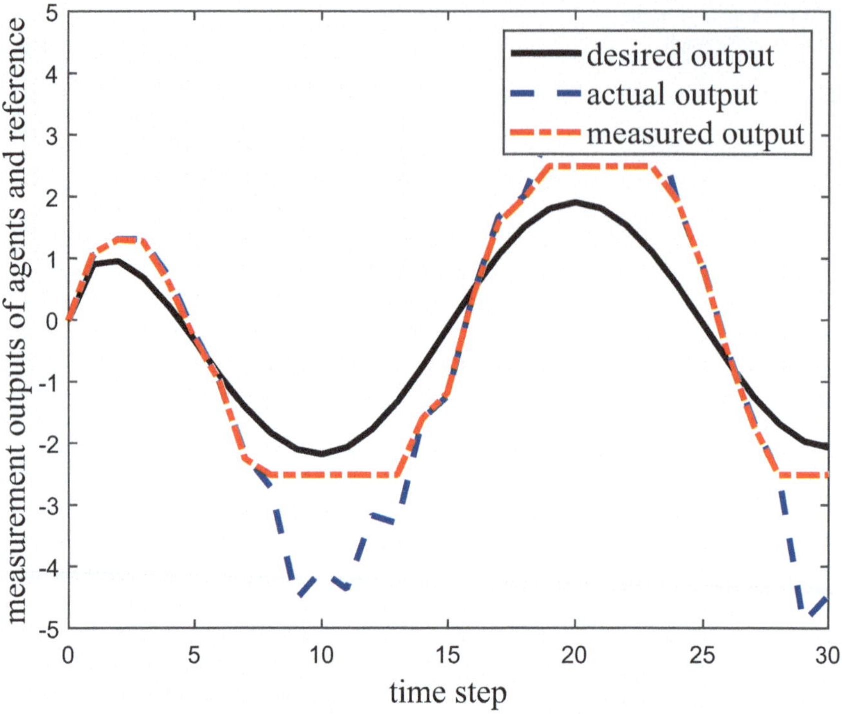

Fig. 3.18 The system output with noises and measurement output at the second iteration (Agent 2)

Moreover, the direction regulation matrix is devised as

$$\Gamma(t) = \begin{bmatrix} 0.8 & 0.5 \\ -0.2 & 1.4 \end{bmatrix}$$

for all t and the decreasing gain λ_k is designed as a piecewise function

$$\lambda_k = \begin{cases} \frac{1}{k}, & 0 < k \leqslant 10 \\ \frac{3}{k}, & k > 10. \end{cases}$$

For the purpose of measurement saturation, in Case 1, Fig. 3.18 illustrates the second dimension of both the actual output $y_{k,2}(t) + \omega_{k,2}(t)$ and the measured output $z_{k,2}(t)$ for Agent 2 during the second iteration. This specific iteration is chosen solely to exemplify measurement saturation. It can be observed that the upper bound for the second dimension is restricted to 2.5, while the lower bound is limited to -2.5. Consequently, only agents whose actual output trajectories fall within this measurable range can be obtained.

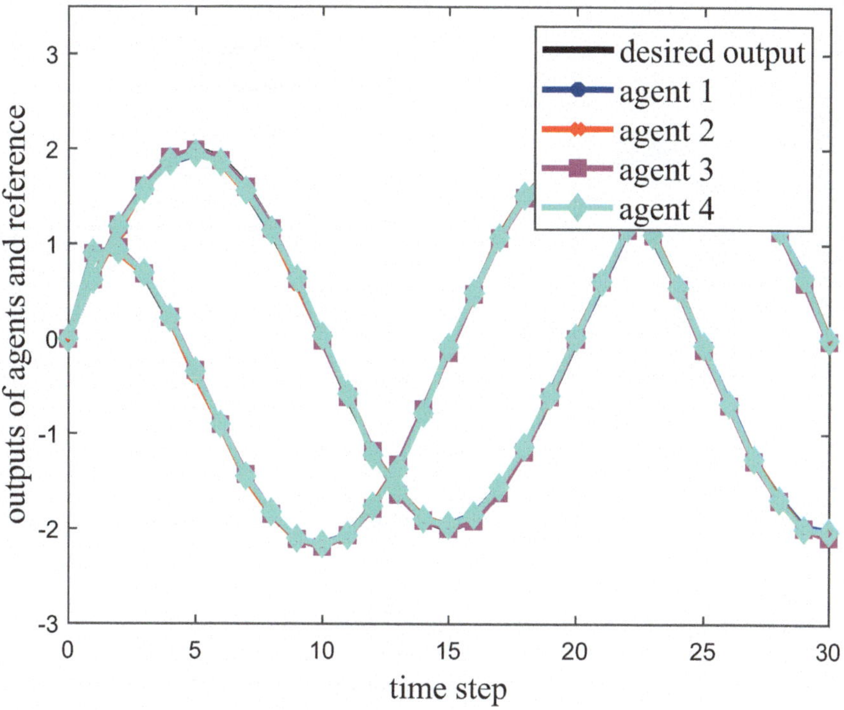

Fig. 3.19 The actual output for all agents (Case 1)

The actual output trajectories at the 40th iteration for Case 1 and 2 are also plotted in Figs. 3.19 and 3.20, respectively. It can be observed that all agents' actual output trajectories remain within the measurement bound. Furthermore, after a sufficient number of learning iterations, all agents successfully track the desired trajectory. This figure effectively demonstrates the efficacy of the proposed learning algorithm (3.106) and (3.125) under conditions of measurement saturation and noise.

The maximum actual tracking error $\max_t \|y_d(t) - y_{k,1}(t)\|$ and the maximum measured tracking error $\max_t \|y_d(t) - z_{k,3}(t)\|$ over the first 40 iterations for Cases 1 and 2 are depicted in Figs. 3.21, 3.22, 3.23, and 3.24 for two dimensions, respectively. These figures clearly demonstrate the successful achievement of the consensus tracking objective. Additionally, we also consider a noise-free scenario where the decreasing gain λ_k is replaced by a constant gain of 0.33, and present the corresponding profiles of maximum tracking errors.

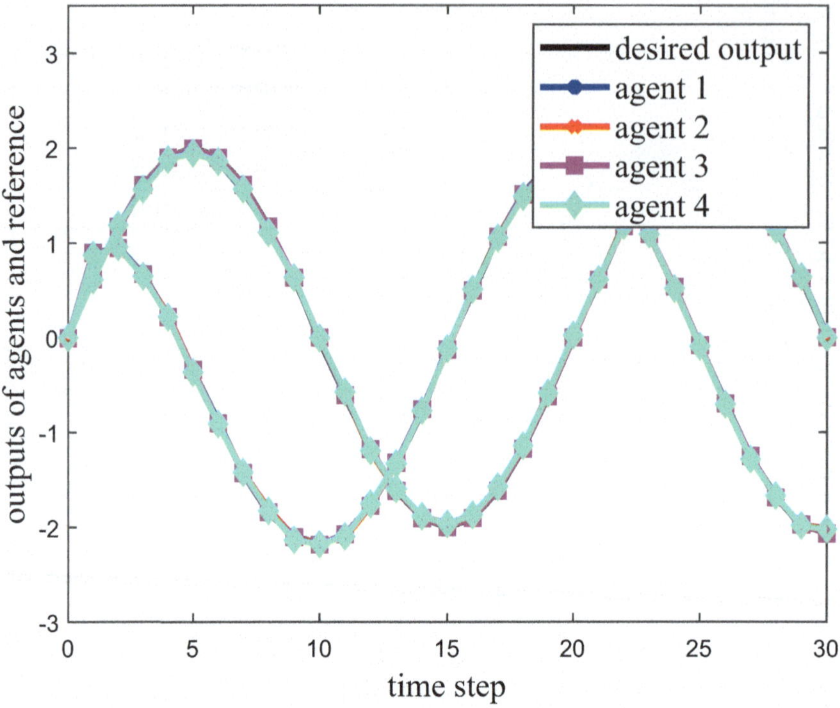

Fig. 3.20 The actual output for all agents (Case 2)

3.4.5.2 MIMO System with Switching Topologies

Case 3 is simulated in this subsection. The system matrices, the initial state, the initial input, the noises and the measurement range are identical to those of the previous cases. In addition, the algorithm runs for 50 iterations.

In this case, the communication graphs among agents are dynamically changing at the time step t. The communication graphs for agents are considered to switch over four states as show in Fig. 3.25, i.e., $\overline{\mathcal{G}}_\alpha = \{G_a, G_b, G_c, G_d\}$, where all of these graphs have a spanning tree with vertex v_0 being the root. Furthermore, in order to simulate the switching of the communication topologies, the switching signal is defined by $\alpha(t)$ varying over the interval $[0, 1]$ as a function of t. We use the MATLAB command '`rand`' to produce the switching signal. The switching rule is described as follows:

- If $\alpha(t) \in [0, 0.25)$, then $\overline{\mathcal{G}}_\alpha = G_a$,
- If $\alpha(t) \in [0.25, 0.5)$, then $\overline{\mathcal{G}}_\alpha = G_b$,
- If $\alpha(t) \in [0.5, 0.75)$, then $\overline{\mathcal{G}}_\alpha = G_c$,
- If $\alpha(t) \in [0.75, 1]$, then $\overline{\mathcal{G}}_\alpha = G_d$.

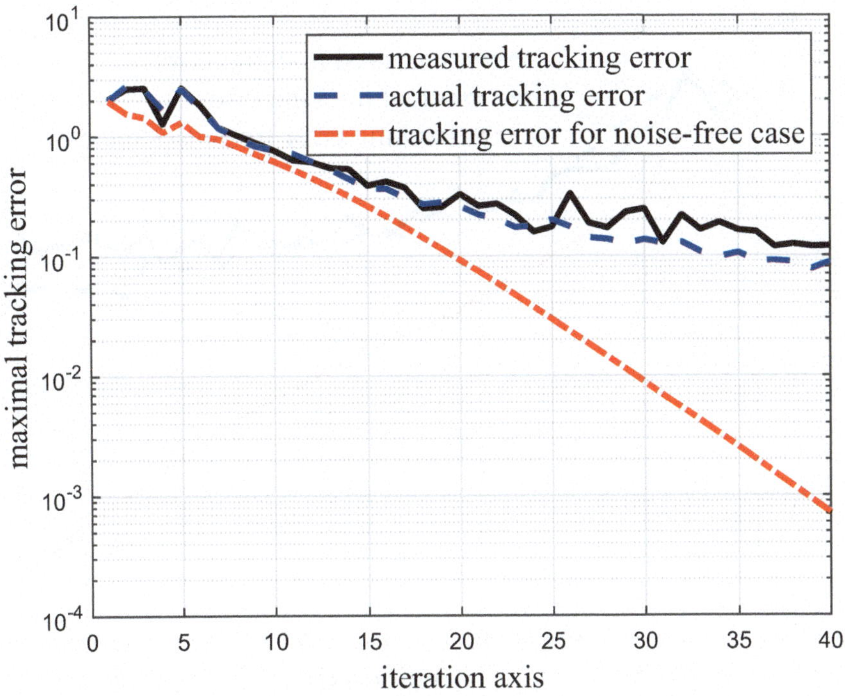

Fig. 3.21 Maximal error profiles of first output (Case 1)

Based on Theorem 3.11, the direction regulation matrix is selected as

$$\Gamma(t) = \begin{bmatrix} 1.2 & 0.3 \\ -0.2 & 1.5 \end{bmatrix}$$

for all t and the decreasing gain λ_k is designed as a piecewise function

$$\lambda_k = \begin{cases} \frac{1}{k}, & 0 < k \leqslant 10 \\ \frac{5}{k}, & k > 10. \end{cases}$$

The actual output trajectories at the 50th iteration are depicted in Fig. 3.26. It can be observed that all agents' actual output trajectories fall within the measurable range. Furthermore, by utilizing (3.133), all agents successfully track the desired trajectory. This figure demonstrates that the proposed learning algorithm (3.133) effectively addresses the consensus tracking problem under measurement saturation and noise disturbances. Additionally, to illustrate measurement saturation, Fig. 3.27 displays both the second dimension of actual output and its corresponding measured output for Agent 3 at the second iteration.

The maximum actual tracking error and the maximum measured tracking error over the first 50 iterations for Agent 3 are depicted in Figs. 3.28 and 3.29, respectively,

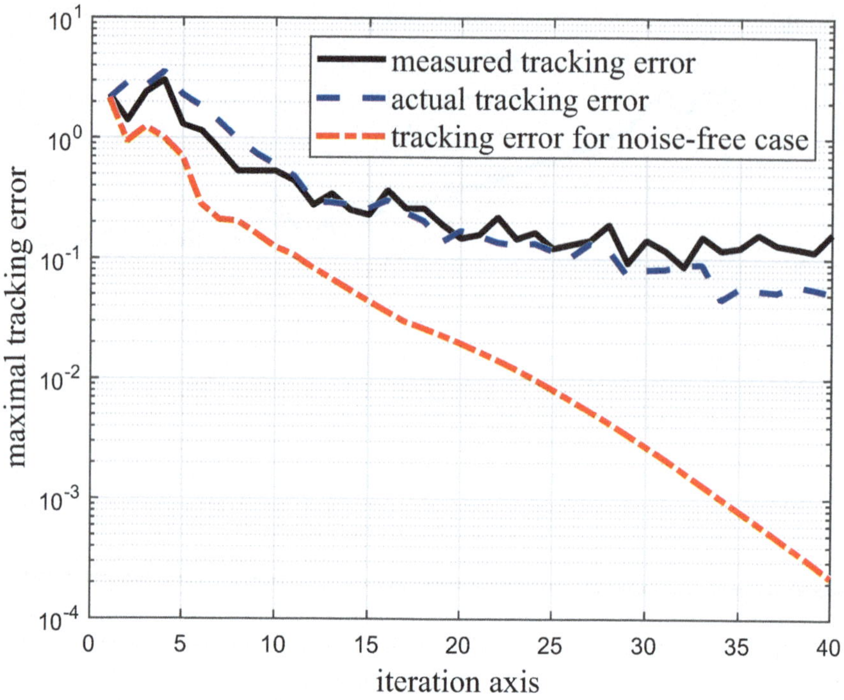

Fig. 3.22 Maximal error profiles of second output (Case 1)

for two dimensions. Additionally, we also conduct simulations on a noise-free system with a constant learning gain of $1/3$.

Evidently, Figs. 3.28 and 3.29 demonstrate the efficacy of employing (3.133) in addressing the consensus tracking problem under dynamic switching topologies. However, due to inherent noise and topology variations, the maximum error profiles exhibit slight irregularities. Nonetheless, it is noteworthy that the achieved tracking performance is already commendable.

3.5 Decreasing Gain Design in Multi-sensor Systems

Multi-sensor systems are extensively employed for achieving comprehensive monitoring and control due to their incorporation of multiple individual sensors/outputs. In such systems, diverse sensors can receive various types of operation signals, including pressure, temperature, and volume. However, the conflicting desired references for different sensors pose a challenge as there is currently no input signal capable of precisely tracking all references simultaneously. This discrepancy has motivated us to address the incompatible multi-objective tracking problem in multi-sensor systems with random process disturbances and measurement noises.

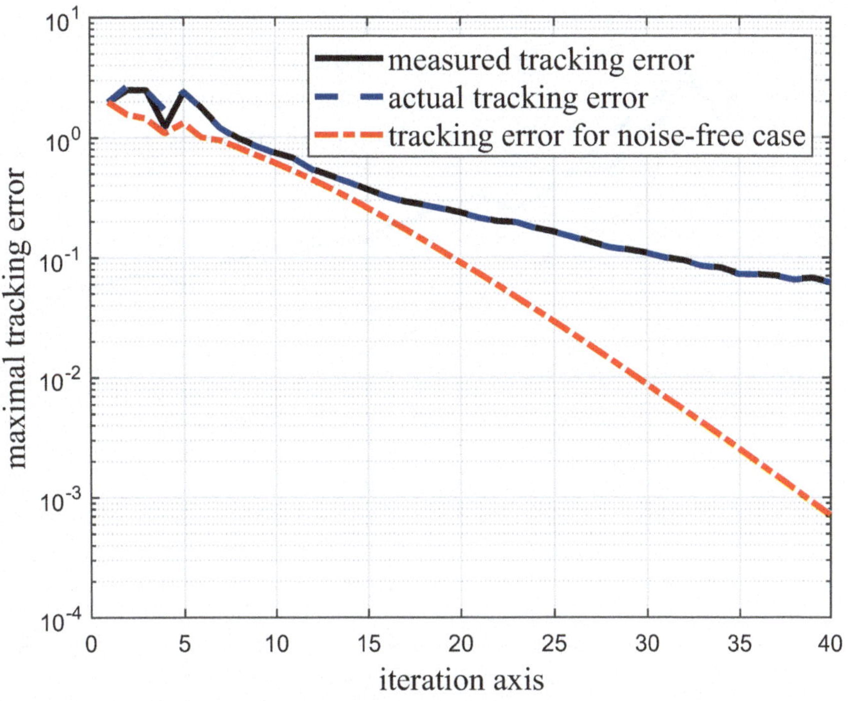

Fig. 3.23 Maximal error profiles of first output (Case 2)

3.5.1 Learning Tracking for Multi-sensor Systems

In recent decades, there has been extensive research on multi-sensor systems for applications in fault detection, target tracking, habitat monitoring, and integrated navigation (Khaleghi et al., 2013). Multi-sensor systems are highly significant in practical scenarios due to their ability to acquire more information of interest through multiple individual outputs or sensors compared to single-sensor systems (Durrant-Whyte, 2012). This rich information can lead to improvements in control performance, such as precision and convergence speed. State estimation or information fusion has emerged as a major focus within the field of multi-sensor systems research (Su et al., 2004; Zhang et al., 2017; Song et al., 2007; Li et al., 2003), with various data fusion and filtering algorithms being central objectives for designing and analyzing controlled systems aiming at achieving better state estimation. A recent paper Sun et al. (2017) provides a comprehensive classification and survey of several distributed fusion estimation algorithms, focusing on random phenomena in networked systems. It is important to note that a multi-sensor system possesses a unique state, which naturally aligns the design objectives across different sensors discussed in the aforementioned papers. These objectives primarily revolve around achieving specific targets by effectively fusing and utilizing diverse information sources, such as iden-

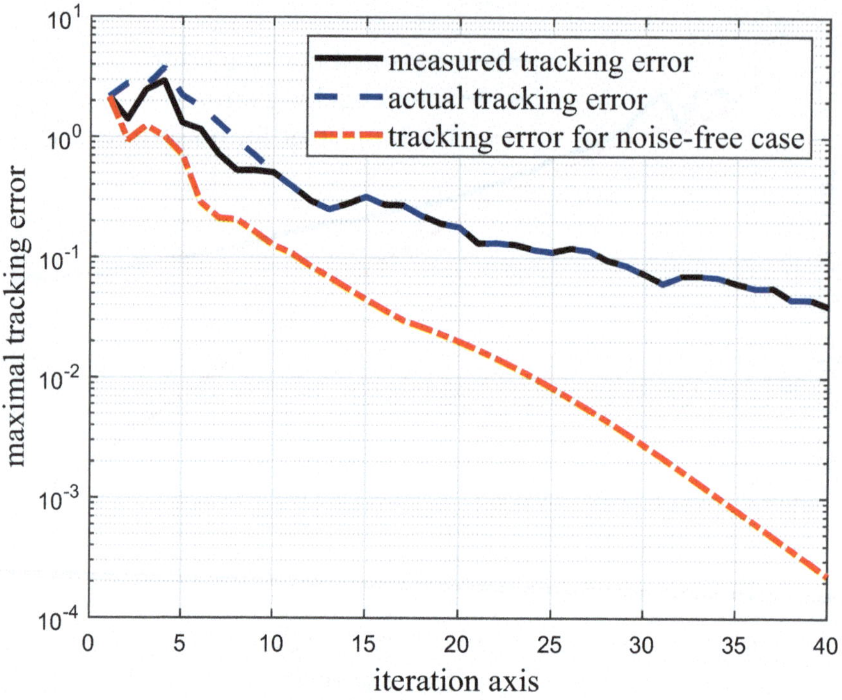

Fig. 3.24 Maximal error profiles of second output (Case 2)

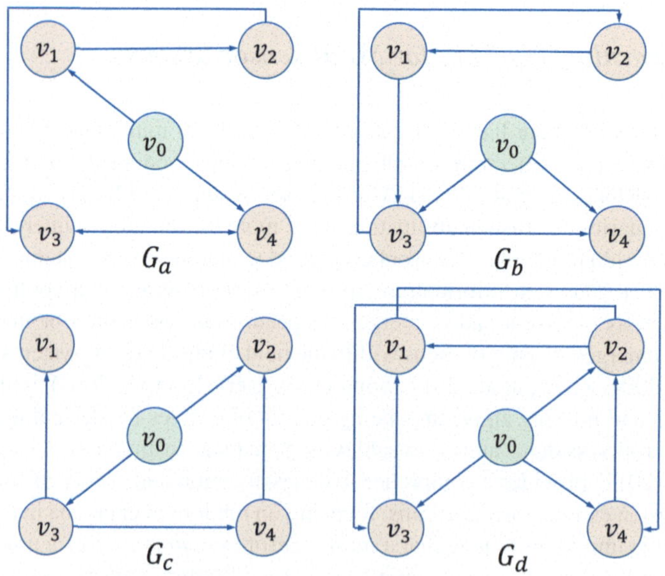

Fig. 3.25 Topologies of the communication graph

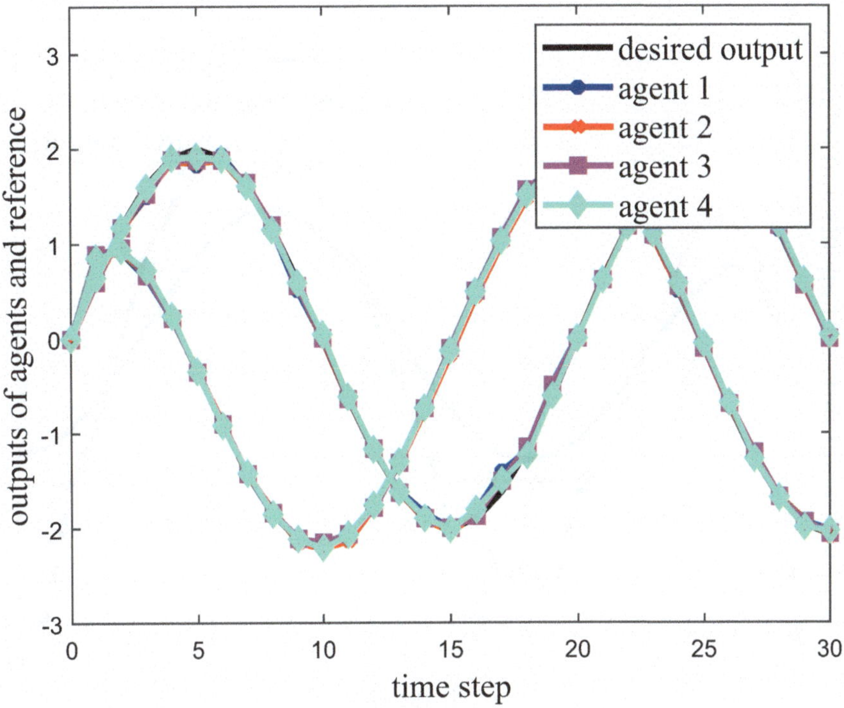

Fig. 3.26 The actual output for all agents (50th iteration)

tifying the most accurate sensor and maintaining balance among multiple sensors. However, limited attention has been given to addressing stabilization and control issues in multi-sensor systems (Johnston and Yuksel, 2014).

In certain applications, a multi-sensor system may necessitate individual trajectory tracking for each sensor; thus, the overall tracking objective comprises multiple trajectories (Ramamurthy et al., 2007). For instance, in monitoring the state of an industrial production process, it is customary to measure various variables such as flow rate, flow volume, pressure, and temperature. Simultaneously, ensuring steady and safe operation typically requires all variables to track predefined references throughout the operation. This scenario represents a multi-objective tracking problem rather than a state estimation problem. Addressing this knowledge gap serves as our primary motivation to explore learning-based solutions for multi-sensor systems with multiple tracking objectives. In this system, each sensor or output has a designated tracking reference and the desired input for a given tracking task may conflict with others. The aim of this section is to propose potential resolutions for the multi-objective tracking problem.

In this section, we consider the multiple tracking objectives to be *compatible* if the associated input signal for each objective is identical to that of other objectives; otherwise, we deem them as *incompatible*. When the objectives are compatible, the

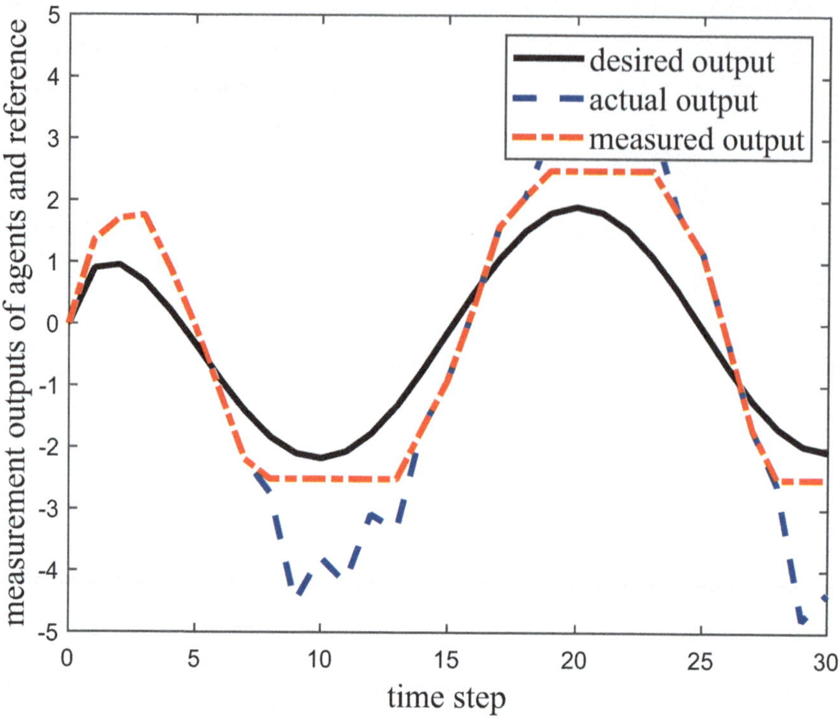

Fig. 3.27 The system output with noises and measurement output at the second iteration (Agent 3)

multi-objective tracking problem can be treated as a conventional tracking problem, which has been successfully addressed in numerous ILC papers. However, when the tracking objectives are incompatible, a trade-off between these objectives becomes necessary. This situation commonly arises when the number of input channels is fewer than that of available sensors. In such cases, engineers must determine which sensor holds greater importance among several options and compute the input primarily based on data from that selected sensor. However, relying solely on subjective decisions may not lead to an optimal solution for solving multi-objective tracking problems.

Generally speaking, two common approaches are employed to tackle this issue. The first approach employs a weighted technique to consolidate the multiple tracking objectives into a single optimization index, incorporating appropriate weights. The optimal input can be uniquely determined based on this weighted index. Subsequently, a learning algorithm can generate an input sequence that progressively approximates the aforementioned computed optimal input as the iteration number increases. The primary challenge in this approach lies in characterizing the optimal input with respect to a specified index, enabling the design of corresponding learning control algorithms. In contrast, the second approach involves directly deriving Pareto

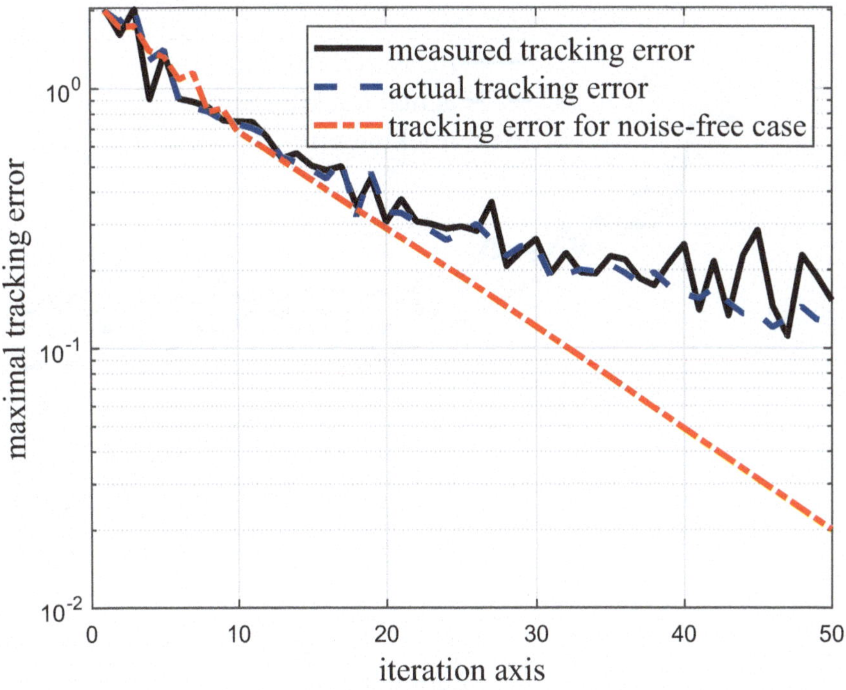

Fig. 3.28 Maximal error profiles of first output (Agent 3)

optimal solutions for the original multi-objective tracking problem by considering an optimization index. This section primarily focuses on elucidating and analyzing the first approach.

3.5.2 Problem Formulation

Consider the following multi-input multi-output discrete-time stochastic system:

$$\begin{aligned} x_k(t+1) &= A_t x_k(t) + B_t u_k(t) + w_k(t), \\ y_{i,k}(t) &= C_{i,t} x_k(t) + v_{i,k}(t), \ i = 1, \ldots, m, \end{aligned} \tag{3.139}$$

where the subscript k is the iteration index: $k = 1, 2, \cdots$; t is the time index: $t = 0, 1, \cdots, T$, where T is the iteration length; and i indicates the sensor index: $i = 1, 2, \cdots, m$, where m is the number of sensors. The expressions $u_k(t) \in \mathbf{R}^p$, $x_k(t) \in \mathbf{R}^n$, and $y_{i,k}(t) \in \mathbf{R}^{q_i}$ are the input, state, and ith output, respectively. The random variables $w_k(t) \in \mathbf{R}^n$ and $v_{i,k}(t) \in \mathbf{R}^{q_i}$ reflect the system disturbances and measurement noises of the ith output/sensor. The time-varying system matrices

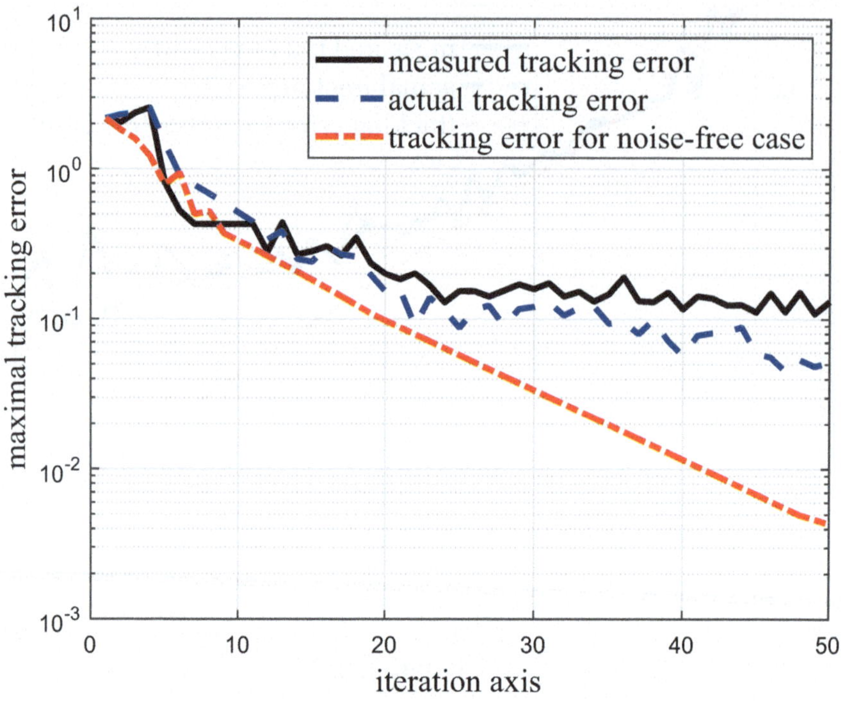

Fig. 3.29 Maximal error profiles of second output (Agent 3)

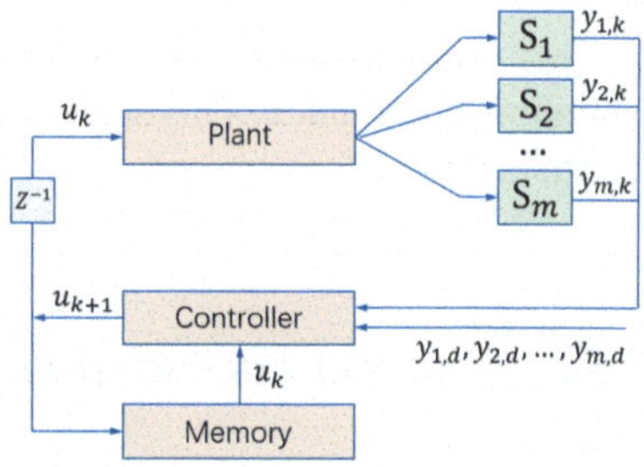

Fig. 3.30 Block diagram of a multi-sensor system

$A_t \in \mathbf{R}^{n\times n}$, $B_t \in \mathbf{R}^{n\times p}$, and $C_{i,t} \in \mathbf{R}^{q_i\times n}$ are compatible with the variables. The control structure of this system is shown in Fig. 3.30, where capital S_i denotes different sensors.

We provide the following assumptions for the systems of interest for further analysis.

Assumption 3.13 Both $w_k(t)$ and $v_{i,k}(t)$ are assumed to be independent in the iteration domain satisfying that $\mathbb{E}w_k(t) = 0$, $\mathbb{E}v_{i,k}(t) = 0$, $\sup_k \mathbb{E}\|w_k(t)\|^2 < \infty$, $\sup_k \mathbb{E}\|v_{i,k}(t)\|^2 < \infty$, $\limsup_{n\to\infty} \frac{1}{n}\sum_{k=1}^{n} w_k(t)w_k^T(t) = R_w^t$, and $\limsup_{n\to\infty} \frac{1}{n}\sum_{k=1}^{n} v_{i,k}(t)w_k^T(t) = R_{v,i}^t$.

Assumption 3.14 The random initial state $x_k(0)$ is mutually independent satisfying that $\mathbb{E}x_k(0) = x_d(0)$, $\sup_k \mathbb{E}\|x_k(0)\|^2 < \infty$, and $\limsup_{n\to\infty} \frac{1}{n}\sum_{k=1}^{n} x_k(0)x_k^T(0) = R_x^0$, where $x_d(0)$ is the desired initial state.

By Assumption 3.14, we can denote $w_k(0) = x_d(0) - x_k(0)$. In Assumptions 3.13 and 3.14, we primarily require iteration-based independence, rather than time-based independence, of all the involved random variables. Although this condition lacks rigor due to the repetitive nature of the process along the iteration axis, relaxing these assumptions would complicate the associated analysis. Hence, we employ these assumptions to emphasize our central concept.

As illustrated in Fig. 3.30, the plant is equipped with m individual sensors/outputs. Existing studies have primarily focused on information fusion or joint estimation of the system state $x_k(t)$, where the fusion kernel combines state estimations from different sensors (Khaleghi et al., 2013; Su et al., 2004; Zhang et al., 2017; Song et al., 2007; Li et al., 2003; Sun et al., 2017). In this section, we address the multi-objective tracking problem by learning, where each output $y_{i,k}(t)$ of the ith sensor aims to track a given reference signal $y_{i,d}(t)$. The main challenge lies in reconciling conflicting tracking objectives that may arise due to diverse desired inputs for different sensors/outputs. Consequently, it becomes difficult to precisely track multiple references simultaneously. This issue motivates us to explore the fundamental tracking capability of a multi-sensor system.

While considering the single-sensor tracking problem, the objective of unified control is to minimize the asymptotically averaged index

$$\limsup_{n\to\infty} \frac{1}{n}\sum_{k=1}^{n}\sum_{t=1}^{T} \|y_{i,k}(t) - y_{i,d}(t)\|^2, \quad \forall t.$$

This index is applicable to cases with completely trackable and partially trackable references for both noise-free and stochastic systems. Here, we take a noise-free system as an example. If a given reference is completely trackable, a desired input $u_d(t)$ generating the reference $y_d(t)$ exists. In this case, the minimum of the index $\limsup_{n\to\infty} \frac{1}{n}\sum_{k=1}^{n}\sum_{t=1}^{T} \|y_{i,k}(t) - y_{i,d}(t)\|^2$ is zero. This minimum can be achieved as long as we find an input sequence $\{u_k(t)\}$ such that $u_k(t) \to u_d(t)$ as $k \to \infty$, $\forall t$. Such a selection of $\{u_k(t)\}$ leads to $\|y_{i,k}(t) - y_{i,d}(t)\|^2 \to 0$ and then $\limsup_{n\to\infty} \frac{1}{n}\sum_{k=1}^{n}\sum_{t=1}^{T} \|y_{i,k}(t) - y_{i,d}(t)\|^2 = 0$. Therefore, our target is to generate an input sequence $\{u_k(t)\}$ satisfying $u_k(t) \to u_d(t), \forall t$. If a given reference is partially trackable, no input signal that can drive the system to generate the given reference exists. However, an input signal, say $u^*(t)$, must exist such that $\|y^*(t) - y_d(t)\|^2$ is minimized; here, $y^*(t)$ is the corresponding output of $u^*(t)$. Then, as long as we can find an input sequence $\{u_k(t)\}$ satisfying $u_k(t) \to u^*(t)$, we can guarantee that the index $\limsup_{n\to\infty} \frac{1}{n}\sum_{k=1}^{n}\sum_{t=1}^{T} \|y_{i,k}(t) - y_{i,d}(t)\|^2$ is minimized.

If the set of references $y_{i,d}(t)$ is compatible for all sensors, there must exist at least one desired input $u_d(t)$ that generates references for all sensors simultaneously. Therefore, any input sequence satisfying $u_k(t) \to u_d(t)$ guarantees $y_{i,k}(t) - y_{i,d}(t) \to 0$, $\forall i$ in a noise-free system. In this case, the same input sequence can be used to minimize the index $\limsup_{n\to\infty} \frac{1}{n}\sum_{k=1}^{n}\sum_{t=1}^{T}\|y_{i,k}(t) - y_{i,d}(t)\|^2$ for a corresponding stochastic system (as per Theorem 2.8 of Chen and Guo (1991)). However, an arbitrarily given set of references is generally incompatible with all sensors. In such cases, the index-minimum solution for one sensor may significantly increase the values of indices for other sensors. This implies that solutions obtained from best achievable inputs are usually distinct across different sensors. Hence, we need to devise a potential strategy to balance tracking indices across multiple sensors.

To address the multi-objective tracking problem (Emmerich and Deutz, 2018), two commonly employed strategies are available. The first strategy involves combining multiple objectives into a single weighted optimization index, allowing for trade-offs between conflicting objectives by assigning appropriate weights. The second strategy directly applies evolutionary algorithms to search for Pareto optimal solutions. As a pioneering work on the multi-objective tracking problem in learning control, we adopt the first strategy to formulate and solve the problem. Specifically, we combine individual tracking indices from all sensors into a weighted optimization index defined as $\sum_{i=1}^{m}\lambda_i\|y_{i,k}(t) - y_{i,d}(t)\|^2$ at any time instant t, where the weights λ_i satisfy $\lambda_i \geq 0$, $1 \leq i \leq m$, and $\sum_{i=1}^{m}\lambda_i = 1$. In this index, greater weight is assigned to sensors that have more influence on input correctness than others.

Then, the asymptotically averaged index for a multi-sensor system is defined as follows:

$$V = \limsup_{n\to\infty} \frac{1}{n}\sum_{k=1}^{n}\sum_{t=1}^{T}\left(\sum_{i=1}^{m}\lambda_i\|y_{i,k}(t) - y_{i,d}(t)\|^2\right). \tag{3.140}$$

As a special case, if the tracking performance of all sensors is treated equally, we have the following optimization index:

$$V = \limsup_{n\to\infty} \frac{1}{n}\sum_{k=1}^{n}\sum_{t=1}^{T}\left(\sum_{i=1}^{m}\|y_{i,k}(t) - y_{i,d}(t)\|^2\right). \tag{3.141}$$

The optimal input signal with respect to (3.140) and (3.141) will be specified below. We select the Euclidean norm as a measure of tracking performance due to its inherent advantages over alternative approaches. Firstly, it quantifies the standard Euclidean distance between two vectors, representing the cumulative tracking error of all sensors throughout the entire time interval. Secondly, computations based on this metric are straightforward and feasible (Du et al., 2013; Deb, 2008).

3.5.3 Learning Tracking Algorithms and Their Convergence

In this section, we present specific learning control algorithms to address the optimization problem described in the previous section. Specifically, we employ the classical P-type update law (Bristow et al., 2006) with constant and decreasing gains. To establish convergence results, we initially project the incompatible reference onto the space spanned by system matrices and subsequently analyze the tracking performance based on the projected trajectory.

To simplify our analysis, we first consider the equal-weighted index (3.141). The learning algorithm for the ith sensor is

$$u_{i,k+1}(t) = u_k(t) + \sigma L_{i,t} e_{i,k}(t+1), \tag{3.142}$$

where $e_{i,k}(t)$ denotes the tracking error of the ith sensor, $e_{i,k}(t) = y_{i,d}(t) - y_{i,k}(t)$, $L_{i,t}$ denotes a direction regulation matrix, and σ is a learning step, $\sigma > 0$. The final input fed to the plant is generated by taking an equal average of all available local inputs:

$$\begin{aligned} u_{k+1}(t) = \frac{1}{m}\sum_{i=1}^{m} u_{i,k+1}(t) &= u_k(t) + \frac{\sigma}{m}\sum_{i=1}^{m} L_{i,t} e_{i,k}(t+1) \\ &= u_k(t) + \frac{\sigma}{m} L^t e_k(t+1), \end{aligned} \tag{3.143}$$

where $L^t \triangleq [L_{1,t}, \cdots, L_{m,t}]$ and $e_k(t+1) = y_d(t+1) - y_k(t+1)$.

Before presenting the convergence analysis of the above algorithms and associated conditions of learning gain matrices, we conduct a strict analysis of the tracking ability in lifting form. We address the multi-sensor tracking problem with incompatible references $y_{i,d}(t)$. It is important to note that simultaneous complete tracking of the lifted reference $y_d(t)$ by all sensors is not achievable. Therefore, we decompose $y_d(t)$ into two components: $y_d(t) = y_r(t) + y_r^+(t)$. Here, $y_r(t)$ represents the best attainable trajectory that can be completely tracked by a noise-free system described by $x_k(t+1) = A_t x_k(t) + B_t u_k(t)$ and $y_k(t) = C_t x_k(t)$. On the other hand, $y_r^+(t) = y_d(t) - y_r(t)$ denotes the residual information which cannot be tracked completely according to the specified index in our problem formulation.

Next, we present a mathematical description of $y_r(t)$. To this end, we stack the input and lifted output as $\boldsymbol{u}_k \triangleq [u_k^T(0), u_k^T(1), \cdots, u_k^T(N-1)]^T$ and $\boldsymbol{y}_k \triangleq [y_k^T(1), y_k^T(2), \cdots, y_k^T(N)]^T$. Similarly, we denote $\boldsymbol{y}_d \triangleq [y_d^T(1), y_d^T(2), \cdots, y_d^T(N)]^T$. Moreover,

$$\mathbf{G} = \begin{bmatrix} C_1 B_0 & 0 & \dots & 0 \\ C_2 A_1 B_0 & C_2 B_1 & \dots & 0 \\ \vdots & \vdots & \ddots & \vdots \\ C_N A_1^{N-1} B_0 & \dots & \dots & C_N B_{N-1} \end{bmatrix}.$$

Then, for the noise-free system, we have $\boldsymbol{y}_k = \mathbf{G}\boldsymbol{u}_k$. According to the incompatible assumption of $y_{i,d}(t)$, a stacked input $\boldsymbol{u}^\circ$ such that $\boldsymbol{y}_d = \mathbf{G}\boldsymbol{u}^\circ$ does not exist. In this case, the achievable trajectory by the noise-free system is the trajectory that solves the optimization problem:

$$\begin{aligned} \mathbf{y}_r &= \arg\min \|\mathbf{y}_d - \mathbf{y}\|^2 \\ \text{s.t.} \quad & \mathbf{y} = \mathbf{G}\boldsymbol{u} \end{aligned} \tag{3.144}$$

according to index (3.141), where $\mathbf{y}_r \triangleq [y_r^T(1), y_r^T(2), \cdots, y_r^T(N)]^T$. Note that any output $\mathbf{y}$ satisfying $\mathbf{y} = \mathbf{G}\boldsymbol{u}$ must be located in the space spanned by all columns of $\mathbf{G}$, denoted by $\mathcal{S}_\mathbf{G}$. Thus, the best achievable reference $\mathbf{y}_r$ is the projection of $\mathbf{y}_d$ on space $\mathcal{S}_\mathbf{G}$. In other words, the optimization problem (3.144) is equivalent to the following relation:

$$\begin{aligned} \mathbf{y}_r &= \arg\min \|\mathbf{y}_d - \mathbf{y}\|^2 \\ \text{s.t.} \quad & \mathbf{y} \in \mathcal{S}_\mathbf{G}. \end{aligned} \tag{3.145}$$

We denote $\mathbf{y}_r = P_\mathbf{G}\mathbf{y}_d$ and $\mathbf{y}_r^+ = \mathbf{y}_d - P_\mathbf{G}\mathbf{y}_d$, where $P_\mathbf{G}$ denotes the projection on the space spanned by columns of $\mathbf{G}$, i.e., $\mathcal{S}_\mathbf{G}$. Then, $\mathbf{G}^T\mathbf{y}_r^+ = 0$ and $\mathbf{y}_r^T\mathbf{y}_r^+ = 0$.

Moreover, index (3.141) is minimized if we can find an input that drives the output $\mathbf{y}_r$. For the completely trackable reference $y_r(t)$, a unique $u_r(t)$ exists such that

$$\begin{aligned} x_r(t+1) &= A_t x_r(t) + B_t u_r(t), \\ y_r(t) &= C_t x_r(t), \end{aligned} \tag{3.146}$$

provided that the coupling matrix $C_{t+1}B_t$ is of full-column rank. Clearly, in the lifting form, we have $\mathbf{y}_r = \mathbf{G}\boldsymbol{u}_r$. Accordingly, we can rewrite (3.143) in a lifting form as follows:

$$\boldsymbol{u}_{k+1} = \boldsymbol{u}_k + \frac{\sigma}{m}\mathbf{L}\boldsymbol{e}_k, \tag{3.147}$$

where $\boldsymbol{e}_k = \mathbf{y}_d - \mathbf{y}_k$, where $\mathbf{y}_k$ is output of the original stochastic system (3.139), and $\mathbf{L} = \text{diag}\{L^0, L^1, \cdots, L^{N-1}\} \in \mathbf{R}^{Np\times N(q_1+\cdots+q_m)}$.

The following theorem summarizes the convergence property.

Theorem 3.12 *Consider a linear discrete-time system* (3.139), *assume that Assumptions 3.13 and 3.14 hold, and apply learning algorithm* (3.143). *Then, the input error $u_r(t) - u_k(t)$ converges to a small bounded zone of zero in the mean-square and almost-sure senses as $k \to \infty$ if the learning step σ is small enough, and the learning gain matrices $L_{i,t}$ satisfy the condition that all eigenvalues of $\sum_{i=1}^m L_{i,t}C_{i,t+1}B_t$ have positive real parts and all of the columns of* $\mathbf{L}^T$ *belong to the space spanned by* $\mathbf{G}$.

Proof. Denote $\Delta x_k(t) = x_r(t) - x_k(t)$ and $\Delta u_k(t) = u_r(t) - u_k(t)$, where $x_r(t)$ and $u_r(t)$ are defined in (3.146). From (3.139) and (3.146), we have

$$\Delta x_k(t) = \sum_{j=0}^{t-1} A_j^{t-2}B_j\Delta u_k(j) - \sum_{j=0}^{t} A_j^{t-1}w_k(j), \tag{3.148}$$

where $A_j^l = A_jA_{j+1}\cdots A_l$, $l \geq j$, and $A_{i+1}^i \triangleq I$.

Then, subtracting both sides of (3.143) from $u_r(t)$ and substituting (3.148) lead to

$$\begin{aligned}
&\Delta u_{k+1}(t) \\
&= \Delta u_k(t) - \frac{\sigma}{m}\sum_{i=1}^{m} L_{i,t} e_{i,k}(t+1) \\
&= \Delta u_k(t) - \frac{\sigma}{m} L^t (y_r(t) - y_k(t) + y_r^+(t)) \\
&= \left(I - \frac{\sigma}{m} L^t C_{t+1} B_t\right) \Delta u_k(t) - \frac{\sigma}{m} L^t C_{t+1} A_t \Delta x_k(t) \\
&+ \frac{\sigma}{m} L^t C_{t+1} w_k(t+1) + \frac{\sigma}{m}\sum_{i=1}^{m} L_{i,t} v_{i,k}(t+1) - \frac{\sigma}{m} L^t y_r^+(t) \\
&= \left(I - \frac{\sigma}{m} L^t C_{t+1} B_t\right) \Delta u_k(t) - \frac{\sigma}{m} L^t C_{t+1} A_t \sum_{j=0}^{t-1} A_j^{t-2} B_j \Delta u_k(j) \\
&+ \frac{\sigma}{m} L^t C_{t+1} A_t \sum_{j=0}^{t} A_j^{t-1} w_k(j) + \frac{\sigma}{m} L^t C_{t+1} w_k(t+1) \\
&+ \frac{\sigma}{m}\sum_{i=1}^{m} L_{i,t} v_{i,k}(t+1) - \frac{\sigma}{m} L^t y_r^+(t),
\end{aligned} \tag{3.149}$$

where $L^t C_{t+1} = \sum_{i=1}^{m} L_{i,t} C_{i,t+1}$.

Denote $\Delta \boldsymbol{u}_k \triangleq [\Delta u_k^T(0), \Delta u_k^T(1), \cdots, \Delta u_k^T(N-1)]^T, \boldsymbol{w}_k \triangleq [w_k^T(1), \cdots, w_k^T(N)]^T$, $v_k(t) \triangleq [v_{1,k}^T(t), v_{2,k}^T(t), \cdots, v_{m,k}^T(t)]^T$, $\boldsymbol{v}_k \triangleq [v_k^T(1), v_k^T(2), \cdots, v_k^T(N)]^T$.

Then, from (3.149),

$$\begin{aligned}
\Delta \boldsymbol{u}_{k+1} &= \left(I - \frac{\sigma}{m}\mathbf{LG}\right)\Delta \boldsymbol{u}_k + \frac{\sigma}{m}\mathbf{L}(\mathbf{H}\boldsymbol{w}_k + \boldsymbol{v}_k) - \frac{\sigma}{m}\mathbf{L}\mathbf{y}_r^+ \\
&= \left(I - \frac{\sigma}{m}\mathbf{LG}\right)\Delta \boldsymbol{u}_k + \frac{\sigma}{m}\mathbf{L}\boldsymbol{\epsilon}_k - \frac{\sigma}{m}\mathbf{L}\mathbf{y}_r^+,
\end{aligned} \tag{3.150}$$

where $\boldsymbol{\epsilon}_k = \mathbf{H}\boldsymbol{w}_k + \boldsymbol{v}_k$ and

$$\mathbf{H} = \begin{bmatrix} C_1 A_0 & C_1 & 0 & \dots & 0 \\ C_2 A_0^1 & C_2 A_1 & C_2 & \dots & 0 \\ \vdots & \vdots & \vdots & \ddots & \vdots \\ C_N A_0^{N-1} & C_N A_1^{N-1} & C_N A_2^{N-1} & \dots & C_N \end{bmatrix}.$$

First, since all columns of $\mathbf{L}^T$ belong to the space spanned by $\mathbf{G}$, each row of $\mathbf{L}$ can be expressed as a linear combination of the rows of $\mathbf{G}^T$. In other words, a matrix M, whose entries are the coefficients of the linear combination, exists such that $\mathbf{L} = M\mathbf{G}^T$. Consequently, $\mathbf{L}\mathbf{y}_r^+ = M\mathbf{G}^T\mathbf{y}_r^+ = 0$, and (3.150) becomes

$$\Delta \boldsymbol{u}_{k+1} = \left(I - \frac{\sigma}{m}\mathbf{LG}\right)\Delta \boldsymbol{u}_k + \frac{\sigma}{m}\mathbf{L}\boldsymbol{\epsilon}_k. \tag{3.151}$$

$\mathbf{G}$ is a block lower-triangular matrix and $\mathbf{L}$ is a block diagonal matrix. Thus, $\mathbf{LG}$ is a block lower-triangular matrix, with its diagonal elements being $L^t C_{t+1} B_t$, $0 \le t \le N-1$. Based on the stack structure of L^t and C_{t+1} we have $L^t C_{t+1} B_t = \sum_{i=1}^m L_{i,t} C_{i,t+1} B_t$. If the learning gain matrices $L_{i,t}$ satisfy the condition that all eigenvalues of $\sum_{i=1}^m L_{i,t} C_{i,t+1} B_t$ have positive real parts, then all eigenvalues of the matrix $\mathbf{LG}$ have positive real parts. According to the Lyapunov equation theory, for any positive-definite matrix $\mathbf{S} > 0$, a unique positive-definite symmetric matrix $\mathbf{Q} > 0$ satisfying $(\mathbf{LG})^T\mathbf{Q} + \mathbf{Q}(\mathbf{LG}) = \mathbf{S}$ exists. Let $\mathbf{S} = I_{Np}$. Then, we have

$$(\mathbf{LG})^T \mathbf{Q} + \mathbf{Q}(\mathbf{LG}) = I_{Np}. \tag{3.152}$$

Define a weighted norm of $\Delta \boldsymbol{u}_k$:

$$V_k = \|\Delta \boldsymbol{u}_k\|_{\mathbf{Q}}^2 = (\Delta \boldsymbol{u}_k)^T \mathbf{Q} \Delta \boldsymbol{u}_k. \tag{3.153}$$

Substituting (3.151) into (3.153) yields

$$\begin{aligned} V_{k+1} &= \left(\Delta \boldsymbol{u}_k - \frac{\sigma}{m}\mathbf{LG}\Delta \boldsymbol{u}_k + \frac{\sigma}{m}\mathbf{L}\boldsymbol{\epsilon}_k\right)^T \mathbf{Q} \\ &\quad \times \left(\Delta \boldsymbol{u}_k - \frac{\sigma}{m}\mathbf{LG}\Delta \boldsymbol{u}_k + \frac{\sigma}{m}\mathbf{L}\boldsymbol{\epsilon}_k\right) \\ &= \|\Delta \boldsymbol{u}_k\|_{\mathbf{Q}}^2 - \frac{\sigma}{m}\|\Delta \boldsymbol{u}_k\|^2 + \frac{\sigma^2}{m^2}\|\mathbf{LG}\Delta \boldsymbol{u}_k\|_{\mathbf{Q}}^2 \\ &\quad + \frac{2\sigma}{m}(\Delta \boldsymbol{u}_k)^T \left(I - \frac{\sigma}{m}\mathbf{LG}\right)^T \mathbf{Q}(\mathbf{L}\boldsymbol{\epsilon}_k) \\ &\quad + \frac{\sigma^2}{m^2}\|\mathbf{L}\boldsymbol{\epsilon}_k\|_{\mathbf{Q}}^2, \end{aligned} \tag{3.154}$$

where (3.152) is applied for the final equality.

Let $\mathcal{F}_k$ be a σ-algebra, where $\mathcal{F}_k \triangleq \sigma\{x_j(t), w_j(t), v_{i,j}(t), 0 \le t \le N, 0 \le j \le k, 1 \le i \le m\}$. On the basis of the update law (3.143), we clearly have $u_k(t) \in \mathcal{F}_{k-1}$, and thus, $\Delta u_k(t) \in \mathcal{F}_{k-1}$ since $u_r(t)$ is iteration-invariant according to the given reference. Note that $w_k(t)$ and $v_{i,k}(t)$ are independent of $\mathcal{F}_{k-1}$. Thus, $\mathbb{E}[w_k(t) \mid \mathcal{F}_{k-1}] = 0$, $\mathbb{E}[v_{i,k}(t) \mid \mathcal{F}_{k-1}] = 0$, $\sup_k \mathbb{E}[\|w_k(t)\|^2 \mid \mathcal{F}_{k-1}] < \infty$, and $\sup_k \mathbb{E}[\|v_k(t)\|^2 \mid \mathcal{F}_{k-1}] < \infty$. Then, $\mathbb{E}\left[(\Delta \boldsymbol{u}_k)^T\left(I - \frac{\sigma}{m}\mathbf{LG}\right)^T \mathbf{Q}(\mathbf{L}\boldsymbol{\epsilon}_k) \,\middle|\, \mathcal{F}_{k-1}\right] = 0$. We first take conditional expectations to (3.154) with respect to $\mathcal{F}_{k-1}$, and then, take the mathematical expectation to the newly generated equation. Since suitable constants $c_1 > 0$, $c_2 > 0$, and $c_3 > 0$ exist such that $I \ge c_1\mathbf{Q}$, $\|\mathbf{LG}\Delta \boldsymbol{u}_k\|_{\mathbf{Q}}^2 \le c_2\|\Delta \boldsymbol{u}_k\|_{\mathbf{Q}}^2$, and $c_3 = \sup_k \mathbb{E}\|\mathbf{L}\boldsymbol{\epsilon}_k\|_{\mathbf{Q}}^2$, we have

$$\begin{aligned}\mathbb{E}[V_{k+1}] =&\mathbb{E}\|\Delta \boldsymbol{u}_k\|_{\mathbf{Q}}^2 - \frac{\sigma}{m}\mathbb{E}\|\Delta \boldsymbol{u}_k\|^2 + \frac{\sigma^2}{m^2}\mathbb{E}\|\mathbf{LG}\Delta \boldsymbol{u}_k\|_{\mathbf{Q}}^2 \\ &+ \frac{\sigma^2}{m^2}\mathbb{E}\|\mathbf{L}\boldsymbol{\epsilon}_k\|_{\mathbf{Q}}^2 \\ \leq& \left(1 - \frac{\sigma}{m}\left(c_1 - \frac{\sigma}{m}c_2\right)\right)\mathbb{E}\|\Delta \boldsymbol{u}_k\|_{\mathbf{Q}}^2 + \frac{\sigma^2}{m^2}c_3 \\ =&\rho_1\mathbb{E}[V_k] + \tau_1, \end{aligned} \tag{3.155}$$

where $\rho_1 = 1 - \frac{\sigma}{m}(c_1 - \frac{\sigma}{m}c_2)$ and $\tau_1 = \frac{\sigma^2}{m^2}c_3$. Then, the condition $\rho_1 < 1$ is valid, if σ is sufficiently small to ensure $c_1 - \frac{\sigma}{m}c_2 > 0$ and $\frac{\sigma}{m}c_1 < 1$. In other words, $\sigma < \min\{mc_1/c_2, m/c_1\}$. Then, $\mathbb{E}[V_k]$ converges to a small bounded zone of zero. Since $\mathbf{Q}$ is a positive-definite matrix, the mean-square convergence of input error $u_r(t) - u_k(t)$ to a small bound of zero is concluded.

Here, we prove the almost sure convergence of the proposed algorithms. Taking the conditional expectations to both sides of (3.154) with respect to $\mathcal{F}_{k-1}$, similar to (3.155), we have

$$\begin{aligned}&\mathbb{E}\left[V_{k+1} \mid \mathcal{F}_{k-1}\right] \\ &= \mathbb{E}\left[\|\Delta \boldsymbol{u}_k\|_{\mathbf{Q}}^2 \mid \mathcal{F}_{k-1}\right] - \frac{\sigma}{m}\mathbb{E}\left[\|\Delta \boldsymbol{u}_k\|^2 \mid \mathcal{F}_{k-1}\right] \\ &+ \frac{\sigma^2}{m^2}\mathbb{E}\left[\|\mathbf{LG}\Delta \boldsymbol{u}_k\|_{\mathbf{Q}}^2 \mid \mathcal{F}_{k-1}\right] + \frac{\sigma^2}{m^2}\mathbb{E}\left[\|\mathbf{L}\boldsymbol{\epsilon}_k\|_{\mathbf{Q}}^2 \mid \mathcal{F}_{k-1}\right] \\ &\leq \rho_1 V_k + \tau_1, \end{aligned} \tag{3.156}$$

where the properties that $\Delta \boldsymbol{u}_k \in \mathcal{F}_{k-1}$ and $\boldsymbol{\epsilon}_k$ are independent of $\mathcal{F}_{k-1}$ are used. Then, we have

$$\mathbb{E}\left[V_{k+1} - \frac{\tau_1}{1-\rho_1} \mid \mathcal{F}_{k-1}\right] \leq \rho_1\left(V_k - \frac{\tau_1}{1-\rho_1}\right). \tag{3.157}$$

As long as $V_k > \frac{\tau_1}{1-\rho_1}$, the sequence $\{V_k - \frac{\tau_1}{1-\rho_1}\}$ is a non-increasing non-negative supermartingale, and the almost sure convergence of V_k to a small bounded zone of zero is obtained. The proof is thus completed.

Remark 3.16 The condition that all columns of $\mathbf{L}^T$ belong to the space spanned by $\mathbf{G}$ implies that the alternative selections of the algorithm are gradient-based cases. A simple choice for the direction regulation matrix $\mathbf{L}$ is the original gradient, i.e., $L_{i,t} = (C_{i,t+1}B_t)^T$. The positive-definiteness of $\sum_{i=1}^m L_{i,t}C_{i,t+1}B_t$ can be ensured if at least one of $C_{i,t+1}B_t$, for all i, has full-column rank. This selection plays a crucial role in eliminating the influence of untrackable components (i.e., $\boldsymbol{y}_r^{\dagger}$) during the learning process of input signals.

Remark 3.17 From (3.155), we observe that the final convergent bound is dependent on the variance of stochastic noises, represented by τ_1. As τ_1 is inversely proportional to m^2, it can be concluded that increasing the number of sensors effectively reduces the upper bound of input error. Furthermore, in case of a noise-free system (i.e., random noises $w_k(t)$ and $v_{i,k}(t)$ are absent from the system model), $\tau_1 = 0$. This observation indicates that exponential convergence of input sequence is guaranteed due to contraction mapping in (3.157), which highlights an advantage of fixed-step σ over its decreasing step counterpart.

The input error $u_r(t) - u_k(t)$, generated by the fixed-gain-type algorithm (3.143), has not been proven to converge to zero; however, it is highly anticipated that zero-error convergence will be achieved in practical applications due to the inability of a fixed gain σ to suppress the influence of stochastic noises $w_k(t)$ and $v_{i,k}(t)$, which continuously impact the input updating process. Consequently, even after a sufficient number of learning iterations, the input sequence will exhibit fluctuations around the desired value. To ensure stable convergence of the sequence $u_k(t)$ towards $u_r(t)$ (i.e., $u_r(t) - u_k(t) \to 0$ as $k \to \infty$), we propose replacing the fixed-step σ with a decreasing-step sequence σ_k.

The learning algorithm (3.143) becomes

$$
\begin{aligned}
u_{k+1}(t) &= u_k(t) + \frac{\sigma_k}{m} \sum_{i=1}^{m} L_{i,t} e_{i,k}(t+1) \\
&= u_k(t) + \frac{\sigma_k}{m} L^t e_k(t+1),
\end{aligned} \tag{3.158}
$$

where the learning step σ_k satisfies $\sigma_k > 0$, $\sigma_k \to 0$, $\sum_{k=1}^{\infty} \sigma_k = \infty$, and $\sum_{k=1}^{\infty} \sigma_k^2 < \infty$.

The following theorem summarizes the convergence property of (3.158).

Theorem 3.13 *Consider a linear discrete-time system* (3.139), *assume that Assumptions 3.13 and 3.14 hold, and apply learning algorithm* (3.158). *Then, the input error $u_r(t) - u_k(t)$ exhibits both mean-square and almost sure convergence to zero as $k \to \infty$ if the learning gain matrices $L_{i,t}$ satisfy the condition that all eigenvalues of $\sum_{i=1}^{m} L_{i,t} C_{i,t+1} B_t$ have positive real parts and all columns of $\mathbf{L}^T$ belong to the space spanned by* **G**.

Proof. Similar to the proof of Theorem 3.12, we can easily derive the following equation:

$$
\Delta \boldsymbol{u}_{k+1} = \left(I - \frac{\sigma_k}{m} \mathbf{LG} \right) \Delta \boldsymbol{u}_k + \frac{\sigma_k}{m} \mathbf{L} \boldsymbol{\epsilon}_k - \frac{\sigma_k}{m} \mathbf{L} \mathbf{y}_r^+. \tag{3.159}
$$

Again, the rows of **L** can be expressed as the linear combination of the rows of $\mathbf{G}^T$. Thus, $\mathbf{L}\mathbf{y}_r^+ = 0$, which leads to

$$
\Delta \boldsymbol{u}_{k+1} = \left(I - \frac{\sigma_k}{m} \mathbf{LG} \right) \Delta \boldsymbol{u}_k + \frac{\sigma_k}{m} \mathbf{L} \boldsymbol{\epsilon}_k. \tag{3.160}
$$

We apply the matrix **Q** given in the proof of Theorem 3.12 and the weighted norm of $\Delta \boldsymbol{u}_k$ with respect to **Q**: $V_k = (\Delta \boldsymbol{u}_k)^T \mathbf{Q} (\Delta \boldsymbol{u}_k)$. From (3.160) we have

$$\begin{aligned}V_{k+1} &= \left(\Delta\boldsymbol{u}_k - \frac{\sigma_k}{m}\mathbf{LG}\Delta\boldsymbol{u}_k + \frac{\sigma_k}{m}\mathbf{L}\boldsymbol{\epsilon}_k\right)^T \mathbf{Q} \\ &\quad\times \left(\Delta\boldsymbol{u}_k - \frac{\sigma_k}{m}\mathbf{LG}\Delta\boldsymbol{u}_k + \frac{\sigma_k}{m}\mathbf{L}\boldsymbol{\epsilon}_k\right) \\ &= \|\Delta\boldsymbol{u}_k\|_{\mathbf{Q}}^2 - \frac{\sigma_k}{m}\|\Delta\boldsymbol{u}_k\|^2 + \frac{\sigma_k^2}{m^2}\|\mathbf{LG}\Delta\boldsymbol{u}_k\|_{\mathbf{Q}}^2 \\ &\quad+ \frac{2\sigma_k}{m}(\Delta\boldsymbol{u}_k)^T\left(I - \frac{\sigma_k}{m}\mathbf{LG}\right)^T \mathbf{Q}(\mathbf{L}\boldsymbol{\epsilon}_k) + \frac{\sigma_k^2}{m^2}\|\mathbf{L}\boldsymbol{\epsilon}_k\|_{\mathbf{Q}}^2. \end{aligned} \tag{3.161}$$

Note that $\mathbb{E}\left[(\Delta\boldsymbol{u}_k)^T\left(I - \frac{\sigma_k}{m}\mathbf{LG}\right)^T \mathbf{Q}(\mathbf{L}\boldsymbol{\epsilon}_k) \mid \mathcal{F}_{k-1}\right] = 0$. Taking the conditional expectations to both sides of (3.161) with respect to $\mathcal{F}_{k-1}$ and then taking the mathematical expectation to both sides of the obtained equation yield

$$\begin{aligned}\mathbb{E}[V_{k+1}] \le& \mathbb{E}[V_k] - \sigma_k c_4 \mathbb{E}[V_k] + \sigma_k^2 c_5 \mathbb{E}[V_k] + c_6\sigma_k^2 \\ =& (1 - c_4\sigma_k + c_5\sigma_k^2)\mathbb{E}[V_k] + c_6\sigma_k^2, \end{aligned} \tag{3.162}$$

where $c_4 = c_1/m$, $c_5 = c_2/m^2$, and $c_6 = c_3/m^2$. Since $\sigma_k \to 0$, we can choose a sufficiently large integer k_0 such that $1 - c_4\sigma_k + c_5\sigma_k^2 < 1$ for all $k \ge k_0$. Then, we have $\mathbb{E}[V_{k+1}] \le \mathbb{E}[V_k] + c_6\sigma_k^2$. From this equation and $\sum_{k=1}^{\infty}\sigma_k^2 < \infty$, $\sup_k \mathbb{E}[V_k] < \infty$ and $\mathbb{E}[V_k]$ converges. On the basis of this boundedness, from (3.162), we have

$$\mathbb{E}[V_{k+1}] \le (1 - c_4\sigma_k)\mathbb{E}[V_k] + c_7\sigma_k^2, \tag{3.163}$$

where $c_7 \ge c_6 + c_5\mathbb{E}[V_k]$. Observing that $\sum_{k=1}^{\infty}\sigma_k = \infty$ and $\sum_{k=1}^{\infty}\sigma_k^2 < \infty$, we conclude that $\lim_{k\to\infty}\mathbb{E}[V_k] = 0$. In other words, $\lim_{k\to\infty}\mathbb{E}[\|\Delta\boldsymbol{u}_k\|^2] = 0$ because $\mathbf{Q}$ is a positive-definite matrix. The mean-square convergence is thus obtained.

Here, we demonstrate the almost sure convergence. By taking the conditional expectation to both sides of (3.161) with respect to $\mathcal{F}_{k-1}$,

$$\mathbb{E}[V_{k+1} \mid \mathcal{F}_{k-1}] \le V_k + \sigma_k^2(c_5 V_k + c_6). \tag{3.164}$$

Using the mean-square convergence result, it is verified that

$$\begin{aligned}&\sum_{k=1}^{\infty}\mathbb{E}\left[\sigma_k^2(c_5 V_k + c_6)\right] \\ &\le \left(c_5 \sup_k \mathbb{E}[V_k] + c_6\right)\sum_{k=1}^{\infty}\sigma_k^2 < \infty, \quad \text{almost surely.}\end{aligned}$$

Then, using Lemma 2.2, V_k can be determined to converge almost surely. Thus far, we have shown that V_k converges to zero based on mean-square convergence. Therefore, the almost sure convergent limitation of V_k should also be zero. The almost sure convergence is thus obtained, and the proof is completed.

Subsequently, we proceed to the general performance index (3.140). In this index, the significance of different tracking objectives is equitably balanced through the assignment of varying weights. Specifically, higher weights are allocated to prioritize tracking performance for more important sensors while comparatively lower weights are assigned to less critical sensors. This approach ensures that due attention is given to the tracking information from vital sensors while minimizing emphasis on less significant ones. In short, for index (3.140), the learning algorithm is formulated as

$$u_{k+1}(t) = u_k(t) + \sigma_k \sum_{i=1}^{m} \lambda_i L_{i,t} e_{i,k}(t+1), \tag{3.165}$$

where λ_i, $1 \le i \le m$, are the weights given in (3.140) with $\lambda_1 + \cdots + \lambda_m = 1$. Here, we apply the decreasing step sequence directly to ensure zero-error tracking performance; analysis of the fixed-step case is omitted here to save space.

Similar to that in the equal-weight case, we rewrite the update law (3.165) as

$$u_{k+1}(t) = u_k(t) + \sigma_k L^t W e_k(t+1), \tag{3.166}$$

where $L^t = [L_{1,t}, \cdots, L_{m,t}]$ and $W = \text{diag}\{\lambda_1 I_{q_1}, \cdots, \lambda_m I_{q_m}\}$. Then, we stack the input and lifted output as above, i.e., we apply the notations $\boldsymbol{u}_k$ and $\boldsymbol{y}_k$. The system evolution does not change, i.e., $\boldsymbol{y}_k = \mathbf{G}\boldsymbol{u}_k$. The achievable trajectory by the noise-free system according to the weighted index (3.140) is a trajectory that solves the following optimization problem:

$$\begin{aligned} \boldsymbol{y}_{wr} &= \arg\min [\boldsymbol{y}_d - \boldsymbol{y}]^T [I_N \otimes W][\boldsymbol{y}_d - \boldsymbol{y}] \\ &= \arg\min \|\boldsymbol{y}_d - \boldsymbol{y}\|^2_{\mathbf{W}_0} \\ &\text{s.t.} \boldsymbol{y} = \mathbf{G}\boldsymbol{u} \end{aligned} \tag{3.167}$$

where $\boldsymbol{y}_{wr}$ denotes the achievable trajectory associated with index (3.140) and $\mathbf{W}_0 \triangleq I_N \otimes W$. Note that $\|\boldsymbol{y}_d - \boldsymbol{y}\|^2_{\mathbf{W}_0} = \|\mathbf{W}_0^{\frac{1}{2}} \boldsymbol{y}_d - \mathbf{W}_0^{\frac{1}{2}} \boldsymbol{y}\|^2$. Thus, the combination solution of the projection of $\boldsymbol{y}_r$ on the space $\mathcal{S}_{\mathbf{G}}$ minimizing (3.167) is identical to that of the projection of $\mathbf{W}_0^{\frac{1}{2}} \boldsymbol{y}_d$ on the space spanned by the columns of $\mathbf{W}_0^{\frac{1}{2}}\mathbf{G}$. In this case, we denote $\boldsymbol{y}_{wr} = P_{\mathbf{G}'} \boldsymbol{y}_d$ and $\boldsymbol{y}^+_{wr} = \boldsymbol{y}_d - P_{\mathbf{G}'} \boldsymbol{y}_d$, where $P_{\mathbf{G}'}$ denotes the projection on the space spanned by the columns of $\mathbf{G}$ while minimizing the weighted index (3.167). Then, we have $\mathbf{G}^T \mathbf{W}_0 \boldsymbol{y}^+_{wr} = 0$ and $\boldsymbol{y}^T_{wr} \boldsymbol{y}^+_{wr} = 0$. A $\boldsymbol{u}_{wr}$ exists such that $\boldsymbol{y}_{wr} = \mathbf{G}\boldsymbol{u}_{wr}$, and our objective is to show the convergence of $\boldsymbol{u}_k$ to $\boldsymbol{u}_{wr}$ as the iteration number increases.

From (3.166),

$$\boldsymbol{u}_{k+1} = \boldsymbol{u}_k + \sigma_k \mathbf{L}\mathbf{W}_0 \boldsymbol{e}_k, \tag{3.168}$$

where $\mathbf{L} = \text{diag}\{L^0, L^1, \cdots, L^{N-1}\}$, $\mathbf{W}_0 = I_N \otimes W$, and $\boldsymbol{e}_k = \boldsymbol{y}_d - \boldsymbol{y}_k$, as defined above.

The convergence results for the weighted case, i.e., learning algorithm (3.165), are summarized in the following theorem.

Theorem 3.14 *Consider a linear discrete-time system* (3.139), *assume that Assumptions 3.13 and 3.14 hold, and apply learning algorithm* (3.165) *according to weighted tracking index* (3.140). *The system input error $u_{wr}(t) - u_k(t)$ exhibits both mean-square and almost sure convergence to zero as $k \to \infty$ if the learning gain matrices $L_{i,t}$ satisfy the condition that all eigenvalues of $\sum_{i=1}^{m} \lambda_i L_{i,t} C_{i,t+1} B_t$ have positive real parts and all columns of $\mathbf{L}^T$ belong to the space spanned by* **G**.

Proof. Similar to the proof of Theorem 3.12, we clearly have

$$\begin{aligned} \Delta \boldsymbol{u}'_{k+1} &= \Delta \boldsymbol{u}'_k - \sigma_k \mathbf{L}\mathbf{W}_0 \left[\mathbf{G}\Delta \boldsymbol{u}'_k - \mathbf{H}\boldsymbol{w}_k - \boldsymbol{v}_k + \boldsymbol{y}^+_{wr}\right] \\ &= \left(I - \sigma_k \mathbf{L}\mathbf{W}_0 \mathbf{G}\right) \Delta \boldsymbol{u}'_k + \sigma_k \mathbf{L}\mathbf{W}_0 \boldsymbol{\epsilon}_k - \sigma_k \mathbf{L}\mathbf{W}_0 \mathbf{y}^+_{wr}, \end{aligned} \tag{3.169}$$

where $\boldsymbol{\epsilon}_k = \mathbf{H}\boldsymbol{w}_k + \boldsymbol{v}_k$ and $\Delta \boldsymbol{u}'_k = \boldsymbol{u}_{wr} - \boldsymbol{u}_k$ with $\boldsymbol{u}_{wr}$ being the desired input according to $\boldsymbol{y}_{wr}$.

Since all columns of $\mathbf{L}^T$ belong to the space spanned by **G**, each row of **L** can be expressed as a linear combination of the rows of $\mathbf{G}^T$. In other words, an M exists such that $\mathbf{L} = M\mathbf{G}^T$. Consequently, $\mathbf{L}\mathbf{W}_0 \mathbf{y}^+_{wr} = M\mathbf{G}^T \mathbf{W}_0 \mathbf{y}^+_{wr} = 0$. Then, (3.169) becomes

$$\Delta \boldsymbol{u}'_{k+1} = \left(I - \sigma_k \mathbf{L}\mathbf{W}_0 \mathbf{G}\right) \Delta \boldsymbol{u}'_k + \sigma_k \mathbf{L}\mathbf{W}_0 \boldsymbol{\epsilon}_k. \tag{3.170}$$

Note that **G** is a block lower-triangular matrix, **L** is a block diagonal matrix, and $\mathbf{W}_0$ is a diagonal matrix. Thus, $\mathbf{L}\mathbf{W}_0\mathbf{G}$ is a block lower-triangular matrix, with its diagonal blocks being $L^t W C_{t+1} B_t$, $0 \le t \le N-1$. From the stack structure of L^t and C_{t+1}, we know that $L^t W C_{t+1} B_t = \sum_{i=1}^{m} \lambda_i L_{i,t} C_{i,t+1} B_t$. According to the design conditions of $L_{i,t}$, we know that the eigenvalues of $L^t W C_{t+1} B_t$ have positive real parts. According to the Lyapunov equation theory, a unique positive-definite matrix $\mathbf{Q}_0 > 0$ exists such that $(\mathbf{L}\mathbf{W}_0\mathbf{G})^T \mathbf{Q}_0 + \mathbf{Q}_0 \mathbf{L}\mathbf{W}_0\mathbf{G} = I_{Np}$.

Similar to the proof of Theorem 3.13, the weighted norm of $\Delta \boldsymbol{u}'_k$ is defined with respect to $\mathbf{Q}_0$, $V_k = \left(\Delta \boldsymbol{u}'_k\right)^T \mathbf{Q}_0 \Delta \boldsymbol{u}'_k = \|\Delta \boldsymbol{u}'_k\|^2_{\mathbf{Q}_0}$. Then, from (3.170),

$$\begin{aligned}V_{k+1} &= \left(\Delta \boldsymbol{u}'_k - \sigma_k \mathbf{L}\mathbf{W}_0\mathbf{G}\Delta \boldsymbol{u}'_k + \sigma_k \mathbf{L}\mathbf{W}_0\boldsymbol{\epsilon}_k\right)^T \mathbf{Q}_0 \\ &\times \left(\Delta \boldsymbol{u}'_k - \sigma_k \mathbf{L}\mathbf{W}_0\mathbf{G}\Delta \boldsymbol{u}'_k + \sigma_k \mathbf{L}\mathbf{W}_0\boldsymbol{\epsilon}_k\right) \\ &= \|\Delta \boldsymbol{u}'_k\|^2_{\mathbf{Q}_0} - \sigma_k \|\Delta \boldsymbol{u}'_k\|^2 + \sigma_k^2 \|\mathbf{L}\mathbf{W}_0\mathbf{G}\Delta \boldsymbol{u}'_k\|^2_{\mathbf{Q}_0} \\ &+ 2\sigma_k (\Delta \boldsymbol{u}'_k)^T (I - \sigma_k \mathbf{L}\mathbf{W}_0\mathbf{G})^T \mathbf{Q}_0 \left(\mathbf{L}\mathbf{W}_0\boldsymbol{\epsilon}_k\right) \\ &+ \sigma_k^2 \|\mathbf{L}\mathbf{W}_0\boldsymbol{\epsilon}_k\|^2_{\mathbf{Q}_0}. \end{aligned} \tag{3.171}$$

Note that $\mathbb{E}\left[(\Delta \boldsymbol{u}'_k)^T (I - \sigma_k \mathbf{L}\mathbf{W}_0\mathbf{G})^T \mathbf{Q}_0 \left(\mathbf{L}\mathbf{W}_0\boldsymbol{\epsilon}_k\right) \mid \mathcal{F}_{k-1}\right] = 0$, where $\mathcal{F}_{k-1}$ is defined as above. Taking the conditional expectation to both sides of (3.171) with respect to $\mathcal{F}_{k-1}$ and then taking the mathematical expectation to both sides of the obtained equation,

$$\begin{aligned}\mathbb{E}[V_{k+1}] \leq &\mathbb{E}[V_k] - \sigma_k c_8 \mathbb{E}[V_k] + \sigma_k^2 c_9 \mathbb{E}[V_k] + c_{10}\sigma_k^2 \\ = &(1 - \sigma_k c_8 + \sigma_k^2 c_9)\mathbb{E}[V_k] + c_{10}\sigma_k^2, \end{aligned} \tag{3.172}$$

where c_8 satisfies $I \geq c_8 \mathbf{Q}_0$, c_9 satisfies $\|\mathbf{L}\mathbf{W}_0\mathbf{G}\Delta \boldsymbol{u}'_k\|^2_{\mathbf{Q}_0} \leq c_9 \|\Delta \boldsymbol{u}'_k\|^2_{\mathbf{Q}_0}$, and $c_{10} = \sup_k \mathbb{E}\|\mathbf{L}\mathbf{W}_0\boldsymbol{\epsilon}_k\|^2_{\mathbf{Q}_0}$. As this equation is identical to (3.162), the steps of this proof are identical to those in the proof of Theorem 3.13. Both the mean-square and almost sure convergence of the input sequence to $\boldsymbol{u}_{wr}$ can be proven, and the proof is completed.

Remark 3.18 Based on the aforementioned derivations, it is observed that the weighted index (3.140) results in imbalanced performance across different sensors. Consequently, the optimal reference trajectory based on incompatible objectives would shift from $\boldsymbol{y}_r$ to $\boldsymbol{y}_{wr}$. In the case where $\lambda_1 = \cdots = \lambda_m$, it becomes evident that both the update law and achievable reference remain identical to those derived for the equal-weight index (3.141). This observation confirms that the update law and convergence outcomes of the equal-weight index (3.141) represent a special weighted instance of (3.140). Furthermore, it should be noted that row selection in matrix $\mathbf{L}$ primarily relies on system matrix $\mathbf{G}$.

3.5.4 *Extensions to the Data Dropout Case*

In the previous section, we established an ILC for multi-sensor systems with both equal-weighted and weighted indices, assuming seamless transmission between sensors and the learning controller. However, in practical applications where wired/wireless networks are used for connectivity between the plant and controller to enhance flexibility, robustness, and implementation ease, random data dropouts may occur due to factors such as data congestion, limited bandwidth, and linkage faults. In this section, we extend the results presented in Sect. 3.5.3 to address the scenario of data dropout. For brevity reasons, we focus on discussing only the equal-weighted index (3.141), while omitting similar steps required for obtaining results related to the weighted index (3.140).

It is worth noting that data transmission can be categorized into two states: successful transmission and loss. Therefore, data dropout is commonly represented by a random binary variable, denoted as $\gamma_{i,k}(t)$, for the ith sensor at the tth time instant of the kth iteration. Specifically, $\gamma_{i,k}(t)$ takes a value of 1 if the corresponding data packet is successfully transmitted; otherwise, it takes a value of 0. In essence, the occurrence of data dropouts can be viewed as a stochastic switch that randomly opens and closes within the network. Generally speaking, the randomness associated with data dropouts can be effectively described using Bernoulli variables. That is, we assume

$$\mathbb{P}(\gamma_{i,k}(t) = 1) = \overline{\gamma}, \quad \mathbb{P}(\gamma_{i,k}(t) = 0) = 1 - \overline{\gamma}, \tag{3.173}$$

where $0 < \overline{\gamma} < 1$ denotes the average level of successful transmission. If $\overline{\gamma} = 0$, the network is completely non-functional, rendering it incapable of transmitting any data packets and consequently preventing any system improvement. Conversely, if $\overline{\gamma} = 1$, the network operates efficiently; this scenario was previously discussed in the preceding subsections. Additionally, we acknowledge that the distribution remains consistent for all i and t in (3.173). This assumption is solely made to simplify the expressions; although extending the identical successful transmission rate case to encompass time- and sensor-varying dropout rates necessitates some effort (alongside tedious notations), it can be accomplished using similar steps as outlined in this section.

We adopt an update law with decreasing steps in this subsection. However, the update law with a fixed step can be addressed by combining the derivations presented in Theorem 3.15 and those in the proof of Theorem 3.12. Specifically, we consider the update law (3.158) discussed in the previous section. In case of data packet loss during transmission, we substitute zero for the tracking error. In this case, the update law is modified as

$$\begin{aligned} u_{k+1}(t) &= u_k(t) + \frac{\sigma_k}{m} \sum_{i=1}^{m} \gamma_{i,k}(t) L_{i,t} e_{i,k}(t+1) \\ &= u_k(t) + \frac{\sigma_k}{m} L^t \Gamma_k(t) e_k(t+1), \end{aligned} \tag{3.174}$$

where the learning step σ_k satisfies $\sigma_k > 0$, $\sigma_k \to 0$, $\sum_{k=1}^{\infty} \sigma_k = \infty$, and $\sum_{k=1}^{\infty} \sigma_k^2 < \infty$. $\Gamma_k(t) = \text{diag}\{\gamma_{1,k}(t) I_{q_1}, \cdots, \gamma_{m,k}(t) I_{q_m}\}$.

We obtain the following convergence result for the learning algorithm with random data dropouts.

Theorem 3.15 *Consider a linear discrete-time system* (3.139), *assume that Assumptions 3.13 and 3.14 hold, and apply learning algorithm* (3.174). *Moreover, the networks from each sensor to the learning controller are not completely broken, i.e.,* $0 < \overline{\gamma} < 1$. *Then, the system input error* $u_r(t) - u_k(t)$ *exhibits both mean-square and almost sure convergence to zero as* $k \to \infty$ *if the learning gain matrices* $L_{i,t}$

satisfy the condition that all eigenvalues of $\sum_{i=1}^{m} L_{i,t}C_{i,t+1}B_t$ *have positive real parts and all columns of* $\mathbf{L}^T$ *belong to the space spanned by* $\mathbf{G}$.

Proof. Similar to the steps in the previous proofs, we stack the input and output of the whole iteration interval into supervectors $\Delta\boldsymbol{u}_k$ and $\Delta\mathbf{y}_k$. The meanings of other stacked supervectors are identical to those in the previous proofs. Then, from (3.174), $\boldsymbol{u}_{k+1} = \boldsymbol{u}_k + \frac{\sigma_k}{m}\mathbf{L}\Gamma_k\boldsymbol{e}_k$, where Γ_k is defined as $\Gamma_k = \text{diag}\{\Gamma_k(0), \cdots, \Gamma_k(N-1)\} \in \mathbf{R}^{N(q_1+\cdots+q_m)\times N(q_1+\cdots+q_m)}$. Clearly, Γ_k is a diagonal matrix.

Subtracting both sides of the last equation from $\boldsymbol{u}_r$ leads to

$$\begin{aligned}\Delta\boldsymbol{u}_{k+1} =& \Delta\boldsymbol{u}_k - \frac{\sigma_k}{m}\mathbf{L}\Gamma_k\mathbf{G}\Delta\boldsymbol{u}_k + \frac{\sigma_k}{m}\mathbf{L}\Gamma_k\boldsymbol{\epsilon}_k + \frac{\sigma_k}{m}\mathbf{L}\mathbf{y}_r^+\\ =& \Delta\boldsymbol{u}_k - \frac{\sigma_k}{m}\mathbf{L}\overline{\Gamma}\mathbf{G}\Delta\boldsymbol{u}_k\\ &+ \frac{\sigma_k}{m}\mathbf{L}(\overline{\Gamma} - \Gamma_k)\mathbf{G}\Delta\boldsymbol{u}_k + \frac{\sigma_k}{m}\mathbf{L}\Gamma_k\boldsymbol{\epsilon}_k,\end{aligned} \tag{3.175}$$

where $\overline{\Gamma} \triangleq \mathbb{E}\Gamma_k$ and $\boldsymbol{\epsilon}_k = \mathbf{H}\boldsymbol{w}_k + \boldsymbol{v}_k$. Clearly, $\overline{\Gamma} = \overline{\gamma} I_{N(q_1+\cdots+q_m)}$ under the homogeneous data dropout rate. Thus, $\mathbf{L}\overline{\Gamma}\mathbf{G} = \overline{\gamma}\mathbf{L}\mathbf{G}$.

Similar to the analysis in the proof of Theorem 3.13, a unique positive-definite matrix $\mathbf{Q}$ exists such that $(\mathbf{L}\mathbf{G})^T\mathbf{Q} + \mathbf{Q}(\mathbf{L}\mathbf{G}) = I_{Np}$. We apply the weighted norm of $\Delta\boldsymbol{u}_k$, $V_k = \|\Delta\boldsymbol{u}_k\|_{\mathbf{Q}}^2 = (\Delta\boldsymbol{u}_k)^T\mathbf{Q}\Delta\boldsymbol{u}_k$, and have

$$\begin{aligned}V_{k+1} =& \|\Delta\boldsymbol{u}_k\|_{\mathbf{Q}}^2 + \frac{\sigma_k^2\overline{\gamma}^2}{m^2}\|\mathbf{L}\mathbf{G}\Delta\boldsymbol{u}_k\|_{\mathbf{Q}}^2 - \frac{\sigma_k}{m}\overline{\gamma}\|\Delta\boldsymbol{u}_k\|^2\\ &+ \frac{\sigma_k^2}{m^2}\|\mathbf{L}(\overline{\Gamma} - \Gamma_k)\mathbf{G}\Delta\boldsymbol{u}_k\|_{\mathbf{Q}}^2 + \frac{\sigma_k^2}{m^2}\|\mathbf{L}\Gamma_k\boldsymbol{\epsilon}_k\|_{\mathbf{Q}}^2\\ &+ \frac{2\sigma_k^2}{m^2}(\Delta\boldsymbol{u}_k)^T\left(\mathbf{L}(\overline{\Gamma} - \Gamma_k)\mathbf{G}\right)^T\mathbf{Q}\mathbf{L}\Gamma_k\boldsymbol{\epsilon}_k\\ &+ \frac{2\sigma_k}{m}(\Delta\boldsymbol{u}_k)^T\left(I - \frac{\sigma_k\overline{\gamma}}{m}\mathbf{L}\mathbf{G}\right)^T\mathbf{Q}\mathbf{L}(\overline{\Gamma} - \Gamma_k)\mathbf{G}\Delta\boldsymbol{u}_k\\ &+ \frac{2\sigma_k}{m}(\Delta\boldsymbol{u}_k)^T\left(I - \frac{\sigma_k\overline{\gamma}}{m}\mathbf{L}\mathbf{G}\right)^T\mathbf{Q}\mathbf{L}\Gamma_k\boldsymbol{\epsilon}_k.\end{aligned} \tag{3.176}$$

The proof can be then completed following the procedures of Theorem 3.12.

Remark 3.19 In update law (3.174), the average coefficient $1/m$ is retained even if the amount of available data packets is less than m when data dropouts occur. In other words, the average coefficient $1/m$ may overly suppress the whole effect of the available tracking errors from the sensors. Thus, in case one considers the average coefficient according to the available packets, (3.174) is modified as

$$u_{k+1}(t) = u_k(t) + \sigma_k\beta_k^{-1}(t)\sum_{i=1}^{m}\gamma_{i,k}(t)L_{i,t}e_{i,k}(t+1), \tag{3.177}$$

where $\beta_k(t) = \sum_{i=1}^{m} \gamma_{i,k}(t)$. Here, $\beta_k(t)$ denotes the number of available packets and thus it is event-triggered. To avoid mathematical irrationality, we must either assume that at least one sensor can successfully transmit its packet for each time instant and each iteration or simply redefine $\beta_k(t) = \max\{\sum_{i=1}^{m} \gamma_{i,k}(t), 1\}$.

3.5.5 Illustrative Simulations

To verify the analysis results, we present numerical simulations according to the following system:

$$
\begin{aligned}
&\boldsymbol{x}_k(t+1) = \\
&\quad \begin{bmatrix} 0.1 - 0.05\sin(0.2t) & 0 & 0.01t \\ 0.1 & -0.01t & -0.02\cos(0.5t) \\ 0 & 0.4 & 0.1\cos(0.2t) \end{bmatrix} \boldsymbol{x}_k(t) \\
&\quad + \begin{bmatrix} 1 - 0.1\sin(0.5\pi t)^2 & 0 \\ 0.1\sin(\pi t) & 0.1 \\ 0.2 & 0.5 + 0.02\cos(\pi t)^3 \end{bmatrix} \boldsymbol{u}_k(t) \\
&\quad + \boldsymbol{\omega}_k(t+1),
\end{aligned}
$$

where the system has three outputs given by

$$
\begin{aligned}
y_{1,k}(t) &= \begin{bmatrix} -2 & 1.2 & 2 \end{bmatrix} \boldsymbol{x}_k(t) + \nu_{1,k}(t), \\
y_{2,k}(t) &= \begin{bmatrix} 1 & 0.04 & 2 \end{bmatrix} \boldsymbol{x}_k(t) + \nu_{2,k}(t), \\
y_{3,k}(t) &= \begin{bmatrix} -0.4 & 0.3 & 3 \end{bmatrix} \boldsymbol{x}_k(t) + \nu_{3,k}(t).
\end{aligned}
$$

The noises are simulated in a Gaussian distribution, $\boldsymbol{\omega}_k(t+1) \sim N(0, 0.05^2 I_3)$ and $\nu_{i,k}(t) \sim N(0, 0.1^2)$ for $i = 1, 2, 3$.

The desired references for the three outputs are defined by $y_{1,d} = \sin(0.1\pi t)$, $y_{2,d} = \cos(0.1\pi t) - \frac{1}{t+1}$, $y_{3,d} = \sin(0.1\pi t) + \frac{1}{2}\cos(0.1\pi t) - \frac{1}{2}$. Since we have only two inputs for three sensors, the desired reference is clearly incompatible, i.e., at least part of the references cannot be precisely tracked.

The iteration length is set to $T = 50$. The learning gain is simply selected as $L_{i,t} = (C_{i,t+1}B_t)^T$ (demonstrated in Remark 3.16). All algorithms are run for 50 iterations. Both fixed and decreasing gains will be discussed in the following subsections. We set the gain as $\sigma_k = \alpha/k^\beta$, $\alpha, \beta \geq 0$. If $\beta = 0$, the gain becomes constant for all iterations. Different selections of α and β will be elaborated to demonstrate the influence of these parameters.

In the subsequent analysis, we sequentially examine three distinct scenarios. Firstly, we rigorously establish the convergence of the equal-weight index case. Subsequently, we elucidate the weighted index case in detail. Lastly, we provide a concise comparative analysis of this scenario incorporating random data dropouts.

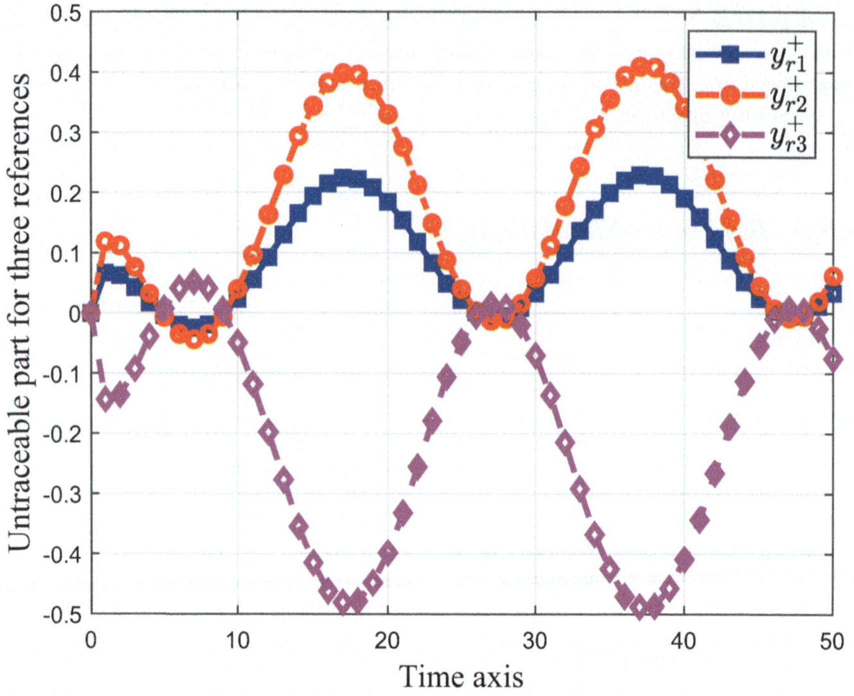

Fig. 3.31 Residual part of the given references for equal-weight index (3.141)

3.5.5.1 Equal-Weight Index Case

For the equal-weight index (3.141), the best achievable trajectory $\mathbf{y}_r$ is computed by $\mathbf{y}_r = P_{\mathbf{G}}\mathbf{y}_d$, where $\mathbf{y}_d$ is lifted by the given references for all three sensors. Then, we have the residual part $\mathbf{y}_r^+ = \mathbf{y}_d - \mathbf{y}_r$. The visible values of $\mathbf{y}_r^+$ are shown in Fig. 3.31 while the achievable reference $\mathbf{y}_r$ is plotted in Fig. 3.33 below. In Fig. 3.31, the dot-dashed line with circles, solid line with squares, and dashed line with diamonds denote residual parts for the three sensors in sequence. Here, $\mathbf{y}_r^+ \neq 0$, which indicates that the three references are incompatible. In other words, we cannot track all three references with zero-error simultaneously. The best achievable reference $\mathbf{y}_r$ describes our objective according to index (3.141).

For a given input $\boldsymbol{u}_k$, we denote the virtual output of the noise-free plant as $\mathbf{y}_k^{\text{v}}$, i.e., $\mathbf{y}_k^{\text{v}} = \mathbf{G}\boldsymbol{u}_k$; then, $\mathbf{y}_k = \mathbf{y}_k^{\text{v}} + \mathbf{H}\boldsymbol{w}_k + \boldsymbol{v}_k$. We call $\mathbf{y}_r - \mathbf{y}_k^{\text{v}}$ a learning error to indicate the part that we can improve by learning control algorithms. The tracking error consists of three parts: current-iteration random noises, a learning error, and the residual part. In particular,

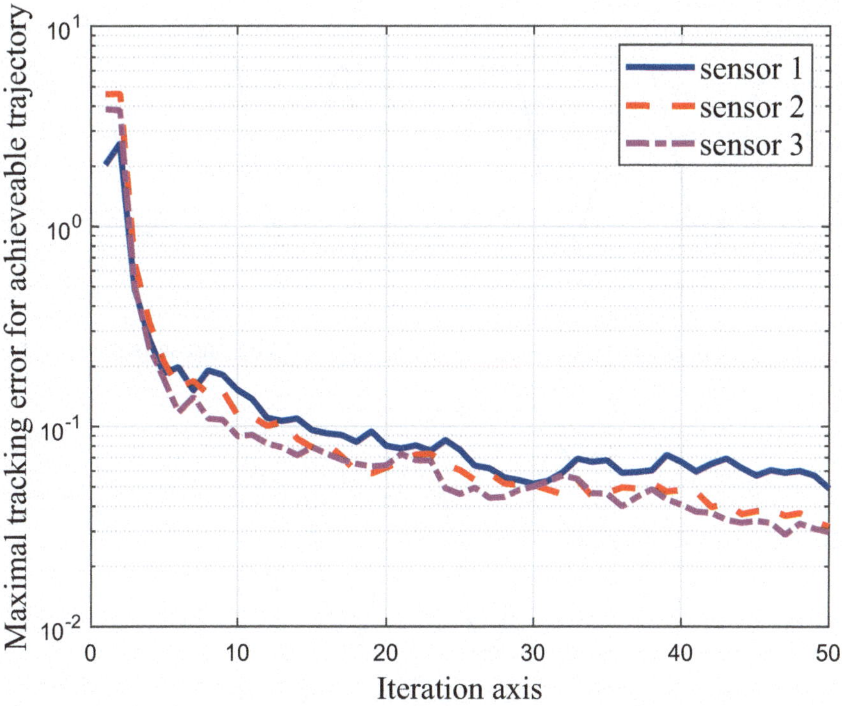

Fig. 3.32 Maximal learning error profiles of three sensors for algorithm (3.158)

$$\begin{aligned} \boldsymbol{e}_k &= \boldsymbol{y}_d - \boldsymbol{y}_k \\ &= \underbrace{(\boldsymbol{y}_d - \boldsymbol{y}_r)}_{\text{residual part}} + \underbrace{(\boldsymbol{y}_r - \boldsymbol{y}_k^{\text{v}})}_{\text{learning error}} - \underbrace{(\mathbf{H}\boldsymbol{w}_k + \boldsymbol{v}_k)}_{\text{random noise}} . \end{aligned}$$

Because both residual part and random noises cannot be improved by any learning control algorithm, the actual objective is to drive the learning error to zero asymptotically along the iteration axis to minimize the index (3.141).

To achieve the above tracking performance, we first apply the learning control algorithm with decreasing gain, i.e., (3.158). The parameters of decreasing gain $\sigma_k = \alpha/k^{\beta}$ are set as $\alpha = 1$ and $\beta = 1$. The maximal learning error profiles for the three sensors are plotted in Fig. 3.32, where the dotted, solid, and dashed lines denote the three sensors in turn. The maximal learning error is defined as $\max_{0\leq t\leq T} |e_{i,k}^{\text{learn}}(t)|$, where $e_{i,k}^{\text{learn}}(t) \triangleq y_{i,r}(t) - y_{i,k}^{\text{v}}(t)$. A continuously decreasing trend is observed in all profiles. This observation verifies the efficiency of the proposed algorithms.

The tracking performance at the 50th iteration is depicted in Fig. 3.33, where each subfigure illustrates the desired reference $\boldsymbol{y}_d$, achievable trajectory $\boldsymbol{y}_r$, and virtual output $\boldsymbol{y}_k^{\text{v}}$ represented by the dotted line, solid line, and dashed line, respectively. It is evident that the outputs precisely follow the achievable trajectories for each

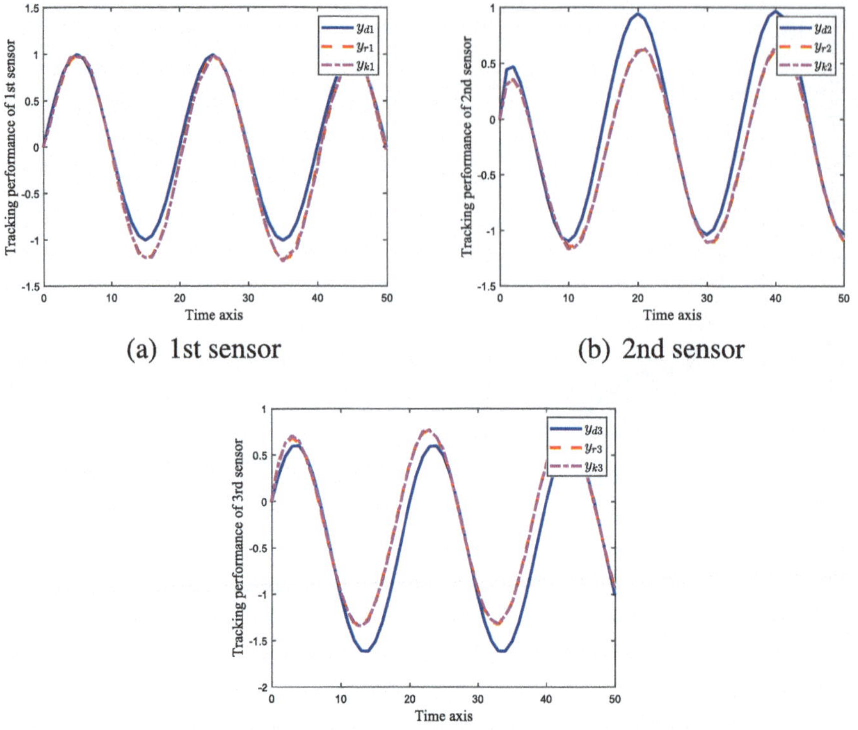

(a) 1st sensor

(b) 2nd sensor

(c) 3rd sensor

Fig. 3.33 Tracking performance at the 50th iteration of three sensors for algorithm (3.158)

sensor; however, they noticeably deviate from the desired references. This figure provides empirical evidence of the satisfactory tracking performance achieved by our proposed algorithm.

To demonstrate the influence of parameters α and β, we consider three cases of algorithm (3.158): $\alpha = 1$, $\beta = 1$; $\alpha = 1/4$, $\beta = 1/4$; and $\alpha = 1/4$, $\beta = 2/3$. The learning index profiles in the iteration domain are shown in Fig. 3.34, where the learning index for each iteration is defined as $\|\mathbf{y}_d - \mathbf{y}_k^{\nu}\|^2$. Since the current-iteration random noise has been removed from this index, the profiles should converge to $\|\mathbf{y}_r^+\|^2$ according to (3.145). The tracking performance can be seen in Fig. 3.34, where the solid line with circles denotes the value of $\|\mathbf{y}_r^+\|^2$. Moreover, we can observe that decreasing α may slow down the convergence speed, while decreasing β would accelerate it. This observation provides a general guideline for parameter setting.

The algorithm is simulated with a fixed gain (3.143) for three cases: $\sigma = 1$, $\sigma = 1/8$, and $\sigma = 1/20$. The obtained tracking performances are similar to those in (3.158), although there are differences in tracking precision. The index profiles are presented in Fig. 3.35. The subplot of the figure demonstrates that the index profiles

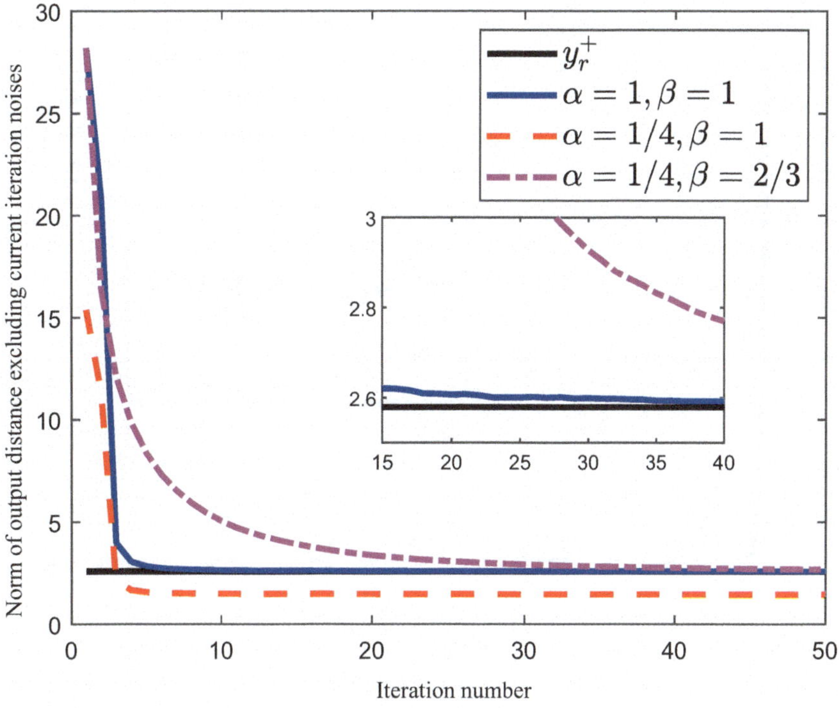

Fig. 3.34 Index profiles for algorithm (3.158)

deviate from the minimal value $\|\mathbf{y}_r^+\|^2$, thus confirming the conclusions of Theorem 3.12. Moreover, caution should be exercised when selecting a constant gain as larger selections may result in greater tracking errors while smaller selections may lead to a slower convergence speed.

3.5.5.2 Weighted Index Case

In this subsection, we consider the weighted index (3.140) with weights being $\lambda_1 = 0.4$, $\lambda_2 = 0.3$, and $\lambda_3 = 0.3$. For this case, the best achievable trajectory is defined according to (3.167). The residual part $\mathbf{y}_{wr}^+$ is shown in Fig. 3.36. Compared to Fig. 3.31, we observe that the residual part for the first sensor has been reduced, which indicates that the larger the weight, the smaller are the residual values.

The weighted algorithm with decreasing gain (3.165) is applied, where $\alpha = 1$ and $\beta = 1$. The maximal learning error profiles for three sensors are illustrated in Fig. 3.37, exhibiting a consistently declining trend. However, the precision of this particular case is comparatively inferior to that of the equal-weight index scenario. This observation presents an intriguing and unresolved issue that necessitates further investigation.

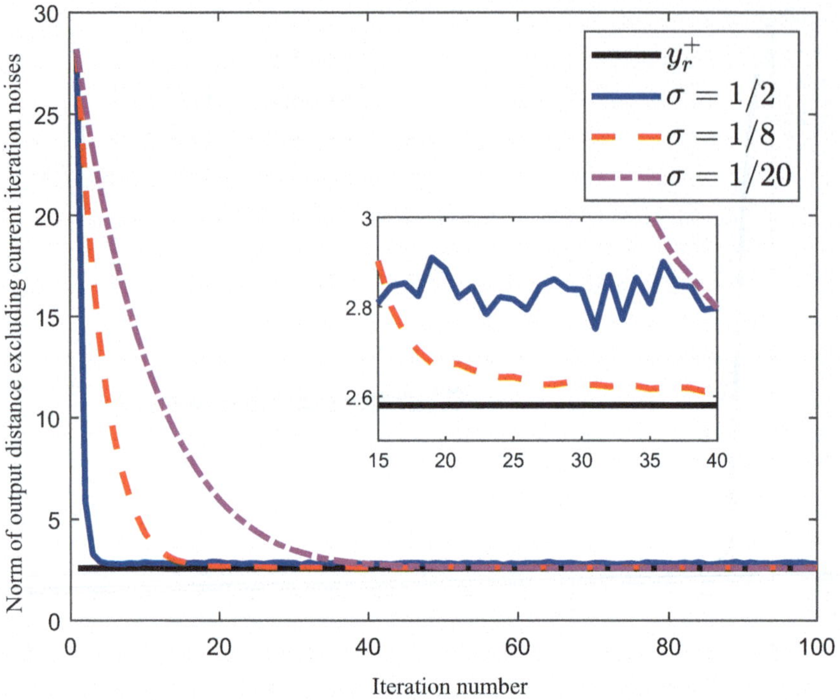

Fig. 3.35 Learning index profiles for algorithm (3.143)

The tracking performance at the 50th iteration is shown in Fig. 3.38; here, the meanings of the lines are identical to those in the equal-weight case in the previous subsection. Acceptable tracking performance can be seen in this figure, which indicates that the proposed scheme is effective in the weighted multi-objective tracking problem. The algorithm is simulated for two additional cases: $\alpha = 1/4$, $\beta = 1$ and $\alpha = 1/4$, $\beta = 2/3$. The learning index profiles in the iteration domain are shown in Fig. 3.39, where the solid line with circles denotes the value of $\|\mathbf{y}_{wr}^{+}\|_{\mathbf{W}_0}^2$ and the learning index for each iteration is defined as $\|\mathbf{y}_d - \mathbf{y}_k^{\nu}\|_{\mathbf{W}_0}^2$. We can see that all profiles converge to the value of $\|\mathbf{y}_{wr}^{+}\|_{\mathbf{W}_0}^2$. Moreover, different selections of α and β can affect the convergence speed similarly to that in the previous subsection. Furthermore, if we replace the decreasing gain in (3.165) with a fixed gain, results similar to those in the previous subsection can be obtained. Here, we omit the details of this analysis to save space.

We check the applicability of algorithm (3.158) for the weighted index case. The learning index profiles for both (3.158) and (3.165) are compared in Fig. 3.40, where the learning index is computed as $\|\mathbf{y}_d - \mathbf{y}_k^{\nu}\|_{\mathbf{W}_0}^2$. The figures illustrates that the algorithm (3.158) cannot drive the learning index $\|\mathbf{y}_d - \mathbf{y}_k^{\nu}\|_{\mathbf{W}_0}^2$ to converge to $\|\mathbf{y}_{wr}^{+}\|_{\mathbf{W}_0}^2$; this result indicates the necessity of introducing weights to the learning control algorithm.

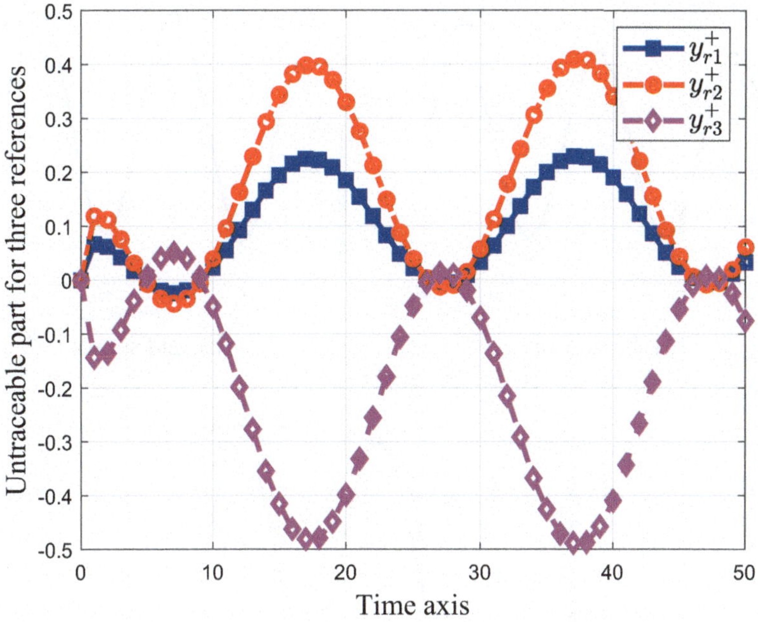

Fig. 3.36 Residual part of the given references for weight index (3.140)

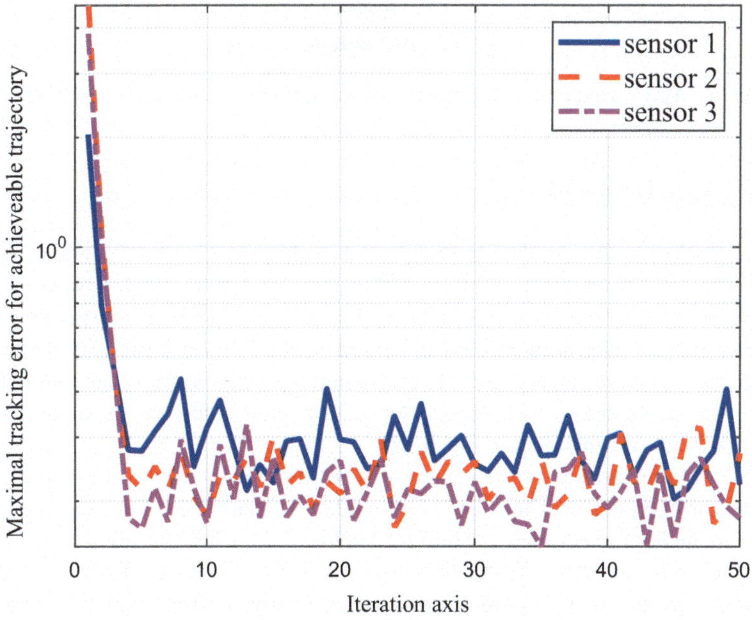

Fig. 3.37 Tracking error profiles of three sensors for algorithm (3.165)

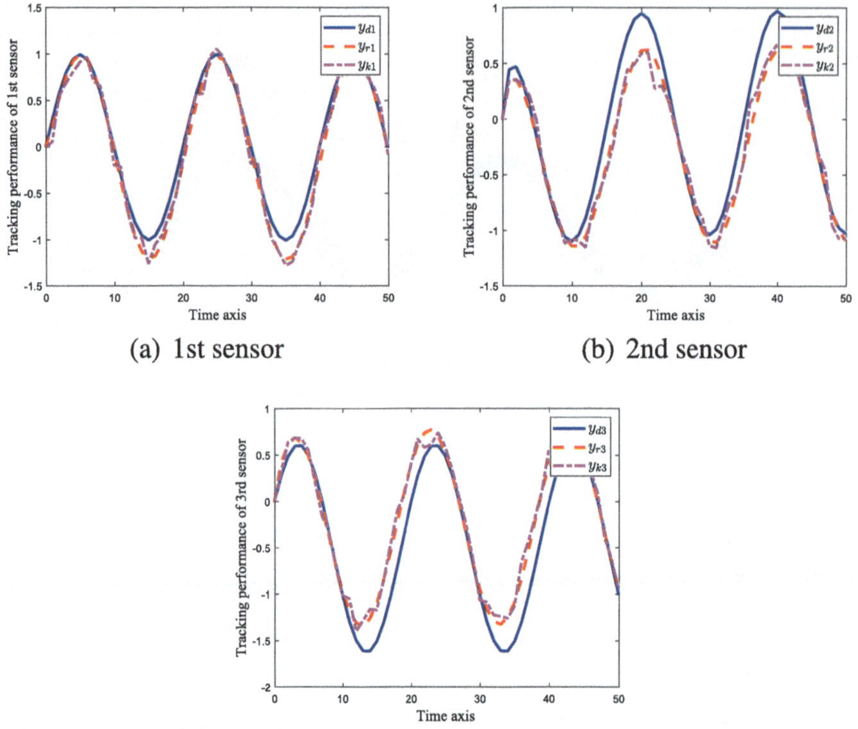

(a) 1st sensor (b) 2nd sensor

(c) 3rd sensor

Fig. 3.38 Tracking performance at the 50th iteration of three sensors for algorithm (3.165)

3.5.5.3 Data Dropout Case

In this subsection, we simulate the random data dropout case for the equal-weight index. The data dropout rate is 30%; that is, $\overline{\gamma} = 0.7$. The update law (3.174) is applied, and its parameters are set as $\alpha = 1$ and $\beta = 1$. The general tracking performance for the data dropout case is similar to the lossless case, except that the convergence speed is much slower. The maximal learning error profiles for the three sensors are shown in Fig. 3.41; here, a decreasing trend is still observed but the rate of decrease is lower than that in Fig. 3.32.

Next, we compare the performance of algorithms with fixed averaging (3.174) and event-triggered averaging (3.177) to validate Remark 3.19. The learning index profiles in the iteration domain are depicted in Fig. 3.42, where the algorithm (3.174) with an arbitrary averaging factor $m = 5$ is also simulated for comparison. Among the tested algorithms, it is evident that the event-triggered averaging mechanism yields superior tracking performance.

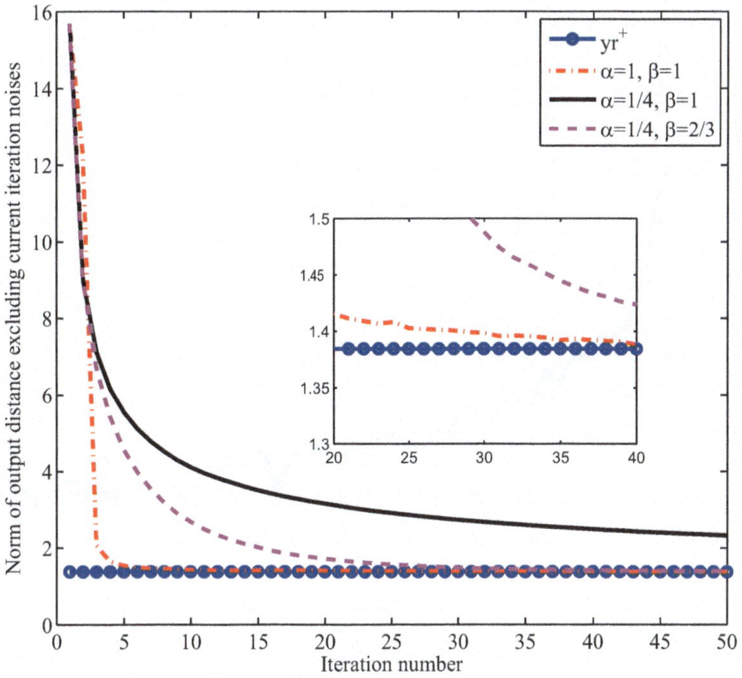

Fig. 3.39 Learning index profiles for algorithm (3.165)

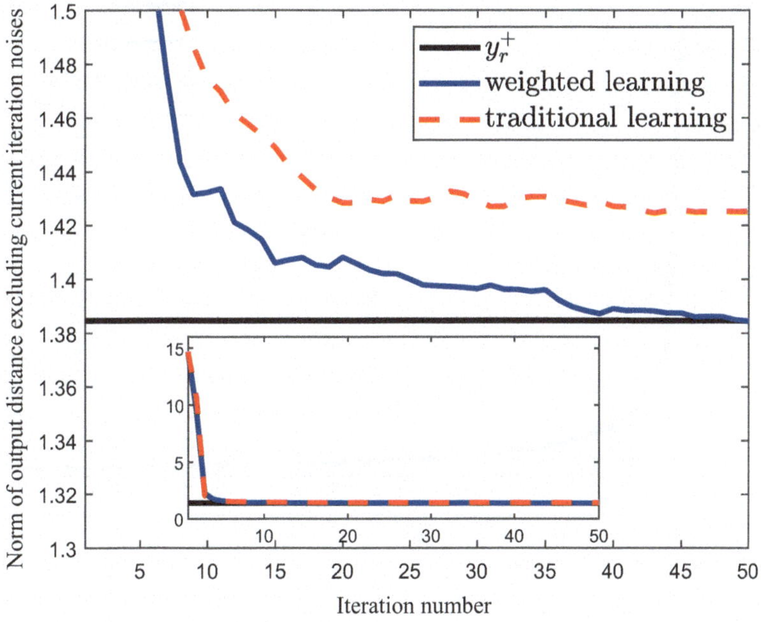

Fig. 3.40 Performance comparison between (3.158) (traditional learning) and (3.165) (weighted learning) under the weighted index (3.140)

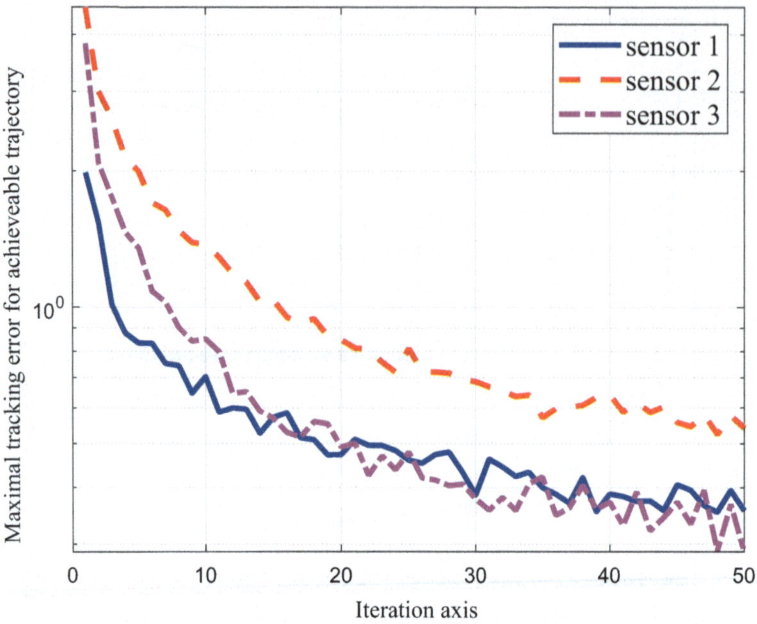

Fig. 3.41 Tracking error profiles of three sensors for algorithm (3.174) with data dropouts

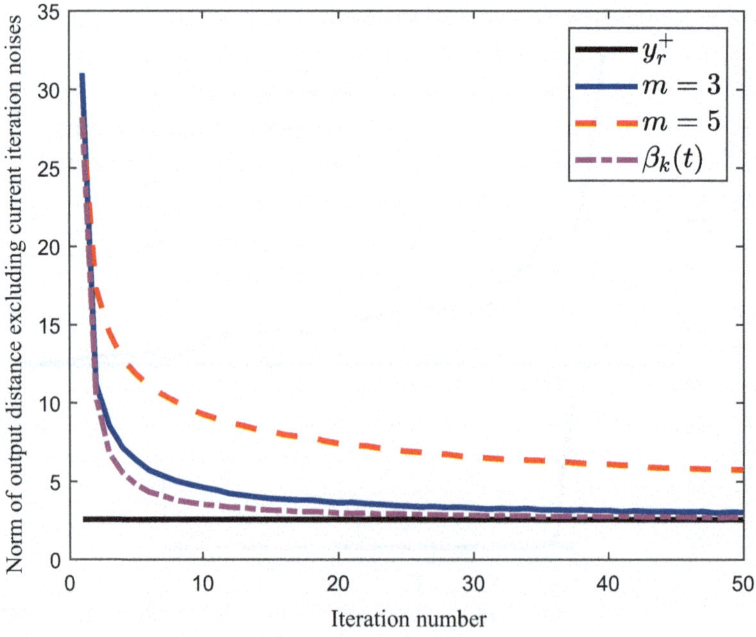

Fig. 3.42 Learning index profiles for (3.174) with $m = 3$, (3.174) with $m = 5$, and (3.177) under data dropouts

Chapter 4
Adaptive Gain Design

4.1 Adaptive Gain Design in SILC

4.1.1 Motivation of Adaptive Gain

In practical applications, stochastic noises are introduced into the system dynamics due to various factors such as mechanical vibration and measurement error. These noises need to be effectively addressed through novel frameworks and techniques. In this regard, ILC for stochastic systems has garnered significant attention (Shen and Wang 2014). One common approach is the application of zero-phase low-pass and linear-phase filters. The Kalman filtering (KF) technique (Cao et al. 2016a, b; Saab 2001a, 2003), which obtains the direction regulation matrix by minimizing the error covariance matrix, has been observed to be highly effective in handling random noises. However, obtaining an optimal gain matrix requires specific knowledge about system matrices and statistical properties of random noises such as covariance matrices. Stochastic approximation (SA) has emerged as a promising alternative for addressing similar problems (Butcher and Karimi 2011; Butcher et al. 2008; Shen 2018; Shen and Xu 2017). The proposed algorithms based on SA provide almost-sure convergence and asymptotic properties (Chen 2003). However, due to the necessity of a decreasing gain sequence for handling stochastic systems, SA-based ILC algorithms exhibit slow convergence. Consequently, the speed limitation inherent in the original SA algorithms also restricts the performance of SA-based ILC laws, as they converge at speeds proportional to a specific power of the decreasing gain sequence (Chen 2003). Furthermore, designing an appropriate decreasing gain sequence prior to the control process poses challenges in optimizing such stochastic systems.

Motivated by these observations, we aim to enhance the practical convergence speed of ILC algorithms for stochastic systems. Firstly, we discover that in a noise-free system (Meng et al. 2015; Pipeleers and Moore 2014; Shen et al. 2016a; Van de Wijdeven et al. 2009), the decreasing gain sequence is eliminated and the convergence speed increases when utilizing an iteration-invariant direction regulation matrix as

D. Shen, *Variable Gain Design in Stochastic Iterative Learning Control*,
Intelligent Control and Learning Systems 13,
https://doi.org/10.1007/978-981-97-8281-9_4

it continuously contracts the input error. Furthermore, in a stochastic system, the tracking error comprises two components: an inaccurate tracking term (caused by an imprecise input signal) and a stochastic noise term. During initial iterations when prior knowledge about the desired input is unavailable, there may be significant deviations from it. In such common scenarios, the inaccurate tracking term dominates the tracking error. To expedite convergence speed, handling incorrectly tracked systems during their learning process can be approached similarly to noise-free cases. After a sufficient number of learning iterations, the inaccurate tracking term will be significantly reduced, and the dominant source of tracking error will be stochastic noise. The impact of these tracking errors can be asymptotically attenuated by implementing a decreasing gain sequence, thereby achieving stable convergence. An intriguing challenge arises: how can we adaptively tune the learning gain along the iteration axis with limited knowledge about the learning process?

To address this issue, we propose a novel ILC algorithm with gain adaptation for stochastic systems. The motivation behind our gain adaptation mechanism stems from several considerations. Firstly, as data-driven methods are well-suited for practical applications, it is crucial to effectively utilize information on tracking errors. Additionally, in scalar cases of noise-free systems, the sign of the tracking error remains unchanged; this property should also hold true in stochastic and MIMO systems when compared to negligible stochastic noise terms. In fact, this conjecture plays a pivotal role in determining the gain adaptation within our proposed algorithm. It is important to highlight that our new ILC algorithm builds upon Kesten's accelerated SA algorithm (Delyon and Juditsky 1993; Kesten 1958).

The principal contribution of this section is a novel ILC algorithm with gain adaptation for discrete-time stochastic systems. This gain adaptation enables the tuning of the direction regulation matrix to be either constant or adaptively decreasing along the iteration axis. The proposed ILC algorithm achieves faster convergence speed and smaller tracking error compared to existing ILC laws for stochastic systems (Butcher and Karimi 2011; Butcher et al. 2008; Shen 2018; Shen and Xu 2017). Furthermore, the proposed gain-adaptation mechanism is entirely data-driven as it is based on the ratio between the inaccurate tracking term and the term representing stochastic noises. Under specific circumstances, this new ILC algorithm can be considered as a variant of the classic P-type ILC law designed for noise-free systems.

4.1.2 Problem Formulation

Consider the following discrete-time stochastic system

$$\begin{aligned} x_k(t+1) =& A_t x_k(t) + B_t u_k(t) + w_k(t), \\ y_k(t) =& C_t x_k(t) + v_k(t), \end{aligned} \tag{4.1}$$

where $t = 0, 1, \cdots, N$ is the time index with N being the final time instant, and $k = 1, 2, \cdots$ is the iteration index. $u_k(t) \in \mathbf{R}^p$, $x_k(t) \in \mathbf{R}^n$, and $y_k(t) \in \mathbf{R}^q$ are the input, state, and output, respectively. Both $w_k(t)$ and $v_k(t)$ are random variables denoting the system noise and measurement noise, respectively. A_t, B_t, and C_t are the system matrices with appropriate dimensions.

The desired reference is denoted by $y_d(t)$, $t = 0, 1, \cdots, N$. In this section, we assume that the reference is realizable; in other words, there is a unique desired input $u_d(t)$ and suitable desired initial state $x_d(0)$ satisfying the following relations:

$$\begin{aligned} x_d(t+1) =& A_t x_d(t) + B_t u_d(t), \\ y_d(t+1) =& C_{t+1} x_d(t+1). \end{aligned} \tag{4.2}$$

In this case, the desired input $u_d(t)$ can be well-defined by the following recursive formula: $u_d(t) = [(C^+ B_t)^T C^+ B_t]^{-1}[y_d(t+1) - C^+ A_t x_d(t)]$ and $x_d(t+1) = A_t x_d(t) + B_t u_d(t)$ with the desired initial value $x_d(0)$, where $C^+ \triangleq C_{t+1}$.

The analysis proceeds under the following assumptions.

Assumption 4.1 The coupling matrix $C_{t+1} B_t$ is of full-column rank, $\forall t$.

Remark 4.1 Assumption 4.1 implies that system (4.1) has a relative degree of one, and it is required that $q \geq p$, where p and q represent the dimensions of the input and output vectors respectively. The necessity of Assumption 4.1 is discussed in Huang et al. (2002). As demonstrated later in the update law (4.3), when the relative degree is one, the updating of $u_k(t)$ relies on the tracking error $e_k(t+1)$. In practical systems, time-delay and system structure can increase the relative degree beyond one. If $\varsigma > 1$, then in (4.3), the tracking error $e_k(t+1)$ must be replaced by $e_k(t+\varsigma)$, along with appropriate modifications to the direction regulation matrix. Our algorithm can be extended to similar cases by introducing additional notations.

Assumption 4.2 The initial state satisfies the identical initial condition. That is, $x_k(0) = x_d(0), \forall k$, where $x_d(0)$ is the desired initial state such that $y_d(0) = C_0 x_d(0)$.

Assumption 4.3 The stochastic noises $w_k(t)$ and $v_k(t)$ are stationary and independent along the iteration axis at different time instants. Moreover, $\mathbb{E} z_k(t) = 0$, $\sup_k \mathbb{E} z_k(t) z_k^T(t) < \infty$, and $\forall \lambda > 0$, $\min_k \mathbb{P}(|z_k(t)| \leq \lambda) > 0$, where $z_k(t)$ denotes $w_k(t)$ or $v_k(t)$.

By considering Assumption 4.3, Assumption 4.2 can be relaxed to the case in which $x_k(0)$ is random with $\mathbb{E} x_k(0) = x_d(0)$. However, because this case is beyond the primary scope of this section, we employ Assumption 4.2 for the sake of brevity.

In system (4.1), an iteration-invariant direction regulation matrix can accumulate the influence of stochastic noises. To eliminate these effects, most methods apply a decreasing gain. The resulting ILC law based on SA algorithms (Butcher and Karimi 2011; Butcher et al. 2008) is

$$u_{k+1}(t) = u_k(t) + a_k L_t e_k(t+1), \tag{4.3}$$

where a_k satisfies $a_k > 0$, $a_k \to 0$, $\sum_{k=1}^{\infty} a_k = \infty$, and $\sum_{k=1}^{\infty} a_k^2 < \infty$. Butcher et al. (2008) and Butcher and Karimi (2011) demonstrated the almost-sure convergence of the ILC law using the conclusions of Robbins-Monro type SA algorithms (Chen 2003). However, a major disadvantage of algorithm (4.3) is its slow convergence speed. In fact, the algorithm converges no faster than the original SA algorithms. In Chen (2003), the convergence speed is shown to be proportional to $\sqrt{a_k}$, which depends on the selection of a_k. As we have little prior knowledge of the system, especially the stochastic noises, selecting the optimal a_k prior to the learning process is difficult.

The tracking error comprises two components: the inaccurately tracked term resulting from inaccurate input and the term representing stochastic noises. The relative ratio between these two components varies throughout the learning process. Initially, the stochastic noise term is typically a small proportion of the tracking error, allowing for approximating the direction regulation matrix as a constant to expedite convergence speed. However, when the stochastic noise term becomes sufficiently large, a decreasing gain can be employed to stabilize convergence. An intriguing question arises: can we devise a gain adaptation mechanism that achieves such learning performance solely based on available information? Based on these observations, we formulate our problem as follows.

Problem Statement: The control objective is to develop a novel ILC algorithm with adaptive gain adjustment for discrete-time systems. In order to expedite convergence, the gain matrix remains constant during the initial iterations and subsequently decreases adaptively after a sufficient number of learning iterations.

4.1.3 New ILC Algorithm with Gain Adaptation

To facilitate the following analysis, first, we express the system in the vector form. Define $U_k = [u_k^T(0), \cdots, u_k^T(N-1)]^T$, $Y_k = [y_k^T(1), \cdots, y_k^T(N)]^T$, $\xi_k = [(v_k(1) + C_1 w_k(0))^T, \cdots, (v_k(N) + \sum_{i=0}^{N-1} C_N A_{i+1}^{N-1} w_k(i))^T]^T$, where $A_i^j \triangleq A_j \cdots A_i$ for $j \geq i$. According to Assumption 4.3, the lifted noise ξ_k is stationary and independent along the iteration axis with zero mean and bounded variance. Without loss of generality, we set $\mathbb{E}\xi_k\xi_k^T \leq \Sigma$, $\forall k$. The associated lifted system matrix is

$$\mathcal{H} = \begin{bmatrix} C_1 B_0 & 0 & \cdots & 0 \\ C_2 A_1 B_0 & C_2 B_1 & \cdots & 0 \\ \vdots & \vdots & \ddots & \vdots \\ C_N A_1^{N-1} B_0 & C_N A_2^{N-1} B_1 & \cdots & C_N B_{N-1} \end{bmatrix}. \tag{4.4}$$

We can now rewrite the original system (4.1) and the desired system (4.2) as $Y_k = \mathcal{H}U_k + \xi_k + Y_{k0}$ and $Y_d = \mathcal{H}U_d + Y_{d0}$ where $Y_{k0} = Mx_k(0)$ and $Y_{d0} = Mx_d(0)$ with $M \triangleq [(C_1 A_0)^T, (C_2 A_0^1)^T, \cdots, (C_N A_1^{N-1})^T]^T$. By Assumption 4.2, we have $Y_{k0} = Y_{d0}$, and the tracking error is then given by

$$E_k \triangleq Y_d - Y_k = \mathcal{H}\Delta U_k - \xi_k, \tag{4.5}$$

where $\Delta U_k \triangleq U_d - U_k$. To achieve the control objective, we define the gain-adaptation parameter s_k as follows:

$$s_k = s_{k-1} + \mathbf{1}_{[E_k^T E_{k-1} < 0]}, \quad k \geq 2 \tag{4.6}$$

with initial value $s_1 = 1$, where $\mathbf{1}_{[event]}$ is an indicator function equal to one if the indicated event is true, and zero otherwise. Obviously, $m_k \triangleq s_k - 1 = \sum_{i=1}^{k-1} \mathbf{1}_{[E_{i+1}^T E_i < 0]}$ denotes the number of times (up to the kth iteration) for which the product of the error vectors in two adjacent iterations is negative. In the following, we will show that m_k is zero in a noise-free system and approaches infinity in a noisy system. Assisted by s_k, the new ILC algorithm is as follows:

$$U_{k+1} = U_k + s_k^{-1} \mathcal{L} E_k, \tag{4.7}$$

where $\mathcal{L}$ is the direction regulation matrix.

The new ILC algorithm with gain adaptation is inspired by Kesten's accelerated SA algorithm (Delyon and Juditsky 1993; Kesten 1958), which enhances the convergence speed of the classical SA algorithm. It should be noted that due to the presence of random noises in the actual output, our focus shifts towards demonstrating the asymptotic convergence of the generated input sequence to the desired input, rather than directly addressing the convergence of tracking errors.

Remark 4.2 The new ILC algorithm with gain adaptation is based on the following concept. The tracking error E_k defined by (4.5) comprises two components: the imprecise tracking term $\mathcal{H}\Delta U_k$ and the stochastic noise term ξ_k. During the initial iterations, the imprecise tracking term $\mathcal{H}\Delta U_k$ may dominate E_k due to significant deviations between the initial input and U_d. In such cases, if $E_k^T E_{k-1}$ remains positive, then s_k will remain invariant. Furthermore, a small value of s_k indicates rapid convergence. After a sufficient number of learning iterations, the proportion of $\mathcal{H}\Delta U_k$ is expected to decrease until the stochastic noise term ξ_k dominates E_k. In this scenario, the occurrence of event $E_k^T E_{k-1} < 0$ can be roughly equated with $\xi_k^T \xi_{k-1} < 0$, indicating a continuous increase in s_k. Eventually, the SA mechanism is activated to ensure stable convergence. Therefore, the new ILC algorithm will execute when $E_k^T E_{k-1} \geq 0$ in a noise-free system or when the contribution of stochastic noises becomes negligible in the tracking error E_k.

Remark 4.3 The gain adaptation strategy in (4.7) is formulated in a simplified manner to elucidate our fundamental concept rather than to guarantee an optimal selection. With the adaptation mechanism defined in (4.6), a generalized form of the ILC algorithm can be expressed as $U_{k+1} = U_k + a_{s_k} \mathcal{L} E_k$, where s_k serves as an event-triggered subscript for the original gain sequence $\{a_k\}$. Given the multitude of choices for gain sequences when considering adaptation, it becomes intriguing and significant to explore a methodology for identifying the optimal one, which presents itself as a promising avenue for further research. Moreover, while the gain adaptation remains time-independent in (4.6) and (4.7), it can be extended to encompass time-dependent scenarios by denoting s_k as s_k^t and updating it based on the event $e_k^T(t+1)e_{k-1}(t+1) < 0$.

4.1.4 Design of the Learning Gain Matrix

Before analyzing the new ILC algorithm in detail, first, we clarify the design conditions of the direction regulation matrix $\mathcal{L}$. Let

$$Q = I - \mathcal{L}\mathcal{H}. \tag{4.8}$$

For asymptotic convergence, the selection of $\mathcal{L}$ is crucial to ensure that Q is quasi-positive definite with a spectral radius $\rho(Q)$ less than 1. This condition serves as an essential requirement for designing a direction regulation matrix. In order to achieve monotonic convergence, it is additionally or alternatively necessary to have the largest singular value $\overline{\sigma}(Q)$ of Q also less than 1. Hereafter, we define a quasi-positive definite matrix Q as one satisfying $z^T Qz > 0$ for any non-zero vector z with compatible dimensions. The symmetry property is not necessarily required since $(Q + Q^T)/2$ always yields a symmetric matrix, and thus we can replace the notation Q in subsequent steps without further modification. In the following sections, we illustrate two specific cases where the direction regulation matrix $\mathcal{L}$ is employed.
(1) *Noncausal Direction Regulation Matrix*. Taking the conditions on Q into account, we intuitively select a special case $Q = \gamma I$ with $0 < \gamma < 1$. From (4.8), $\mathcal{L}$ is designed as follows:

$$\mathcal{L} = (1-\gamma)(\mathcal{H}^T\mathcal{H})^{-1}\mathcal{H}^T. \tag{4.9}$$

Note that $\mathcal{H}$ is a lower-triangular block matrix but $\mathcal{H}^T\mathcal{H}$ is not lower-triangular, indicating that $\mathcal{L}$ is a general matrix. With this direction regulation matrix, the update law (4.7) is noncausal, and the input updating must use the forward tracking error. Thus, the ILC algorithm (4.7) must be in a lifted vector form, risking a dimensional problem for large iteration length N. Therefore, we are interested in decomposing (4.7) into a time-based form.
(2) *Block Diagonal Direction Regulation Matrix*. The computation of $\mathcal{L}$ in (4.9) is costly for a large iteration length N. For practical purposes, we decompose (4.7) into the following time-based form:

$$u_{k+1}(t) = u_k(t) + s_k^{-1}L_t e_k(t+1) \tag{4.10}$$

where the direction regulation matrix is block diagonal:

$$\mathcal{L} = \sigma \mathcal{L}_0, \quad \mathcal{L}_0 = \text{diag}\{L_0, \ L_1, \cdots, L_{N-1}\} \tag{4.11}$$

and $L_i \in \mathbf{R}^{p\times q}, 0 \le i \le N-1$, must satisfy the positive-definiteness of the matrix $L_t C_{t+1} B_t$. Additionally, we define a sufficiently small positive constant σ.

When $\mathcal{L}$ is block diagonal, $\mathcal{LH}$ is clearly a lower-triangular block matrix with all positive eigenvalues. For any vector z of suitable dimension, we have $z^T Q z = z^T(I - \mathcal{LH})z = z^T[I - \frac{\mathcal{LH}+(\mathcal{LH})^T}{2}]z = z^T z - \sigma z^T \frac{\mathcal{L}_0\mathcal{H}+\mathcal{H}^T\mathcal{L}_0^T}{2} z$. It should be noted that $(\mathcal{L}_0\mathcal{H} + \mathcal{H}^T\mathcal{L}_0^T)/2$ is a symmetric matrix with all positive eigenvalues, thereby establishing its positive definiteness. Consequently, for any non-zero vector z and sufficiently small value of σ, we can assert that $z^T Q z > 0$. In the following subsection, we will provide further elaboration on the selection of parameter σ. The introduction of a suitably small constant σ is intriguing; however, it is solely introduced to facilitate the analysis of a noise-free system (refer to Sect. 4.1.5.1), as in stochastic systems, s_k tends to increase towards infinity as k increases (see Sect. 4.1.5.2). Therefore, the convergence of $1/s_k$ towards zero naturally suppresses the impact of $\mathcal{LH}$.

In the aforementioned context, we introduce a novel parameter σ to ensure the positive-definiteness of the Q matrix when σ is sufficiently small. However, during systematic controller design, our focus lies on determining an appropriate range for σ. Herein, we elaborate on the existence of such a range in the single-input-single-output case of (4.1) to achieve concise expression.

Note that because $\mathcal{H}$ is lower triangular and $\mathcal{L}_0$ is diagonal, the product $\mathcal{L}_0\mathcal{H}$ is also a lower-triangular matrix. To clarify the notations, we let

$$\mathcal{L}_0\mathcal{H} = \begin{bmatrix} a_{11} & 0 & 0 & \cdots & 0 \\ a_{21} & a_{22} & 0 & \cdots & 0 \\ a_{31} & a_{32} & a_{33} & \cdots & 0 \\ \vdots & \vdots & \vdots & \ddots & \vdots \\ a_{N1} & a_{N2} & a_{N3} & \cdots & a_{NN} \end{bmatrix},$$

where the diagonal elements satisfy $a_{ii} > 0$ for $1 \le i \le N$ by the design of the learning gain L_t.

For any N-dimensional vector $z = [z_1, z_2, \cdots, z_N]^T$,

$$\begin{aligned} z^T Q z =& z^T(I - \sigma\mathcal{L}_0\mathcal{H})z \\ =& \sum_{i=1}^{N} z_i^2 - \sigma\left(\sum_{i=1}^{N} a_{ii} z_i^2 + \sum_{1\le j<i\le N} a_{ij} z_i z_j\right) \\ =& \sum_{i=1}^{N}(1 - \sigma a_{ii}) z_i^2 - \sigma \sum_{1\le j<i\le N} a_{ij} z_i z_j. \end{aligned} \tag{4.12}$$

Note that

$$
\begin{aligned}
\sum_{1\le j<i\le N} a_{ij} z_i z_j \le& \sum_{1\le j<i\le N} |a_{ij}| \frac{z_i^2+z_j^2}{2} \\
=& \sum_{j=1}^{N-1}\sum_{i=j+1}^{N} \frac{|a_{ij}|}{2} z_j^2 + \sum_{j=2}^{N}\sum_{i=1}^{j-1} \frac{|a_{ji}|}{2} z_j^2 \\
=& \sum_{j=1}^{N-1} z_j^2 \left(\sum_{i=j+1}^{N} \frac{|a_{ij}|}{2}\right) + \sum_{j=2}^{N} z_j^2 \left(\sum_{i=1}^{j-1} \frac{|a_{ji}|}{2}\right) \\
=& z_1^2 \left(\sum_{i=2}^{N} \frac{|a_{i1}|}{2}\right) + \sum_{j=2}^{N-1} z_j^2 \left(\sum_{i=j+1}^{N} \frac{|a_{ij}|}{2} + \sum_{i=1}^{j-1} \frac{|a_{ji}|}{2}\right) \\
&+ z_N^2 \left(\sum_{i=1}^{N-1} \frac{|a_{Ni}|}{2}\right).
\end{aligned}
\tag{4.13}
$$

To ensure the positiveness of Q, that is, to ensure $z^T Qz > 0$ for any $z \neq 0$, we substitute (4.13) into (4.12) and compare the z_i^2 coefficients. This process yields the following inequalities:

$$
\begin{aligned}
1-\sigma a_{11} >& \sigma \left(\sum_{i=2}^{N} \frac{|a_{i1}|}{2}\right), \quad 1-\sigma a_{NN} > \sigma \left(\sum_{i=1}^{N-1} \frac{|a_{Ni}|}{2}\right), \\
1-\sigma a_{jj} >& \sigma \left(\sum_{i=j+1}^{N} \frac{|a_{ij}|}{2} + \sum_{i=1}^{j-1} \frac{|a_{ji}|}{2}\right), \quad 2 \le j \le N-1.
\end{aligned}
$$

From these inequalities,

$$
\begin{aligned}
&\sigma < \left(a_{11} + \sum_{i=2}^{N} \frac{|a_{i1}|}{2}\right)^{-1}, \quad \sigma < \left(a_{NN} + \sum_{i=1}^{N-1} \frac{|a_{Ni}|}{2}\right)^{-1}, \\
&\sigma < \left(\sum_{i=j}^{N} \frac{|a_{ij}|}{2} + \sum_{i=1}^{j} \frac{|a_{ji}|}{2}\right)^{-1}, \quad 2 \le j \le N-1.
\end{aligned}
\tag{4.14}
$$

Thus we can select a positive constant σ that satisfies all of the inequalities in (4.14).

4.1.5 Convergence Analysis

The first part of this section discusses the noise-free system and verifies that the learning gain remains invariant to iterations when the stochastic noises in the tracking error are negligible. Subsequently, it establishes the convergence of the novel ILC algorithm for stochastic systems.

4.1.5.1 Analysis for Noise-Free Systems

Analysis of the noise-free system reveals the evolution property of s_k. We expect that s_k is constant for all k. Note that $s_1 = 1$. Then, for $k = 1$, the new ILC algorithm (4.7) becomes $U_2 = U_1 + \mathcal{L}E_1$. If we can guarantee $E_2^T E_1 \geq 0$, then $s_2 = 1$. For $k = 2$, algorithm (4.7) becomes $U_3 = U_2 + \mathcal{L}E_2$. In fact, provided that $E_k^T E_{k-1} \geq 0$ for all k, algorithm (4.7) becomes the traditional P-type ILC law, i.e.,

$$U_{k+1} = U_k + \mathcal{L}E_k. \tag{4.15}$$

Alternatively, we can show by mathematical induction that both the ILC update law (4.15) with an iteration-invariant learning gain and the inequality $E_k^T E_{k-1} \geq 0$ hold. To prove this, it is sufficient to show that $E_k^T E_{k-1} \geq 0$ holds for all k, where U_k is generated by (4.15), as clarified in the following theorem.

Theorem 4.1 *Suppose that system* (4.1) *is noise-free, i.e., the noise terms $w_k(t)$ and $v_k(t)$ are removed. Assume also that Assumptions 4.1 and 4.2 both hold, and that the ILC algorithm* (4.15) *is applied. If the direction regulation matrix $\mathcal{L}$ satisfies quasi-positive definiteness of $Q = I - \mathcal{L}\mathcal{H}$, then $E_{k+1}^T E_k \geq 0$ for all k.*

Proof. Subtracting both sides of (4.15) from the desired input U_d yields $\Delta U_{k+1} = \Delta U_k - \mathcal{L}E_k = (I - \mathcal{L}\mathcal{H})\Delta U_k$. The product $E_{k+1}^T E_k$, $\forall k$, is then expanded as follows:

$$\begin{aligned} E_{k+1}^T E_k =& \Delta U_{k+1}^T \mathcal{H}^T \mathcal{H} \Delta U_k \\ =& \Delta U_k^T (I - \mathcal{L}\mathcal{H})^T \mathcal{H}^T \mathcal{H} \Delta U_k \\ =& \Delta U_k^T [\mathcal{H}^T \mathcal{H} - \mathcal{H}^T \mathcal{L}^T \mathcal{H}^T \mathcal{H}] \Delta U_k \\ =& \Delta U_k^T \mathcal{H}^T (I - \mathcal{L}^T \mathcal{H}^T) \mathcal{H} \Delta U_k. \end{aligned} \tag{4.16}$$

Note that $I - \mathcal{L}^T \mathcal{H}^T$ is quasi-positive definite when $\mathcal{L}$ is designed as in Sect. 4.1.4. Therefore, $E_{k+1}^T E_k \geq 0$ for all k.

From Theorem 4.1, we deduce two key observations. Firstly, since $s_k = 1$ holds for all k, the classical algorithm (4.15) designed for the noise-free system can be regarded as a special case of the novel ILC algorithm (4.7), and under the condition that $\overline{\sigma}(Q) < 1$, the tracking error E_k exhibits exponential convergence towards zero.

Secondly, if the stochastic noise terms in the stochastic system (4.1) constitute a relatively small proportion of the overall tracking error, it is highly likely that the inequality $E_{k+1}^T E_k \geq 0$ will still hold with a high probability. However, as the proportion of stochastic noises relative to the tracking error increases, there is a decrease in probability for event $E_{k+1}^T E_k \geq 0$.

4.1.5.2 Analysis for Stochastic Systems

Returning to algorithm (4.7), we propose the following convergence theorem.

Theorem 4.2 *Consider the discrete-time stochastic system* (4.1) *and assume that Assumptions 4.1-4.3 hold. If the direction regulation matrix $\mathcal{L}$ in the new ILC algorithm* (4.7) *satisfies the condition that all the eigenvalues of $\mathcal{LH}$ are positive real numbers, then the following statements hold: (i) the input error ΔU_k converges asymptotically to zero, a.s., and (ii) the scale variable s_k satisfies $s_k/k \to \mathbb{P}(\xi_1^T \xi_2 < 0)$ as $k \to \infty$, a.s.*

The new algorithm for ILC in stochastic systems is based on Kesten's accelerated SA algorithm (Delyon and Juditsky 1993; Kesten 1958). Consequently, the obtained results can be effectively applied in the convergence analysis. To enhance clarity and convenience, we initially restate the theorem in a high-dimensional scenario (Delyon and Juditsky 1993) as the following lemma.

Lemma 4.1 *Consider the following process*

$$\bar{y}_k = \varphi(\bar{x}_k) + \epsilon_k, \quad \bar{x}_{k+1} = \bar{x}_k - \bar{y}_k/s_k, \tag{4.17}$$

where $\varphi(\cdot) : \boldsymbol{R}^n \to \boldsymbol{R}^n$ with $\bar{x}^*$ being the root and $s_k = 1 + \sum_{i=1}^{k-1} \mathbf{1}_{[\bar{y}_{i+1}^T \bar{y}_i < 0]}$. Assume that the following conditions hold.

Assumption 4.4 ϵ_k is a sequence of random variables independent of $\bar{x}_k$ and $\mathbb{E}(\epsilon_k) = 0$, $\mathbb{E}(\epsilon_k \epsilon_k^T) \leq M$ for some M. Moreover, for any $\lambda > 0$, $\min_k \mathbb{P}(|\epsilon_k| \leq \lambda) > 0$.

Assumption 4.5 $\varphi(\bar{x})$ is a continuous function of $\bar{x}$. There exists $\alpha > 0$ such that for any $\gamma \leq \alpha$ and any starting point $\bar{x}_0$, the deterministic sequence $\bar{x}_{k+1} = \bar{x}_k - \gamma\varphi(\bar{x}_k)$ converges to $\bar{x}^*$, and there also exists a function $V(\bar{x}) : \mathbf{R}^n \to \mathbf{R}^+$, some positive β and $\overline{R}$, and a matrix $J > 0$ such that $V(\bar{x}^*) = 0$, and

$$\nabla^2 V(\bar{x}) \leq J, \text{ for all } \bar{x}, \tag{4.18}$$

$$\varphi(\bar{x})^T \nabla V(\bar{x}) \geq \frac{\alpha}{2}\left[\varphi(\bar{x})^T J \varphi(\bar{x}) + \mathrm{tr}(MJ)\right] + \beta, \tag{4.19}$$

for any $\bar{x}$ such that $\|\bar{x} - \bar{x}^*\| \geq \overline{R}$. Moreover, for any $\bar{x} \neq \bar{x}^*$, $\varphi(\bar{x})^T \nabla V(\bar{x}) > 0$.

The sequence generated by (4.17) satisfies $\bar{x}_k \to \bar{x}^*$ and $\lim_{k\to\infty} \frac{s_k}{k} \to \mathbb{P}(\epsilon_1^T \epsilon_2 < 0)$, a.s., where $\mathbb{P}(\cdot)$ is the probability of its indicated event.

We would like to emphasize that Assumption 4.3 in (4.1) can be further relaxed, as suggested by Delyon and Juditsky (1993), although this would lead to a more intricate analysis. Assumption 4.5 delineates the requirements on a Lyapunov function, allowing for some flexibility through the parameters α and β.

Proof. Subtracting both sides of (4.7) from U_d yields

$$\Delta U_{k+1} = \Delta U_k - s_k^{-1}(\mathcal{L}\mathcal{H}\Delta U_k - \xi_k), \tag{4.20}$$

where (4.5) has been used. The quantities ΔU_k, $\mathcal{L}\mathcal{H}\Delta U_k$, and $-\xi_k$ correspond to $\bar{x}_k$, $\varphi(\bar{x}_k)$, and ϵ_k in Lemma 4.1, respectively. That is, the regression function in this case is $\varphi(x) = \mathcal{L}\mathcal{H}x$. To validate the theorem, it is sufficient to verify Assumptions 4.4 and 4.5.

First, as the matrix $\mathcal{L}\mathcal{H}$ is of full rank, the root of the regression function $\varphi(x) = \mathcal{L}\mathcal{H}x$ is $x = 0$. In other words, the convergence limit of ΔU_k is zero.

In Assumption 4.4, the noise term ξ_k is stationary and independent along the iteration axis with zero mean and a finite variance matrix (according to A4.3). Furthermore, the matrix M in Assumption 4.4 is equal to the matrix Σ, with the result that Assumption 4.4 holds.

We now consider Assumption 4.5. Theorem 4.1 clarifies that the deterministic recursion $\Delta U_{k+1} = \Delta U_k - \gamma\mathcal{L}\mathcal{H}\Delta U_k$ always converges to zero, $\forall\gamma > 0$. Moreover, note that $\mathcal{L}\mathcal{H}$ is a square matrix and all of its eigenvalues are positive. There is a positive matrix Γ such that $(\mathcal{L}\mathcal{H})^T\Gamma + \Gamma\mathcal{L}\mathcal{H} = I$. Define the Lyapunov function $V(x) = x^T\Gamma x$. As Γ is positive, $V(x)$ is minimized only at $x = 0$, which defines the root of $\mathcal{L}\mathcal{H}x$. Meanwhile, $\nabla^2 V(x) = 2\Gamma$; therefore, the matrix J in (4.18) corresponds to 2Γ. Now let us check the validity of (4.19). The left-hand side is

$$\begin{aligned}\varphi(x)^T\nabla V(x) =& 2x^T(\mathcal{L}\mathcal{H})^T\Gamma x = x^T(\mathcal{L}\mathcal{H})^T\Gamma x + x^T\Gamma(\mathcal{L}\mathcal{H})x \\ =& x^T[(\mathcal{L}\mathcal{H})^T\Gamma + \Gamma(\mathcal{L}\mathcal{H})]x = x^Tx.\end{aligned} \tag{4.21}$$

Furthermore,

$$\begin{aligned}\varphi(x)^TJ\varphi(x) =& x^T(\mathcal{L}\mathcal{H})^T 2J\mathcal{L}\mathcal{H}x \\ \le& 4\overline{\sigma}((\mathcal{L}\mathcal{H})^T\Gamma\mathcal{L}\mathcal{H})x^Tx.\end{aligned} \tag{4.22}$$

Therefore, by selecting $\alpha < [4\overline{\sigma}((\mathcal{L}\mathcal{H})^T\Gamma\mathcal{L}\mathcal{H})]^{-1}$, we obtain

$$\alpha\varphi(x)^TJ\varphi(x) \le x^Tx. \tag{4.23}$$

To validate (4.19), we combine (4.21)–(4.23) and show that there exist ρ and β such that $x^Tx \ge 2\alpha\mathrm{tr}(\Sigma\Gamma) + 2\beta$ for any x satisfying $\|x\| \ge \overline{R}$. This inequality holds when α and β are sufficiently small. Consequently, Assumption 4.5 is also verified. Using Lemma 4.1 completes the proof.

In addition, the noise ξ_k is assumed to be symmetrically distributed around zero. Thus, s_k can be estimated more specifically by the following corollary.

Corollary 4.1 *Apply the new ILC algorithm* (4.7) *to the stochastic system* (4.1) *under Assumptions 4.1–4.3. If the distribution function of ξ_k is symmetric about zero, then*

$$\lim_{k\to\infty}\frac{s_k}{k}=\frac{1}{2}. \tag{4.24}$$

Proof. $s_k/k \to \mathbb{P}(\xi_1^T\xi_2<0)$ by Theorem 4.2. Therefore, we can prove the corollary by showing that $\mathbb{P}(\xi_1^T\xi_2<0)=1/2$. As ξ_k is independent along the iteration axis and $\mathbb{E}\xi_k=0$ for all k, we have $\mathbb{E}\xi_1\xi_2=0$. In addition, as the distribution function of ξ_k is symmetric about zero, the joint distribution function $\tilde{F}$ of $\xi_1^T\xi_2$ is symmetric about zero, and, thus, the median point of $\tilde{F}$ is actually the expectation, i.e., zero. Therefore,

$$\mathbb{P}(\xi_1^T\xi_2<0)=\tilde{F}(0)=\frac{1}{2}. \tag{4.25}$$

This completes the proof.

This corollary connects the proposed algorithm (4.7) and the classic SA-based ILC law. Specifically, the classic SA-based ILC law with $a_k=1/k$ can be formulated as $U_{k+1}=U_k+k^{-1}\mathcal{L}E_k$. Corollary 4.1 specifically relates the two cases by estimating s_k/k.

4.1.5.3 Comparisons of the New and Conventional ILC Algorithms

In this subsection, we briefly compare the new ILC algorithm (4.7), the classical ILC update laws (4.15), and a common case of (4.3), namely

$$U_{k+1}=U_k+k^{-1}\mathcal{L}E_k. \tag{4.26}$$

In a noise-free system, all stochastic noises are eliminated and U_k converges in the three algorithms (4.7), (4.15), and (4.26). However, the SA-based algorithm (4.26) exhibits significantly slower convergence compared to the other algorithms. Specifically, the convergence speed is inversely proportional to k in (4.26), while it is exponential in both (4.7) and (4.15). Notably, under these circumstances (see Sect. 4.1.5.1), both (4.7) and (4.15) yield identical results.

In a stochastic system, both (4.7) and (4.26) guarantee almost-sure convergence of the input error ΔU_k to zero. However, (4.7) converges more rapidly than (4.26) because it retains a larger learning gain during the first few iterations, when the stochastic noises are overwhelmed by the inaccurate tracking term $\mathcal{H}\Delta U_k$. Conversely, (4.15) fails to ensure a stable convergence of U_k as a result of the stochastic noises. To show this point, if U_k converges stably to a limit, then the lifted

Table 4.1 Convergence comparisons of the three tested algorithms

ILC algorithm	Noise-free case	Noise-involved case
P-type ILC (4.15)	Exponential	Non-convergent
SA-based ILC (4.26)	$O(k^{-1})$	$O(k^{-\frac{1}{2}})$
New ILC (4.7)	Identical to (4.15)	Faster than (4.26)

form of (4.15) yields $\lim_{k\to\infty} U_{k+1} = \lim_{k\to\infty} U_k + \lim_{k\to\infty} \mathcal{L}E_k$ or equivalently $\lim_{k\to\infty} \mathcal{L}E_k = 0$, contradicting the system formulation.

Table 4.1 provides a clear overview of these comparisons.

In summary, the novel ILC algorithm dynamically adjusts its direction regulation matrix from a constant value during the initial iterations to a decreasing type after a sufficient number of learning iterations. This adaptive process is solely based on tracking information in stochastic systems and employs a data-driven approach to enhance convergence speed for practical applications. Simulation results demonstrate that this gain adaptation effectively addresses the challenge of iteration-varying stochastic noise levels. In noise-free systems, the proposed ILC algorithm reverts to the classical P-type ILC law.

4.1.6 Illustrative Example

To verify the theoretical results, we adopt the permanent magnet synchronous motor example in Xu and Jin (2013), which is a MIMO model. The system is described as follows:

$$\begin{aligned}\dot{x}_k &= Ax_k + Bu_k,\\ y_k &= Cx_k + \xi_k,\end{aligned}$$

where $x = [i_r, i_s]^T$ are the r-axis and s-axis currents to be controlled, and

$$\begin{aligned}A = \begin{bmatrix} -\frac{R}{L_r} & \frac{L_s\omega}{L_r} \\ -\frac{L_r\omega}{L_s} & -\frac{R}{L_s}\end{bmatrix}, \quad B = \begin{bmatrix} \frac{1}{L_r} & 0 \\ 0 & \frac{1}{L_s}\end{bmatrix},\\ u_k = [u_r,\ u_s]^T, \quad C = I,\end{aligned}$$

where L_r, L_s, and R are the inductances and resistance of the stator, and u_k is the control input voltages. Thus, we have $p = q = 2$ regarding (4.11). The reference current to be tracked is $0.5\sin(2\pi t/T)$ and $1.5\sin(2\pi t/T)$ for the first and second dimensions of the output, respectively, where $T = 10$ s is the iteration length. In this simulation, the sampling period is $h = 0.04$ s and, thus, $N = 250$. The resistance R is assumed to be time varying, $R = R_0(1.4 - 0.4e^{-t})$, $L_r = L_{r0}/1 + 0.1$, and $L_s = L_{s0}/1 + 0.1$, where $R_0 = 0.5\,\Omega$ and $L_{r0} = L_{s0} = 1$ H are as in Xu and Jin (2013).

Let $\omega = 0.1$. For simplicity, the stochastic noise ξ_k is assumed to follow a normal distribution with covariance τ^2, where τ reflects the level of the stochastic noises. The noise level is varied as $\tau = 0$, 0.01, 0.05, and 0.3. Note that $\tau = 0$ implies a noise-free system. The learning gain L_t is set to $0.6I$, which satisfies the condition in Theorem 4.2. The new ILC algorithm (4.7) and the SA-based algorithm (4.26) (i.e., (4.3) with $a_k = 1/k$) are run for 100 iterations. We omit the repetitive simulations to save space because the noncausal direction regulation matrix would provide similar tracking performance to the block diagonal direction regulation matrix.

The following simulations clarify four aspects of the proposed algorithm. First, we verify that (4.7) and (4.15) are identical in a noise-free system (Figs. 4.1 and 4.2).

Fig. 4.1 Averaged tracking error profiles at different noise levels. The initial values are different in the two plots because of the independent operation and random noises

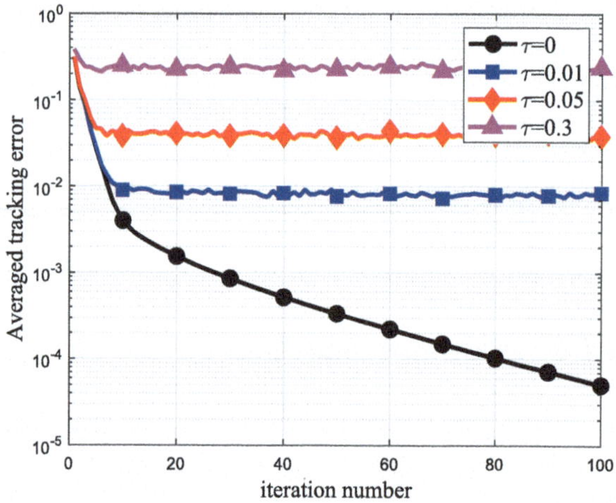

(a) The first output i_r

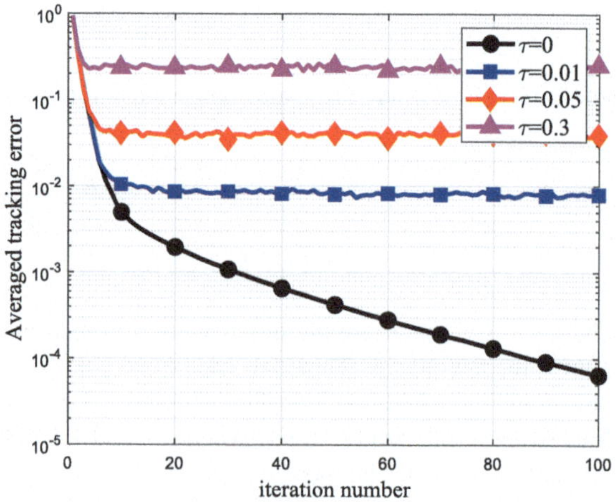

(b) The second output i_s

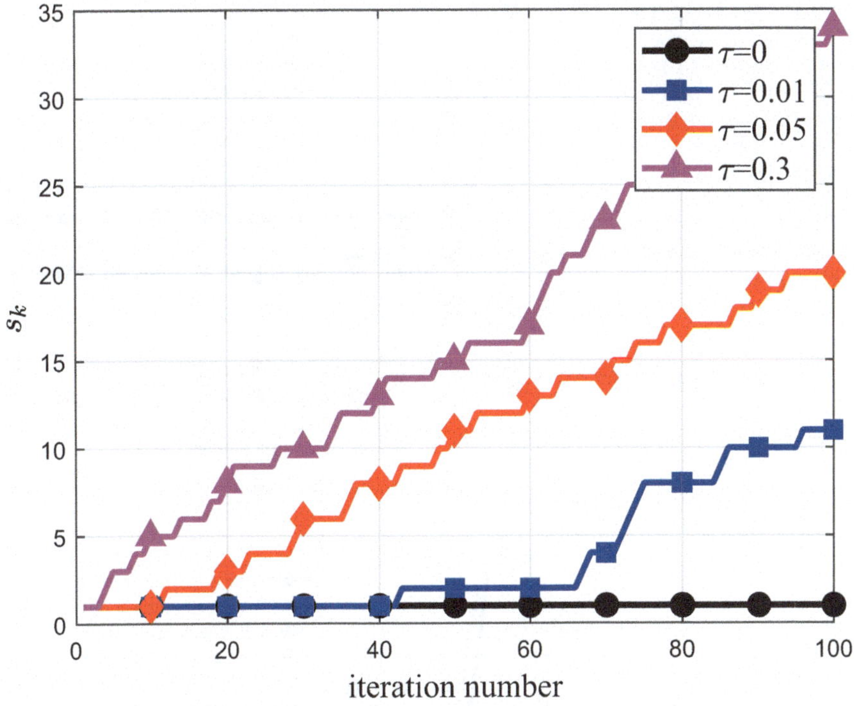

Fig. 4.2 s_k profiles with different noise levels

Second, we reveal the behaviors of (4.7) and the gain adaptation at different noise levels (Figs. 4.1 and 4.2). Third, we compare the tracking performances of (4.7), (4.26), and the KF algorithm proposed in Saab (2001a) (Fig. 4.3). Finally, we demonstrate the performance adaptation while varying the noise level throughout the iterations (Fig. 4.4).

In the absence of noise, the new ILC algorithm (4.7) exhibits identical behavior to the classical ILC law (4.15). For the noiseless scenario ($\tau = 0$, solid line with green circles), it can be observed from the linear-log plot in Fig. 4.1 that the error decreases linearly with iteration number after an initial few iterations, indicating exponential convergence speed. However, this trend is not evident in other cases. Additionally, Fig. 4.2 demonstrates that s_k remains to be one for all iterations in the noise-free case (solid line with green circles). This further confirms that (4.7) is equivalent to (4.15) when operating in a noise-free system.

The learning algorithm involves stochastic noises when $\tau \neq 0$. Here the noise level is varied as $\tau = 0.01$, 0.05, and 0.3. Because these noises are unpredictable, they cannot be eliminated by any learning algorithm. Thus, the tracking performance is successful when the inaccurate tracking term $\mathcal{H}\Delta U_k \to 0$. In this case, the asymptotic tracking error consists only of the stochastic noises. Figure 4.1 illustrates the tracking performance of (4.7). The fluctuation levels in the converged tracking

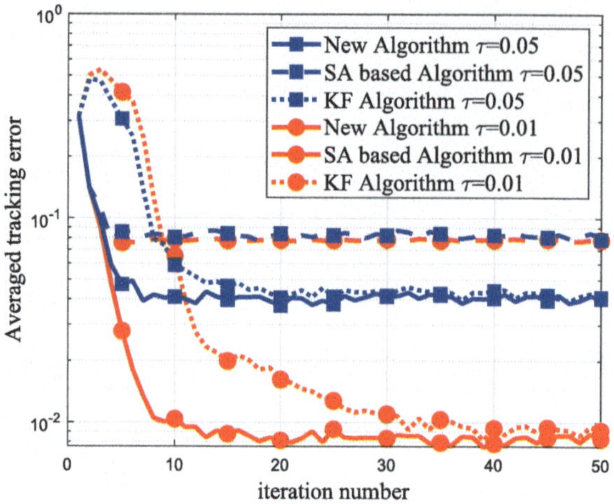

(a) The first output i_r

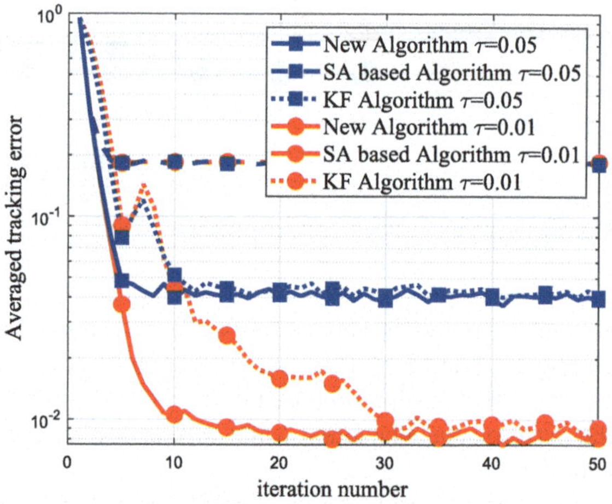

(b) The second output i_s

Fig. 4.3 Averaged tracking error profiles of the new ILC algorithm and the SA-based algorithm for $\tau = 0.01$ and $\tau = 0.05$. The solid, dashed, and dotted lines are generated by the new algorithm (4.7), the SA-based algorithm (4.26), and KF algorithm in Saab (2001a), respectively

error profiles are identical to those of the stochastic noises. Figure 4.2 shows the convergence of $s_k/k \to 1/2$ as $k \to \infty$. Across all three noise levels, the value of s_k consistently increases with each iteration. As the noise level τ becomes higher, the proportion s_k/k approaches a closer approximation to 1/2 after 100 iterations. This is due to the fact that the sign of $E_{k+1}^T E_k$ does not turn negative until the ratio between stochastic noises and tracking error reaches a sufficiently large threshold. Consequently, a larger value of τ leads to an earlier initiation of gain adaptation and ultimately results in a more accurate approximation effect.

To demonstrate the superiority of the new ILC algorithm (4.7) over both the SA-based algorithm and the KF-based algorithm, we conducted simulations using

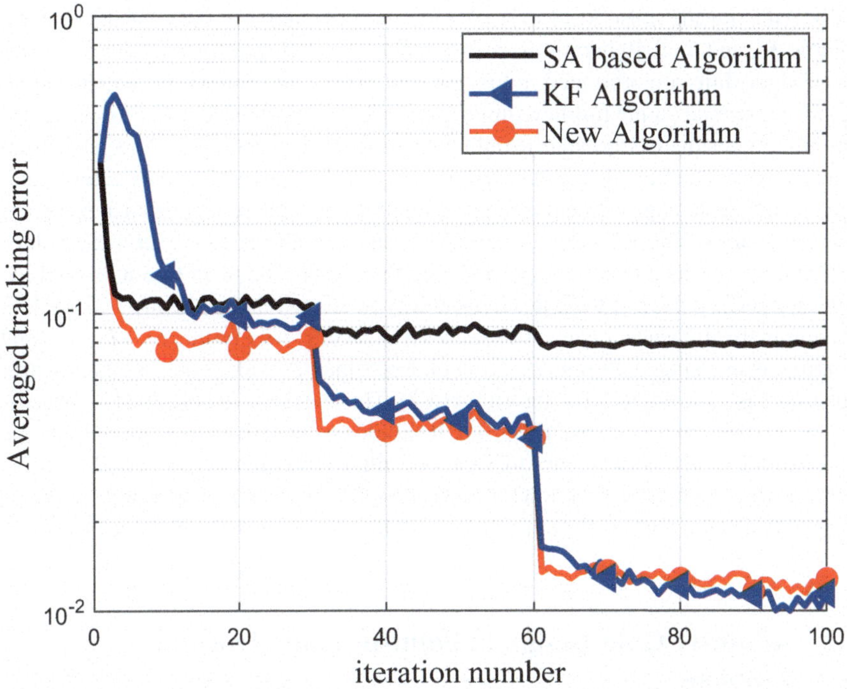

Fig. 4.4 Averaged tracking error profiles for the first output of the new ILC algorithm (4.7), the SA-based algorithm (4.26), and the KF algorithm in Saab (2001a) when the noise level is varied throughout the iterations. During the first 30 iterations, subsequent 30 iterations, and final 40 iterations, the noise level is $\tau = 0.1$, $\tau = 0.05$, and $\tau = 0.01$, respectively

algorithms (4.7), (4.26), and the KF-based algorithm from Saab (2001a) with $\tau = 0.01$ and $\tau = 0.05$. The KF-based algorithm in Saab (2001a) requires complete plant dynamics. Figure 4.3 illustrates the results obtained from these simulations. While both our proposed algorithm and the KF-based algorithm converge within the range of noise-leveled tracking errors, it is observed that after a few iterations, the SA-based algorithm exhibits slower convergence speed compared to other two algorithms. In other words, its tracking precision is inferior to that demonstrated by remaining two algorithms. Furthermore, although both our proposed algorithm and the KF-based algorithm reduce tracking errors relative to theoretical noise-leveled range, our proposed method ensures significantly faster convergence speed than that achieved by the KF-based algorithm. In summary, (4.7) outperforms equation (4.26) in terms of tracking performance while surpassing even the convergence speed exhibited by KF-based algorithm of Saab (2001a).

Since the tracking performance is heavily influenced by the noise level τ, we are interested in investigating how the tracking behavior varies when the stochastic noise level changes during iterations. To achieve this, we conduct simulations of algorithms (4.7), (4.26), and the KF-based algorithm from Saab (2001a) over 100 iterations, gradually reducing the noise level to $\tau = 0.1$, $\tau = 0.05$, and $\tau = 0.01$ for the first 30, middle 30, and final 40 iterations, respectively. The averaged tracking error profiles of all algorithms are presented in Fig. 4.4 (only showing results for the first output; results for second output follow a similar pattern but are omitted here due to space constraints). When there is a change in unknown noise levels, our proposed ILC algorithm quickly adapts and exhibits improved tracking performance after only a few iterations, while little improvement is observed with the SA-based algorithm as the noise level decreases. The KF-based algorithm behaves similarly to our proposed algorithm; however, its adaptation speed is slightly slower. The main reason behind the superior performance of (4.7) lies in its ability to effectively handle lower noise levels dominated by inaccuracies in tracking term $\mathcal{H}\Delta U_k$. As a result, $E_{k+1}^T E_k > 0$ and s_k remains unchanged during transition phases which provides constant gain matrix ensuring fast convergence for algorithm (4.7). This mechanism is verified by observing significant improvements in Fig. 4.4. It should be noted that our proposed scheme may exhibit similar performance to SA-based algorithms after longer periods of time. A promising alternative direction for future research involves incorporating an event-triggered scheme with well-defined events indicating dominant influence of noise.

4.2 Adaptive Gain Design in Point-to-Point Tracking Problem

4.2.1 Stochastic Point-to-Point Tracking Problem

Point-to-point tracking systems (PTSs) are commonly employed in various industrial processes (Chi et al. 2022; Ding and Wu 2007; Van de Wijdeven and Bosgra 2008). Instead of continuously tracking a reference throughout the entire time domain, PTSs aim to achieve precise tracking at critical time instances while allowing system dynamics to remain unconstrained during the remaining time period. Consequently, point-to-point tracking problems ensure optimal performance only at specific designated points (Chen 1991). To illustrate this concept, consider basketball shooting where the primary objective is to accurately shoot the ball into the basket rather than following a particular trajectory. Therefore, only the final point-the position of the basket-needs to be precisely targeted. Similarly, in pick-and-place tasks performed by industrial manipulators, the main goal is to accurately grasp an object from a specified location and place it elsewhere (Li et al. 2013), potentially passing through multiple predetermined positions. Numerous other examples can exemplify such tasks including production line automation, robot arms operation, crane positioning, and high-speed train control.

Various control approaches have been employed in point-to-point tracking problems, among which ILC has demonstrated its advantage in achieving high tracking performance for repetitive systems. Numerous papers have contributed to the field of ILC for PTSs. However, none of these studies have taken into consideration any stochastic elements within the investigated systems. Preliminary results on stochastic systems were explored in Shen et al. (2017); Xu et al. (2017) using stochastic approximation algorithms. The achieved tracking performance exhibited superior performance indices for specified positions despite the presence of system and measurement noise; however, this was at the expense of a relatively slow convergence rate. Despite this valuable contribution, there remains a lack of comprehensive exploration regarding learning control for PTSs.

For the point-to-point tracking problem, only a subset of the reference trajectory (referred to as subreference hereafter) is expected to precisely follow the predicted path, while disregarding the remaining trajectory. By formulating this problem as a mapping that utilizes inputs from the entire time interval and produces outputs at specified time instants, we have discovered that there may exist numerous (potentially infinite) input solutions capable of generating the desired subreference. Consequently, an important question arises: Does the input sequence generated by our proposed learning control algorithm converge towards a limit? Furthermore, if it does converge, will this limit remain invariant across different experiments given a fixed initial input? This issue has not been adequately addressed in existing literature and serves as the primary motivation for this section.

In practical applications, the presence of random noise, including dynamics disturbance and measurement noise, is inevitable. Consequently, tracking errors are consistently affected by this randomness, necessitating its consideration in algorithm design. In order to mitigate the impact of randomness as the iteration number increases, a decreasing gain is introduced. However, a drawback associated with this approach is that it often leads to a slow convergence rate since it is typically implemented independently of the actual tracking performance. Therefore, our objective in this section is to investigate whether introducing performance-based adaptive gains can accelerate the convergence rate.

In this section, we investigate a discrete-time linear system subjected to random noise and focus on the point-to-point tracking problem. The tracking problem is formulated as a linear mapping from input to output with additional noise, where the input has higher dimensions than the output. To address this issue, we propose an accelerated gradient algorithm that incorporates an adaptive gain mechanism based on adjacent tracking errors. Initially, the learning gain remains constant to achieve rapid reduction in tracking errors; subsequently, it gradually decreases after a certain number of iterations to mitigate the impact of noise. Furthermore, we prove that our accelerated gradient algorithm converges to the invariant limit closest to the initial input in terms of Euclidean norm among all possible solutions. Additionally, we introduce a generalized accelerated algorithm by incorporating a direction regulation matrix concept which eliminates the need for gradient information. The convergence limit of this algorithm depends not only on the initial input but also on the direction regulation matrix as it determines the improvement space for the input sequence. We provide a direct analysis of tracking error and complete asymptotic analysis of our proposed control scheme. Detailed comparisons between our proposed algorithms and traditional approaches are presented along with simulation results using a permanent magnet synchronous motor model to validate our theoretical findings.

4.2.2 Problem Formulation

Consider the discrete time-varying linear systems

$$\begin{aligned} x_k(t+1) &= A_t x_k(t) + B_t u_k(t), \\ y_k(t) &= C_t x_k(t) + v_k(t), \end{aligned} \tag{4.27}$$

where $A_t \in \mathbf{R}^{n\times n}$, $B_t \in \mathbf{R}^{n\times p}$, and $C_t \in \mathbf{R}^{q\times n}$ are the system matrices at time t, $t = 0, 1, \ldots, N$. In addition, $x_k(t) \in \mathbf{R}^n$, $u_k(t) \in \mathbf{R}^p$ and $y_k(t) \in \mathbf{R}^q$ denote the system state, input, and output at time instant t of iteration k, respectively, where $k = 1, 2, \ldots$. Further, $v_k(t) \in \mathbf{R}^q$ denotes the measurement noise subject to a normal distribution with zero mean and bounded covariance.

The following assumptions are required for the analysis:

Assumption 4.6 The input/output coupling matrix $C_{t+1}B_t$ is of full row rank.

Assumption 4.7 The measurement noise $v_k(t)$ is independent with respect to the labels k and t. Further, $v_k(t) \sim \mathcal{N}(0, R_t)$, where R_t denotes the covariance matrix.

We lift the system quantities of the entire iteration into supervectors. In particular,

$$\begin{aligned} \boldsymbol{y}_k &= [y_k^T(1), y_k^T(2), \ldots, y_k^T(N)]^T, \\ \boldsymbol{u}_k &= [u_k^T(0), u_k^T(2), \ldots, u_k^T(N-1)]^T, \\ \boldsymbol{y}_0 &= \left[(C_1A_0)^T, (C_2A_1A_0)^T, \ldots, \left(C_N \prod_{j=0}^{N-1} A_j \right)^T \right]^T x_k(0), \end{aligned}$$

where $\prod_{k=i}^{j} A_k \triangleq A_jA_{j-1}\cdots A_i$ for $j \geq i$.

Then, the lifting formulation for (4.27) can be expressed as

$$\boldsymbol{y}_k = H\boldsymbol{u}_k + \boldsymbol{y}_0 + \boldsymbol{\varpi}_k, \tag{4.28}$$

where

$$H = \begin{bmatrix} C_1B_0 & 0 & \cdots & 0 \\ C_2A_1B_0 & C_2B_1 & \cdots & 0 \\ \vdots & \vdots & \ddots & \vdots \\ C_N\prod_{j=1}^{N-1}A_jB_0 & C_N\prod_{j=1}^{N-1}A_jB_1 & \cdots & C_NB_{N-1} \end{bmatrix}$$

and $\boldsymbol{\varpi}_k = [v_k^T(1), v_k^T(2), \ldots, v_k^T(N)]^T$. By Assumption 4.7, we can reasonably assume that $\boldsymbol{\varpi}_k$ follows a normal distribution $\mathcal{N}(0, Q)$, where $Q \in \mathbf{R}^{nq\times nq}$ is a suitable positive-definite matrix. From the lifting model (4.28), it is clear that the case with additive noise in the state dynamics of (4.27) is a straightforward extension because the noise can be integrated into $\boldsymbol{\varpi}_k$.

In the standard ILC framework, the desired reference is a complete curve defined over the entire iteration interval $\{1, 2, \ldots, N\}$, i.e., $\boldsymbol{y}_d = [y_d^T(1), y_d^T(2), \ldots, y_d^T(N)] \in \mathbf{R}^{qN}$. The objective of standard ILC is to design an appropriate learning control scheme that enables precise tracking of $\boldsymbol{y}_d$ by generating inputs capable of driving the system output of (4.28). However, in various scenarios such as those involving pick-and-place robots or complex chemical processes, it may only be necessary to accurately track a subset of the complete reference $\boldsymbol{y}_d$, while allowing other parts to remain unconstrained. Thus, accurate tracking can be achieved by focusing on a sub-reference, which gives rise to what is commonly known as the point-to-point tracking problem.

Remark 4.4 In traditional ILC schemes, the desired reference $\boldsymbol{y}_d$ is typically assumed to be realizable, meaning that there exists a *unique* input $\boldsymbol{u}_d$ such that $\boldsymbol{y}_d = H\boldsymbol{u}_d$. Consequently, the convergence of the output $\boldsymbol{y}_k$ towards the desired reference $\boldsymbol{y}_d$ can be established by demonstrating that $\boldsymbol{u}_k \to \boldsymbol{u}_d$. However, in this section, Assumption 4.6 allows for any reference to be realizable without imposing uniqueness. Therefore, a more meticulous investigation into the asymptotic behavior of the input sequence is required.

Without any loss of generality, we consider the initial state to be zero, i.e., $x_k(0) = 0$ for all k, to simplify the derivations. Thus, $\boldsymbol{y}_0$ is zero and (4.28) becomes

$$\boldsymbol{y}_k = H\boldsymbol{u}_k + \boldsymbol{\varpi}_k. \tag{4.29}$$

Remark 4.5 Indeed, the initial state can be set to a fixed value or randomly varied around a specific value for all iterations. In the former case, the initial state can be adjusted as part of the desired reference; in the latter case, any random initialization error can be incorporated into $\boldsymbol{\varpi}_k$. Consequently, these extensions of the reinitialization condition will not impact the subsequent results.

The point-to-point tracking problem suggests that precise realization of only a subset, rather than all, of the reference points is required. This formulation allows for increased flexibility in control design.

We assume that $\boldsymbol{y}_r \in \mathbf{R}^m$ is the desired sub-reference of $\boldsymbol{y}_d$, which represents an ordered subset of entries from the vector $\boldsymbol{y}_d$. Here, m denotes the dimensions for tracking, where $m < qN$. At any given time instant, $\boldsymbol{y}_r$ can only contain the output of a specific dimension. In other words, simultaneous tracking of all elements in $y_d(t)$ is not necessary. The relationship between $\boldsymbol{y}_r$ and $\boldsymbol{y}_d$ is modeled as described in Freeman et al. (2011); Freeman (2012); Freeman and Tan (2013), which is given as follows:

$$\boldsymbol{y}_r = \Psi\boldsymbol{y}_d, \tag{4.30}$$

where the transformation matrix $\Psi \in \mathbf{R}^{m\times qN}$ is a binary matrix with m nonzero entries scattered among different rows and columns. These nonzero entries are equal to one, i.e., $\Psi(i, j)$ is either 1 or 0. Further, the matrix Ψ should satisfy that $\sum_{j=1}^{qN} \Psi(i, j) = 1, i = 1, 2, \ldots, m$, and $\sum_{i=1}^{m} \Psi(i, j) \leq 1, j = 1, 2, \ldots, qN$.

Based on the given definition, each row of matrix Ψ contains only one entry equal to one, and this entry's label corresponds to the dimension that needs to be precisely tracked in vector $\mathbf{y}_d$. Additionally, for any two row labels i_0 and i_1, we denote the column labels of the nonzero entries in the i_0th and i_1th rows as j_0 and j_1, respectively (i.e., $\Psi(i_0, j_0) = 1$ and $\Psi(i_1, j_1) = 1$). It follows that if $i_1 > i_0$, then it must be true that $j_1 > j_0$, indicating that the order of entries in vector $\mathbf{y}_m$ preserves the original order given in vector $\mathbf{y}_d$. Therefore, matrix Ψ acts as a selection matrix determining the dimensions of the entire output vector.

The dimension m of the sub-reference $\mathbf{y}_r$ is typically much smaller than that of the entire reference $\mathbf{y}_d$. As a result, the transformation matrix Ψ possesses full row rank. Leveraging Assumption 4.6, the subsequent lemma establishes a crucial rank property concerning the matrix product of Ψ and H.

Lemma 4.2 *(Shen et al. (2017)) The matrix ΨH is of full row rank, if $C_{t+1}B_t$ is of full row rank for all time instants.*

When analyzing the convergence of any learning control scheme, existing literature primarily focuses on demonstrating the convergence of $\mathbb{E}[\|\mathbf{y}_d - \Psi\mathbf{y}_k\|^2]$ to a limit as $k \to \infty$. Instead of proving the convergence of the input sequence, emphasis is placed on showing that the tracking error converges with increasing iteration number. This approach stems from the nonuniqueness associated with potential input solutions based on a given reference. Specifically, under Assumption 4.6 and Lemma 4.2, for any desired tracking objective $\mathbf{y}_r$, there exists an infinite number of inputs $\boldsymbol{u} \in \mathbf{R}^{np}$ that satisfy $\mathbf{y}_r = \Psi H\boldsymbol{u}$.

However, the convergence of the tracking error does not necessarily imply that the input sequence asymptotically approaches a limit. Considering that the learning control scheme successively generates the input sequence in the iteration domain, it becomes crucial to reveal the asymptotic performance of the generated input sequence. Several unresolved issues remain.

- The set of possible inputs satisfying the tracking reference (i.e., $\mathbf{y}_r = \Psi H\boldsymbol{u}$) should be considered instead of a single candidate. Therefore, for the designed learning control scheme for point-to-point tracking problems, it is necessary to determine whether the generated input sequence converges to a limit along the iteration axis.
- Moreover, if the generated input sequence converges to a limit as the iteration number increases, it is imperative to ascertain whether this limit is uniquely determined for a given initial input value. In case of an affirmative response, it becomes essential to inquire about explicit methods for describing this limit and elucidating its specific properties.
- Additionally, the application of a decreasing gain to counteract the impact of random noise is common practice. However, this predefined sequence can impede convergence speed. Therefore, it is worth exploring practical methods to accelerate convergence rate based on tracking performance.

The control objective of this section aims to address the aforementioned issues pertaining to conventional learning control schemes. These issues are brought to light through the proposal of an accelerated gradient algorithm, and the findings are further extended to encompass a broader design scenario.

The set of all possible solutions to the desired tracking reference $\boldsymbol{y}_r$ is denoted as $\mathcal{S}_u = \{\boldsymbol{u} : \boldsymbol{y}_r = \Psi H \boldsymbol{u}\}$. In order to achieve favorable tracking performance, it is necessary for the input sequence generated by the learning control scheme to converge towards $\mathcal{S}_u$, i.e., $\boldsymbol{u}_k \to \mathcal{S}_u$. This section provides a proof that, under the proposed assumptions, for any given initial input $\boldsymbol{u}_0$, the input sequence $\{\boldsymbol{u}_k\}$ converges to a specific limit within the set $\mathcal{S}_u$, addressing the first issue. Furthermore, we denote this limit of the input sequence as $\boldsymbol{u}^*$, which belongs to $\mathcal{S}_u$. Among all candidates in the set $\mathcal{S}_u$, $\boldsymbol{u}^* = \arg\min_{\boldsymbol{u}\in\mathcal{S}_u} \|\boldsymbol{u} - \boldsymbol{u}_0\|^2$. This addresses the second issue by minimizing the distance between initial input $\boldsymbol{u}_0$ and limit $\boldsymbol{u}^*$. Additionally, we introduce a gain design technique based on tracking error in order to accelerate convergence rate. The newly designed gain remains unchanged if tracking error is dominated by input bias and starts decreasing if tracking error is dominated by measurement noise. These contributions propose an accelerated learning control scheme that addresses the third issue. All these findings contribute towards a profound understanding of learning control problems for PTSs.

4.2.3 Accelerated Gradient Algorithm

4.2.3.1 Algorithm Design

The practical tracking reference is $\boldsymbol{y}_r$. Thus, only a subset of the dimensions of the output $\boldsymbol{y}_k$ is required to track the reference. To this end, we apply the transformation matrix Ψ to the lifting system dynamics (4.29):

$$\Psi \boldsymbol{y}_k = \Psi H \boldsymbol{u}_k + \Psi \boldsymbol{\varpi}_k. \tag{4.31}$$

We define the practical tracking error $\boldsymbol{e}_k = \boldsymbol{y}_r - \Psi \boldsymbol{y}_k \in \mathbf{R}^m$. Then, the accelerated gradient algorithm is given as follows:

$$\boldsymbol{u}_{k+1} = \boldsymbol{u}_k + r_k^{-1} H^T \Psi^T \boldsymbol{e}_k, \tag{4.32}$$

where r_k is a gain adaptation parameter used for accelerating, $r_k = 1 + \sum_{i=1}^{k-1} \mathbf{1}_{[\boldsymbol{e}_{i+1}^T \boldsymbol{e}_i < 0]}$. In particular, this parameter is initially set to 1, i.e., $r_1 = 1$, and then incremented by 1 whenever the indicated event occurs. Thus, the update process of r_k is random and data-driven. The corresponding recursive formula is given by $r_k = r_{k-1} + \mathbf{1}_{[\boldsymbol{e}_k^T \boldsymbol{e}_{k-1} < 0]}$.

Here, we present an intuitive explanation of the accelerated gradient algorithm given in (4.32). Firstly, the tracking error $\boldsymbol{e}_k$ is expressed as a linear function of the input $\boldsymbol{u}_k$, where the gradient is denoted by ΨH. To ensure dimensional compatibility between the tracking error and input, we employ the transpose of this gradient as the direction regulation matrix. The utilization of complete system information H is essential for this algorithm.

To comprehend the principle of acceleration in the proposed scheme, we initially revisit conventional P-type algorithms mentioned in previous literature such as (Shen et al. 2022). It has been observed that a constant gain can lead to rapid convergence speed. However, due to random noise, the input sequence fails to converge towards a limit under this circumstance. To gradually eliminate the impact of random noise and ensure steady convergence towards a limit, it is necessary to employ a decreasing gain sequence that approaches zero at an appropriate rate. Nevertheless, this approach often results in slow convergence speed. This observation motivated us to investigate whether maintaining a constant gain value during the early stages of the learning process could expedite convergence speed and subsequently decrease after a certain number of iterations for improved tracking performance. The design of an appropriate switching condition based on available tracking information is crucial for implementing this mechanism. In this section, we achieve this mechanism by evaluating the sign of successive tracking errors.

Now we proceed to the principle of the acceleration mechanism in (4.32). We note that the tracking error $\boldsymbol{e}_k$ consists of two parts: $\boldsymbol{e}_k = \boldsymbol{y}_r - \Psi H\boldsymbol{u}_k - \Psi\boldsymbol{\varpi}_k$, where the first part $P_1 \triangleq \boldsymbol{y}_r - \Psi H\boldsymbol{u}_k$ denotes the input bias, and the second part $P_2 \triangleq \Psi\boldsymbol{\varpi}_k$ denotes the noise fluctuation. In the initial phase of the learning process, we consider that the generated input $\boldsymbol{u}_k$ is still distant from $\mathcal{S}_u$, thereby resulting in a tracking error primarily dominated by the first component P_1. In such cases, the product of consecutive tracking errors remains positive, leading to an unchanged value of r_k. As iterations progress, it is anticipated that the input will improve and approach proximity to the set $\mathcal{S}_u$, while significantly reducing the influence of P_1. Consequently, random noise becomes dominant and contributes to the second component P_2 as a major source of tracking error. Under these circumstances, there exists a certain probability for successive tracking errors to yield negative products and consequently increase r_k. As a result, within Algorithm (4.32), the gain factor $1/r_k$ gradually diminishes towards zero with increasing iteration number.

A related detailed analysis is given in Sect. 4.2.3.3. In addition, if the system is free of random noise, i.e., $\boldsymbol{\varpi}_k = 0$, one can prove that the event $\boldsymbol{e}_{i+1}^T\boldsymbol{e}_i < 0$ will never occur; thus, the adaptive parameter r_k is always equal to 1. Therefore, the proposed algorithm (4.32) turns into the conventional learning control scheme (Bristow et al. 2006). This is verified by simulations in Sect. 4.2.5.

4.2.3.2 Limit of the Input Sequence

In this subsection, we outline the limits of the input sequence prior to presenting a comprehensive convergence analysis. Specifically, the feasible solutions for $\boldsymbol{y}_r = \Psi H\boldsymbol{u}$ form a set denoted as $\mathcal{S}_u$. To ensure optimal tracking performance, it suffices to guarantee that $\boldsymbol{u}_k$ converges to a point within $\mathcal{S}_u$. In such cases, the limit of the tracking error $\boldsymbol{e}_k$ is given by $\boldsymbol{e}^* = \boldsymbol{y}_r - \Psi H \boldsymbol{u}|_{\boldsymbol{u} \in \mathcal{S}_u} - \Psi \boldsymbol{\varpi} = -\Psi \boldsymbol{\varpi}$. The mean square value of the tracking error can be expressed as $\mathbb{E}[\|\boldsymbol{e}^*\|^2] = \mathbb{E}[\|\Psi \boldsymbol{\varpi}\|^2]$. Enhancing this metric further poses challenges due to its dependence solely on measurement noise in each iteration, which cannot be predicted or eliminated using any learning control scheme.

Motivated by these observations, most existing studies have demonstrated the convergence of the auxiliary error $\boldsymbol{y}_r - \Psi H \boldsymbol{u}_k$ to zero with respect to iterations. However, little attention has been paid to the iteration-wise performance of the generated input sequence $\boldsymbol{u}_k$. In contrast, we provide clear evidence that the input sequence generated by the accelerated gradient algorithm (4.32) converges to a fixed limit $\boldsymbol{u}^*$ within $\mathcal{S}_u$, regardless of any given initial input. In particular, we will strictly prove that the limit $\boldsymbol{u}^*$ of the generated input sequence is closest to the initial input $\boldsymbol{u}_0$ in the sense that

$$\boldsymbol{u}^* = \arg\min_{\boldsymbol{u} \in \mathcal{S}_u} \|\boldsymbol{u} - \boldsymbol{u}_0\|^2, \tag{4.33}$$

or equivalently, $\boldsymbol{u}^*$ is the solution of the optimization problem:

$$\begin{aligned} &\min_{\boldsymbol{u} \in \mathbf{R}^{pN}} \|\boldsymbol{u} - \boldsymbol{u}_0\|^2, \\ &\text{s.t. } \boldsymbol{y}_r = \Psi H \boldsymbol{u}. \end{aligned}$$

The given problem can be formulated as a constrained convex optimization problem. By employing elementary derivations, specifically by defining the Lagrangian function and utilizing the well-established Karush-Kuhn-Tucker (KKT) conditions (Polyak 1987), we are able to obtain the solution to this optimization problem:

$$\boldsymbol{u}^* = \boldsymbol{u}_0 + H^T \Psi^T (\Psi H H^T \Psi^T)^{-1} (\boldsymbol{y}_r - \Psi H \boldsymbol{u}_0). \tag{4.34}$$

As a special case, if we begin the accelerated gradient algorithm (4.32) from zero, i.e., $\boldsymbol{u}_0 = 0$, we obtain a concise limit

$$\boldsymbol{u}^* = H^T \Psi^T (\Psi H H^T \Psi^T)^{-1} \boldsymbol{y}_r.$$

The limit satisfying (4.33) can be considered as an energy-efficient convergence route of the input sequence to a certain extent. It is evident from (4.34) that the convergence limit is dependent on the initial input. However, we will not consider selecting an optimal initial input based on some index since it is equivalent to finding the desired input. Indeed, in practical applications, the initial input is arbitrarily given and our proposed scheme always achieves perfect tracking performance.

4.2.3.3 Convergence Analysis

In this section, we present a convergence analysis for the accelerated gradient algorithm (4.32). Specifically, we demonstrate that the input sequence generated by (4.32) converges asymptotically to a limit for any initial input. Furthermore, this limit corresponds to the nearest point in set $\mathcal{S}_u$ to the given initial input under the Euclidean norm. These findings are summarized in the theorem.

Theorem 4.3 *Consider the system* (4.27) *under Assumptions 4.6 and 4.7. For the point-to-point tracking reference* $\mathbf{y}_r$, *apply the accelerated gradient algorithm* (4.32). *Starting from the initial input* $\boldsymbol{u}_0$, *the input sequence* $\{\boldsymbol{u}_k\}$ *converges to* $\boldsymbol{u}^*$, *which is defined by* (4.34).

We first provide a concise overview of the analysis to enhance comprehension before presenting a detailed proof. The proof comprises two main components: firstly, we transform algorithm (4.32) into a recursion of vectors within the space spanned by $H^T\Psi^T$. This transformation effectively resolves the challenge of nonuniqueness encountered when proving convergence limits for the input sequence. Secondly, we accomplish the convergence analysis by employing Lemma 4.1.

Proof. We first establish an intrinsic relationship within the input sequence $\{\boldsymbol{u}_k\}$ using the mathematical induction method. To this end, we start from the initial input. By the algorithm (4.32), we have $\boldsymbol{u}_1 = \boldsymbol{u}_0 + H^T\Psi^T\boldsymbol{e}_0$, where $\boldsymbol{e}_0$ denotes the tracking error driven by the input $\boldsymbol{u}_0$. Notably, $\boldsymbol{e}_0$ is a vector, indicating that $\boldsymbol{u}_1 - \boldsymbol{u}_0 \in \text{Span}\{H^T\Psi^T\}$. In other words, the difference between $\boldsymbol{u}_1$ and $\boldsymbol{u}_0$ is a linear combination of the columns of $H^T\Psi^T$.

By the principle of induction method, assume that $\boldsymbol{u}_k - \boldsymbol{u}_0 \in \text{Span}\{H^T\Psi^T\}$. From the algorithm (4.32), we obtain

$$(\boldsymbol{u}_{k+1} - \boldsymbol{u}_0) = (\boldsymbol{u}_k - \boldsymbol{u}_0) + r_k^{-1}H^T\Psi^T\boldsymbol{e}_k. \tag{4.35}$$

Considering that $1/r_k$ is a real number, it is evident that $\boldsymbol{u}_{k+1} - \boldsymbol{u}_0 \in \text{Span}\{H^T\Psi^T\}$. Thus, for the input sequence $\{\boldsymbol{u}_k\}$ generated by (4.32), it is determined that $\boldsymbol{u}_k - \boldsymbol{u}_0$ belongs to the space spanned by the columns of $H^T\Psi^T$ for all iterations.

By Lemma 4.2, $H^T\Psi^T$ is of full column rank under Assumption 4.6. Therefore, for each term $\boldsymbol{u}_k - \boldsymbol{u}_0$, there must exist a unique vector $\phi_k \in \mathbf{R}^m$ such that

$$\boldsymbol{u}_k - \boldsymbol{u}_0 = H^T\Psi^T\phi_k. \tag{4.36}$$

That is, the linear combination of the columns of $H^T\Psi^T$ for the term $\boldsymbol{u}_k - \boldsymbol{u}_0$ is unique.

Substituting (4.36) into (4.35), we have

$$H^T\Psi^T\phi_{k+1} = H^T\Psi^T\phi_k + r_k^{-1}H^T\Psi^T\boldsymbol{e}_k. \tag{4.37}$$

Again, by the full column rank property of $H^T\Psi^T$, we obtain

$$\phi_{k+1} = \phi_k + r_k^{-1}\boldsymbol{e}_k. \tag{4.38}$$

Therefore, to prove the theorem, it is sufficient to show that ϕ_k converges to $(\Psi HH^T\Psi^T)^{-1}(\boldsymbol{y}_r - \Psi H\boldsymbol{u}_0)$. This step can be verified by using Lemma 4.1. To this end, we substitute the expression of the error $\boldsymbol{e}_k$ into (4.38):

$$\begin{aligned}
\phi_{k+1} &= \phi_k + r_k^{-1}(\boldsymbol{y}_r - \Psi\boldsymbol{y}_k)\\
&= \phi_k + r_k^{-1}(\boldsymbol{y}_r - \Psi H\boldsymbol{u}_k - \Psi\boldsymbol{\varpi}_k)\\
&= \phi_k - r_k^{-1}[\Psi HH^T\Psi^T\phi_k - (\boldsymbol{y}_r - \Psi H\boldsymbol{u}_0)]\\
&\quad - r_k^{-1}\Psi\boldsymbol{\varpi}_k,
\end{aligned} \tag{4.39}$$

where (4.36) is substituted for the last equality.

Denote $\Gamma = \Psi HH^T\Psi^T$ and $\eta = \boldsymbol{y}_r - \Psi H\boldsymbol{u}_0$. Based on this, we rewrite (4.39) as follows:

$$\underbrace{\phi_{k+1} - \Gamma^{-1}\eta}_{\triangleq\theta_{k+1}} = \underbrace{\phi_k - \Gamma^{-1}\eta}_{\triangleq\theta_k} - \frac{1}{r_k}[\Gamma\phi_k - \eta] - \frac{1}{r_k}\Psi\boldsymbol{\varpi}_k. \tag{4.40}$$

The quantities in (4.40) correspond to those in Lemma 4.1: θ_k corresponds to $\bar{x}_k$, $\Gamma\phi_k - \eta = \Gamma\theta_k$ corresponds to the function $\varphi(\bar{x}_k)$, and $\Psi\boldsymbol{\varpi}_k$ corresponds to ϵ_k. The limit $\bar{x}^*$ in Lemma 4.1 is 0 for (4.40). Thus, verifying the conditions of Lemma 4.1 is sufficient to complete the proof, i.e., Assumptions 4.4 and 4.5.

We first check Assumption 4.4. Notably, $\Psi\boldsymbol{\varpi}_k$ is a condensed subvetor for $\boldsymbol{\varpi}_k$. Thus, by Assumption 4.7, we obtain $\Psi\boldsymbol{\varpi}_k \sim N(0, \Psi Q\Psi^T)$. Then, it is evident that $\mathbb{E}[\Psi\boldsymbol{\varpi}_k] = 0$, and $\mathbb{E}[\Psi\boldsymbol{\varpi}_k\boldsymbol{\varpi}_k^T\Psi^T] \le M$, where $M = \Psi Q\Psi^T$.

Next, we proceed to check Assumption 4.5. Corresponding to the deterministic sequence $\bar{x}_{k+1} = \bar{x}_k - \gamma\varphi(\bar{x}_k)$ and based on (4.40), we obtain

$$\theta_{k+1} = \theta_k - \gamma\Gamma\theta_k = (I - \gamma\Gamma)\theta_k.$$

By Lemma 4.2, ΨH is of full row rank, meaning that Γ is a positive definite matrix. Therefore, for a sufficiently small $\alpha > 0$ and $\gamma \le \alpha$, one can ensure that $I - \gamma\Gamma$ is a positive definite matrix with all its eigenvalues within the interval (0, 1). This guarantees $\theta_k \to 0$.

In addition, we define $V(\theta) = \theta^T\theta$. Then, $\nabla^2 V(\theta) = I_m \le J$ for all θ, where J is set to $2I_m$. Further, $\varphi(\theta)^T\nabla V(\theta) = 2\theta^T\Gamma\theta$ and $\varphi(\theta)^T J\varphi(\theta) = 2\theta^T\Gamma^2\theta \le 2\lambda_{\max}(\Gamma)\theta^T\Gamma\theta$, where $\lambda_{\max}(\Gamma)$ is the largest eigenvalue of Γ.

For $\|\theta\| \ge \bar{R}$, we obtain $\theta^T\Gamma\theta \ge \lambda_{\min}(\Gamma)\theta^T\theta \ge \lambda_{\min}(\Gamma)\bar{R}^2$, where $\lambda_{\min}(\Gamma)$ is the smallest eigenvalue of Γ. Let $\alpha < \min\left\{\frac{\lambda_{\min}(\Gamma)\bar{R}^2}{\text{trace}(2\Psi Q\Psi^T)}, \frac{1}{2\lambda_{\max}(\Gamma)}\right\}$ and $\beta < \frac{\lambda_{\min}(\Gamma)\bar{R}^2}{2}$. We thus have

$$\begin{aligned}
&\frac{\alpha}{2}[\varphi(\theta)^T J\varphi(\theta) + \mathrm{tr}(MJ)] + \beta \\
&\leq \frac{1}{2}\theta^T \Gamma\theta + \lambda_{\min}(\Gamma)\bar{R}^2 \\
&< 2\theta^T \Gamma\theta = \varphi(\theta)^T \nabla V(\theta).
\end{aligned}$$

In other words, the inequality (4.19) holds.

Therefore, all conditions of Lemma 4.1 hold for (4.40), and we can conclude that $\theta_k \to 0$, i.e., $\lim_{k\to\infty} \phi_k = \Gamma^{-1}\eta$. Thus,

$$\begin{aligned}
\boldsymbol{u}_k - \boldsymbol{u}_0 &\to H^T \Psi^T \Gamma^{-1}\eta \\
&= H^T \Psi^T (\Psi H H^T \Psi^T)^{-1}(\boldsymbol{y}_r - \Psi H \boldsymbol{u}_0).
\end{aligned} \tag{4.41}$$

The proof is thus completed.

Theorem 4.3 establishes the asymptotic convergence of the input sequence generated by the accelerated gradient algorithm (4.32), demonstrating a significant advancement compared to existing results. We rigorously prove the convergence of the input sequence for PTSs, which directly implies that $\Psi H \boldsymbol{u}_k \to \boldsymbol{y}_r$. Consequently, our proposed learning control scheme successfully tracks the given reference $\boldsymbol{y}_r$, and it can be inferred that existing convergence results regarding tracking errors are a direct consequence of Theorem 4.3.

In practical applications, the eigenvalues of matrix Γ may exhibit significant magnitudes, leading to undesirable overshooting during the initial iterations of the learning process. To address this issue, an additional parameter ζ is commonly employed to adjust these eigenvalues. In other words, algorithm (4.32) is replaced with $\boldsymbol{u}_{k+1} = \boldsymbol{u}_k + r_k^{-1}\zeta H^T \Psi^T \boldsymbol{e}_k$, where ζ represents a suitable constant value. It can be observed from the proof that incorporating such a modification does not affect the convergence analysis since ζ can be considered as part of matrix Γ.

In contrast to asymptotic convergence, achieving monotonic convergence is desirable as it ensures a decline in input errors in a specific norm from trial to trial. Monotonic convergence is particularly appealing for deterministic systems. However, due to the presence of random noise, it becomes challenging, if not impossible, to establish the monotonic convergence of the transformed vector θ_k. By leveraging the positive-definiteness of Γ and the vanishing property of r_k^{-1} as stated in Lemma 4.1, we can deduce that $\mathbb{E}[\theta_k]$ exhibits a monotonically decreasing trend after several iterations when taking expectation to both sides of (4.40).

4.2.4 Generalized Accelerated Algorithm

The asymptotic convergence of the input sequence for the accelerated gradient algorithm was established in the previous subsection. To formulate the learning control scheme, precise information about the system is necessary, including all matrices A_t, B_t, and C_t to construct the lifting matrix H. In this section, we extend these results to a generalized algorithm.

4.2.4.1 Accelerated Algorithm with Direction Regulation Matrix

The generalized learning control scheme is given as follows:

$$\boldsymbol{u}_{k+1} = \boldsymbol{u}_k + r_k^{-1} L \boldsymbol{e}_k, \tag{4.42}$$

where $L \in R^{pN \times m}$ is a direction regulation matrix.

The matrix L serves two purposes in the algorithm. Firstly, it regulates the dimensional discrepancy between the input and tracking error, thus earning its name as the direction regulation matrix. Secondly, it establishes the learning space for the input sequence. To illustrate this point, we set the initial input $\boldsymbol{u}_0 = 0$. It is evident that all elements of the input sequence must lie within the subspace spanned by the columns of L. In other words, the input sequence approaches a solution set $\mathcal{S}_u$ within this subspace spanned by L. Consequently, although not converging to $\boldsymbol{u}^*$, it is still ensured that under certain conditions, the input sequence will converge to a specific point in $\mathcal{S}_u$, which depends on both initial input $\boldsymbol{u}_0$ and direction regulation matrix L.

Convergence conditions for direction regulation matrix L (C1): The matrix L is designed such that $\Psi H L$ is a square matrix with only positive eigenvalues.

The gradient $H^T \Psi^T$ used in (4.32) satisfies the condition (C1). Under the condition (C1), the limit $\boldsymbol{u}^\circ$ of the input sequence generated by (4.42) can be characterized by

$$\boldsymbol{u}^\circ = \boldsymbol{u}_0 + L(\Psi H L)^{-1}(\boldsymbol{y}_r - \Psi H \boldsymbol{u}_0) \tag{4.43}$$

for any initial input $\boldsymbol{u}_0$. Clearly, $\Psi H \boldsymbol{u}^\circ = \boldsymbol{y}_r$, i.e., $\boldsymbol{u}^\circ \in \mathcal{S}_u$.

The verification of the condition (C1) may still necessitate complete system information. In this regard, we propose a practical design for the matrix L. By leveraging Assumption 4.6, we can construct a matrix L_t at each time instant t such that all eigenvalues of $C_{t+1} B_t L_t$ are positive. A straightforward choice is to set $L_t = (C_{t+1} B_t)^T$. Consequently, the direction regulation matrix L is given by $L = \text{diag}\{L_0, L_1, \ldots, L_{N-1}\} \Psi^T$.

4.2.4.2 Convergence Analysis for Input Sequence

The convergence of the generalized scheme (4.42) is provided in the following theorem.

Theorem 4.4 *Consider the system* (4.27) *under Assumptions 4.6 and 4.7. For the point-to-point tracking reference* $\boldsymbol{y}_r$*, apply the accelerated learning control scheme*

(4.42), *in which the direction regulation matrix L satisfies the condition (C1). Starting from the initial input* $\boldsymbol{u}_0$, *the input sequence* $\{\boldsymbol{u}_k\}$ *converges to* $\boldsymbol{u}^\circ$, *which is defined by* (4.43).

Proof. This theorem can be proven in a similar manner to Theorem 4.3. Therefore, we present only the key steps and highlight the significant modifications. Firstly, following a similar approach as the proof of Theorem 4.3, we employ mathematical induction to establish that $\boldsymbol{u}_k - \boldsymbol{u}_0$ belongs to the column space of L for all iterations, i.e., $\boldsymbol{u}_k - \boldsymbol{u}_0 \in \text{Span}\{L\}, \forall k$.

By the condition (C1), the matrix ΨHL is of full rank and thus L must be of full column rank. Then, by the linear representation theory, there exists a unique vector $\bar{\phi}_k \in \mathbf{R}^m$ such that $\boldsymbol{u}_k - \boldsymbol{u}_0 = L\bar{\phi}_k$. Substituting it into (4.42) yields

$$\bar{\phi}_{k+1} = \bar{\phi}_k + r_k^{-1}\boldsymbol{e}_k, \tag{4.44}$$

where the full column rank property of L is applied again. Then, substituting the tracking error expression yields

$$\begin{aligned}\bar{\phi}_{k+1} &= \bar{\phi}_k + r_k^{-1}(\boldsymbol{y}_r - \Psi \boldsymbol{y}_k)\\ &= \bar{\phi}_k - r_k^{-1}[\Psi HL\bar{\phi}_k - (\boldsymbol{y}_r - \Psi H\boldsymbol{u}_0)] - r_k^{-1}\Psi\boldsymbol{\varpi}_k,\end{aligned}$$

which differs from (4.39) in that $H^T\Psi^T$ is replaced by L.

We define $\Gamma_e = \Psi HL$ and $\bar{\theta}_k = \bar{\phi}_k - \Gamma_e^{-1}\eta$. Then,

$$\bar{\theta}_{k+1} = \bar{\theta}_k - r_k^{-1}\Gamma_e\bar{\theta}_k - r_k^{-1}\Psi\boldsymbol{\varpi}_k. \tag{4.45}$$

The following steps are similar to those of Theorem 4.3 by using Lemma 4.1. The verification of Assumption 4.4 is identical, but the verification of Assumption 4.5 is modified as follows because Γ_e is no longer a positive definite matrix.

First, $\bar{\theta}_k$, $\Gamma_e\bar{\theta}_k$, and $\Psi\boldsymbol{\varpi}_k$ correspond to $\bar{x}_k$, $\varphi(\bar{x}_k)$, and ϵ_k in Lemma 4.1, respectively. For the deterministic recursion, we have $\bar{\theta}_{k+1} = (I - \gamma\Gamma_e)\bar{\theta}_k$. Γ_e is not positive definite. However, for a sufficiently small α, it is clear that the eigenvalues of $I - \gamma\Gamma_e$ are positive and smaller than 1. Then, $\bar{\theta}_k \to 0$.

Then, define a Lyapunov function $V(\bar{\theta}) = \bar{\theta}^T\Xi\bar{\theta}$, where Ξ is a positive-definite matrix such that $(\Psi HL)^T\Xi + \Xi\Psi HL = I$. The existence and uniqueness of Ξ are ensured by Lyapunov equation theory under the condition (C1). We can easily verify that $\nabla^2 V(\bar{\theta}) = I_m \leq J$ with $J = 2\Xi$ for all $\bar{\theta}$. Moreover,

$$\begin{aligned}\varphi(\bar{\theta})^T\nabla V(\bar{\theta}) &= 2\bar{\theta}^T(\Gamma_e)^T\Xi\bar{\theta}\\ &= \bar{\theta}^T(\Gamma_e)^T\Xi\bar{\theta} + \bar{\theta}^T\Xi\Gamma_e\bar{\theta} = \bar{\theta}^T\bar{\theta}.\end{aligned}$$

Note that $\varphi(\bar{\theta})^T J\varphi(\bar{\theta}) = 2\bar{\theta}^T\Gamma_e^T\Xi\Gamma_e\bar{\theta} \leq 2\lambda_{\max}(\Gamma_e^T\Xi\Gamma_e)\cdot\bar{\theta}^T\bar{\theta}$. For $\|\bar{\theta}\| \geq \bar{R}$, we have $\bar{\theta}^T\bar{\theta} \geq \bar{R}^2$. Thus, we let $\alpha < \min\left\{\frac{\bar{R}^2}{\text{trace}(2\Psi Q\Psi^T)}, \frac{1}{4\lambda_{\max}(\Gamma_e^T\Xi\Gamma_e)}\right\}$ and $\beta < \frac{\bar{R}^2}{4}$.

The inequalities can then be verified. Thus, by Lemma 4.1, we conclude that $\bar{\theta}_k \to 0$, implying that $\boldsymbol{u}_k \to \boldsymbol{u}^\circ$.

The proof sketch consists of two parts: the former part elucidates the characteristics of the input increment at each iteration, while the latter part addresses the analytical distinction caused by relaxing condition (C1). Theorem 4.4 demonstrates the effectiveness of the generalized learning control scheme (4.42), where the gradient matrix in (4.32) is replaced with a direction regulation matrix L. While retaining convergence to the desired set $\mathcal{S}_u$, this relaxation simultaneously alters the iteration path of the input sequence. Specifically, the input sequence $\boldsymbol{u}_k - \boldsymbol{u}_0$ generated by (4.42) remains confined within a subspace spanned by the direction regulation matrix. In other words, under this constraint, the convergence limit $\boldsymbol{u}^\circ$ is closest to the initial input $\boldsymbol{u}_0$. Among all possible convergence limits with respect to the direction regulation matrix L, $\boldsymbol{u}^*$ contributes minimally to distance. This represents an advantage of utilizing accelerated gradient algorithm when having full knowledge of system dynamics.

4.2.4.3 Alternative Analysis of the Tracking Error

According to Theorem 4.4, it can be directly inferred that the mean square of tracking errors will converge as the iteration number increases. Specifically, $\mathbb{E}[\|\boldsymbol{e}_k\|^2] \to \text{trace}(\Psi Q \Psi^T)$, where the latter represents optimal tracking performance due to its dependence solely on the measurement noise of the current iteration. In this subsection, we present an alternative analysis technique that omits checking input convergence, thereby completing the asymptotic analysis of the proposed learning control schemes. Introducing an auxiliary error term $\boldsymbol{e}'_k \triangleq \boldsymbol{y}_r - \Psi H \boldsymbol{u}_k$, we have $\boldsymbol{e}_k = \boldsymbol{e}'_k + \Psi \boldsymbol{\varpi}_k$. Therefore, it suffices to demonstrate that $\mathbb{E}[\|\boldsymbol{e}'_k\|^2] \to 0$.

Corollary 4.2 *Consider the system* (4.27) *under Assumptions 4.6 and 4.7. For the point-to-point tracking reference* $\boldsymbol{y}_r$, *apply the accelerated learning control scheme* (4.42), *in which L satisfies the condition (C1). Then, the auxiliary tracking error* $\boldsymbol{e}'_k$ *converges to zero in the mean-square sense.*

Proof. Multiplying both sides of (4.42) from the left by ΨH and subtracting them from $\boldsymbol{y}_r$, we obtain

$$\boldsymbol{e}'_{k+1} = \boldsymbol{e}'_k - r_k^{-1} \Psi H L (\boldsymbol{e}'_k - \Psi \boldsymbol{\varpi}_k). \tag{4.46}$$

Note that Lemma 4.1 cannot be applied to this recursion because the gain adaptation parameter is not generated by the innovation term $\Psi H L \boldsymbol{e}_k$, which essentially differs from (4.17). However, by Lemma 4.1 and Theorem 4.4, it can be determined that $\lim_{k\to\infty} \frac{r_k}{k} \to \mathbb{P}(\boldsymbol{\varpi}_1^T \Psi^T \Psi \boldsymbol{\varpi}_2 < 0) > 0$ almost surely, i.e., the ratio of r_k over the iteration number k converges to a constant. Therefore, $1/r_k > 0$, $1/r_k \to 0$ and $\sum_{k=1}^{\infty} 1/r_k = \infty$. Denote $\rho_k = 1/r_k$. The proof of this proposition is completed by constructing a Lyapunov function and establishing the contraction inequality directly.

We define a Lyapunov function $\mathcal{V}_k = (\boldsymbol{e}'_k)^T \Xi (\boldsymbol{e}'_k)$, where Ξ satisfies $(\Psi HL)^T \Xi + \Xi \Psi HL = I$. Denote $\Phi \triangleq \Psi HL$. Then,

$$\begin{aligned}\mathcal{V}_{k+1} &= (\boldsymbol{e}'_{k+1})^T \Xi (\boldsymbol{e}'_{k+1}) \\ &= \mathcal{V}_k + \rho_k^2 (\boldsymbol{e}'_k)^T \Phi^T \Xi \Phi (\boldsymbol{e}'_k) + \rho_k^2 \boldsymbol{\varpi}_k^T \Psi^T \Phi^T \Xi \Phi \Psi \boldsymbol{\varpi}_k \\ &\quad - \rho_k (\boldsymbol{e}'_k)^T (\Phi^T \Xi + \Xi \Phi)(\boldsymbol{e}'_k) + 2\rho_k (\boldsymbol{e}'_k)^T \Xi \Psi \boldsymbol{\varpi}_k \\ &\quad - 2\rho_k^2 (\boldsymbol{e}'_k)^T \Phi^T \Xi \Phi \Psi \boldsymbol{\varpi}_k. \end{aligned} \tag{4.47}$$

We introduce an increasing σ-algebra $\mathcal{F}_k = \sigma\{\boldsymbol{u}_0, x_i(0), \boldsymbol{\varpi}_i, 1 \le i \le k\}$, indicating the set of all possible events up to the kth iterations. Based on the scheme (4.42), it is evident that $\boldsymbol{u}_k \in \mathcal{F}_{k-1}$, i.e., $\mathbb{E}[\boldsymbol{u}_k|\mathcal{F}_{k-1}] = \boldsymbol{u}_k$ and $\mathbb{E}[\boldsymbol{e}'_k|\mathcal{F}_{k-1}] = \boldsymbol{e}'_k$. In addition, by Assumption 4.7, $\boldsymbol{\varpi}_k$ is independent of $\mathcal{F}_{k-1}$. Thus, $\mathbb{E}[\boldsymbol{\varpi}_k|\mathcal{F}_{k-1}] = 0$. Applying this result, we obtain

$$\mathbb{E}[(\boldsymbol{e}'_k)^T \Phi^T \Xi \Phi (\boldsymbol{e}'_k)|\mathcal{F}_{k-1}] \le d_1 \mathcal{V}_k, \tag{4.48}$$

$$\mathbb{E}[\boldsymbol{\varpi}_k^T \Psi^T \Phi^T \Xi \Phi \Psi \boldsymbol{\varpi}_k|\mathcal{F}_{k-1}] \le d_2, \tag{4.49}$$

$$\mathbb{E}[(\boldsymbol{e}'_k)^T \Xi \Psi \boldsymbol{\varpi}_k|\mathcal{F}_{k-1}] = 0, \tag{4.50}$$

$$\mathbb{E}[(\boldsymbol{e}'_k)^T \Phi^T \Xi \Phi \Psi \boldsymbol{\varpi}_k|\mathcal{F}_{k-1}] = 0, \tag{4.51}$$

where $d_1 > 0$ satisfies $\Phi^T \Xi \Phi \le c_1 \Xi$, and $d_2 > 0$ is a constant. Moreover, there is a suitable constant $d_3 > 0$ such that $d_3 \Xi \le I$. Thus, we have

$$\begin{aligned}&\mathbb{E}[(\boldsymbol{e}'_k)^T (\Phi^T \Xi + \Xi \Phi)(\boldsymbol{e}'_k)|\mathcal{F}_{k-1}] \\ &\qquad = (\boldsymbol{e}'_k)^T (\boldsymbol{e}'_k) \ge d_3 \mathcal{V}_k. \end{aligned} \tag{4.52}$$

Taking the conditional expectation to both sides of (4.47) with respect to $\mathcal{F}_{k-1}$ and substituting (4.48)-(4.52) into the resulted inequality, we obtain $\mathbb{E}[\mathcal{V}_{k+1}|\mathcal{F}_{k-1}] \le (1 - d_3\rho_k + d_1\rho_k^2)\mathcal{V}_k + d_2\rho_k^2$. Taking the mathematical expectation to both sides of this inequality yields $\mathbb{E}[\mathcal{V}_{k+1}] \le (1 - d_3\rho_k + d_1\rho_k^2)\mathbb{E}[\mathcal{V}_k] + d_2\rho_k^2$. Corresponding $\mathbb{E}[\mathcal{V}_k]$, $d_3\rho_k - d_1\rho_k^2$, and $d_2\rho_k^2$ to a_k, b_k, and c_k in Lemma 2.1, respectively, we can verify the conditions of the lemma without any difficulty. Thus, by Lemma 2.1, we obtain $\mathbb{E}[\mathcal{V}_k] \to 0$. Further, $\mathbb{E}[\|\boldsymbol{e}'_k\|^2] \le \lambda_{\max}(\Xi)[\mathcal{V}_k]$, which leads to $\mathbb{E}[\|\boldsymbol{e}'_k\|^2] \to 0$. The proof is completed.

The learning control theory of the point-to-point tracking problem is completed by Corollary 4.2, which directly demonstrates the zero-error convergence of the auxiliary tracking errors. Notably, this convergence is independent of the analysis of the input sequence. While this perspective towards performance analysis has been considered in existing literature, it has not yet been applied to stochastic point-to-point tracking problems. By summarizing the results from Sects. 4.2.3 and 4.2.4, this subsection successfully proposes an integrated analysis framework for point-to-point tracking problems using an accelerated scheme.

4.2.4.4 Discussions on the Schemes

In this subsection, we illustrate the connections and distinctions between the accelerated gradient algorithm and the generalized one, which reflect the impact of both the gradient matrix $H^T\Psi^T$ and direction regulation matrix L. Specifically, as matrix $H^T\Psi^T$ satisfies condition (C1), the accelerated gradient algorithm can be regarded as a special case of the generalized scheme.

The illustration of the abstract convergence routes for the proposed schemes is presented in Fig. 4.5. The plane represents the space $\mathbf{R}^{pN}$, which corresponds to the lifted input vector $\boldsymbol{u}_k$. All input vectors, including the initial input $\boldsymbol{u}_0$ and the limit $\boldsymbol{u}^*$, are represented as points on this plane. The shaded zone indicates the solution set $\mathcal{S}_u$. The lines connecting from the initial input $\boldsymbol{u}_0$ to the solution set $\mathcal{S}_u$ represent the convergence routes of schemes (4.32) and (4.42), respectively. Each line signifies a potential evolutionary path within this specified space. In other words, each iteration's incremental part must lie within the subspace spanned by its corresponding scheme's direction regulation matrix.

We first explain the accelerated gradient algorithm. Section 4.2.3 explains that the convergence limit $\boldsymbol{u}^*$ has the minimal distance from the initial input, i.e., $\|\boldsymbol{u}^* - \boldsymbol{u}_0\|^2 = \min_{\boldsymbol{u}\in\mathcal{S}_u} \|\boldsymbol{u} - \boldsymbol{u}_0\|^2$. This is demonstrated by the straight line from $\boldsymbol{u}_0$ to $\boldsymbol{u}^*$ in Fig. 4.5 and illustrates the main advantage of the gradient algorithm. The incremental part of the accelerated gradient algorithm (4.32) always belongs to the space spanned by $H^T\Psi^T$.

The generalized scheme (4.42) is illustrated by the curve from $\boldsymbol{u}_0$ to $\boldsymbol{u}^*$. This curve is determined by the direct regulation matrix L in the sense that the incremental part of each iteration must belong to the space spanned by L. Thus, the gradient $H^T\Psi^T$ is a special case of L. Further, if the spaces spanned by L and $H^T\Psi^T$ are nonidentical, the convergence route of (4.42) certainly differs from that of (4.32). Accordingly,

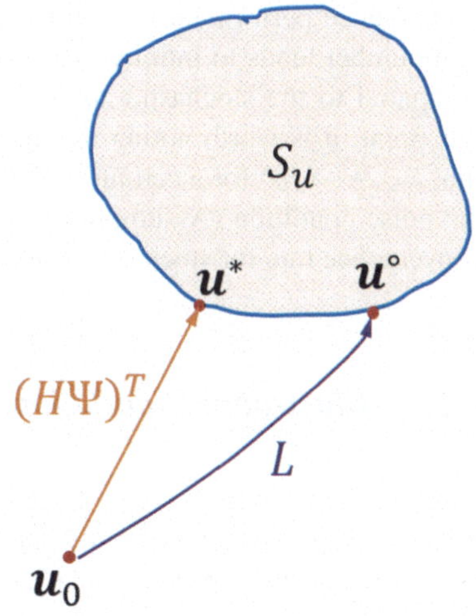

Fig. 4.5 Convergence routes for different schemes

the convergence limit $\boldsymbol{u}^{\circ}$ differs from $\boldsymbol{u}^*$. This observation explains how the matrix L plays the role of control direction regulation. Meanwhile, it reflects the relationship between the accelerated gradient algorithm and the generalized scheme.

In addition, the distance $\|\boldsymbol{u}^{\circ} - \boldsymbol{u}_0\|^2$ cannot be smaller than $\|\boldsymbol{u}^* - \boldsymbol{u}_0\|^2$ because the latter is the minimal distance. A natural problem arises: why we still study the generalized scheme (4.42) since the distance is not minimized? In fact, the distance between the initial input and convergence limit does not correspond to the convergence rate of the proposed algorithms. Although using the gradient $H^T\Psi^T$ leads to the minimal distance, the convergence rate of the accelerated gradient algorithm (4.32) may be lower than that of the generalized scheme (4.42) when the direction regulation matrix L is well tuned. This is because the convergence rate is not only determined by the gain sequence r_k but also by the matrix ΨHL. Roughly speaking, if $H^T\Psi^T$ is applied, the eigenvalues of $\Psi HH^T\Psi^T$ are primarily determined by the system matrices and reference positions. They may be considerably large or small; in both cases, this can cause the matrix to have a very large conditional number. Accordingly, the convergence rate of (4.32) can be difficult to tune according to the application. In contrast, for the generalized scheme (4.42), an appropriately selected matrix L can improve the eigenvalue distribution of ΨHL, which may improve the convergence rate.

The accelerated gradient algorithm (4.32) guarantees the convergence of the generated input sequence to the point in $\mathcal{S}_u$ that is closest to the initial input, thereby potentially reducing energy consumption. In contrast, the generalized scheme (4.42) does not guarantee minimal distance of the input. However, by designing a direction regulation matrix L, it is possible to significantly improve the convergence rate.

Before concluding this subsection, we would like to provide some remarks on the convergence speed of the proposed schemes. Compared to the stochastic approximation scheme with a prior given decreasing gain sequence (Xu et al. 2017), our approach significantly accelerates the convergence speed during early iterations. In other words, by introducing the adaptive gain r_k, we have improved the transient convergence performance. As for the asymptotic convergence rate when the iteration number tends to infinity, our proposed scheme achieves superior performance compared to the stochastic approximation scheme (Xu et al. 2017). To illustrate this point, it is worth noting that according to Lemma 4.1 and the main theorem, $\lim_{k\to\infty}\frac{r_k}{k} \to p^*$ for a certain probability $p^* > 0$, implying that $r_k = O(k)$. Under the noise condition (Assumption 4.7), our approach attains an optimal asymptotic convergence rate for stochastic control.

4.2.5 Illustrative Simulations

In this section, we apply the proposed schemes to the permanent magnet synchronous motor example, which is a multi-input-multi-output model. The system is given by the following state-space model (Xu and Jin 2013):

$$\begin{aligned}\dot{x}_k &= A_t x_k + B u_k,\\ y_k &= C x_k + \xi_k,\end{aligned}$$

where $x = [i_r, i_s]^T$ are the r-axis and s-axis currents to be controlled. In addition,

$$\begin{aligned}A_t &= \begin{bmatrix} -\frac{R_t}{L_r} & \frac{L_s\omega}{L_r} \\ -\frac{L_r\omega}{L_s} & -\frac{R_t}{L_s}\end{bmatrix}, \quad B = \begin{bmatrix} \frac{1}{L_r} & 0 \\ 0 & \frac{1}{L_s}\end{bmatrix},\\ u_k &= [u_r, u_s]^T, \quad C = I,\end{aligned}$$

where L_r, L_s, and R_t are the inductances and resistance of the stator, and u_k is the control input voltage. Thus, the system dimension is $p = q = 2$. The simulation settings are given as follows: The iteration length is $T = 10\,\mathrm{s}$, and the sampling period is $h = 0.04\,\mathrm{s}$. Thus, $N = 250$. Similar to Xu and Jin (2013), a time-varying resistance R_t is considered $R_t = R_0(1.4 - 0.4e^{-t})$, where $R_0 = 0.5\,\Omega$. Further, $L_r = L_{r0}/1 + 0.1$ and $L_s = L_{s0}/1 + 0.1$, where $L_{r0} = L_{s0} = 1\,\mathrm{H}$. Next, we set $\omega = 0.1$. The measurement noise ξ_k follows a normal distribution with covariance τ^2. Here, τ reflects the noise level and considerably affects the learning process.

The full reference currents to be tracked are represented by $0.5\sin(2\pi t/T)$ and $1.5\sin(2\pi t/T)$ for the first and second dimensions of the output, respectively. However, for point-to-point tracking problems, only a subset of the references need to be tracked. We randomly select 100 integers from $\{1, 2, \ldots, N\}$, indicating the positions that are required for the first dimension of the output. Then, we perform the same operation for the second dimension. Thus, the actual tracking reference is a set of 100 points for each dimension.

4.2.5.1 Accelerated Gradient Algorithm

In this subsection, we apply the accelerated gradient algorithm (4.32) and compare it with the traditional stochastic approximation-type algorithm given in Shen et al. (2017); Shen and Qu (2020); Shen et al. (2021).

As stated above, the point-to-point tracking references have 100 points for each dimension. Thus, the lifting sub-reference $\boldsymbol{y}_r$ has dimension 200. Applying the accelerated gradient algorithm (4.32) and focusing on the recursion of (4.39), it can be seen that $\Psi H H^T \Psi^T$ is a square matrix with dimension 200. After calculation, we find that the eigenvalues of this matrix are distributed from approximately 0.2 to 110, i.e., the condition number of $\Psi H H^T \Psi^T$ is very large. Meanwhile, a large eigenvalue (approximately 110) can yield an undesired learning process. Thus, we add an additional gain ζ to algorithm (4.32) to ensure an appropriate learning process. It is merely a practical technique for implementation without affecting the theoretical results. Generally, additional gain ζ is required to make all eigenvalues of $\zeta \Psi H H^T \Psi^T$ be approximately within the range (0, 1). In particular, we use $\boldsymbol{u}_{k+1} = \boldsymbol{u}_k + \frac{\zeta}{r_k} H^T \Psi^T \boldsymbol{e}_k$, where $\zeta = 0.01$ is employed to avoid an unacceptable overshoot of the algorithm in

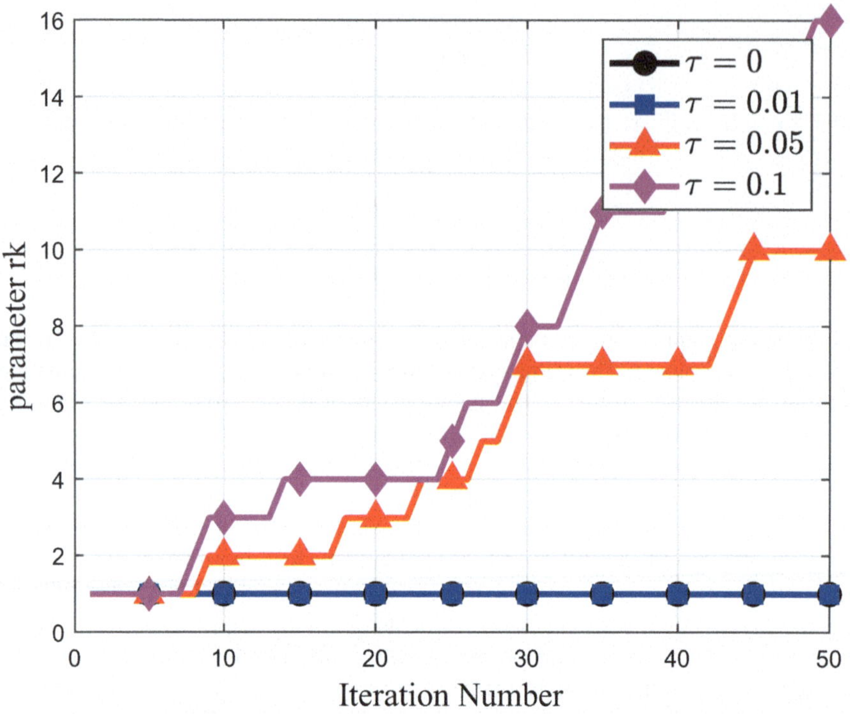

Fig. 4.6 Evolution of the adaptive parameter r_k: τ denotes the noise level

the iteration domain. For comparison, we also apply the stochastic approximation algorithm $\boldsymbol{u}_{k+1} = \boldsymbol{u}_k + \frac{\zeta}{k} H^T \Psi^T \boldsymbol{e}_k$, in which the decreasing gain $1/k$ is independent of the practical tracking performance, and the traditional gradient algorithm $\boldsymbol{u}_{k+1} = \boldsymbol{u}_k + \zeta H^T \Psi^T \boldsymbol{e}_k$. Each algorithm is run for 50 iterations. Four noise levels are considered: τ =0.01, 0.05, 0.1, and 0.

We first demonstrate the principle of the acceleration mechanism. Figure 4.6 illustrates the evolution of the adaptive parameter r_k within the iteration domain. When $\tau = 0$, indicating absence of measurement noise, r_k remains constantly equal to 1. This signifies that the acceleration mechanism is never triggered and the proposed algorithm functions as a traditional contraction mapping algorithm. In this scenario, exponential convergence towards zero tracking error is observed. However, when $\tau \neq 0$, the acceleration mechanism is activated at a specific iteration depending on the level of noise present. Notably, lower noise levels result in an increase in r_k starting from a larger number of iterations, implying that during its initial stages, the proposed algorithm maintains a constant gain and exhibits rapid learning speed. Conversely, higher noise levels lead to an earlier initiation of the acceleration mechanism. By evaluating input bias and measurement noise ratio, we can ascertain how much dependence exists between practical learning performance and activation of the acceleration mechanism.

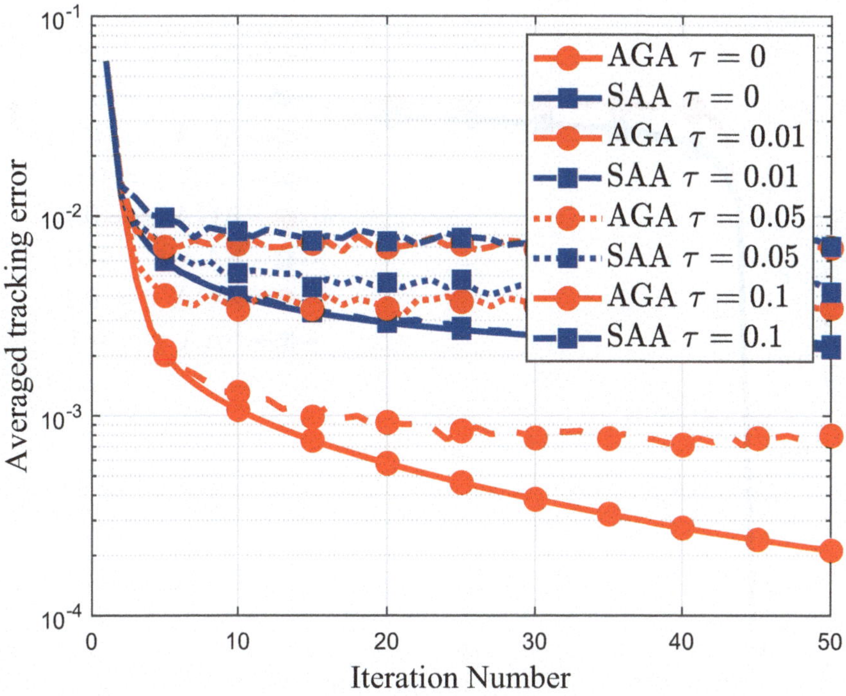

Fig. 4.7 Averaged tracking error profiles: AGA, accelerated gradient algorithm; SAA, stochastic approximation algorithm

The averaged tracking error profiles for both the accelerated gradient and stochastic approximation algorithms are shown in Fig. 4.7. Here, the averaged tracking error is defined as $\|\boldsymbol{e}_k\|^2/m$, where $m = 200$ represents the dimension of lifting sub-reference $\boldsymbol{y}_r$. Based on this figure, two observations can be made. Firstly, under the same noise level, the accelerated gradient algorithm exhibits faster convergence compared to the stochastic approximation algorithm. Secondly, for each noise level, the accelerated gradient algorithm ensures that the fluctuation levels of tracking errors align with those of measurement noise. Consequently, it can be concluded that the proposed algorithm achieves optimal performance.

The input distance from the initial input, defined as $\|\boldsymbol{u}_k - \boldsymbol{u}_0\|^2$ for each iteration, is presented in Fig. 4.8. The proposed algorithm exhibits a larger distance compared to the stochastic approximation algorithm, indicating its superior ability to improve the input. Furthermore, the distance profiles show significant overlap across different noise levels, suggesting that both algorithms have effectively enhanced the input.

In these simulations, an additional gain ζ is introduced to regulate the range of eigenvalues of $\Psi H H^T \Psi^T$ in order to prevent excessive overshoot during early iterations. The selection of ζ requires prior knowledge of the system information H. However, if the precise model is unavailable, a conservative gain can be assigned

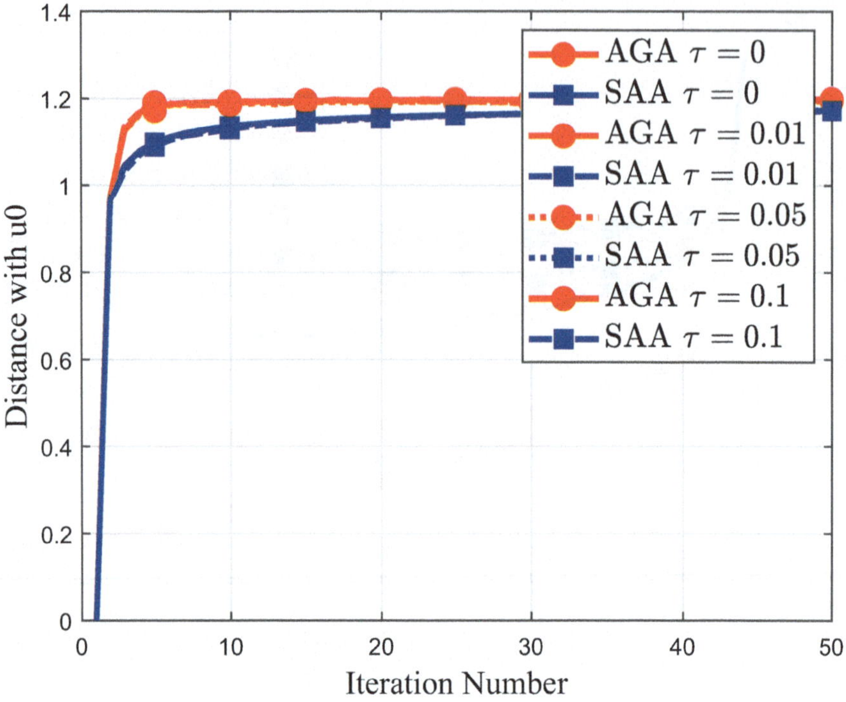

Fig. 4.8 Distance of the inputs $\boldsymbol{u}_k$ from the initial input $\boldsymbol{u}_0$: AGA, accelerated gradient algorithm; SAA, stochastic approximation algorithm

based on a rough estimate of H. To investigate the impact of ζ on convergence performance, we conducted simulations for four scenarios with different values of ζ, namely $0.01, 0.005, 0.001$, and 0.03, respectively. The averaged tracking error profiles of the accelerated gradient algorithms under a noise level ($\tau = 0.01$) are presented in Fig. 4.9. It can be observed that as ζ decreases from 0.01, the convergence speed slows down; however, adopting a large value for ζ may lead to overshoot during early iterations.

We note that the accelerated gradient algorithm is a variant of the traditional gradient algorithm $\boldsymbol{u}_{k+1} = \boldsymbol{u}_k + \zeta H^T \Psi^T \boldsymbol{e}_k$ by introducing an adaptive parameter to improve the performance. One may be interested in what promising strengths the proposed scheme may have compared with tradition gradient algorithm. Indeed, there are two advantages. First, the adaptive parameter can help relax the selection range of ζ, which is beneficial for the imprecise model case. To seen this point, we simulate both algorithms for two scenarios of ζ: $\zeta = 0.01$ and 0.02. The averaged tracking error profiles are given in Fig. 4.10. For the scenario $\zeta = 0.01$ in Fig. 4.10a where the eigenvalues are well tuned, the proposed scheme achieves the same convergence rate to the traditional gradient algorithm. Especially, for the scenario $\tau = 0$, the profiles of both algorithms completely overlap with each other, which confirms the conclusion

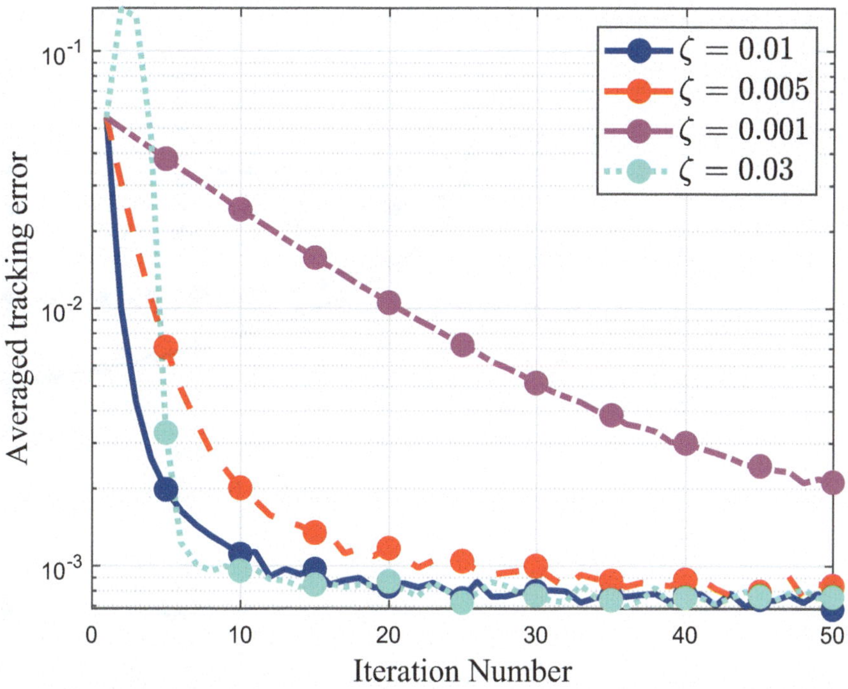

Fig. 4.9 Averaged tracking error profiles for the accelerated gradient algorithm with different ζ

that r_k is constant in the absence of measurement noise. In contrast, for the scenario $\zeta = 0.02$ in Fig. 4.10b, the traditional gradient algorithm yields fast divergence of the tracking errors, while the proposed scheme can still ensure perfect convergence and favorite tracking performance. Second, the adaptive parameter can improve the input evolutionary behavior in the iteration domain. To demonstrate this point, we present the input distance profiles of both algorithms in Fig. 4.11 for the scenario $\zeta = 0.01$ where the tracking performance are almost the same. From the amplified subplot in Fig. 4.11, it is seen that the input profiles generated by the traditional gradient algorithm fluctuate significantly larger than those of the proposed scheme. In other words, the proposed scheme achieves a more steady improvement of the input sequence.

4.2.5.2 Generalized Scheme

In this subsection, we consider the generalized scheme (4.42) and compare it with (4.32). For the generalized scheme (4.42), L is set to $L = \text{diag}\{L_0, L_1, \ldots, L_{N-1}\}\Psi^T$ with $L_t = 0.2(C_{t+1}B_t)^T$; then, the eigenvalues of ΦHL are equal to 0.1653. Due to the continuity of the eigenvalue function with respect to its matrix entries, a rough estimate of the coupling matrix $C_{t+1}B_t$ will not affect the performance significantly.

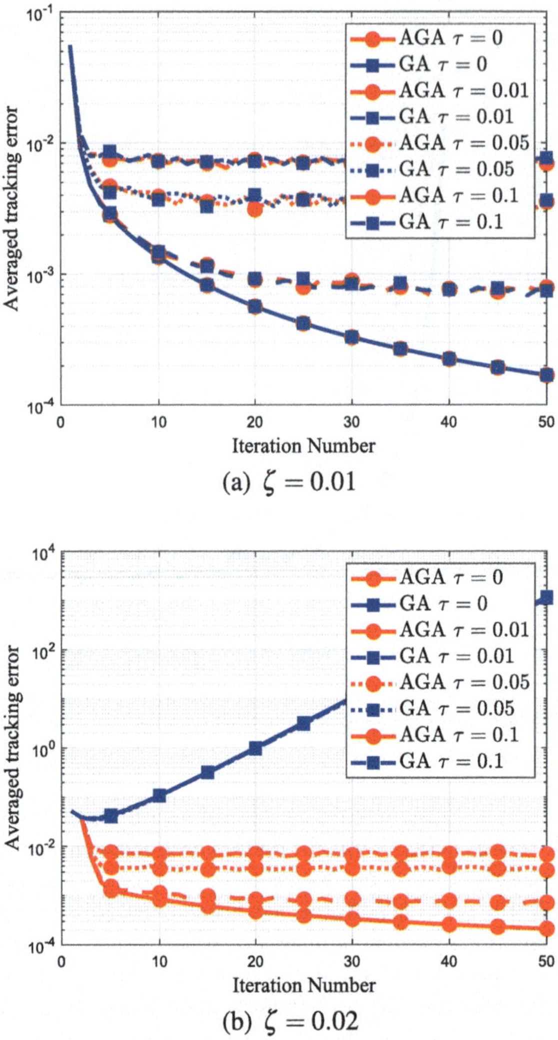

Fig. 4.10 Averaged tracking error profiles: AGA, accelerated gradient algorithm; GA, gradient algorithm

The accelerated gradient algorithm formulated in the previous subsection is employed for a comparison. Each algorithm is run for 50 iterations. The noise levels are 0.01, 0.05, and 0.1.

The averaged tracking error profiles of both algorithms for three noise levels are shown in Fig. 4.12. Both algorithms demonstrate similar tracking performance after a few iterations as they converge to the solution set $\mathcal{S}_u$. For lower noise levels, the generalized scheme achieves superior tracking performance after several learning iterations, potentially due to the effective eigenvalue regulation by matrix L. In general, both algorithms can achieve satisfactory tracking performance for various applications.

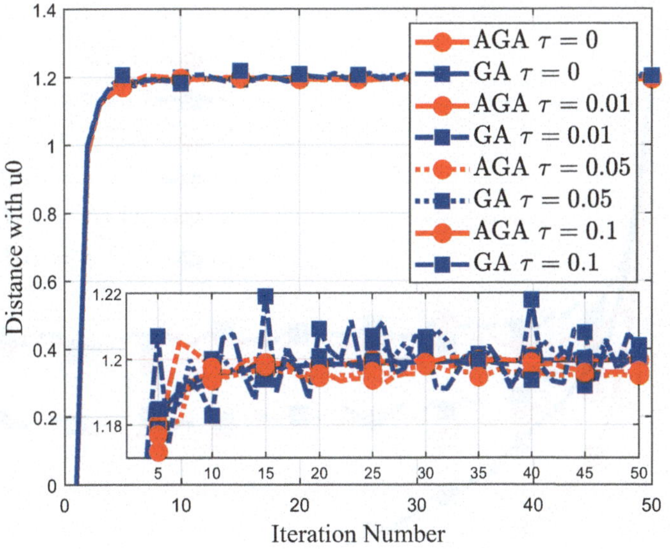

Fig. 4.11 Distance of the input $\boldsymbol{u}_k$ from the initial input $\boldsymbol{u}_0$: AGA, accelerated gradient algorithm; GA, gradient algorithm

The input distance from the initial input is demonstrated in Fig. 4.13. Several observations can be made. Firstly, it is evident that the generalized scheme has a larger input distance compared to the accelerated gradient algorithm, thus confirming the conclusion that the latter converges to the closest limit point in set $\mathcal{S}_u$. Furthermore, regardless of different noise levels, the input distance profiles remain almost identical. This indicates that improvements in input are primarily determined by the algorithm or evolutionary space and are seldom affected by random noise.

4.3 Adaptive Gain Design in Markov Nonuniform Trial Lengths Problem

4.3.1 Markov Nonuniform Trial Lengths Problem

In various practical systems such as industrial processes, rehabilitation systems, cruise surveillance, and robots, the experiences from previous iterations can be utilized to enhance the performance of subsequent iterations. However, existing literature often assumes an invariant iteration length for all iterations, assuming complete information availability in each iteration. Nevertheless, in real-world applications, this assumption is often unattainable due to unknown uncertainties and safety considerations. For instance, adaptive ILC has been successfully employed in func-

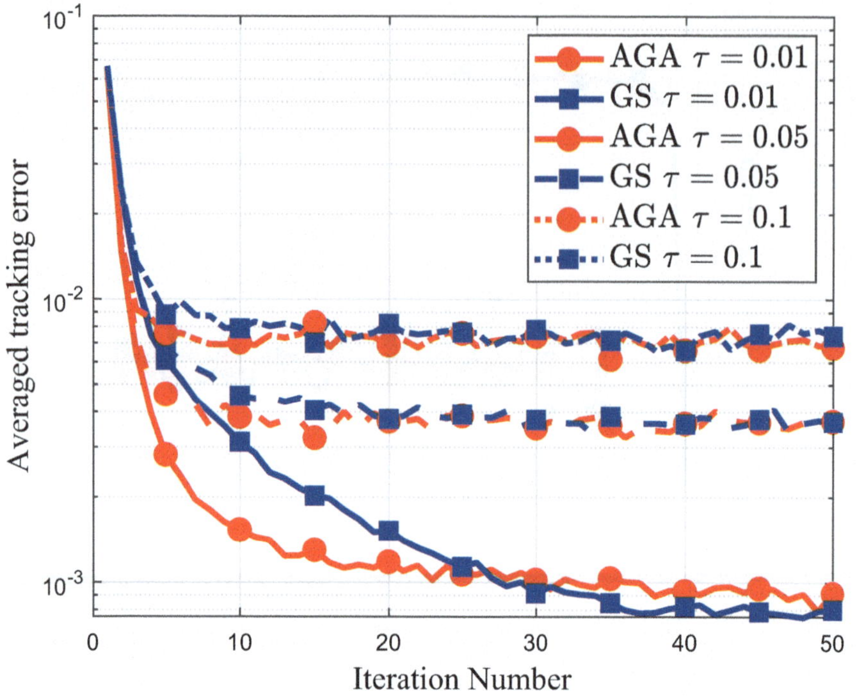

Fig. 4.12 Averaged tracking error profiles: AGA, accelerated gradient algorithm; GS, generalized scheme

tional electrical stimulation for upper limb movement and gait assistance (Seel et al. 2016b, a), where experimental results demonstrate that the actual operation process may prematurely terminate when the output significantly deviates from the desired reference. Another example is provided in Longman and Mombaur (2006) for humanoid and biped walking robots featuring periodic or quasi-periodic gaits with varying phase durations between iterations. These examples highlight that iteration lengths can vary randomly in practice, posing a challenging problem. Specifically, if an iteration ends earlier than intended lengthwise, there will be missing information required for updating purposes which necessitates appropriate compensation.

In addressing this issue, it is crucial to incorporate the stochastic nature of varying lengths. Most existing literature employs an iteration-independent random variable that may follow a specific probability distribution. By virtue of its independence, the randomness associated with varying lengths can be eliminated by applying an expectation to relevant quantities. Numerous approaches have been proposed to tackle this problem.

The first approach is the conventional contraction principle (Li and Shen 2017; Li et al. 2014, 2015; Meng and Zhang 2018; Shen et al. 2016b; Shi et al. 2016), where several studies require prior statistical knowledge of the random lengths (Li

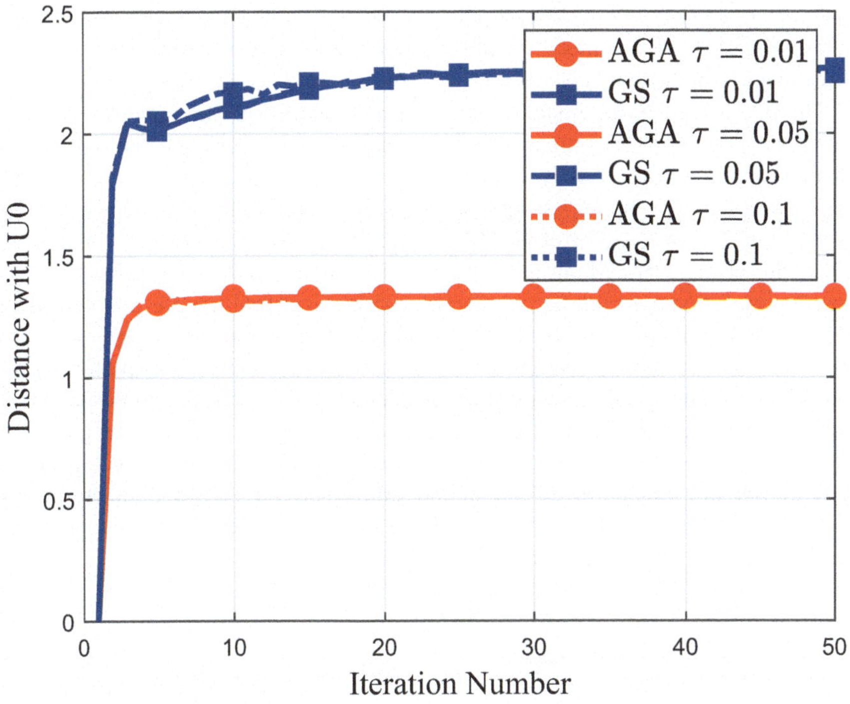

Fig. 4.13 Distance of the inputs $\boldsymbol{u}_k$ from the initial input $\boldsymbol{u}_0$: AGA, accelerated gradient algorithm; GS, generalized scheme

et al. 2014, 2015; Shi et al. 2016). However, this approach can be costly and unsafe without effective control, making it generally impractical. Averaging operators are often introduced to compensate for the lack of information (Li and Shen 2017; Li et al. 2014, 2015; Shi et al. 2016). Furthermore, we consider linear and affine systems with bounded noise and establish a bounded convergence property for tracking errors (Meng and Zhang 2018; Shi et al. 2016).

The second approach, proposed in Shen et al. (2016a), introduces a novel recursive computation method for determining the mean and variance matrix of the input error. This enables the convergence of tracking errors to zero in various contexts; however, it is primarily suitable for noise-free linear systems. The third approach applies a modified composite energy function to both parametric and nonparametric continuous-time nonlinear systems (Shen and Xu 2019a, b), where the random varying trial length is modeled using a general distribution function. Summarizing existing results reveals that iteration-independence of varying lengths is necessary; otherwise, additional requirements must be imposed to ensure full-length occurrence. However, applications such as Seel et al. (2016b, a); Longman and Mombaur (2006) demonstrate that nonuniform trial length may depend on system states or controller performance (Seel et al. 2011); in other words, trial length can be iteration-dependent.

In this section, we employ a Markov chain model to describe this circumstance since it has proven effective in modeling limited dependence across various environments (Meyn and Tweedie 2012).

The ILC for stochastic systems with Markov nonuniform trial lengths is investigated in this section. In comparison to existing literature, our contributions can be summarized as follows: Firstly, a Markov chain is introduced to model the randomness in trial lengths, encompassing both random and deterministic models proposed in previous works (e.g., Li et al. 2014, 2015; Shen et al. 2016a, b; Li and Shen 2017; Shi et al. 2016; Meng and Zhang 2018; Shen and Xu 2019a, b). Notably, the proposed model accommodates iteration-dependent trial lengths and considers random measurement noise. Secondly, iteration-varying gains are incorporated into the conventional P-type algorithm for controller design to enhance its performance. Specifically, update- and performance-dependent adaptive gains are proposed to further improve tracking accuracy. Convergence conditions for both types of gains are also derived. Lastly, a Lyapunov function-based approach is employed for analysis purposes to establish mean-square and almost-sure convergence under various sources of randomness. The presence of iteration-dependency in relevant quantities poses significant challenges during the analysis process.

4.3.2 Problem Formulation

Consider the following linear time-varying system

$$\begin{aligned} x_k(t+1) &= A_t x_k(t) + B_t u_k(t), \\ y_k(t) &= C_t x_k(t) + w_k(t), \end{aligned} \tag{4.53}$$

where the subscript k denotes the iteration number, $k = 1, 2, \ldots$, and t denotes the time iteration, $t = 0, \ldots, N$. Here, N denotes the expected length of each iteration. The variables $x_k(t) \in \mathbf{R}^n$, $u_k(t) \in \mathbf{R}^p$, and $y_k(t) \in \mathbf{R}^q$ represent the system state, input, and measured output, respectively. $w_k(t)$ denotes the random measurement noise. A_t, B_t, and C_t are time-varying system matrices with appropriate dimensions.

The tracking reference is given by $y_r(t)$, $t = 0, 1, \ldots, N$, which is realizable in the sense that there exist an initial state $x_r(0)$ and input $u_r(t)$ that satisfy the following equation:

$$\begin{aligned} x_r(t+1) &= A_t x_r(t) + B_t u_r(t), \\ y_r(t) &= C_t x_r(t). \end{aligned} \tag{4.54}$$

We first present and rationalize the assumptions employed in our analytical study.

Assumption 4.8 The identical initialization condition holds, i.e., $x_k(0) = x_r(0), \forall k$.

Potential extensions of Assumption 4.8 are feasible. However, that would cost a lot of space. Thus, we employ this condition to highlight our novelty.

Assumption 4.9 The input/output coupling matrix $C_{t+1}B_t$ is of full-column rank.

Assumption 4.9 implicitly imposes a constraint that the number of inputs should not exceed the number of outputs. This constraint ensures the uniqueness of the desired input with respect to (4.54), provided that the tracking reference is feasible. Specifically, it can be expressed as $u_r(t) = [C_B^T C_B]^{-1} C_B^T [y_r(t+1) - C_{t+1}A_t x_r(t)]$, where $C_B \triangleq C_{t+1}B_t$. Relaxing this assumption would have no impact on the subsequent results but would complicate the analysis.

Assumption 4.10 For any t, the noise sequence $\{w_k(t), k \in \mathbf{N}\}$ forms a martingale difference sequence with respect to the iteration index k; that is, $\mathbb{E}\{w_{k+1}(t) \mid \mathcal{F}_k\} = 0$, where the σ-algebra is defined as $\mathcal{F}_k = \sigma\{w_i(t), 1 \le i \le k, 0 \le t \le N_k\}$ (i.e., the set of all events induced by the available randomness up to the kth iteration). Additionally, $\mathbb{E}w_k(t) < \infty$ and $\sup_k \mathbb{E}\{\|w_{k+1}(t)\|^2 \mid \mathcal{F}_k\} < \infty, \forall t, k$.

Assumption 4.10 is imposed within the iteration domain rather than in the time domain. It is evident that the conventional zero-mean white noise satisfies this assumption, which holds true in numerous applications (Campi et al. 2008; Li et al. 2019).

In this section, we assume that the actual output for tracking is given by $z_k(t) = C_t x_k(t)$, while only the noisy output $y_k(t)$ is available for learning. Therefore, the control objective is to design an iterative learning law that enables $z_k(t)$ to track the reference signal $y_r(t)$ as the iteration number increases. Indeed, by considering $y_k(t)$ as the output, we can formulate the control objective to minimize the averaged norm of noisy tracking errors as defined by $J_t = \limsup_{k\to\infty} \frac{1}{k}\sum_{i=1}^{k} \|y_r(t) - y_i(t)\|^2, \forall t$. It can be easily verified that these two objectives are equivalent; thus, enabling us to apply our results in this section to stochastic system's learning control. To achieve this latter objective, it suffices to demonstrate that the generated input sequence satisfies $u_k(t) \to u_r(t)$ for all time instances. Henceforth, our design and analysis goals revolve around ensuring convergence of $u_k(t) \to u_r(t)$ as $k \to \infty$.

In practice, the actual iteration length may terminate prematurely, resulting in a nonuniform trial length. To capture this stochasticity, we introduce a corresponding mathematical model. The objective of the system is to complete a given tracking task within the time interval $[0, N]$. Therefore, it is reasonable to assume the existence of a lower bound on the actual iteration length denoted by $N_\circ$, where $N_\circ < N$. This implies that the actual operation length can take values from $N_\circ$ to N. Let $\theta \triangleq N - N_\circ + 1$ represent the number of possible trial lengths. We denote the actual length of the kth iteration as N_k. It is evident that $N_\circ \le N_k \le N$, and there are θ possible values for N_k. In previous literature, it has been commonly assumed that the nonuniform length N_k is independent across iterations; however, this assumption may not hold under practical conditions where variable lengths could depend on system states or randomly occurring events (Seel et al. 2011). In other words, variable lengths can be dependent on iterations. Henceforth, we employ Markov chain modeling to capture this dependence.

In particular, let $\mathcal{A}_i$ denote the event that the actual iteration length is $N_\circ + i - 1$, $1 \le i \le \theta$. It is assumed that the jump among the events $\mathcal{A}_i$, $1 \le i \le \theta$, forms a finite-state Markov chain, where the state transition probability matrix is $P = [p_{ij}]$, $p_{ij} \triangleq \mathbb{P}(\mathcal{A}_i \to \mathcal{A}_j)$, $1 \le i, j \le \theta$. In other words, the occurrence of all possible trial lengths forms a Markov chain along the iteration axis. We present the following assumption for further analysis.

Assumption 4.11 The Markov chain is iteration homogeneous, irreducible, recurrent, and independent of the measurement noise $w_k(t)$. Moreover, a finite summation of N_k exists such that its expectation is independent of the iteration number.

We remark this assumption as follows. The iteration-homogeneous property ensures that the transition probability remains independent of k. The recurrent condition of state $\mathcal{A}_\theta$, which signifies that the trial length equals the desired length N, plays a crucial role in achieving precise tracking throughout the entire time interval. If state $\mathcal{A}_\theta$ is transient, it will never reach the desired length N after a finite number of iterations, thereby rendering any improvement in tracking performance at time instant N unattainable. Consequently, asymptotically accurate tracking over $[0, N]$ becomes unachievable. Indeed, the partial states in $\{\mathcal{A}_i, 1 \le i \le \theta - 1\}$ can be transient (resulting in a reducible Markov chain). In this scenario, the subsequent results can still be derived similarly based on the subset of left recurrent states. However, including this treatment significantly complicates the mathematical derivations without introducing any novel concepts; therefore, we choose to omit this case. A special scenario of the recurrence of $\mathcal{A}_\theta$ is when all Markov states $\mathcal{A}_i$ can transition to state $\mathcal{A}_\theta$ with a positive probability ($\min_{1 \le i \le \theta} p_{i\theta} > 0$). In such cases, the subsequent analysis can be simplified. If the Markov chain is stationary, then the expectation of a finite summation of N_k remains independent of the iteration number. This condition is crucial for convergence analysis purposes. It should be emphasized that designing the controller does not require statistical information about the Markov chain model.

The recurrent nature of the full trial length $\mathcal{A}_\theta$ suggests a straightforward approach to updating the control only on a subset of trials where $\mathcal{A}_\theta$ occurs. However, this update scheme may prove highly inefficient when the probability of $\mathcal{A}_\theta$ is low. Moreover, disregarding these incomplete trials results in valuable information being wasted.

4.3.3 Learning Controller Design

The actual length of the trial varies within the set $\mathcal{N}^c \triangleq \{N_\circ, \ldots, N\}$, which is worth noting. When the trial length equals the desired length N, tracking errors for the entire trial can be accessed. However, if it falls short of N, tracking errors at missing time instants cannot be utilized to update input signals. In this section, absent output information is supplemented by tracking references, resulting in zero tracking errors at missing time instants. The definition of tracking errors is as follows:

$$e_k(t) = \begin{cases} y_r(t) - y_k(t), & 1 \le t \le N_k \\ 0, & N_k + 1 \le t \le N. \end{cases} \tag{4.55}$$

Now we provide our learning control algorithm as follows:

$$u_{k+1}(t) = u_k(t) + a_k L_t e_k(t+1),\ 0 \le t \le N-1, \tag{4.56}$$

where L_t denotes the regulatory matrix and $\{a_k\}$ is a sequence of variable gains. We have the following conditions:

$$\text{(C1)}\quad a_k > 0, a_k \to 0, \sum_{k=1}^{\infty} a_k = \infty, \tag{4.57}$$

$$\text{(C2)}\quad \exists \alpha_1 > 0 \text{ such that } \alpha_1 a_k \le a_{k+1} \le a_k, \tag{4.58}$$

$$\text{(C3)}\quad \sum_{k=1}^{\infty} a_k^2 < \infty. \tag{4.59}$$

Remark 4.6 The regulatory matrix L_t is incorporated into the algorithm to regulate the control direction. To mitigate the impact of random noise and ensure convergence, a variable gain sequence $\{a_k\}$ is introduced. Without this mitigation, continuous disturbance from measurement noise would prevent the input sequence from converging. If the random noise $w_k(t)$ is absent in system (4.53), it can be omitted from (4.56), transforming the learning algorithm (4.56) into a traditional one.

Remark 4.7 It is evident that $a_k = \rho/k$ with $\rho > 0$ satisfies conditions (C1)–(C3), where ρ serves as a tunable parameter for regulating the transient performance of the iteration. Generally, a smaller value of ρ would result in slower convergence but with reduced overshoot in the iteration domain, while a larger value of ρ can lead to faster convergence albeit with increased overshoot.

The learning gain a_k in (4.56) is typically predetermined prior to the operational process, irrespective of the specific evolution of system performance. Hence, the efficacy of learning heavily relies on the initial selection of a_k, which necessitates some general knowledge about the system. Consequently, it becomes imperative for the algorithm to dynamically adapt and fine-tune itself based on tracking performance. In this section, we propose two practical adaptive alternatives for determining the learning gain.

- Update-dependent adaptive mechanism: The input signal is updated by the following adaptive algorithm,

$$u_{k+1}(t) = u_k(t) + a_{\tau_t^k} L_t e_k(t+1), \tag{4.60}$$

where the update index τ_t^k is recursively defined as:

$$\tau_t^k = \begin{cases} \tau_t^{k-1} + 1, & \text{if } t \leq N_k \\ \tau_t^{k-1}, & \text{otherwise} \end{cases} \tag{4.61}$$

with the initial values being $\tau_t^1 = 1$.

- Performance-dependent adaptive mechanism: The input signal is updated by the following adaptive algorithm:

$$u_{k+1}(t) = u_k(t) + \rho \zeta_k L_t e_k(t+1), \tag{4.62}$$

where $\rho > 0$ is a sufficiently small constant (tunable parameter) and the learning gain ζ_k is recursively defined as:

$$\zeta_k = \begin{cases} (\zeta_{k-1}^{-1} + 1)^{-1}, & \text{if } \sum_{t=1}^{N} e_k^T(t) e_{k-1}(t) < 0 \\ \zeta_{k-1}, & \text{otherwise} \end{cases} \tag{4.63}$$

with the initial values being $\zeta_1 = 1$.

The first alternative is referred to as the update-dependent adaptive mechanism, as the index τ_t^k increases only when both the corresponding tracking information is available and an update of the input signal occurs. It should be noted that in the same iteration, the decreasing sequence $a_{\tau_t^k}$ is time-varying. At any given time within set $\mathcal{N}^c$, due to the recurrence of the Markov chain, practical updates of the input (i.e., deleting repeated trials) form a continuous updating process in the iteration domain. We term the second alternative as a performance-dependent adaptive mechanism since changes in index ζ_k occur solely based on satisfaction of performance conditions throughout iterations. Both adaptive mechanisms contribute to enhancing convergence speed through self-tuning principles.

Moreover, an intuitive comprehension of the performance-dependent adaptive mechanism can be presented. It should be noted that the tracking error consists of two components: one caused by unsatisfactory control input, $C\Delta x_k(t) \triangleq C(x_r(t) - x_k(t))$, and random noise $w_k(t)$. During the initial learning stage, the input may deviate significantly from the desired input, leading to a dominance of $C\Delta x_k(t)$ in the tracking error. Under mild conditions, this would result in a positive value for $\sum_{t=1}^{N} \Delta x_k^T(t) C^T C \Delta x_{k+1}(t)$ and make $\sum_{t=1}^{N} e_k^T(t) e_{k-1}(t)$ positive as well; consequently, there would be no triggering of performance condition (4.63). After a sufficient number of iterations, it is believed that the input has been significantly improved and the tracking error is primarily influenced by the noise term $w_k(t)$. In this scenario, according to Assumption 4.10, there exists a certain probability for the inner product of the noise vector to be negative. Consequently, depending on the specific distribution of $w_k(t)$, there will be a certain probability for condition (4.63) to be triggered.

4.3.4 Convergence Analysis

In this section, a strict convergence analysis is provided for the proposed algorithm (4.56) in both the mean-square and almost-sure senses while considering the stochastic noise. To make a concise expression, the signals of each iteration are lifted, assuming that the full trial length N is achieved: $\boldsymbol{u}_\star = [u_\star^T(0), u_\star^T(1), \ldots, u_\star^T(N-1)]^T \in \mathbf{R}^{pN}$ and $\boldsymbol{y}_\star = [y_\star^T(1), y_\star^T(1), \ldots, y_\star^T(N)]^T \in \mathbf{R}^{qN}$, where the subscript $\star$ is "r" or "k" denoting the reference signal or actual signal, respectively, from the kth iteration. Then, we have

$$\boldsymbol{y}_k = G\boldsymbol{u}_k + Mx_k(0) + \xi_k, \tag{4.64}$$

$$\boldsymbol{y}_r = G\boldsymbol{u}_r + Mx_r(0), \tag{4.65}$$

where

$$G = \begin{bmatrix} C_1B_0 & 0 & \ldots & 0 \\ C_2A_1B_0 & C_2B_1 & \ldots & 0 \\ \vdots & \vdots & \ddots & \vdots \\ C_NA_1^{N-1}B_0 & C_NA_2^{N-1} & \ldots & C_NB_{N-1} \end{bmatrix}, \tag{4.66}$$

$$M = [(C_1A_0)^T, (C_2A_0^1)^T, \ldots, (C_NA_0^{N-1})^T]^T, \tag{4.67}$$

and $\xi_k = [w_k^T(1), \ldots, w_k^T(N)]^T$ with $A_i^j \triangleq A_j \ldots A_i$, $j \geq i$. It is evident that $\{\xi_k\}$ is a martingale difference sequence with bounded conditional second moments.

However, due to the problem of the nonuniform trial lengths, the entire output trajectory $\boldsymbol{y}_k$ is generally not available for the update, but only the measured part can be accessed. To fill the gap in notations, we introduce the following matrix

$$Q_{N_k} = \begin{bmatrix} I_{N_k} \otimes I_q & 0 \\ 0 & \mathbf{0}_{(N-N_k)} \otimes I_q \end{bmatrix}, \quad N_\circ \leq N_k \leq N \tag{4.68}$$

where I_l and $\mathbf{0}_l$ denote the unit matrix and zero matrix with the dimension of $l \times l$, respectively. Then, for the kth iteration with the trial length being N_k, the actual lifted output, whose missing part is complemented by the tracking reference, is given by:

$$\boldsymbol{y}_k^* = Q_{N_k}\boldsymbol{y}_k + (I_{qN} - Q_{N_k})\boldsymbol{y}_r. \tag{4.69}$$

Similarly, the lifted tracking error is denoted by $\boldsymbol{e}_k = [e_k^T(1), e_k^T(2), \ldots, e_k^T(N-1)]^T$, where $e_k(t)$ is defined in (4.55). It is apparent that

$$\boldsymbol{e}_k = \boldsymbol{y}_r - \boldsymbol{y}_k^* = Q_{N_k}(\boldsymbol{y}_r - \boldsymbol{y}_k). \tag{4.70}$$

Now, the update law (4.56) could be rewritten in the lifted form

$$\boldsymbol{u}_{k+1} = \boldsymbol{u}_k + a_kL\boldsymbol{e}_k = \boldsymbol{u}_k + a_kLQ_{N_k}(\boldsymbol{y}_r - \boldsymbol{y}_k), \tag{4.71}$$

where $L = \text{diag}\{L_0, L_1, \ldots, L_{N-1}\}$ is the block-diagonal matrix consisting of the regulation matrices $L_t, 0 \le t \le N-1$.

Substituting (4.64) and (4.65) into (4.71) leads to

$$\boldsymbol{u}_{k+1} = \boldsymbol{u}_k + a_k L Q_{N_k} G \Delta \boldsymbol{u}_k - a_k L Q_{N_k} \xi_k, \tag{4.72}$$

where $\Delta \boldsymbol{u}_k \triangleq \boldsymbol{u}_r - \boldsymbol{u}_k$ and the identical initialization condition $x_k(0) = x_r(0)$ is used.

Before proceeding to the main results, it should be emphasized that the lifted formulations above are only used to facilitate the convergence analysis. The input signals are actually updated by the original algorithms (i.e., (4.56) or its alternatives (4.60) and (4.62)). Consequently, this learning update effectively addresses the dimensional problem associated with handling long trial lengths N. The subsequent theorem presents the convergence of the proposed algorithm (4.56).

Theorem 4.5 *Consider system* (4.53) *with Assumptions 4.8-4.11. For any t, select the regulatory matrix L_t such that $-L_t C_{t+1} B_t$ is Hurwitz. Let the learning gain a_k satisfy conditions (C1)–(C2), then the input sequence generated by algorithm* (4.56) *converges to the reference input in the mean-square sense, i.e.,*

$$\lim_{k\to\infty} \mathbb{E}\{\|u_r(t) - u_k(t)\|^2\} = 0, \quad 0 \le t \le N-1. \tag{4.73}$$

Additionally, if condition (C3) is satisfied, then the input sequence also converges in the almost-sure (a.s.) sense, i.e.,

$$\lim_{k\to\infty} u_k(t) = u_r(t), \quad a.s. \quad 0 \le t \le N-1. \tag{4.74}$$

Proof. Reminding $\Delta \boldsymbol{u}_k = \boldsymbol{u}_r - \boldsymbol{u}_k$ and subtracting both sides of (4.72) from $\boldsymbol{u}_r$, we have:

$$\Delta \boldsymbol{u}_{k+1} = \Delta \boldsymbol{u}_k - a_k L Q_{N_k} G \Delta \boldsymbol{u}_k + a_k L Q_{N_k} \xi_k. \tag{4.75}$$

Backward iterating the last equation for h times, where h is a positive integer describing the length of successive iterations, we have:

$$\Delta \boldsymbol{u}_{(n+1)h} = \Phi((n+1)h, nh) \Delta \boldsymbol{u}_{nh} + \Omega_{nh}, \tag{4.76}$$

where $\Phi(i+1, j) = (I - a_i L Q_{N_i} G) \ldots (I - a_j L Q_{N_j} G)$, $\forall i \ge j$, $\Phi(i,i) = I_{pN}$, and $\Omega_k = \sum_{j=k}^{k+h-1} \Phi(k+h-1, j) a_j L Q_{N_j} \xi_j$.

It should be noted that L is a block-diagonal matrix and G is a block-lower-triangular matrix; thus, LG is a block-lower-triangular matrix. The diagonal blocks of LG are $L_t C_{t+1} B_t, 0 \le t \le N-1$. The regulatory matrix L_t ensures that $-L_t C_{t+1} B_t$ is Hurwitz (i.e., all eigenvalues of $-L_t C_{t+1} B_t$ are with negative real parts). By noticing the definition of Q_{N_k} in (4.68), $L Q_{N_k} G$ is a block-lower-triangular matrix with its diagonal blocks being either $L_t C_{t+1} B_t$ or $\mathbf{0}_q$.

By Assumption 4.11, there exists an integer $h > 0$ such that $\mathbb{E}(\sum_{i=k}^{k+h-1} Q_{N_i})$ is independent of the iteration number, where the expectation is taken in respect of the probability space constructed by the events of the variable trial lengths. Meanwhile, the full length scenario is recurrent; thus, the diagonal blocks of $\Xi \triangleq -\mathbb{E}(\sum_{i=k}^{k+h-1} \Lambda_i)$ cannot be zero, where $\Lambda_i = LQ_{N_i}G$. In other words, all eigenvalues of Ξ are with negative real parts. This ensures the existence of a positive-definite matrix S such that $\Xi^T S + S\Xi = I_{pN}$. For the positive-definite matrix S, a positive constant $c_0 > 0$ such that $c_0\lambda_{\max}(S) < 1$ can guarantee that $I_{pN} \geq c_0 S$, where $\lambda_{\max}(S)$ denotes the largest eigenvalue of S. Therefore, we obtain:

$$\Xi^T S + S\Xi \geq c_3 S, \quad \text{for some } c_0 > 0. \tag{4.77}$$

A Lyapunov function $V_k = (\Delta \boldsymbol{u}_k)^T S(\Delta \boldsymbol{u}_k)$ is defined. Then, by (4.76), we have

$$\begin{aligned} V_{(n+1)h} &= (\Delta \boldsymbol{u}_{(n+1)h})^T S(\Delta \boldsymbol{u}_{(n+1)h}) \\ &= (\Delta \boldsymbol{u}_{nh})^T \Phi((n+1)h, nh)^T S\Phi((n+1)h, nh)\Delta \boldsymbol{u}_{nh} \\ &\quad + 2(\Delta \boldsymbol{u}_{nh})^T \Phi((n+1)h, nh)^T S\Omega_{nh} \\ &\quad + \Omega_{nh}^T S\Omega_{nh}. \end{aligned} \tag{4.78}$$

By condition (C2), there exist constants $c_1 > 0$ and $c_2 > 0$ such that

$$\begin{aligned} &\Phi((n+1)h, nh)^T S\Phi((n+1)h, nh) \\ &\leq S - \sum_{i=nh}^{(n+1)h-1} a_i \left(\Lambda_i^T S + S\Lambda_i\right) + c_1 a_{nh}^2 S \\ &\leq S - c_2 a_{nh} H_{nh} + c_1 a_{nh}^2 S, \end{aligned} \tag{4.79}$$

where $H_{nh} = \sum_{i=nh}^{(n+1)h-1} \left(\Lambda_i^T S + S\Lambda_i\right)$.

Substituting (4.79) into (4.78) yields

$$\begin{aligned} &V_{(n+1)h} \\ &\leq V_{nh} - c_2 a_{nh}(\Delta \boldsymbol{u}_{nh})^T H_{nh}(\Delta \boldsymbol{u}_{nh}) + c_1 a_{nh}^2 V_{nh} \\ &\quad + \Omega_{nh}^T S\Omega_{nh} + 2(\Delta \boldsymbol{u}_{nh})^T \Phi((n+1)h, nh)^T S\Omega_{nh}. \end{aligned} \tag{4.80}$$

Let $\mathcal{G}_k = \sigma\{w_i(t), N_i, 1 \leq i \leq k-1, 0 \leq t \leq N_i\}$. Then, by Assumption 4.10 and the independence between nonuniform trial length and the measurement noise in Assumption 4.11,

$$\mathbb{E}\{(\Delta \boldsymbol{u}_{nh})^T \Phi((n+1)h, nh)^T S\Omega_{nh} \mid \mathcal{G}_{nh}\} = 0. \tag{4.81}$$

By the independence in Assumption 4.11 and (4.77) with a sufficiently large integer h, a constant $c_3 > 0$ exists such that

$$\mathbb{E}\{H_{nh} \mid \mathcal{G}_{nh}\} \geq c_3 S. \tag{4.82}$$

By the independence between the random matrix Q_{N_k} and the noise term ξ_k again, and remarking that the matrix Λ_i is bounded, we have:

$$\mathbb{E}\{\Omega_{nh}^T S \Omega_{nh} \mid \mathcal{G}_{nh}\} \leq c_4 \sum_{j=nh}^{(n+1)h-1} a_j^2 \mathbb{E}\{\|\xi_j\|^2 \mid \mathcal{G}_{nh}\}, \tag{4.83}$$

where $c_4 > 0$ is a suitable constant and Minkowski's inequality is applied.

Taking conditional expectation to both sides of (4.80) with respect to $\mathcal{G}_{nh}$ and substituting (4.81)-(4.83), we have:

$$\begin{aligned}\mathbb{E}\{V_{(n+1)h} \mid \mathcal{G}_{nh}\} \leq &(1 - c_2 c_3 a_{nh} + c_1 a_{nh}^2) V_{nh} \\ &+ c_4 \sum_{j=nh}^{(n+1)h-1} a_j^2 \mathbb{E}\{\|\xi_j\|^2 \mid \mathcal{G}_{nh}\}.\end{aligned} \tag{4.84}$$

Taking expectation to (4.84), we have:

$$\mathbb{E}\{V_{(n+1)h}\} \leq (1 - c_2 c_3 a_{nh} + c_1 a_{nh}^2) \mathbb{E}\{V_{nh}\} + c_5 a_{nh}^2, \tag{4.85}$$

where the boundedness of $\mathbb{E}\{\|\xi_j\|^2 \mid \mathcal{G}_{nh}\}$ and condition (C3) are applied, and $c_5 \triangleq c_4 h \sup_k \mathbb{E}\{\|\xi_k\|^2 \mid \mathcal{F}\}$.

Define $\mu_n = c_2 c_3 a_{nh} - c_1 a_{nh}^2$ and $\nu_n = c_5 a_{nh}^2$. It is clear that we have $\sum_{n=1}^{\infty} \mu_n = \infty$ and $\lim_{n\to\infty} \nu_n / \mu_n = 0$ by conditions (C1) and (C2). It follows from (4.85) that

$$\begin{aligned}\beta_{n+1} &\leq (1 - \mu_n)\beta_n + \nu_n \\ &\leq \left(\prod_{l=m}^{n}(1 - \mu_l)\right)\beta_m + \sum_{l=m}^{n}\left(\prod_{j=l+1}^{n}(1 - \mu_j)\right)\nu_l \\ &= \left(\prod_{l=m}^{n}(1 - \mu_l)\right)\beta_m + \sum_{l=m}^{n}\left(\prod_{j=l+1}^{n}(1 - \mu_j)\right)\mu_l \cdot \frac{\nu_l}{\mu_l} \\ &\leq \left(\prod_{l=m}^{n}(1 - \mu_l)\right)\beta_m \\ &\quad + \left(\sup_{m\leq l\leq n}\frac{\nu_l}{\mu_l}\right) \times \sum_{l=m}^{n}\left(\prod_{j=l+1}^{n}(1 - \mu_j)\mu_l\right) \\ &= \left(\prod_{l=m}^{n}(1 - \mu_l)\right)\beta_m \\ &\quad + \left(\sup_{m\leq l\leq n}\frac{\nu_l}{\mu_l}\right) \times \left(1 - \prod_{l=m}^{n}(1 - \mu_l)\right),\end{aligned}$$

where $\beta_n \triangleq \mathbb{E}\{V_{nh}\}$. Therefore, we have

$$\limsup_{n\to\infty} \beta_n \le \limsup_{n\to\infty} \frac{\nu_n}{\mu_n} = 0.$$

That is, we conclude that $\lim_{n\to\infty} \mathbb{E}\{V_{nh}\} = 0$. To complete the proof, the gap needs to be filled to show the convergence for an arbitrary iteration number rather than the subsequence of subscripts nh. To this end, from (4.75), for $k = nh + 1, \ldots, (n+1)h - 1$:

$$\Delta \boldsymbol{u}_k = \Phi(k, nh)\Delta \boldsymbol{u}_{nh} + \Omega_{k,nh}, \tag{4.86}$$

where $\Omega_{k,nh} = \sum_{j=nh}^{k-1} \Phi(k, j+1) a_j L Q_{N_j} \xi_j$. Then, we have:

$$\begin{aligned}
\mathbb{E}\{V_k\} &\le 2\mathbb{E}\left\{\|\Phi(k, nh)\|^2 V_{nh}\right\} + 2\mathbb{E}\left\{\Omega_{k,nh}^T S \Omega_{k,nh}\right\} \\
&\le c_6 \mathbb{E}\{V_{nh}\} + c_7 \sum_{j=nh}^{k-1} a_j^2 \mathbb{E}\{\|\xi_j\|^2\} \\
&\le c_6 \mathbb{E}\{V_{nh}\} + c_8 a_{nh}^2.
\end{aligned} \tag{4.87}$$

Consequently, it is clear that $\lim_{k\to\infty} \mathbb{E}\{V_k\} = 0$. Furthermore, as S is a positive-definite matrix, we have that $\lim_{k\to\infty} \mathbb{E}\{\|\Delta \boldsymbol{u}_k\|^2\} = 0$. This completes the proof of the mean-square convergence.

To show the convergence in the almost-sure sense, we divide the iteration number into h subsequences $\{nh + j, n = 0, 1, \ldots\}$, $j = 1, \ldots, h$. For each subsequence, similar to (4.84) we have

$$\mathbb{E}\{V_{(n+1)h+j} \mid \mathcal{G}_{nh+j}\} \le V_{nh+j} - \eta_n + \chi_n, \tag{4.88}$$

where $\eta_n \triangleq (c_2 c_3 a_{nh+j} - c_1 a_{nh+j}^2) V_{nh+j}$ and $\chi_n \triangleq c_4 \sum_{i=nh+j}^{(n+1)h-1+j} a_i^2 \mathbb{E}\{\|\xi_i\|^2 | \mathcal{G}_{nh}\}$.

By condition (C3), we have $\sum_{n=0}^{\infty} a_{nh+j}^2 < \infty$ and $\sum_{n=0}^{\infty} \sum_{i=nh+j}^{(n+1)h-1+j} a_i^2 < \infty$. Using Assumption 4.10,

$$\sum_{n=0}^{\infty} \chi_n = c_4 \sum_{n=0}^{\infty} \sum_{i=nh+j}^{(n+1)h-1+j} a_i^2 \mathbb{E}\{\|\xi_i\|^2 \mid \mathcal{G}_{nh}\} < \infty.$$

Clearly, $\mathbb{E}\{|V_j|\} < \infty$. Therefore, by Lemma 2.2, it follows that V_{nh+j} converges almost surely as $n \to \infty$, $1 \le j \le h$. From Theorem 4.5 we already have $\lim_{k\to\infty} \mathbb{E}\{V_k\} = 0$; thus, $\lim_{n\to\infty} V_{nh+j} = 0$, $1 \le j \le h$, further implying that $\lim_{k\to\infty} V_k = 0$ in the almost-sure sense. Noting that S is a positive-definite matrix, we obtain $\lim_{k\to\infty} \boldsymbol{u}_k = \boldsymbol{u}_r$, a.s. The proof is completed.

Remark 4.8 In this proof, the decaying trend of the Lyapunov function V_k is demonstrated through a series of successive iterations using a general Markov chain model.

Assuming that all Markov states $\mathcal{A}_i$ can transition to state $\mathcal{A}_\theta$ (corresponding to the full trial length case) with positive probability, i.e., $\min_{1\le i\le\theta} p_{i\theta} > 0$, simplifies the above proofs and shows the decaying trend of V_k in each iteration. In this scenario, it holds for every iteration number k that $\mathbb{E}\{\Lambda_k^T S + S\Lambda_k \mid \mathcal{G}_{k-1}\} \ge c_5 S$, where $c_5 > 0$ is an appropriate constant.

Remark 4.9 The results presented herein can be extended to accommodate the inclusion of process noise in the state equation of (4.53). In such cases, the introduction of state noise contributes an additional term to $\boldsymbol{y}_k$, which can be considered as a component of ξ_k. The convergence analysis remains valid; however, it should be noted that the actual output $z_k(t) = C_t x_k(t)$ may no longer accurately track $y_r(t)$.

Remark 4.10 We examine the terms on the right-hand side of (4.76). The first term, namely $\Phi((n+1)h, nh)\Delta\boldsymbol{u}_{nh}$, involves incomplete trials that can expedite convergence due to considering all iterations. However, it simultaneously introduces additional noise terms as indicated by the last term in (4.76), i.e., Ω_{nh}. Consequently, we observe a trade-off between the convergence rate and the impact of noise. This matter remains unresolved and is left for future investigation.

In Theorem 4.5, the convergence of the input error is established strictly in both the mean-square and the almost-sure senses. According to (4.53), (4.54), and Assumption 4.8, $z_k(t)$ and $y_r(t)$ have the same dynamics except that they are driven by $u_k(t)$ and $u_r(t)$, respectively. Thus, the convergence of $u_k(t)$ to $u_r(t)$ yields that $z_k(t)$ converges to $y_r(t)$ in both stochastic senses. However, due to the existence of random noise $w_k(t)$, it is impossible to precisely track the desired reference. Indeed, we have $\mathbb{E}[\|y_k(t) - y_r(t)\|^2] - \mathbb{E}[\|w_k(t)\|^2] \to 0$ by direct calculations. Therefore, the index $\limsup_{k\to\infty} J_k$ is determined only by the noise term $w_k(t)$.

Before ending this section, we clarify the convergence for the two adaptive alternatives (4.60) and (4.62) with learning gains $a_{\tau_t^k}$ and ζ_k. The update-dependent gain $a_{\tau_t^k}$ remains unchanged if the corresponding tracking error is absent due to a short trial length. Therefore, for any fixed time instant, if all the iterations in which no update occurs are removed, the remainder is actually identical to (4.56). For the performance-dependent gain ζ_k, as long as the random noise exists, the inverse ζ_k^{-1} will increase to infinity similarly to the positive integer sequence. Further details are provided in Sect. 4.1. To summarize, the conditions (C1)–(C3) are satisfied by $a_{\tau_t^k}$ and ζ_k.

4.3.5 Numerical Simulations

The effectiveness of the proposed controller on a 4 DoF robot arm with only joint position measurements is demonstrated in this example. Furthermore, it is shown that ILC can effectively handle non-repeatable dynamics by utilizing relatively smaller controller gains. In contrast to most non-iterative controllers designed for robot manipulator tracking, such as the deterministic PD controller (Loria 2016), the stochastic

PD controller (Saab and Ghanem 2018), and sliding mode controllers (Mitra et al. 2016), which rely on gravity compensation, Coriolis and centripetal compensation models, partial friction force compensation models, and/or inertia matrix models; the ILC controller used in this example treats the plant as a black box without utilizing any part of its model. Additionally, the zero control signal is employed during the initial iteration.

The performance of the proposed controller is numerically evaluated on a model of a cable-driven 4 DoF Barret whole arm manipulator (WAM). To account for joint position measurement errors based on Barrett WAM sensor specifications, zero-mean uniformly distributed measurement noise with a variance of $4 \times 10^{-8}/12$ is added to the joint position measurements. The employed WAM model, fully described in Saab and Ghanem (2018), is based on supplier parameters.

As in Loria (2016); Saab and Ghanem (2018); Mitra et al. (2016), the Euler-Lagrange model that describes the continuous-time dynamics of a rigid 4-DOF serial non-redundant robot manipulator is considered, where all the actuated revolute joints are described in joint coordinates as follows

$$u = M(\theta)\ddot{\theta} + C(\theta, \dot{\theta})\dot{\theta} + G(\theta) + F(\theta, \dot{\theta}),$$

where θ, $\dot{\theta}$, and $\ddot{\theta} \in \mathbf{R}^4$ represent the joint angles, velocities, and accelerations, respectively, $M(\theta) \in \mathbf{R}^{4\times4}$ is the symmetric positive-definite inertia matrix, $C(\theta, \dot{\theta}) \in \mathbf{R}^{4\times4}$ denotes the Coriolis and centripetal matrix, $G(\theta) \in \mathbf{R}^4$ represents the gravity vector, $F(\theta, \dot{\theta}) \in \mathbf{R}^4$ denotes the joint friction, and $u \in \mathbf{R}^4$ denotes the joint torque input. The joint velocities, $\hat{\dot{\theta}}_k(t)$, are estimated from the noisy measurement of the joint positions using the backward Euler method. Subsequently, the estimated joint velocities are treated as the plant output. Consequently, it can be shown that (a) $C_{t+1}B_t$ is nothing but $M^{-1}(\theta)$ for the plant under consideration, and (b) there exists a positive and constant diagonal regulatory matrix $L \in \mathbf{R}^{4\times4}$ such that $0 < \lambda(I - LM^{-1}(\theta)) < 1$ and $0 < \lambda(I - a_k LM^{-1}(\theta)) < 1$, where the sequence $\{a_k\}$ is chosen such that $0 < a_k < 1$ for all k.

The learning control algorithm (4.60) is represented as follows:

$$u_{k+1}(t) = u_k(t) + a_{\tau_t^k} L_t \dot{e}_k(t+1),$$

where $\dot{e}_k(t) \triangleq \dot{\theta}_d(t) - \hat{\dot{\theta}}_k(t)$. The sequence $\{a_k\}$ is selected such that a_k decreases at a very low rate within the first 100 iterations; in particular, $a_k = 1/\sqrt[4]{k}$, and for $k > 100$, we can set $a_k = 1/k$ so that $\sum_{k=1}^{\infty} a_k^2 < \infty$. The recursive algorithm that updates τ_t^k is as described in (4.61). With some tuning, the regulatory matrix, $L = \text{diag}\{4, 4.6, 0.6, 0.5\}$ is selected.

The numerical simulations were conducted at a sampling frequency of 125 Hz and a time interval of two seconds, resulting in $N = 250$. The employed Markov chain generating the nonuniform trial lengths is a one-dimensional random walk process with length variation, $\Delta N_{k+1} = \Delta N_k + \sigma_v v_k$, where the random sequence v_k represents zero-mean independent identically distributed uniformly distributed

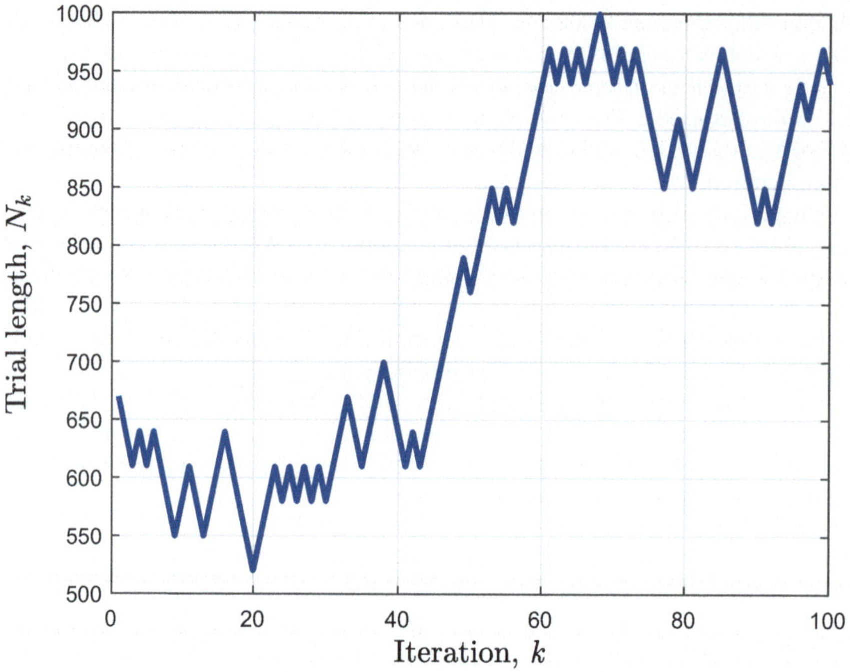

Fig. 4.14 The nonuniform trial lengths, N_k

noise with unity variance. We set $N_k = N + \Delta N_k, \sigma_v = 30$, and select corresponding seeds to ensure that $0 \leq N_k \leq N$. The nonuniform trial lengths, denoted as N_k, are illustrated in Fig. 4.14.

Figure 4.15 shows the time trajectories of all the joint positions corresponding to the 0th, 10th and 100th iterations. It can be seen that the system outputs track the desired references remarkably well. However, when the nonuniform trial lengths shown in Fig. 4.14, where $N_k \leq N$ are used, only two learning iterations, namely $k = 80$ and $k = 83$ are used to learn from the entire time span of our reference trajectories. By contrast, for $0 \leq t \leq 200$, there are 83 iterations to learn from. That is, it is anticipated that there will be larger tracking errors as $t \to N$.

To evaluate the performance of our proposed method, three scenarios that use the same regulatory matrix, $L = \text{diag}\{4, 4.6, 0.6, 0.5\}$ are considered. In the first scenario, the same proposed sequence of variable gains, $a_{\tau_t^k}$ is considered but with full trial lengths; that is $N_k = N$ for all k. In the second scenario, a non-adaptive ILC is used, that is, $a_k = 1$ for all k, and $N_k = N$ for all k. In the third scenario, a non-adaptive ILC is considered with the same nonuniform trial lengths shown in Fig. 4.14.

Scenario 1: First the same sequence of variable gains while using full trial length is considered; that is:

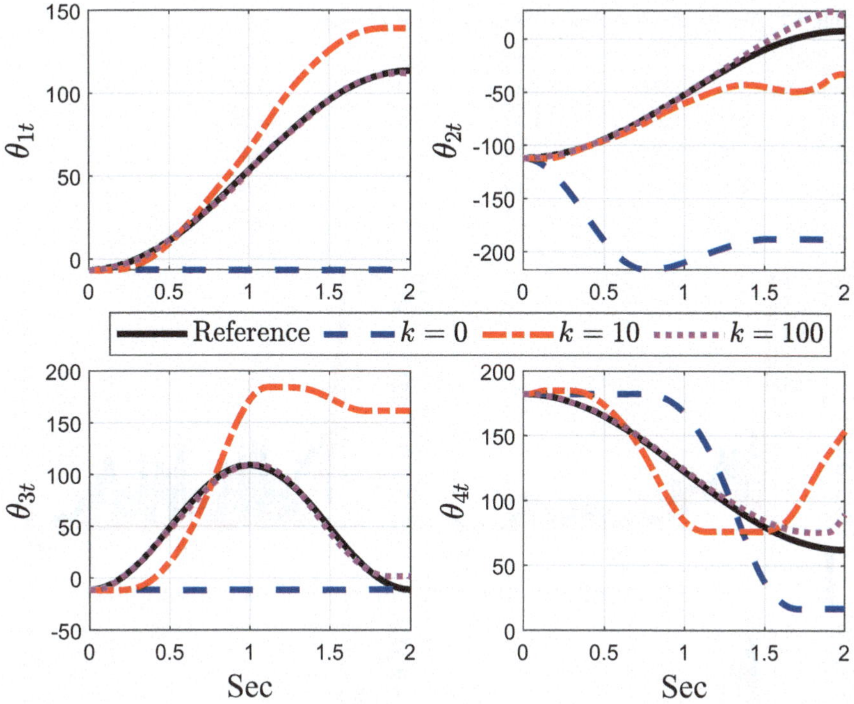

Fig. 4.15 Joint position trajectories using the proposed method for the iterative index $k = 0, k = 10$, and $k = 100$

$$u_{k+1}(t) = u_k(t) + a_{\tau_t^k} L e_k(t+1) \tag{4.89}$$

where $e_k(t) = \dot{\theta}_d(t) - \dot{\hat{\theta}}_k(t), 0 \leq t \leq N$, and $a_{\tau_t^k}$ is the same sequence employed in our ILC update law.

Scenario 2: A non-adaptive strategy while using full trial length is then considered; that is:

$$u_{k+1}(t) = u_k(t) + L e_k(t+1) \tag{4.90}$$

where $e_k(t) = \dot{\theta}_d(t) - \dot{\hat{\theta}}_k(t), 0 \leq t \leq N$.

Scenario 3: A non-adaptive strategy while using the same nonuniform trial length is also considered; that is:

$$u_{k+1}(t) = u_k(t) + L e_k(t+1) \tag{4.91}$$

where $e_k(t) = \dot{\theta}_d(t) - \dot{\hat{\theta}}_k(t), 0 \leq t \leq N_k$ and $e_k(t) = 0, N_{k+1} \leq t \leq N$.

Table 4.2 Average absolute error, $AAE \times 10^{-2}$, for all joints

Method	Joint 1	Joint 2	Joint 3	Joint 4	Average
Proposed	0.77	2.44	1.75	3.27	2.06
Scenario 1 (4.89)	0.67	1.77	1.49	1.75	1.42
Scenario 2 (4.90)	2.73	7.79	4.91	6.45	5.47
Scenario 3 (4.91)	1.98	6.19	5.46	8.90	5.63

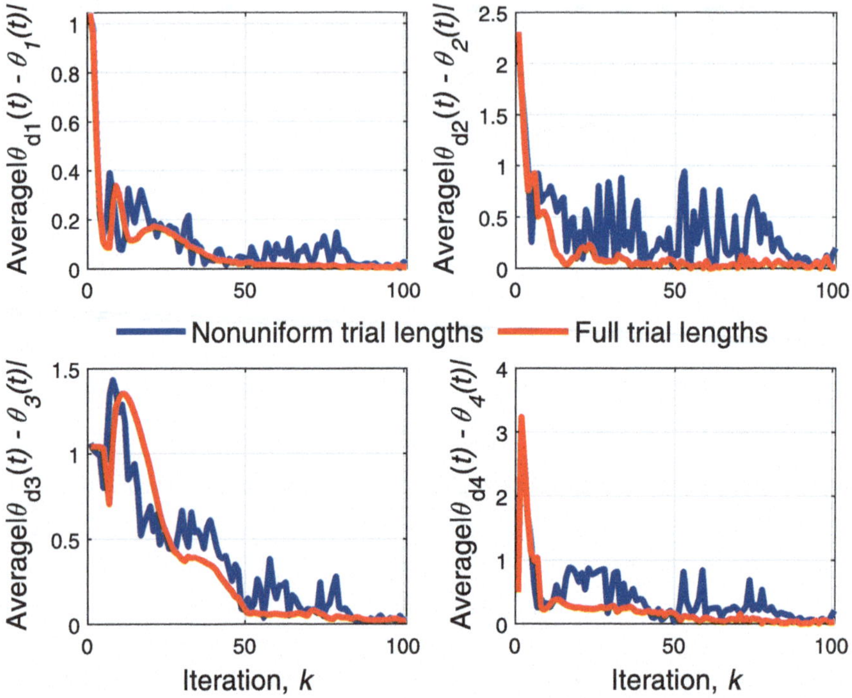

Fig. 4.16 Time average of absolute error versus number of iterations using the proposed variable gains: nonuniform trial lengths in blue and Scenario 1 (4.89) using full trial lengths is in red

The average absolute error (AAE) is evaluated at the 100th iteration to assess the steady-state performance of each ILC scheme in the iterative domain, that is, $AAE_i = avg_{0\leq t\leq N}|\theta_{i,d}(t) - \theta_{i,k=100}(t)|$. Ten different independent experiments are conducted and the average of AAE_i, $1 \leq i \leq 4$ is taken. The results are listed in Table 4.2.

Figure 4.16 shows the propagation of $avg_{0\leq t\leq N}|\theta_{i,d}(t) - \theta_{i,k}(t)|$ versus the number of iterations for the proposed method, as well as the scheme of Scenario 1 (4.89).

Examining Table 4.2 and Fig. 4.16, the following observations can be made:

- As expected, Scenario 1 (4.89) that uses the proposed sequence of variable gains, $a_{\tau_i^k}$ but takes advantage of the full-trial lengths performs the best. Figure 4.16

shows accelerated convergence within the first 50 iterations but subsequently, the proposed method with nonuniform trial lengths performs very similarly.
- Whenever the proposed sequence of variable gains is employed with nonuniform trial lengths, the overall errors are 2.65 times smaller than those of Scenarios 2 and 3, in which non-adaptive variable gains are employed. This observation illustrates the effectiveness of the proposed variable gains.

4.4 Adaptive Gain Design in Temperature Control for Large-Scale Buildings

4.4.1 Temperature Control for Large-Scale Buildings

Heating, ventilation, and air conditioning (HVAC) systems are utilized to maintain an optimal temperature in large-scale buildings. Buildings consume approximately 20–40% of global energy (Perez-Lombard et al. 2008). Given the imperative task of achieving carbon neutrality, enhancing HVAC control systems for human comfort and energy conservation is inevitable (Kolarik et al. 2011; Peeters et al. 2008). Typically, a large-scale building comprises multiple units and zones where the temperature of one zone is influenced by its neighboring zones and surrounding environment. Therefore, implementing a distributed control strategy holds significant importance for practical HVAC systems (Yang et al. 2020). Moreover, considering disturbances from outdoor weather conditions and indoor human activities, the dynamics of HVAC systems inherently exhibit stochastic behavior with various sources of randomness. Consequently, a major challenge arises: How can each zone be effectively controlled to attain the desired temperature amidst random noise?

Various approaches have been developed to address the high energy consumption of HVAC systems (Mei et al. 2021; Peng et al. 2016; Thomas et al. 2005). In order to enhance indoor thermal comfort and energy efficiency with a time-invariant temperature setpoint, a guaranteed cost periodically intermittent-switch control has been proposed (Mei et al. 2021). The works (Peng et al. 2016; Thomas et al. 2005) integrate feedback and feedforward control techniques for temperature regulation in HVAC systems. Feedback control reduces sensitivity to parameter changes, while feedforward control utilizes historical information from the database to generate prior estimates for unknown variables.

Another commonly employed strategy is model predictive control (MPC) (Bejarano et al. 2020; Kelman and Borrelli 2011; Venkat et al. 2008; Wang and Hu 2018). MPC utilizes the system model to predict states and generate an input signal for optimizing a control index (Kelman and Borrelli 2011). However, due to the complexity of building materials, stored items, and human activities, it is challenging to obtain an accurate system model in practical scenarios. To address this issue, a distributed MPC scheme has been proposed by partitioning the centralized system into multiple subsystems (Venkat et al. 2008). Nevertheless, this approach heavily relies on the precision of the system model and exhibits limited robustness for real-world applications. Furthermore, MPC necessitates significant computational resources to solve given tasks.

The objective of HVAC systems is to ensure that the temperature in each zone closely follows a time-varying reference. The control task is performed on a daily basis, with the specified temperature reference being repeated every 24 h. This means that HVAC systems consistently maintain the temperature of all zones in accordance with the given reference on a daily basis. However, none of the aforementioned studies have taken into account this inherent repeatability characteristic for control purposes. Considering the importance of repeatability in control, it is reasonable to explore input and output information from previous iterations in order to gradually enhance control performance (Bensidhoum and Bouakrif 2020). This suggests that ILC can be applied effectively for temperature regulation in HVAC systems.

There is a limited amount of literature on the application of ILC in HVAC systems (Chi et al. 2020, 2022; Dhar et al. 2020; Peng et al. 2017; Van Pham and Nguyen 2017). As an initial attempt, an interconnected dynamic equation was used to model the HVAC system and a D-type decentralized ILC algorithm was proposed (Van Pham and Nguyen 2017). The implementation of this algorithm for 5000 iterations resulted in zero tracking errors. To enhance the efficiency of HVAC systems, an event-triggered near-optimal controller was designed for the interconnected system which conserved communication, computation and energy resources (Dhar et al. 2020).

Furthermore, the HVAC system was characterized using a multi-agent system approach (Chi et al. 2020, 2022; Peng et al. 2017). To address the challenges posed by its distributed nature, a distributed cooperative ILC algorithm was proposed by decomposing the centralized system into multiple subsystems (Peng et al. 2017). Additionally, a data-driven distributed iterative learning temperature control algorithm was developed for large-scale HVAC systems by reformulating the linear data model (Chi et al. 2020), representing an early attempt to consider both repetitive characteristics and distributed topology. Subsequently, this approach was extended to handle nonrepetitive uncertainties as well (Chi et al. 2022). Despite existing efforts in developing distributed ILC algorithms for HVAC systems, further investigation is still required. Currently, limited progress has been made towards enhancing the convergence speed of temperature control algorithms.

In this section, we address the distributed control problem for stochastic HVAC systems by formulating the learning temperature control task. We fully exploit the repeatability of HVAC systems to achieve exceptional performance, while introducing a decreasing learning gain sequence to effectively eliminate noise effects. Consequently, we propose a distributed learning control (DLC) scheme with a decreasing learning gain for HVAC systems to ensure accurate tracking of reference temperatures in each zone. The convergence is rigorously proven in terms of mean-square sense.

However, the decreasing gain is given independently of the learning process. Consequently, an inappropriate selection of this gain may impede the convergence speed. To address this issue, we analyze the impact of noise to design an optimal learning gain. We observe that noise has a negligible effect during the initial stages of the learning process but becomes dominant as iterations increase. Based on our analysis, we propose a novel mechanism for generating an adaptive gain sequence. This adaptive gain remains constant when noise is smaller than the input error and decreases when noise exceeds it. By employing this novel mechanism, we establish an accelerated DLC scheme that ensures both convergence and accelerated convergence speed in practice.

4.4.2 Problem Formulation

Graph Theory: Define $\mathcal{G} = \{\mathcal{V}, \mathcal{E}\}$ as a directed graph, describing the topology of communication among n zones. The vertex and edge set are $\mathcal{V} = (v_1, v_2, \ldots, v_2)$ and $\mathcal{E} \subset \mathcal{V} \times \mathcal{V}$, respectively. Every vertex v_i represents a zone, and the edge (v_i, v_j) indicates that zone v_i can receive the information of zone v_j. Moreover, $\mathcal{N}_i = \{v_j \in \mathcal{V} : (v_i, v_j) \in \mathcal{E}\}$ is the adjacent vertex set. The adjacency matrix is denoted by $\mathcal{A} \triangleq [a_{i,j}]$, where $a_{i,j} = 1$ if $(v_i, v_j) \in \mathcal{E}$; otherwise, $a_{i,j} = 0$. The degree matrix is $\mathcal{O} \triangleq diag\{o_1, o_2, \ldots, o_n\}$, where $o_i = \sum_{j \in \mathcal{N}_i} a_{i,j}$. The Laplacian matrix $\mathcal{L}$ is defined as $\mathcal{L} = \mathcal{O} - \mathcal{A}$.

Consider a large-scale building comprising multiple zones, as illustrated in Fig. 4.17. The HVAC system structure consists of an Air Handling Unit (AHU) and Variable Air Volume (VAV) boxes, which serve as the primary components (Yang et al. 2021). The AHU comprises a damper, a cooling/heating coil, and a supply fan. The damper mixes outside fresh air with return air, while the cooling/heating coil adjusts the temperature of the mixed air. Subsequently, the supply fan facilitates distribution of the mixed air to each zone. Additionally, each zone is equipped with a VAV box that regulates airflow rate to control individual zone temperatures.

The temperature of each zone is affected by its neighbor zones and the external environments. The thermal dynamics model of zone i is described using the resistance-capacitance network (Yang et al. 2021):

$$
\begin{aligned}
C_i \dot{T}_i &= \sum_{j \in \mathcal{N}_i} \frac{1}{r_{ij}}(T_j - T_i) + \frac{1}{r_{0i}}(T_e - T_i) + u_i + \omega_i, \\
u_i &= c_a m_i (T_s - T_i)
\end{aligned}
\tag{4.92}
$$

Here, $i, j \in \{1, 2, \ldots, n\}$ represents the zone index; $T_i \in \mathbf{R}$ is the temperature of the ith zone, i.e., the system output; $T_e \in \mathbf{R}$ is the outdoor temperature; $\omega_i \in \mathbf{R}$ is the random noise denoting internal and external disturbances; $u_i \in \mathbf{R}$ is the control input; c_a is the specific heat capacity of air; m_i is the flow rate of the supply air; T_s is the supply air temperature; $\mathcal{N}_i$ is the set of all neighbor zones of ith zone; $C_i \in \mathbf{R}$, representing thermal capacitance, is the energy required for temperature change of $1°C$. The thermal resistance $r_{ij} \in \mathbf{R}$ describes the thermal transmission between zones i and j, and $r_{ij} = r_{ji}$. Additionally, the thermal transmission between the external environment and zone i is described as r_{0i}.

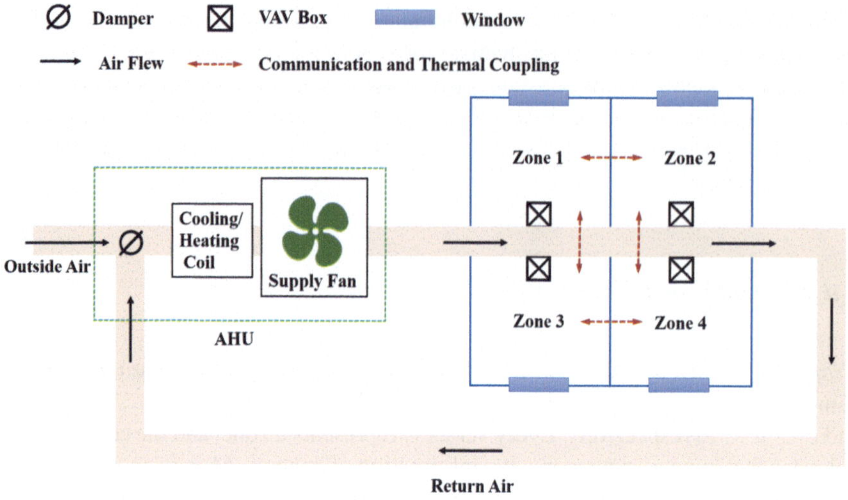

Fig. 4.17 Schematic of the HVAC system

The desired temperature $T_{d,i}$ of the ith zone, which is set for a 24-hour period, can be denoted as $T_{d,i} \in \mathbf{R}$. To achieve the desired temperature $T_{d,i}$, the adjustable variables are the airflow rate m_i and the supply air temperature T_s. In practical applications, we generally set $T_s > T_{d,i}$ for heating and $T_s < T_{d,i}$ for cooling. Subsequently, a suitable input value of $u_{d,i}$ is generated by adjusting the value of m_i. However, due to difficulties in obtaining accurate system parameters such as C_i and r_{ij} in practice, it becomes challenging to directly calculate an appropriate input value of $u_{d,i}$.

To enhance control performance, we incorporate a learning concept into the temperature control system, which undergoes repeated cycles every 24 h. A data-driven strategy is employed to iteratively update the current input based on previous information. Consequently, we introduce an iteration index to account for the dynamics of the system. In particular, (4.92) is rewritten as follows:

$$C_i \frac{dT_{k,i}}{dt} = \sum_{j \in \mathcal{N}_i} \frac{1}{r_{ij}} (T_{k,j}(t) - T_{k,i}(t)) + u_{k,i}(t) + x_{k,i}(t),$$
$$x_{k,i}(t) = \frac{1}{r_{0i}} (T_{k,e}(t) - T_{k,i}(t)) + \omega_{k,i}(t), \tag{4.93}$$

where k and t represent the iteration and time indices, respectively. The time interval of an iteration is 24 h. $x_i(t) \in \mathbf{R}$ indicates the impact from the outdoor temperature and random noise.

The outdoor temperature $T_{k,e}(t)$ varies throughout the learning process, and weather patterns may repeat annually (Minakais et al. 2019). Consequently, the mean daily outdoor temperature $T_e(t)$ is employed as a substitute for the actual outdoor temperature $T_{k,e}(t)$ in system dynamics (4.93), while the uncertain discrepancy between $T_e(t)$ and $T_{k,e}(t)$ is represented by random variables $\omega_{k,i}(t)$ (Peng et al. 2016). Thus, an approximate model of (4.93) can be formulated as follows:

$$
\begin{aligned}
C_i \frac{dT_{k,i}}{dt} &= \sum_{j \in \mathcal{N}_i} \frac{1}{r_{ij}} (T_{k,j}(t) - T_{k,i}(t)) + u_{k,i}(t) + x_{k,i}(t), \\
x_{k,i}(t) &= \frac{1}{r_{0i}} (T_e(t) - T_{k,i}(t)) + \omega_{k,i}(t),
\end{aligned} \tag{4.94}
$$

For further analysis, the following assumptions are required.

Assumption 4.12 For each zone, the initial temperature $T_{k,i}(0)$ is identical to the desired initial value $T_{d,i}(0)$ for all iterations, i.e., $T_{k,i}(0) = T_{d,i}(0), \forall k, i$.

Assumption 4.13 The random variable $\omega_{k,i}(t)$ is independent and stationary with respect to iteration index k, time index t, and zone number i. Moreover, $\mathbb{E}\omega_{k,i}(t) = 0$, $\sup_k \mathbb{E}\|\omega_{k,i}(t)\|^2 < \infty$, and $\forall \lambda > 0$, $\min_k \mathbb{P}(|\omega_{k,i}(t)| \leq \lambda) > 0, \forall i \in \{1, 2, \ldots, n\}$.

Assumption 4.12 is referred to as identical initialization condition. In the case of HVAC systems, where the control task and objective are repetitive, this assumption holds significant relevance. Assumption 4.13 introduces a fundamental requirement for random noise. Given that the learning process can be repeated along the iterative axis, it is reasonable to accept Assumption 4.13 even when dealing with unknown random noise. An example that satisfies this assumption is commonly encountered Gaussian noise.

In practical implementation, it is acknowledged that only a subset of zones attains the desired temperature from a multitude of zones. The control objective entails ensuring consistent achievement of the desired temperature across all zones through inter-zone communication. Each zone acquires temperature information from its neighboring zones by deploying sensors, facilitated by Internet of Things devices (Papadopoulos et al. 2020).

Remark 4.11 Compared to centralized frameworks, distributed schemes offer advantages such as reduced computation, enhanced flexibility, and improved stability. In contrast to decentralized frameworks lacking communication capabilities, distributed schemes prove more effective in jointly tracking the desired temperature. However, practical applications often encounter data damage due to signal transmission through the communication network. This issue is extensively discussed within the field of network control.

A virtual zone v_0 is introduced to represent the desired temperature. For the communication network $\bar{\mathcal{G}} = \{\mathcal{V} \cup 0, \bar{\mathcal{E}}\}$, the edge $(v_i, v_j) \in \mathcal{E}$ indicates that zone v_i obtains the temperature of zone v_j. Additionally, define $\mathcal{D} \triangleq \text{diag}\{d_1, d_2, \ldots, d_n\}$. If zone v_i is connected to the virtual zone v_0, then $d_i = 1$, indicating that the desired temperature is available for zone v_i; otherwise, $d_i = 0$.

Assumption 4.14 The graph $\bar{\mathcal{G}} = \{\mathcal{V} \bigcup 0, \bar{\mathcal{E}}\}$ contains a spanning tree, and the virtual zone is the root of the spanning tree.

Assumption 4.14 guarantees that the desired temperature is globally accessible. It means that each zone obtains the desired temperature through multiple iterations.

Lemma 4.3 *(Hu and Hong (2007)) The eigenvalues of $\mathcal{L} + \mathcal{D}$ have positive real parts when the virtual agent is the root of the spanning tree of the graph $\bar{\mathcal{G}}$.*

Problem Statement: The control objective is to establish a distributed control scheme with gain adaptation for stochastic HVAC systems, ensuring that all zones achieve the desired temperature. The convergence speed is of paramount importance in practical applications. Moreover, operational data can be effectively utilized to practically accelerate the convergence speed.

The temperature dynamics are influenced by the intricate nature of building materials, stored items, and human activities, resulting in variations across different zones. This inherent characteristic poses practical challenges in designing and analyzing distributed learning temperature control systems (de Dear et al. 2013).

4.4.3 Distributed Learning Control Scheme

The DLC scheme for HVAC systems is established in this subsection, incorporating a pre-defined decreasing gain sequence to effectively mitigate the impact of random noise. Furthermore, an analysis on the convergence of the distributed control scheme is conducted.

For all zones satisfying (4.92), we lift the variables to have

$$
\begin{aligned}
\mathbf{T}_k(t) &= [T_{k,1}(t), T_{k,2}(t), \ldots, T_{k,n}(t)]^T \in \mathbf{R}^{n\times 1},\\
\mathbf{u}_k(t) &= [u_{k,1}(t), u_{k,2}(t), \ldots, u_{k,n}(t)]^T \in \mathbf{R}^{n\times 1},\\
\mathbf{T}_E(t) &= 1_n \otimes T_e(t) \in \mathbf{R}^{n\times 1},\\
\omega_k(t) &= [\omega_{k,1}(t), \omega_{k,2}(t), \ldots, \omega_{k,n}(t)]^T \in \mathbf{R}^{n\times 1},
\end{aligned}
$$

where $1_n = [1, 1, \ldots, 1]^T \in \mathbf{R}^{n\times 1}$.

The HVAC systems (4.92) are transformed into a compact form as follows:

$$
C_T \frac{d\mathbf{T}_k(t)}{dt} = (R_T - R_0)\mathbf{T}_k(t) + \mathbf{u}_k(t) + R_0\mathbf{T}_E(t) + \omega_k(t). \tag{4.95}
$$

Here, $C_T = \mathrm{diag}\{C_1, C_2, \ldots, C_n\} \in \mathbf{R}^{n\times n}$, and $R_0 = \mathrm{diag}\{\frac{1}{r_{01}}, \frac{1}{r_{02}}, \ldots, \frac{1}{r_{0n}}\} \in \mathbf{R}^{n\times n}$ represents the thermal interactions between zones and the outside environment; R_T is the thermal resistance matrix, including the information of thermal transmission:

$$R_T = \begin{pmatrix} -\sum_{j\in\mathcal{N}_1}\frac{1}{r_{1j}} & \frac{1}{r_{12}} & \cdots & \frac{1}{r_{1n}} \\ \frac{1}{r_{21}} & -\sum_{j\in\mathcal{N}_2}\frac{1}{r_{2j}} & \cdots & \frac{1}{r_{2n}} \\ \vdots & \vdots & \ddots & \vdots \\ \frac{1}{r_{n1}} & \frac{1}{r_{n2}} & \cdots & -\sum_{j\in\mathcal{N}_n}\frac{1}{r_{nj}} \end{pmatrix}.$$

By defining the sampling period as Δt, the dynamic equation (4.95) is discretized as follows:

$$\begin{aligned} \mathbf{T}_k(t+1) &= A_p(t)\mathbf{T}_k(t) + B_p(t)(\mathbf{u}_k(t) + R_0\mathbf{T}_E(t) + \omega_k(t)), \\ \mathbf{y}_k(t) &= C_p(t)\mathbf{T}_k(t). \end{aligned} \tag{4.96}$$

Here, $t = 0, 1, 2, \ldots, N$ is the time index, and N is the length of an iteration corresponding to the period of 24 h; $A_p(t) \in \mathbf{R}^{n\times n}$ $B_p(t) \in \mathbf{R}^{n\times n}$ and $C_p(t) \in \mathbf{R}^{n\times n}$ are the system matrices, where $A_p(t) = I + \Delta t {C_T}^{-1}(R_T - R_0)$, $B_p(t) = \Delta t {C_T}^{-1}$, $C_p(t) = I$. Since $C_T = \text{diag}\{C_1, C_2, \ldots, C_n\}$, we have $B_p(t) = \text{diag}\{\Delta t/C_1, \Delta t/C_2, \ldots, \Delta t/C_n\}$.

Denote $\mathbf{y}_d(t) = [T_{d,1}(t), T_{d,2}(t), T_{d,3}(t), \ldots, T_{d,n}(t)]^T \in \mathbf{R}^{n\times 1}$ as the desired temperature reference for all zones. The thermal capacitance C_i is a positive real number, with $C_i > 1$. Thus, the square matrix $B_p(t)$ is a diagonal matrix with diagonal element $\Delta t/C_i > 0$. The desired temperature is realizable because the square matrix $C_p(t)B_p(t)$ is of full rank. It means there are opposite initial state $\mathbf{T}_d(0)$ and desired input $\mathbf{u}_d(t)$ such that

$$\begin{aligned} \mathbf{T}_d(t+1) &= A_p(t)\mathbf{T}_d(t) + B_p(t)(\mathbf{u}_d(t) + R_0\mathbf{T}_E(t)), \\ \mathbf{y}_d(t) &= C_p(t)\mathbf{T}_d(t). \end{aligned} \tag{4.97}$$

The DLC scheme is given as follows:

$$u_{k+1,j}(t) = u_{k,j}(t) + \lambda_k \Gamma(t)\xi_{k,j}(t+1), \tag{4.98}$$

where

$$\xi_{k,j}(t) = \sum_{i\in N_j} a_{j,i}(T_{k,i}(t) - T_{k,j}(t)) + d_j(T_{d,j}(t) - T_{k,j}(t)).$$

The matrix $\Gamma(t)$ is the direction regulation matrix (DRM), and λ_k is a decreasing gain satisfying $\lambda_k > 0$, $\lambda_k \to 0$, $\sum_{k=1}^{\infty}\lambda_k = \infty$, and $\sum_{k=1}^{\infty}\lambda_k^2 < \infty$.

Define the output error $E_{k,j}(t) \triangleq T_{d,j}(t) - T_{k,j}(t)$. Then, the learning scheme (4.98) is written as

$$\begin{aligned}
&u_{k+1,j}(t)\\
&= u_{k,j}(t) + \lambda_k \Gamma(t)\Big[\sum_{i\in N_j} a_{j,i}(E_{k,j}(t+1) - E_{k,i}(t+1)).\\
&\quad + d_j E_{k,j}(t+1)\Big].
\end{aligned} \tag{4.99}$$

Define $\mathbf{E}_k(t) \triangleq \mathbf{T}_d(t) - \mathbf{T}_k(t)$. Then, the learning scheme (4.98) is given in the lifting form as follows:

$$\mathbf{u}_{k+1}(t) = \mathbf{u}_k(t) + \lambda_k[(\mathcal{L} + \mathcal{D}) \otimes \Gamma(t)]\mathbf{E}_k(t+1), \tag{4.100}$$

where $\mathcal{L}$ is the Laplacian matrix and $\mathcal{D} \triangleq \mathrm{diag}\{d_1, d_2, \ldots, d_n\}$ represents the access ability of the desired temperature $T_{d,i(t)}$ by the ith zone.

Before proceeding to the convergence of the DLC scheme (4.98), we further lift the vectors over the entire iteration interval, i.e.,

$$\begin{aligned}
\mathbf{U}_k &= [\mathbf{u}_k(0)^T, \mathbf{u}_k(1)^T, \cdots, \mathbf{u}_k(N-1)^T]^T,\\
\mathbf{T}_k &= [\mathbf{T}_k(1)^T, \mathbf{T}_k(2)^T, \cdots, \mathbf{T}_k(N)^T]^T,\\
\bar{\mathbf{T}}_e &= \left[(B_p(0)R_0\mathbf{T}_E(0))^T \cdots \left(\sum_{i=0}^{N-1} A_{i+1}^{N-1} B_p(i) R_0 \mathbf{T}_E(0)\right)^T\right]^T,\\
\zeta_k &= \left[(B_p(0)\omega_k(0))^T, \cdots, \left(\sum_{i=0}^{N-1} A_{i+1}^{N-1} B_p(i)\omega_k(i)\right)^T\right]^T,
\end{aligned}$$

where $A_i^j \triangleq A_p(i)A_p(i+1)\ldots A_p(j)$ for $j \geq i$. Denote $\mathcal{H}$ as the lifted system matrix:

$$\mathcal{H} = \begin{pmatrix}
B_p(0) & 0 & \cdots & 0\\
A_p(1)B_p(0) & B_p(1) & \cdots & 0\\
\vdots & \vdots & \ddots & \vdots\\
A_1^{N-1}B_p(0) & A_2^{N-1}B_p(0) & \cdots & B_p(N-1)
\end{pmatrix}.$$

The dynamics (4.96) becomes

$$\mathbf{T}_k = \mathcal{H}\mathbf{U}_k + \mathbf{T}_{k0} + \bar{\mathbf{T}}_e + \zeta_k, \tag{4.101}$$

where $\mathbf{T}_{k0} = [(A_p(0)\mathbf{T}_k(0))^T, \cdots, (A_0^{N-1}\mathbf{T}_k(0))^T]^T$.

Similarly, for the desired temperature reference T_d, we have

$$\mathbf{T}_d = \mathcal{H}\mathbf{U}_d + \mathbf{T}_{d0} + \bar{\mathbf{T}}_e, \tag{4.102}$$

where $\mathbf{T}_{d0} = [(A_p(0)\mathbf{T}_d(0))^T, \cdots, (A_0^{N-1}\mathbf{T}_d(0))^T]^T$. According to Assum ption 4.12, we have $\mathbf{T}_{d0} = \mathbf{T}_{k0}, \forall k$.

Define $\Delta\mathbf{U}_k = \mathbf{U}_d - \mathbf{U}_k$. From (4.101) and (4.102), the tracking error is expressed as follows:

$$\mathbf{E}_k \triangleq \mathbf{T}_d - \mathbf{T}_k = \mathcal{H}\Delta\mathbf{U}_k - \zeta_k. \tag{4.103}$$

Denote

$$\Phi = \mathrm{diag}\{(\mathcal{L}+\mathcal{D}) \otimes \Gamma(0), \cdots, (\mathcal{L}+\mathcal{D}) \otimes \Gamma(N-1)\}. \tag{4.104}$$

Here, the matrix Φ represents the information flow among zones.

The compact form of (4.100) is given by

$$\mathbf{U}_{k+1} = \mathbf{U}_k + \lambda_k \Phi \mathbf{E}_k. \tag{4.105}$$

Indeed, the actual temperature of each zone is affected by various random noises. Thus, it is not easy to realize $\mathbf{T}_k - \mathbf{T}_d \to 0$. Instead, we prove the convergence of input error $\Delta\mathbf{U}_k$, i.e., $\Delta\mathbf{U}_k \to 0$.

Theorem 4.6 *Consider the HVAC system* (4.93) *under Assumptions 4.12–4.14. For the DLC scheme* (4.98)*, if DRM $\Gamma(t)$ satisfies that all eigenvalues of* $[(\mathcal{L}+\mathcal{D}) \otimes \Gamma(t)]B_p(t)$ *are with positive real parts for all time instants and the decreasing gain λ_k satisfies $\lambda_k > 0$, $\lambda_k \to 0$ and $\sum_{k=1}^{\infty} \lambda_k = \infty$, then the input error $\Delta\mathbf{U}_k$ converges to zero in the mean-square sense, i.e.,* $\lim_{k\to\infty} \mathbb{E}[\|\mathbf{U}_d - \mathbf{U}_k\|^2] = 0$.

Corollary 4.3 *Consider the HVAC system* (4.93) *under the conditions of Theorem 4.6. Applying the DLC scheme* (4.98)*, if the decreasing gain λ_k satisfies an additional requirement $\sum_{k=1}^{\infty} \lambda_k^2 < \infty$, then the input error $\Delta\mathbf{U}_k$ converges to zero in the almost-sure sense, i.e.,* $\mathbb{P}(\lim_{k\to\infty} \Delta\mathbf{U}_k = 0) = 1$.

The proofs of Theorem 4.6 and Corollary 4.3 can be established by following the procedures and techniques provided in the previous chapter, as the learning gain sequence is given prior to the learning process. These results summarize the convergence property of the DLC scheme (4.98), which relies on a decreasing learning gain sequence $\{\lambda_k\}$. However, selecting such a sequence $\{\lambda_k\}$ based on experience and trial rather than operational data poses challenges in practice. Finding a suitable sequence is not straightforward, and an inappropriate selection could significantly slow down convergence speed, particularly during early stages of learning. This naturally raises the question: Can we design an adaptive mechanism that utilizes available information to automatically generate an adaptive learning gain according to some performance index? We address this question in the next subsection.

4.4.4 Accelerated Distributed Learning Control Scheme Using Gain Adaptation

In this subsection, we initially investigate the role of random noise in the learning process. Subsequently, we propose a gain adaptation mechanism to enhance convergence speed during the early stages of learning. Finally, we introduce and analyze the accelerated DLC (ADLC) scheme.

According to (4.103), the tracking error e_k comprises of an inaccurate tracking term $\mathcal{H}\Delta\mathbf{U}_k$ and a noise term ζ_k. The process of learning $\mathbf{E}_k \to 0$ can be divided into two stages. In the early stage, the output error $\mathbf{E}_k$ is primarily influenced by $\mathcal{H}\Delta\mathbf{U}_k$, while the impact of noise is negligible. This is attributed to the initial input $\mathbf{U}_k$ having a significant deviation from the desired input $\mathbf{U}_d$. Consequently, for most iterations in this stage, $\mathbf{E}_k^T\mathbf{E}_{k-1} > 0$. Subsequently, there is a rapid reduction in the inaccurate tracking term leading to its diminished influence on the tracking error. As a result, the noise term becomes dominant and marks the transition into the second stage of learning. The occurrence of event ${\zeta_k}^T\zeta_{k-1} < 0$ will consequently lead to event $\mathbf{E}_k^T\mathbf{E}_{k-1} < 0$. Therefore, it becomes necessary to introduce a mechanism that eliminates or mitigates any adverse effects caused by noise.

Motivated by the above observations, we define an adaptive gain parameter s_k as follows:

$$s_k = s_{k-1} + \mathbf{1}_{[\mathbf{E}_k^T\mathbf{E}_{k-1}<0]}, \quad k \geq 2, \tag{4.106}$$

where $\mathbf{1}_{[\mathbf{E}_k^T\mathbf{E}_{k-1}<0]} = 1$ if the event $\mathbf{E}_k^T\mathbf{E}_{k-1} < 0$ is true, otherwise $\mathbf{1}_{[\mathbf{E}_k^T\mathbf{E}_{k-1}<0]} = 0$. The initial value is $s_1 = 1$. From (4.106), we obtain $s_k = 1 + \sum_{i=1}^{k-1} \mathbf{1}_{[\mathbf{E}_k^T\mathbf{E}_{k-1}<0]}$.

Remark 4.12 The calculation of s_k in (4.106) requires the output error from all zones. Each zone is equipped with a temperature sensor to monitor its temperature (Bhaskoro et al. 2013). Subsequently, the central processing unit processes information gathered from these sensors. The adaptive gain s_k is then determined and disseminated by the central processing unit (Gholamzadehmir et al. 2020).

Based on the adaptive gain parameter s_k, the proposed ADLC algorithm is given as follows:

$$u_{k+1,j}(t) = u_{k,j}(t) + s_k^{-1}\Gamma(t)\xi_{k,j}(t+1), \tag{4.107}$$

where

$$\xi_{k,j}(t) = \sum_{i\in N_j} a_{j,i}(T_{k,i}(t) - T_{k,j}(t)) + d_j(T_{d,j}(t) - T_{k,j}(t)).$$

Compared with the original distributed algorithm (4.98), the decreasing gain λ_k is replaced by an adaptive gain s_k^{-1} in the novel algorithm.

By the definition of Φ in (4.104), the proposed ADLC scheme (4.107) is rewritten as follows:

$$\mathbf{U}_{k+1} = \mathbf{U}_k + s_k^{-1}\Phi\mathbf{E}_k. \tag{4.108}$$

Remark 4.13 The adaptive selection of the learning gain is a formulation of a simple event-trigger mechanism that utilizes the sign of $\mathbf{E}_k^T\mathbf{E}_{k-1}$ to reflect the impact of random noise. Consequently, the value of s_k will experience random increases, resulting in a sequence of decreasing gains s_k^{-1} that are contingent upon practical tracking performance. The adaptive learning gain s_k^{-1} represents a compromise between constant and decreasing gains and can be considered equivalent to a constant gain during the initial stages.

We summarize the main theorem as follows.

Theorem 4.7 *Consider the HVAC system* (4.93) *under Assumptions 4.12-4.14. Apply the proposed ADLC scheme* (4.107)*, where the gain sequence are generate by* (4.106)*. If DRM $\Gamma(t)$ satisfies that all eigenvalues of $[(\mathcal{L}+\mathcal{D})\otimes\Gamma(t)]B_p(t)$ have positive real parts for all time instants, then the input error $\Delta\mathbf{U}_k$ converges to zero, a.s., and the scale variable s_k satisfies $s_k/k \to \mathbb{P}(\zeta_1^T\zeta_2 < 0)$ as $k\to\infty$ a.s.*

Proof. The proof is based on the Lemma 4.1. We mainly verify that Assumptions 4.4 and 4.5 are valid.

The input error is expressed by

$$\Delta\mathbf{U}_{k+1} = \Delta\mathbf{U}_k - s_k^{-1}(\Phi\mathcal{H}\Delta\mathbf{U}_k - \Phi\zeta_k).$$

Here, $\Delta\mathbf{U}_{k+1}$, $\Phi\mathcal{H}\Delta\mathbf{U}_k$ and $-\Phi\zeta_k$ correspond to $\bar{x}_{k+1}$, $\varphi(\bar{x}_k)$ and ϵ_k in Lemma 4.1, respectively. Meanwhile, the continuous function $\varphi(\cdot)$ in Lemma 4.1 corresponds to $\Phi\mathcal{H}x$.

As stated in the proof of Theorem 4.6, $\Phi\mathcal{H}$ is a low-triangular matrix with diagonal blocks $[(\mathcal{L}+\mathcal{D})\otimes\Gamma(t)]B_p(t)$, which are of full rank because all eigenvalues of $[(\mathcal{L}+\mathcal{D})\otimes\Gamma(t)]B_p(t)$ are nonzero. This implies that $\Phi\mathcal{H}$ is of full rank, indicating that $x=0$ is the only root of continuous function $\varphi(x)=\Phi\mathcal{H}x$.

Next, we verify Assumption 4.4. According to Assumption 4.13, the random variable ζ_k is independent and identical with respect to the iteration index k, such that $\mathbb{E}\zeta_k = 0$, $\mathbb{E}\zeta_k(t)\zeta_k^T(t) < \infty$, and $\forall\lambda > 0$, $\min_k \mathbb{P}(|\zeta_k| \le \lambda) > 0$. It means that Assumption 4.4 holds.

For Assumption 4.5, we first show that the input error $\Delta\mathbf{U}_{k+1} = \Delta\mathbf{U}_k - \gamma\Phi\mathcal{H}\Delta\mathbf{U}_k$ converges to zero for all $\gamma > 0$. Since $\Phi\mathcal{H}$ is a low-triangular matrix, $I - \gamma\Phi\mathcal{H}$ is a low-triangular matrix, whose eigenvalues are the group of eigenvalues of diagonal blocks $1-\gamma[(\mathcal{L}+\mathcal{D})\otimes\Gamma(t)]B_p(t)$, $t=0,1,\ldots,N-1$.

To ensure $\Delta\mathbf{U}_k \to 0$, the spectral radius of the matrix Φ should satisfy $\rho(I-\gamma\Phi\mathcal{H}) < 1$. In other words, $\Gamma(t)$ must satisfy the following inequality

$$\rho(1-\gamma[(\mathcal{L}+\mathcal{D})\otimes\Gamma(t)]B_p(t)) < 1 \quad \forall t\in\{0,1,\ldots,N-1\}.$$

Meanwhile, $B_p(t) = \text{diag}[\frac{\Delta t}{C_1}, \frac{\Delta t}{C_2}, \ldots, \frac{\Delta t}{C_n}]$, where C_i is the thermal capacitance, and Δt representing sampling period is smaller than C_i. Clearly, the eigenvalues of $B_p(t)$ are its diagonal elements, $\rho(B_p(t)) < 1$ for $\forall t \in \{0, 1, \ldots, N-1\}$

$\Gamma(t)$ is designed by $\Gamma(t) = \tau\Gamma_0(t)$, where τ is a sufficiently small positive constant, and $\Gamma_0(t)$ is a positive function of t. By Lemma 4.3, the eigenvalues of $\mathcal{L} + \mathcal{D}$ have positive real parts. From $\rho(B_p(t)) < 1$ and for sufficiently small constant τ, the condition $\rho(1 - \gamma[(\mathcal{L} + \mathcal{D}) \otimes \Gamma(t)]B_p(t)) < 1\ \forall t \in \{0, 1, \ldots, N-1\}$ holds.

Additionally, all eigenvalues of $\Phi\mathcal{H}$ are positive real numbers. Thus, there exists a positive-definite matrix P such that $(\Phi\mathcal{H})^T P + P\Phi\mathcal{H} = I$. Define $\mathbb{V}(x) = x^T Px$, where $x = 0$ is the root of $\mathbb{V}(x) = 0$. Moreover, $\nabla^2\mathbb{V}(x) = 2P$, then the matrix J in (4.18) is considered as $2P$. Next, we verify the establishment of (4.19). The left side of inequality is

$$\begin{aligned}\varphi(x)^T \nabla\mathbb{V}(x) &= 2x^T(\Phi\mathcal{H})^T Px \\ &= x^T(\Phi\mathcal{H})^T Px + x^T P(\Phi\mathcal{H})x \\ &= x^T[(\Phi\mathcal{H})^T P + P(\Phi\mathcal{H})]x = x^T x.\end{aligned} \tag{4.109}$$

Meanwhile, according to $J = 2P$, we have

$$\begin{aligned}\varphi(x)^T J\varphi(x) &= 2x^T(\Phi\mathcal{H})^T J(\Phi\mathcal{H})x \\ &\le 4\bar{\sigma}((\Phi\mathcal{H})^T P(\Phi\mathcal{H}))x^T x.\end{aligned} \tag{4.110}$$

Hence, selecting $\alpha < [4\bar{\sigma}((\Phi\mathcal{H})^T P(\Phi\mathcal{H}))]^{-1}$, we have

$$\alpha\varphi(x)^T J\varphi(x) \le x^T x. \tag{4.111}$$

By making α and β sufficiently small, we obtain the following inequality

$$\frac{1}{2}x^T x \ge \alpha tr(MJ) + \beta \quad for \quad \|x\| \ge \bar{R}. \tag{4.112}$$

Considering (4.110)–(4.112), we obtain

$$x^T x = \frac{1}{2}x^T x + \frac{1}{2}x^T x \ge \frac{\alpha}{2}\varphi(x)^T J\varphi(x) + \alpha tr(MJ) + \beta.$$

The proof is completed according to Lemma 4.1, whose conditions are satisfied.

If the distribution function of ζ_k is symmetric about zero, then the distribution function F of $\zeta_1^T\zeta_2$ is symmetric about zero. Thus, by Lemma 4.1, we have $\mathbb{P}(\zeta_1^T\zeta_2 < 0) = F(0) = 1/2$. Then, $s_k^{-1} - \frac{1}{2k} \to 0$, i.e., s_k^{-1} is of the same order of $\frac{1}{2k}$. Thus, s_k^{-1} satisfies $s_k^{-1} > 0$, $s_k^{-1} \to 0$, $\sum_{k=1}^{\infty} s_k^{-1} = \infty$ and $\sum_{k=1}^{\infty} \frac{1}{s_k^2} < \infty$. By Theorem 4.6, we obtain that the input error $\Delta\mathbf{U}_k$ converges to zero.

Remark 4.14 The adaptive learning gain s_k^{-1} in the ADLC scheme is generated based on the output error, as opposed to being given as prior information like in DLC scheme (4.98). This adaptation of s_k^{-1} according to practical tracking performance significantly enhances the convergence speed at early stages, where it remains constant. In terms of asymptotic behavior, the adaptive learning gain s_k^{-1} follows an order of $\frac{1}{2k}$. Consequently, the ADLC scheme achieves exponential convergence with $\|\Delta \mathbf{U}_k\|^2 \to 0$, specifically characterized by a decay rate of $O(k^{-1}) \to 0$.

The specific procedure is provided as follows:

Algorithm 4.1 The ADLC scheme for HVAC systems

Data: operation length N, initial temperature $T_{k,j}(0)$, the stopping criterion (maximum iteration or precision demand);
Result: Optimal input $u_{k,j}(t)$;
Step 1: Initialize $k := 1$, $t := 0$, $s_1 = 1$, $u_{1,j}(t) = 0$;
Step 2: For $t = 0$ to N, using $u_{k,j}(t)$, run HVAC system (4.93) and compute the tracking error $E_{k,j}(t)$;
Step 3: Using equation (4.106), update the value of s_k;
Step 4: Using equation (4.107), update input $u_{k+1,j}(t)$;
Step 5: If the stopping criterion is not satisfied, let $k := k + 1$ and go to **Step 2**; otherwise, end the algorithm.

4.4.5 *Numerical Results*

In this section, we simulate a building HVAC system to demonstrate the effectiveness of the proposed schemes using a four-zone building (Yang et al. 2020; Yu et al. 2018). The system dynamics parameters are adopted from Li et al. (2016), as presented in Table 4.3. The sampling period is set at $\Delta t = 1\text{h}$, and the iteration length N is 24.

Figure 4.18 shows the communication topology, where v_0 represents the virtual zone. The zones v_2 and v_3 directly access the tracking reference, whereas v_1 and v_4 obtain information from v_3 and v_2. According to the communication topology, we have

$$\mathcal{L} = \begin{bmatrix} 1 & 0 & -1 & 0 \\ 0 & 1 & 0 & -1 \\ 0 & 0 & 0 & 0 \\ 0 & 0 & -1 & 1 \end{bmatrix}, \quad \mathcal{D} = \begin{bmatrix} 0 & 0 & 0 & 0 \\ 0 & 1 & 0 & 0 \\ 0 & 0 & 1 & 0 \\ 0 & 0 & 0 & 0 \end{bmatrix}.$$

The hourly mean outdoor temperature T_e of the summer is calculated from the historical temperature data of Beijing. The data are collected from the European Centre for Medium-Range Weather Forecasts[1]. Figure 4.19. shows the hourly mean outdoor temperature T_e and the desired temperature T_d, depicting the significance of HVAC systems. For simplicity, the random noise ω_k is assumed to follow a normal

[1] [Online]. Available: https://cds.climate.copernicus.eu.

Table 4.3 Simulation parameters

Parameter	Value	Unit
C_i	20	kJ/°C
$r_{12}, r_{13}, r_{14}, r_{23}, r_{24}, r_{34}$	23	°C/k W
$r_{01}, r_{02}, r_{03}, r_{04}$	15	°C/k W
Δt	1	Hour

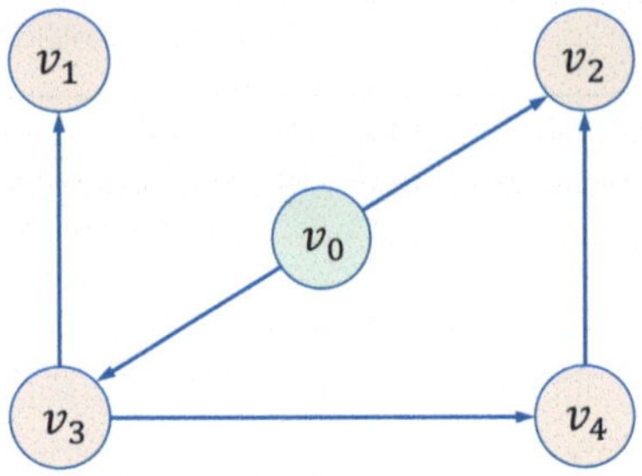

Fig. 4.18 Topology of the communication graph

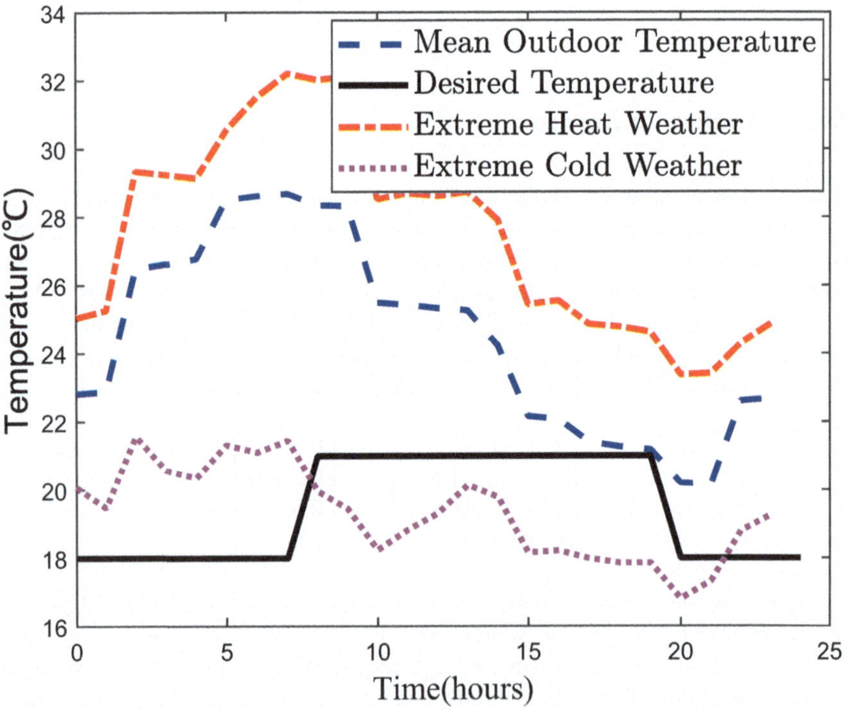

Fig. 4.19 Outdoor temperature and tracking reference

distribution with covariance $\sigma^2 = 0.1^2$. The learning matrix $\Gamma(t)$ is set to $5I$ satisfying the conditions in Theorem 4.6 and 4.7. The DLC scheme (4.98) and ADLC scheme (4.107) are implemented for 120 iterations.

The following numerical results explain several aspects of the proposed schemes to verify the theoretical results.

First, to demonstrate the effectiveness of the DLC scheme, Fig. 4.20 presents the tracking performance of four zones at the 15th and 60th iterations. The decreasing gain is set as $\lambda_k = 3.5/k^{0.55}$. It can be observed that the tracking performance is unsatisfactory at the 15th iteration but becomes acceptable by the 60th iteration.

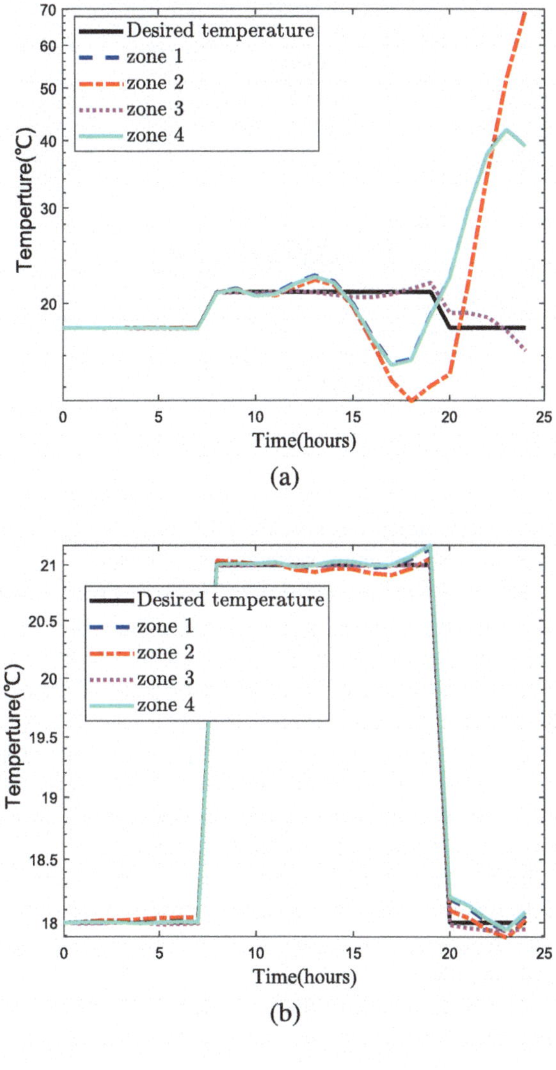

Fig. 4.20 Profiles for temperature of all zones for the DLC scheme: **a** the tracking performance at the 15th iteration. **b** the tracking performance at the 60th iteration

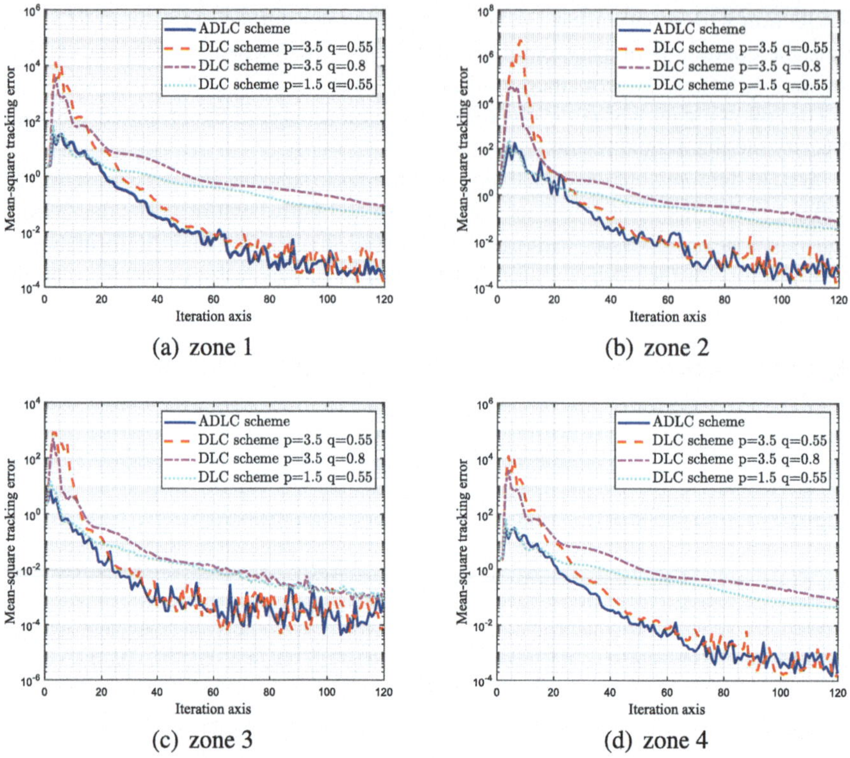

(a) zone 1 (b) zone 2

(c) zone 3 (d) zone 4

Fig. 4.21 Mean-square tracking error profiles of the DLC and ADLC schemes

This clearly illustrates how the DLC scheme (4.98) enhances tracking performance over iterations, thereby validating its efficacy.

By satisfying the condition of Theorem 4.6, a suitable selection for the decreasing gain can be expressed as $\lambda_k = p/k^q$, where $0 < p$ and $0.5 < q \leq 1$. However, in practical applications, it is challenging to determine an appropriate decreasing gain prior to the learning process, while the generation of s_k remains adaptive throughout the learning process.

The mean-square tracking error profiles of the proposed schemes are illustrated in Fig. 4.21, where the mean-square tracking error is defined as $\|e_k\|^2/N$. For the DLC scheme, three selections of the decreasing gain are provided based on empirical knowledge. Specifically, the dash-dotted, dotted, and dashed lines correspond to $\lambda_k = 3.5/k^{0.55}$, $\lambda_k = 3.5/k^{0.8}$, and $\lambda_k = 1.5/k^{0.55}$ respectively. As depicted in Fig. 4.21, a decrease in convergence rate is observed as p decreases or q increases. Thus, employing a larger decreasing gain is anticipated to enhance tracking precision during finite iterations.

The performance of the proposed DLC scheme (4.98) is compared with that of the ADLC scheme (4.107). In Fig. 4.21, the solid line represents the proposed ADLC

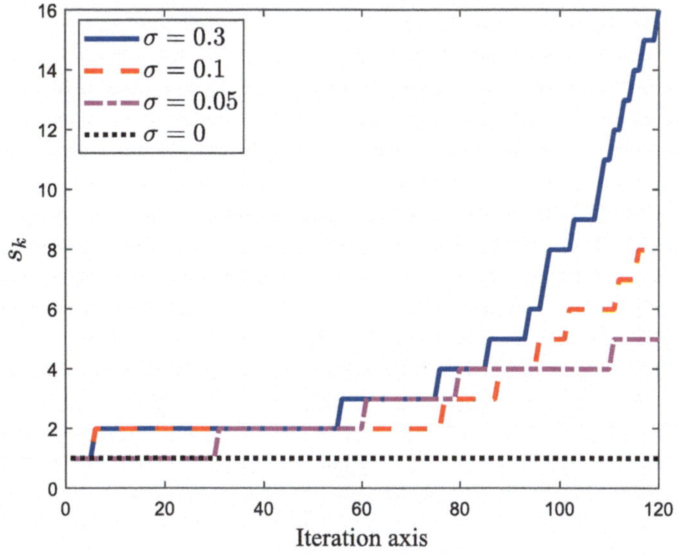

Fig. 4.22 Value of s_k at different noise levels, the covariance of noise is varied as $\sigma = 0, 0.05, 0.1$ and 0.3

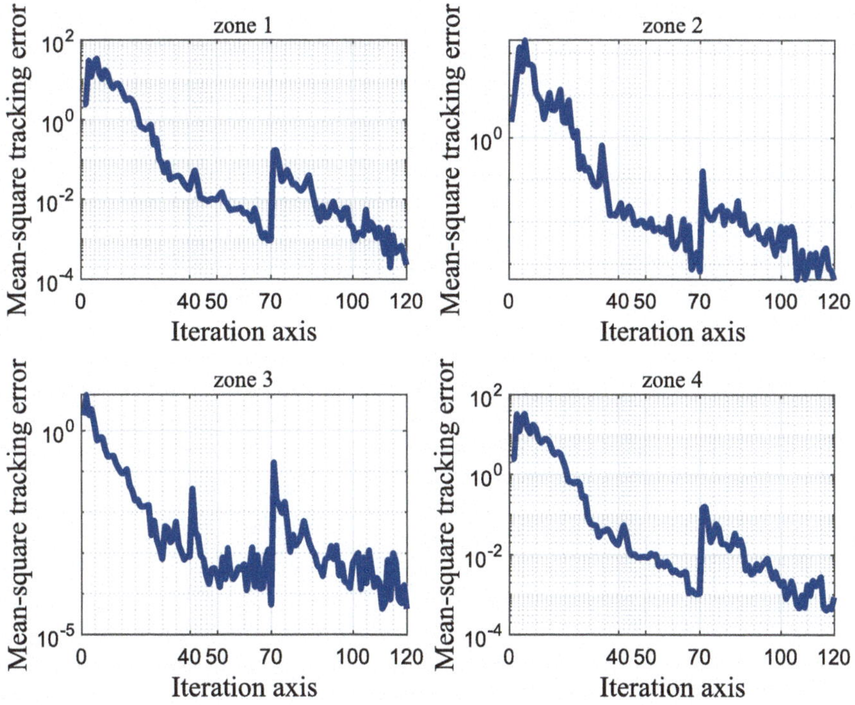

Fig. 4.23 Tracking error for the ADLC scheme when the extreme weathers appear. The extreme heat weather and extreme cold weather appear at the 40th and 70th iteration, respectively

scheme. It can be observed that both DLC and ADLC schemes converge to a small error range when appropriate decreasing gains for DLC schemes are selected. The ADLC scheme exhibits improved convergence speed at the early stages, highlighting the significance of introducing decreasing gain and gain adaptation mechanisms.

The adaptive gain behaviors at different noise levels are illustrated in Fig. 4.22, where the level of noise is represented by the covariance σ^2. Here, we set σ to be σ =0, 0.05, 0.1, and 0.3. From Fig. 4.22, it can be observed that for all noise levels, the value of s_k remains constant during the early stage of the trials and then intermittently increases as the iteration number increases. A higher noise level σ corresponds to an earlier onset of s_k increase. These observations validate the adaptability of the learning gain $1/s_k$ with respect to varying noise levels.

Finally, the impact of extreme weather conditions is taken into consideration as they heavily influence the temperature of different zones. Two instances of extreme weather are depicted in Fig. 4.19. We conduct simulations using the proposed ADLC scheme when extreme heat and cold weather occur at the 40th and 70th iterations respectively. Figure 4.23 illustrates the mean-square tracking error for all zones during extreme weather events. It can be observed that when such events occur, there is a rapid increase in tracking errors across all zones; however, this increase remains within an acceptable range for maintaining precision. Moreover, it is noteworthy that our proposed ADLC scheme effectively mitigates these errors soon after encountering extreme weather conditions.

Chapter 5
Event-Triggering Gain Design

5.1 Event-Triggering Gain Design in SILC

5.1.1 Motivation of Event-Triggering Gain Design

The most prevalent form of ILC is the P-type scheme, wherein the input for the subsequent iteration is a linear combination of the current iteration's input and its corresponding tracking error (Bristow et al. 2006). In this scheme, the input signal is enhanced by multiplying the tracking error with a constant gain factor. This formulation represents a feedback control structure in the iteration domain rather than a feedforward control structure in the time domain. One commonly employed approach for convergence analysis in this methodology relies on the contraction mapping principle (Moore 2012). For deterministic systems, under certain design conditions of the learning gain matrix, it can be observed that with increasing learning iterations, the input error contracts and eventually converges asymptotically to zero.

In practical applications, stochastic factors are inevitable and diverse. For instance, the operation process of a given system may encompass random disturbances, while the measurement of the output signal can be susceptible to contamination from random noise and inaccurate measurements. Furthermore, in response to the rapid advancement of communication technology, decentralized and distributed structures have been widely adopted in control systems to enhance overall facilitation, flexibility, and robustness (Zhang et al. 2016, 2017). However, these exchange networks also introduce randomness through phenomena like data dropout and quantization. Henceforth, it is imperative to investigate learning control methods for stochastic systems.

In the case of stochastic systems, the traditional P-type scheme with a constant learning gain fails to achieve zero-error tracking (Shen and Wang 2014). This is attributed to the inability of a constant gain to ensure continuous improvement of the input due to inherent randomness. Consequently, the input converges within a bounded region near the desired input determined by the tracking reference, where

D. Shen, *Variable Gain Design in Stochastic Iterative Learning Control*,
Intelligent Control and Learning Systems 13,
https://doi.org/10.1007/978-981-97-8281-9_5

manipulation of the constant gain allows for tunable upper bounds. Furthermore, convergence cannot be guaranteed by applying deterministic analysis methods as they are inadequate in handling situations involving unbounded random variables.

To address this issue, several innovative learning control schemes have been proposed. One approach involves modifying the conventional P-type learning control scheme by replacing the constant gain with a decreasing gain sequence (Shen and Wang 2014). In this scheme, the generated input sequence converges to a limit for systems affected by random noise. However, while enhancing tracking performance, there is a trade-off in terms of reduced convergence speed of the learning control scheme. Specifically, although it suppresses the impact of random noise, a decreasing gain sequence also limits the potential improvement in tracking error. Moreover, it should be noted that the establishment of a decreasing gain sequence typically occurs prior to initiating the learning control process and is independent of observed performance. A more promising technique involves introducing a performance-dependent gain calculation. In Saab (2001a), an approach based on Kalman filtering (KF) was proposed to determine an optimal learning gain matrix by minimizing the trace of input error covariance matrix. However, this method requires complete information about both system dynamics and inherent randomness within it. Some data-driven mechanisms for adaptive gain adaptation have also been suggested. For instance, in Shen and Xu (2020), the learning gain was determined based on sign changes in tracking errors between adjacent iterations; this choice drew inspiration from Kesten's accelerated algorithm (Delyon and Juditsky 1993). These studies represent initial efforts towards reconciling conflicting objectives related to achieving superior tracking performance and high convergence speed; nevertheless, this problem remains unresolved as yet unaddressed fully effectively through adaptive design approaches which have demonstrated effectiveness in improving control performance under uncertain conditions (Li et al. 2021; Tong et al. 2018).

The conflicting objectives of achieving high-precision tracking performance and fast convergence speed are thoroughly investigated in this chapter, leading to the proposal of a novel multi-stage adaptive gain strategy for stochastic systems. This approach aims to strike a balance between maintaining a constant gain, which accelerates convergence speed, and employing a vanishing sequence that enhances tracking performance. In the proposed strategy, the iteration domain is divided into multiple stages consisting of successive iterations. Within each stage, the learning gain remains constant but decreases during transitions between stages to expedite convergence. A crucial aspect of this strategy lies in determining the switching iteration at which the learning gain is reduced. In this study, we determine the switching iteration based on both the ratio of initial input error at the beginning of a stage and consideration of noise drift influence.

The stochastic linear system under consideration incorporates both system and measurement noises. In this study, we apply the conventional P-type scheme with a variable learning gain and thoroughly investigate the generation of the gain sequence. Our approach is inspired by the adaptive steplength method proposed in Yousefian et al. (2012). Firstly, we evaluate the evolution of the mean square of the input error for a constant learning gain, which is estimated using two explicitly defined terms:

contracted initial input error and accumulated noise drift. Subsequently, we determine the switching iteration for a given learning gain by observing when the contracted initial input error becomes smaller than the accumulated noise drift for the first time. To reduce the learning gain, we multiply it with a contracting factor that is less than one; its value can be determined by users. Additionally, to enhance effectiveness and flexibility, we introduce longitudinal stretch parameter and transverse stretch parameter to expand candidate range of learning gains within our contraction function. Compared to Yousefian et al. (2012), our mechanism significantly improves both learning gain sequence and tracking performance. We provide an extensive analysis on newly generated gain sequence as well as convergence properties of our proposed scheme.

5.1.2 Problem Formulation

Consider the following discrete-time stochastic system

$$\begin{cases} x_k(t+1) = A_t x_k(t) + B_t u_k(t) + \omega_k(t), \\ y_k(t) = C_t x_k(t) + \nu_k(t), \end{cases} \tag{5.1}$$

where t and k denote the time and iteration label, respectively. In the above expressions, $t \in \{0, 1, \cdots, N\}$, where N is the iteration length and $k \in \mathbf{N}$. The vectors $x_k(t) \in \mathbf{R}^n$, $u_k(t) \in \mathbf{R}^p$, and $y_k(t) \in \mathbf{R}^q$ denote the system state, input, and output, respectively. The random variables $\omega_k(t)$ and $\nu_k(t)$ represent the system noise and measurement noise, respectively. Time-varying matrices A_t, B_t, and C_t are of appropriate dimensions.

We denote the desired tracking reference by $y_d(t)$. Without loss of generality, the reference is assumed to be realizable, i.e., there exist a unique control $u_d(t)$ and an initial value $x_d(0)$ that satisfies the following equations:

$$\begin{cases} x_d(t+1) = A_t x_d(t) + B_t u_d(t) \\ y_d(t) = C_t x_d(t) \end{cases} \tag{5.2}$$

This requirement guarantees the existence of an appropriate input such that the output can precisely track the desired tracking reference.

To permit further analysis, the following assumptions are imposed.

Assumption 5.1 The input/output coupling matrix $C_{t+1}B_t$ is of full-column rank, $\forall t$.

Assumption 5.2 The state initialization in every iteration is equal to $x_d(0)$, i.e., $x_k(0) = x_d(0)$, $\forall k$, where $x_d(0)$ is the desired initial state, which satisfies $y_d(0) = C_0 x_d(0)$.

Assumption 5.3 The noise terms $\omega_k(t)$ and $\nu_k(t)$ are independent for different time indices and iteration numbers. Furthermore, $\omega_k(t)$ and $\nu_k(t)$ satisfy the following expressions:

$$\mathbb{E}[\omega_k(t)] = \mathbb{E}[\nu_k(t)] = 0, \tag{5.3}$$

$$\mathbb{E}[\|\omega_k(t)\|^2] \le \sigma_\omega^2, \quad \mathbb{E}[\|\nu_k(t)\|^2] \le \sigma_\nu^2. \tag{5.4}$$

Remark 5.1 Assumption 5.1 implies that $p \le q$, where p and q are the input and output dimensions. The assumption ensures the existence of a unique input. Under Assumption 5.1, the desired input, $u_d(t)$, can be defined using the recursive expression $u_d(t) = [(C^+B_t)^T C^+B_t]^{-1}[y_d(t+1) - C^+A_t x_d(t)]$ and $x_d(t+1) = A_t x_d(t) + B_t u_d(t)$ with the desired initial value $x_d(0)$, where $C^+ \triangleq C_{t+1}$. If Assumption 5.1 is violated, i.e., no input exists to drive the output to perfectly track the desired trajectory, the proposed scheme can drive the plant output to converge to a trajectory that is closest to the desired reference (in the Euclidian norm sense). Assumption 5.2 is a necessary condition to guarantee the perfect tracking performance for all time instants. The assumption can be generalized to various scenarios (Shen et al. 2021; Xu and Yan 2005; Zhang and Meng 2023). For example, the initial state can vary randomly but satisfy $\mathbb{E}x_k(0) = x_d(0)$ and $\mathbb{E}[{x_k}^2(0)] \le \infty$. In this case, the effect of initial state errors is completely equivalent to the random noise $\omega_k(t)$ and $\nu_k(t)$. Assumption 5.3 constrains both the system and measurement noises to be of zero mean and have a bounded variance. It is evident that Gaussian white noise satisfies this condition.

The state-space model given in (5.1) is first transformed into the lifting formulation to provide a more concise analysis. By utilizing the lifting formulation, we can focus on designing the learning mechanism along the iteration axis. The input, output, and two forms of system noise are denoted by their respective lifting vectors as follows:

$$\begin{aligned}
\boldsymbol{u}_k &= [u_k(0)^T, u_k(1)^T, \cdots, u_k(N-1)^T]^T,\\
\boldsymbol{y}_k &= [y_k(1)^T, y_k(2)^T, \cdots, y_k(N)^T]^T,\\
\boldsymbol{\omega}_k &= [\omega_k(0)^T, \cdots, \omega_k(N-1)^T]^T,\\
\boldsymbol{\nu}_k &= [\nu_k(1)^T, \cdots, \nu_k(N)^T]^T.
\end{aligned}$$

We denote the lifting vectors of the desired tracking reference and input by $\boldsymbol{y}_d = [y_d(1)^T, y_d(2)^T, \cdots, y_d(N)^T]^T$ and $\boldsymbol{u}_d = [u_d(0)^T, u_d(1)^T, \cdots, u_d(N-1)^T]^T$. We define $A_{i,j} \triangleq A(i)\cdots A(j)$ for $i \le j$ and $A_{j,j+1} = I$.

From (5.1) and (5.2) and Assumption 5.2, we see that

$$\boldsymbol{y}_k = \mathcal{H}\boldsymbol{u}_k + \mathcal{G}\boldsymbol{\omega}_k + \boldsymbol{\nu}_k + \mathcal{M}x_k(0), \tag{5.5}$$

$$\boldsymbol{y}_d = \mathcal{H}\boldsymbol{u}_d + \mathcal{M}x_d(0). \tag{5.6}$$

where

$$\mathcal{H} = \begin{bmatrix} C_1 A_{0,1} B_0 & 0 & 0 & \cdots & 0 \\ C_2 A_{1,1} B_0 & C_2 A_{1,2} B_1 & 0 & \cdots & 0 \\ \vdots & \vdots & \vdots & \ddots & \vdots \\ C_N A_{N-1,1} B_0 & C_N A_{N-1,2} B_1 & \cdots & \cdots & C_N A_{N-1,N} B_{N-1} \end{bmatrix},$$

$$\mathcal{G} = \begin{bmatrix} C_1 A_{0,1} & 0 & 0 & \cdots & 0 \\ C_2 A_{1,1} & C_2 A_{1,2} & 0 & \cdots & 0 \\ \vdots & \vdots & \vdots & \ddots & \vdots \\ C_N A_{N-1,1} & C_N A_{N-1,2} & \cdots & \cdots & C_N A_{N-1,N} \end{bmatrix},$$

$$\mathcal{M} = [(C_1 A_0)^T, (C_2 A_1 A_0)^T, \cdots, (C_N A_{1,N-1})^T]^T.$$

We define the tracking error as $e_k(t) \triangleq y_d(t) - y_k(t)$. Using this definition, it is evident that,

$$\boldsymbol{e}_k \triangleq \boldsymbol{y}_d - \boldsymbol{y}_k = \mathcal{H}\Delta\boldsymbol{u}_k - \mathcal{G}\boldsymbol{\omega}_k - \boldsymbol{v}_k, \tag{5.7}$$

where $\Delta\boldsymbol{u}_k \triangleq \boldsymbol{u}_d - \boldsymbol{u}_k$.

The learning control is typically designed to iteratively enhance the input based on historical information; specifically, these methodologies heavily rely on the tracking error information, denoted as $e_k(t)$. Therefore, (5.7) plays a crucial role in analyzing the convergence of the proposed method. As per (5.7), the tracking error arises from both inaccuracies in the input, $\Delta\boldsymbol{u}_k$, and system noise, $\mathcal{G}\boldsymbol{\omega}_k + \boldsymbol{v}_k$. The ratio between these two components is utilized for controlling the multi-stage learning (MSL) mechanism proposed subsequently.

In noiseless systems, the objective of learning control is to ensure that $\boldsymbol{y}_d - \boldsymbol{y}_k$ approaches zero as k tends to infinity. To achieve this objective, a sufficient condition is to ensure that $\boldsymbol{u}_d - \boldsymbol{u}_k$ tends to zero as k increases indefinitely. It has been demonstrated in existing literature that the latter condition can be guaranteed by implementing the following P-type learning control scheme:

$$\boldsymbol{u}_{k+1} = \boldsymbol{u}_k + \gamma\mathcal{L}\boldsymbol{e}_k, \tag{5.8}$$

where γ and $\mathcal{L}$ are the learning gain and direction regulation matrix (DRM), respectively. The DRM is used to bridge the dimension mismatch between the input and output vectors.

When noise is present in the system, represented by non-zero values of $\boldsymbol{\omega}_k$ and $\boldsymbol{v}_k$, (5.8) fails to guarantee the precise convergence of $\boldsymbol{y}_k$ to $\boldsymbol{y}_d$, and instead, $\boldsymbol{u}_k$ converges within a bounded region around $\boldsymbol{u}_d$. To enhance the tracking performance of this method, an alternative approach can be employed in lieu of the constant learning gain method, which utilizes a decreasing gain sequence γ_k, i.e.,

$$\boldsymbol{u}_{k+1} = \boldsymbol{u}_k + \gamma_k\mathcal{L}\boldsymbol{e}_k, \tag{5.9}$$

where $\gamma_k \longrightarrow 0$. To ensure the asymptotic convergence of $\boldsymbol{u}_k$ to the desired input, a common selection of γ_k satisfies the following conditions:

$$\gamma_k > 0, \quad \sum_{k=1}^{\infty} \gamma_k = \infty, \quad \sum_{k=1}^{\infty} \gamma_k^2 < \infty. \tag{5.10}$$

The convergence results are given by Theorem 5.1 below.

The decreasing gain sequence is employed to asymptotically mitigate the impact of noise in the iteration domain. This gain sequence is typically predetermined prior to algorithm implementation. For instance, a commonly utilized gain sequence can be expressed as $\gamma_k = \frac{a}{k}$, where $a > 0$ represents an adjustable parameter. However, it should be noted that $\gamma_k = \frac{a}{k}$ does not incorporate any information pertaining to the system or learning process. Consequently, this may lead to unsatisfactory algorithm performance for certain applications and undesirable characteristics such as rapid initial increase in tracking error or persistently large tracking error even after numerous iterations. Therefore, it is hypothesized that appropriately designing the decreasing gain sequence, denoted as γ_k, could significantly enhance transient learning performance.

The objective of this section is to establish a novel MSL scheme by introducing a multi-stage-based gain sequence. Specifically, the iteration axis is divided into a series of stages, which are jointly determined by the input error, $\Delta\boldsymbol{u}_k$, and the system noise, $\mathcal{G}\boldsymbol{\omega}_k + \boldsymbol{v}_k$. In each stage, the gain remains constant to facilitate rapid learning and decreases during transitions between stages. The fundamental principle of the MSL mechanism and the computation of iterations corresponding to stage transitions are elaborated in the subsequent subsection.

In this work, our focus lies on a P-type scheme presented in (5.8) to emphasize the fundamental concepts of the MSL methodology. It is worth noting that the P-type learning control scheme is widely used in ILC and its proportional structure offers inherent advantages for achieving effective learning effects (Bristow et al. 2006). Therefore, it holds significant importance to investigate and enhance the fundamental learning capability of the P-type scheme under random noise conditions, which serves as a motivation for proposing the MSL mechanism. The subsequent results can be extended to encompass more general structures such as PD-type and PID-type schemes by employing a similar approach of comparing input error variances with random noise.

5.1.3 Multi-stage Learning Gain Design

5.1.3.1 Design of the DRM

As our primary objective is to establish a MSL method by introducing a stage-based gain sequence, we consider a general design of the DRM $\mathcal{L}$.

To avoid the computational complexity caused by the high dimension of the lifted matrices, we consider a block diagonal DRM as follows:

$$\mathcal{L}_0 = \text{diag}\{L_0, L_1, ..., L_{N-1}\}, \tag{5.11}$$

where $L_i \in \mathbf{R}^{p\times q}$ $(0 \le i \le N-1)$ is such that all the eigenvalues of $L_t C_{t+1} B_t$ have positive real parts. The matrix $\mathcal{L}_0\mathcal{H}$ is block lower triangular because $\mathcal{H}$ has this property. Thus, the set of eigenvalues of $\mathcal{L}_0\mathcal{H}$ is a combination of the eigenvalues of $L_t C_{t+1} B_t$, and thus, all the eigenvalues of $\mathcal{L}_0\mathcal{H}$ have real parts that are greater than zero.

Then, for any positive-definite matrix $\mathcal{S}$, the equation $(\mathcal{L}_0\mathcal{H})^T\Phi + \Phi(\mathcal{L}_0\mathcal{H}) = \mathcal{S}$ has a unique solution for Φ, which is a positive-definite matrix. We can rewrite $\mathcal{L}$ as $\mathcal{L} = \Phi\mathcal{L}_0$, and then, for any $\mathbf{x} \neq 0$ and $x \in \mathbf{R}^{p\times N}$, we can obtain

$$\begin{aligned}
\mathbf{x}^T\mathcal{L}\mathcal{H}\mathbf{x} &= \mathbf{x}^T\left(\frac{(\mathcal{L}\mathcal{H})^T + (\mathcal{L}\mathcal{H})}{2}\right)\mathbf{x} \\
&= \mathbf{x}^T\left(\frac{(\mathcal{L}_0\mathcal{H})^T\Phi + \Phi(\mathcal{L}_0\mathcal{H})}{2}\right)\mathbf{x} \\
&= \frac{\mathbf{x}^T\mathcal{S}\mathbf{x}}{2} > 0.
\end{aligned}$$

In this case, the matrix $\mathcal{L}\mathcal{H}$ is called a quasi-positive-definite matrix, i.e., $\mathbf{x}^T\mathcal{L}\mathcal{H}\mathbf{x} > 0$, $\forall\mathbf{x} \neq 0$. We note that it is not necessary for $\mathcal{L}\mathcal{H}$ to be symmetric. The following lemma demonstrates that the matrix Q is quasi-positive-definite.

Lemma 5.1 *Assume Q is a quasi-positive-definite matrix. Then, it can be shown that*

$$\lambda_{\max}^{\frac{1}{2}}(Q^TQ) \ge \lambda_{\max}(Q) \ge \lambda_{\min}(Q) \ge \lambda_{\min}\left(\frac{Q^T+Q}{2}\right). \tag{5.12}$$

Proof. The first inequality holds because the maximum singular value of a matrix cannot be smaller than the spectral radius of that matrix. The second inequality is clear. We now prove the last inequality.

We denote the smallest eigenvalue of $\frac{Q^T+Q}{2}$ as λ_0 and its corresponding eigenvector as $\mathbf{z}_0$. For any $\mathbf{z} \in \mathbf{C}^n$, we can write

$$\lambda_{\min}\left(\frac{Q^T+Q}{2}\right) = \frac{\mathbf{z}_0^H(Q^T+Q)\mathbf{z}_0}{2\mathbf{z}_0^H\mathbf{z}_0} \le \frac{\mathbf{z}^H(Q^T+Q)\mathbf{z}}{2\mathbf{z}^H\mathbf{z}}, \tag{5.13}$$

where $\mathbf{z}^H$ represents the conjugate transpose of $\mathbf{z}$.

Suppose that λ_1 is the eigenvalue with the smallest modulus among the eigenvalues of Q, and we denote its corresponding eigenvector as $\mathbf{z}_1$. Then, we obtain

$$\lambda_{\min}(\frac{Q^T+Q}{2}) \leq \frac{\mathbf{z}_1{}^H(Q^T+Q)\mathbf{z}_1}{2\mathbf{z}_1{}^H\mathbf{z}_1} = Re(\lambda_1) \leq \lambda_{\min}(Q), \tag{5.14}$$

where $Re(\lambda_1)$ represents the real part of λ_1.

5.1.3.2 Motivation of the MSL Mechanism

To understand the principle of the MSL mechanism, we first investigate the performance of (5.8) using a constant gain. Subtracting both sides of (5.8) from $\boldsymbol{u}_d$ leads to

$$\Delta\boldsymbol{u}_{k+1} = \Delta\boldsymbol{u}_k - \gamma\mathcal{L}\boldsymbol{e}_k. \tag{5.15}$$

Substituting (5.7) into (5.15) leads to

$$\begin{aligned}\Delta\boldsymbol{u}_{k+1} &= \Delta\boldsymbol{u}_k - \gamma\mathcal{L}[\mathcal{H}\Delta\boldsymbol{u}_k - \xi_k] \\ &= (I-\gamma\mathcal{L}\mathcal{H})\Delta\boldsymbol{u}_k + \gamma\mathcal{L}\xi_k,\end{aligned} \tag{5.16}$$

where $\xi_k = \mathcal{G}\boldsymbol{w}_k + \boldsymbol{v}_k$.

Taking the vector norm to both sides of (5.16), we find that

$$\begin{aligned}\|\Delta\boldsymbol{u}_{k+1}\|^2 &= \Delta\boldsymbol{u}_{k+1}^T\Delta\boldsymbol{u}_{k+1} \\ &= \Delta\boldsymbol{u}_k^T(I-\gamma\mathcal{L}\mathcal{H})^T(I-\gamma\mathcal{L}\mathcal{H})\Delta\boldsymbol{u}_k \\ &\quad + 2\gamma\Delta\boldsymbol{u}_k^T(I-\gamma\mathcal{L}\mathcal{H})^T\mathcal{L}\xi_k \\ &\quad + \gamma^2\xi_k^T\mathcal{L}^T\mathcal{L}\xi_k^T.\end{aligned} \tag{5.17}$$

We denote a σ-algebra $\mathcal{F}_k$ to indicate the history of the random factors up to iteration k, i.e., $\mathcal{F}_k \triangleq \sigma\{x_d(0), u_1(t), \omega_i(t-1), v_i(t), 1 \leq i \leq k, 1 \leq t \leq N\}$. We note that ξ_k is independent of $\mathcal{F}_{k-1}$ and thus, $\mathbb{E}[\xi_k|\mathcal{F}_{k-1}] = 0$. Furthermore, $\Delta\boldsymbol{u}_k$ depends only on the information of the $(k-1)$th iteration; thus, we see that $\Delta\boldsymbol{u}_k \in \mathcal{F}_{k-1}$. Taking the conditional expectation of (5.17) with respect to $\mathcal{F}_{k-1}$, we obtain

$$\begin{aligned}\mathbb{E}[\|\Delta\boldsymbol{u}_{k+1}\|^2|\mathcal{F}_{k-1}] &= \Delta\boldsymbol{u}_k^T(I-\gamma\mathcal{L}\mathcal{H})^T(I-\gamma\mathcal{L}\mathcal{H})\Delta\boldsymbol{u}_k \\ &\quad + \gamma^2\mathbb{E}[\xi_k^T\mathcal{L}^T\mathcal{L}\xi_k].\end{aligned} \tag{5.18}$$

We define $\lambda_1 = \lambda_{\max}(\mathcal{H}^T\mathcal{L}^T\mathcal{L}\mathcal{H})$ and $\lambda_2 = \lambda_{\min}\left(\frac{\mathcal{H}^T\mathcal{L}^T+\mathcal{L}\mathcal{H}}{2}\right)$. By the design of $\mathcal{L}$ given in Sect. 5.1.3, we see $\lambda_2 > 0$ and $\Delta\boldsymbol{u}_k^T(\mathcal{H}^T\mathcal{L}^T+\mathcal{L}\mathcal{H})\Delta\boldsymbol{u}_k \geq \lambda_2\|\Delta\boldsymbol{u}_k\|^2$. Furthermore, we see that $\Delta\boldsymbol{u}_k^T\mathcal{H}^T\mathcal{L}^T\mathcal{L}\mathcal{H}\Delta\boldsymbol{u}_k \leq \lambda_1\|\Delta\boldsymbol{u}_k\|^2$. We can thus obtain

$$\begin{aligned}
&\Delta \boldsymbol{u}_k^T (I - \gamma \mathcal{L}\mathcal{H})^T (I - \gamma \mathcal{L}\mathcal{H}) \Delta \boldsymbol{u}_k \\
&= \Delta \boldsymbol{u}_k^T \Delta \boldsymbol{u}_k + \gamma^2 \Delta \boldsymbol{u}_k^T \mathcal{H}^T \mathcal{L}^T \mathcal{L}\mathcal{H} \Delta \boldsymbol{u}_k \\
&\quad - \gamma \Delta \boldsymbol{u}_k^T (\mathcal{H}^T \mathcal{L}^T + \mathcal{L}\mathcal{H}) \Delta \boldsymbol{u}_k \\
&\leq (1 - 2\lambda_2 \gamma + \lambda_1 \gamma^2) \| \Delta \boldsymbol{u}_k \|^2.
\end{aligned} \tag{5.19}$$

Considering Assumption 5.3, we obtain

$$\begin{aligned}
&\mathbb{E}[\xi_k^T \mathcal{L}^T \mathcal{L} \xi_k] \\
&= \mathbb{E}[\boldsymbol{v}_k^T \mathcal{L}^T \mathcal{L} \boldsymbol{v}_k] + \mathbb{E}[\boldsymbol{w}_k^T \mathcal{G}^T \mathcal{L}^T \mathcal{L}\mathcal{G} \boldsymbol{w}_k] \\
&\leq \lambda_{\max}(\mathcal{L}^T \mathcal{L}) q N \sigma_\nu^2 + \lambda_{\max}(\mathcal{G}^T \mathcal{L}^T \mathcal{L}\mathcal{G}) p N \sigma_\omega^2.
\end{aligned} \tag{5.20}$$

Here, we use the notation $\sigma_\xi^2 = \lambda_{\max}(\mathcal{L}^T \mathcal{L}) q N \sigma_\nu^2 + \lambda_{\max}(\mathcal{G}^T \mathcal{L}^T \mathcal{L}\mathcal{G}) p N \sigma_\omega^2$. Then, substituting (5.19) and (5.20) into (5.18), we find

$$\begin{aligned}
&\mathbb{E}[\| \Delta \boldsymbol{u}_{k+1} \|^2 \mid \mathcal{F}_{k-1}] \\
&\leq (1 - 2\lambda_2 \gamma + \lambda_1 \gamma^2) \| \Delta \boldsymbol{u}_k \|^2 + \gamma^2 \sigma_\xi^2.
\end{aligned} \tag{5.21}$$

Taking mathematical expectation to the above inequality leads to

$$\begin{aligned}
&\mathbb{E}[\| \Delta \boldsymbol{u}_{k+1} \|^2] \\
&\leq (1 - 2\lambda_2 \gamma + \lambda_1 \gamma^2) \mathbb{E}[\| \Delta \boldsymbol{u}_k \|^2] + \gamma^2 \sigma_\xi^2.
\end{aligned} \tag{5.22}$$

Thus,

$$\mathbb{E}[\| \Delta \boldsymbol{u}_{k+1} \|^2] \leq q(\gamma)^k \mathbb{E}[\| \Delta \boldsymbol{u}_1 \|^2] + \frac{\gamma^2}{1 - q(\gamma)} \sigma_\xi^2, \tag{5.23}$$

where $q(\gamma) \triangleq 1 - 2\lambda_2 \gamma + \lambda_1 \gamma^2$ denotes the contraction function.

From (5.23), it can be observed that the upper bound of the norm of the input error, $\mathbb{E}\| \Delta \boldsymbol{u}_{k+1} \|^2$, comprises two components: the contracted initial input error, $q(\gamma)^k \mathbb{E}[\| \Delta \boldsymbol{u}_1 \|^2]$, and the accumulated noise drift, $\frac{\gamma^2}{1-q(\gamma)} \sigma_\xi^2$. During the initial stages of the learning process, it is likely for the contracted initial input error to surpass the accumulated noise drift, thereby dominating the overall input error. In such cases, employing an appropriate constant gain can facilitate achieving a higher convergence speed. As iterations progress, however, there will be a decrease in the contracted initial input error until it becomes comparable to the accumulated noise drift. This prompts us to consider suppressing any influence caused by system noise. Consequently, reducing the learning gain becomes necessary in order to mitigate noise drift. Figure 5.1 visually demonstrates how an adaptive gain strategy can be employed by comparing these two terms contributing to input error. Based on this analysis, a question arises: How can we design an appropriate mechanism that determines when it is suitable to reduce this gain?

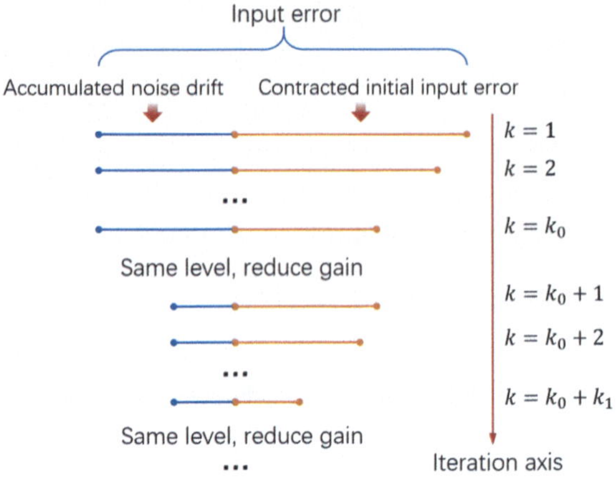

Fig. 5.1 Schematic of the event-triggering gain methodology

The question at hand necessitates the development of a robust metric to gauge the ratio between the contracted initial input error and the accumulated noise drift. With regards to the accumulation function $f(\gamma) = \frac{\gamma^2}{1-q(\gamma)}$, we present the following lemma:

Lemma 5.2 *The function* $f(\gamma) = \frac{\gamma^2}{1-q(\gamma)}$, *where* $q(\gamma) = 1 - 2\lambda_2\gamma + \lambda_1\gamma^2$, *is monotonically increasing over the interval* $\left(0, \frac{2\lambda_2}{\lambda_1}\right)$.

Proof. By differentiation, we obtain $f'(\gamma) = \frac{2\lambda_2}{(2\lambda_2-\lambda_1\gamma)^2} > 0$ for all $\gamma \neq \frac{2\lambda_2}{\lambda_1}$. Thus, $f(\gamma) = \frac{\gamma^2}{1-q(\gamma)}$ is monotonically increasing over the interval $\left(0, \frac{2\lambda_2}{\lambda_1}\right)$.

By direct calculation, we see that the quadratic function $q(\gamma) = 1 - 2\lambda_2\gamma + \lambda_1\gamma^2$ has a minimum at $\left(\frac{\lambda_2}{\lambda_1}, 1 - \frac{\lambda_2^2}{\lambda_1}\right)$. If the learning gain γ is restricted to be within the interval $\left(0, \frac{2\lambda_2}{\lambda_1}\right)$, as a result of Lemma 5.1, we see that $0 < q(\gamma) < 1$. Therefore, if the gain is restricted to be within the interval $\left(0, \frac{\lambda_2}{\lambda_1}\right)$, $q(\gamma)$ is decreasing.

5.1.3.3 The New MSL Mechanism

The MSL mechanism originates from the analysis of (5.23) and is detailed as follows. We consider first a constant gain $\gamma \in (0, \gamma^*)$, where $\gamma^* \triangleq \frac{\lambda_2}{\lambda_1}$. If the contracted initial input error, $q(\gamma)^k\mathbb{E}[\|\Delta \boldsymbol{u}_1\|^2]$, is larger than the accumulated noise drift, $\frac{\gamma^2}{1-q(\gamma)}\sigma_\xi^2$, the constant gain is retained in order to ensure a quick reduction of the former; otherwise, the gain value is decreased to reduce the value of the latter. In other words, the learning gain remains unchanged until the effect of the contracted initial input error is reduced as the same level to the accumulated noise drift.

We denote K as the number of iterations in a given stage; in a given stage, the learning gain is set to be a constant value, γ, for each iteration. Then, for the initial input error, $\|\Delta \boldsymbol{u}_1\|$, K should satisfy the expression

$$q(\gamma)^K \|\Delta \boldsymbol{u}_1\|^2 > \frac{\gamma^2}{1-q(\gamma)}\sigma_\xi^2, \tag{5.24}$$

$$q(\gamma)^{K+1} \|\Delta \boldsymbol{u}_1\|^2 \le \frac{\gamma^2}{1-q(\gamma)}\sigma_\xi^2. \tag{5.25}$$

To obtain the appropriate value of K, we introduce an event trigger as follows:

$$K = \min\left\{ k \in \mathbf{N} \,\middle|\, q(\gamma)^{k+1} \|\Delta \boldsymbol{u}_1\|^2 \le \frac{\gamma^2}{1-q(\gamma)}\sigma_\xi^2 \right\}. \tag{5.26}$$

Based on this event trigger for generating the critical value of K, we define the length of an stage. Then, after K iterations, the gain γ is reduced by a factor of θ $(0 < \theta < 1)$. Meanwhile, when the event is triggered, by considering (5.24), the corresponding input error is estimated to be

$$\mathbb{E}[\|\Delta \boldsymbol{u}_{K+1}\|_2^2] \le 2q(\gamma)^K \mathbb{E}[\|\Delta \boldsymbol{u}_1\|^2]. \tag{5.27}$$

By considering the upper bound of $\mathbb{E}[\|\Delta \boldsymbol{u}_{K+1}\|_2^2]$ as a new starting point, which serves a similar purpose to $\mathbb{E}[\|\Delta \boldsymbol{u}_1\|^2]$ in the initial stage, we derive a second critical value and initiate a subsequent stage. Through iteration of this process, it becomes possible to construct a series of stages that correspond to multiple learning stages. This establishes what is known as the MSL process. Specifically, K_i represents the length of the ith stage and $\overline{E_i}$ denotes the estimated upper bound of input error at the beginning of this stage. The initial learning gain is denoted by γ_1. The MSL mechanism of the learning gain is given by

$$\overline{E_1} = \mathbb{E}[\|\Delta \boldsymbol{u}_1\|^2], \tag{5.28}$$

$$K_i = \min\left\{ k \in \mathbf{N} \,\middle|\, q(\gamma_i)^{k+1}\overline{E}_i \le \frac{{\gamma_i}^2}{1-q(\gamma_i)}\sigma_\xi^2 \right\}, \tag{5.29}$$

$$\overline{E_{i+1}} = 2q(\gamma_i)^{K_i}\overline{E_i}. \tag{5.30}$$

where $\gamma_i = \gamma_1\theta^{i-1}$, $\gamma_1 \in (0, \gamma^*)$, and $\theta \in (0, 1)$. In the above expression, $q(\gamma_i) = 1 - 2\lambda_2\gamma_i + \lambda_1\gamma_i^2$. Then, the gain sequence can be defined by

$$\left\{ \gamma_k \,\middle|\, \gamma_k = \gamma_1\theta^{i-1}, \sum_{i=1}^{i-1} K_i < k \le \sum_{i=1}^{i} K_i, i, k \in \mathbf{N} \right\}. \tag{5.31}$$

Next, we demonstrate the feasibility of the proposed methodology. That is, we show that K_i is well defined, i.e., K_i is finite. We have the following lemma:

Lemma 5.3 *Given the initial value* $\overline{E}_1$ *and* γ_1, *the stage length,* K_i, *calculated using* (5.28)–(5.30) *is well defined.*

Proof. For any $i \in \mathbf{N}$, the inequality $0 < q(\gamma_i) < 1$ holds. Thus, there must exist a finite K_i such that $q(\gamma_i)^{K_i+1}\overline{E}_i < \frac{\gamma_i{}^2}{1-q(\gamma_i)}\sigma_\xi^2$ for any given $\overline{E}_i$ and γ_i. It must then be shown that

$$\overline{E}_i > \frac{\gamma_i{}^2}{1-q(\gamma_i)}\sigma_\xi^2, \forall i \in \mathbf{N}. \tag{5.32}$$

Here, the validity of (5.32) indicates that the vanishing of the learning gain sequence will launch a new stage.

We prove (5.32) using the principle of induction. For the initial case K_1, we assume that $\|\Delta \boldsymbol{u}_1\|^2$ is larger than σ_ξ^2 and $\overline{E}_1 > \frac{\gamma_1^2}{1-q(\gamma_1)}\sigma_\xi^2$. Otherwise, the initial learning gain, γ_1, can be reduced.

Suppose (5.32) holds for the index i. According to the definition of K_i, we have

$$q(\gamma_i)^{K_i}\overline{E}_i > \frac{\gamma_i{}^2}{1-q(\gamma_i)}\sigma_\xi^2. \tag{5.33}$$

Therefore, by (5.30), we obtain

$$\overline{E}_{i+1} = 2q(\gamma_i)^{K_i}\overline{E}_i > q(\gamma_i)^{K_i}\overline{E}_i > \frac{\gamma_i{}^2}{1-q(\gamma_i)}\sigma_\xi^2. \tag{5.34}$$

We note that $\gamma_{i+1} < \gamma_i$. By Lemma 5.2, we see

$$\frac{\gamma_i{}^2}{1-q(\gamma_i)}\sigma_\xi^2 > \frac{\gamma_{i+1}^2}{1-q(\gamma_{i+1})}\sigma_\xi^2. \tag{5.35}$$

Thus, (5.32) holds for index $i+1$. This completes the proof.

5.1.3.4 An Improved MSL Mechanism

The previous subsection proposed an event-triggering-type mechanism for determining the transition between learning stages based on a comparison of the contracted initial input error and accumulated noise drift. However, several shortcomings have been observed in practice. Firstly, to ensure the contraction requirement in the initial stage, the initial learning gain γ_1 may need to be set to a very small value, which can significantly decrease convergence speed. Additionally, if γ_i is small, the value of $q(\gamma_i)$ approaches 1, resulting in a potentially large length for each stage (K_i). In other words, there can be a significant iteration-delay for reducing the learning gain.

To address these issues, we propose a revised expression for $q(\gamma)$ by introducing additional parameters. It is worth noting that the expression of $q(\gamma)$ in (5.19) serves as an upper bound for $\lambda_{\max}(I-\gamma\mathcal{L}\mathcal{H})^T(I-\gamma\mathcal{L}\mathcal{H})$. However, it is generally observed that the value of $q(\gamma)$ significantly exceeds the latter quantity. A promising approach to mitigate the aforementioned limitations of our proposed method involves minimizing the disparity between $q(\gamma)$ and $\lambda_{\max}(I - \gamma\mathcal{LH})^T(I - \gamma\mathcal{LH})$ across all values of γ.

To this end, we introduce two parameters: $\alpha > 0$ and $\beta > 0$; we use these parameters to define a variable equivalent to $q(\gamma)$. In particular, we propose an alternative to $q(\gamma)$, denoted by $\overline{q}(\gamma)$; this new variable is given by

$$\overline{q}(\gamma) = 1 - 2\alpha\beta\lambda_2\gamma + \alpha\beta^2\lambda_1\gamma^2, \tag{5.36}$$

where $1 < \alpha < \frac{\lambda_1}{\lambda_2^2}$ and $0 < \beta < 1$.

The function $\overline{q}(\gamma)$ is employed as the multi-stage generation function. The parameters α and β are denoted as the longitudinal stretch and transverse stretch parameters, respectively. It is evident that $\overline{q}(\gamma)$ exhibits a quadratic behavior; its minimum value occurs at $(\frac{\lambda_2}{\beta\lambda_1}, 1 - \frac{\alpha\lambda_2^2}{\lambda_1})$. Clearly, the learning gain γ should be constrained within the interval $(0, \overline{\gamma}^*)$, where $\overline{\gamma}^* = \frac{\lambda_2}{\beta\lambda_1}$. Figure 5.2 illustrates the impact of parameters α and β on the multi-stage generation function.

Remark 5.2 We explain the effect of the parameters α and β on the control performance to provide a selection guidance for these parameters. The primary shortcoming of the original mechanism given in (5.28)–(5.30) is that the value of $q(\gamma)$ is generally larger than $\lambda_{\max}(I - \gamma\mathcal{LH})^T(I - \gamma\mathcal{LH})$, as shown in Fig. 5.2. Therefore, the parameters α and β are introduced to generate an alternative function, $\overline{q}(\gamma)$, to approximate $\lambda_{\max}(I - \gamma\mathcal{LH})^T(I - \gamma\mathcal{LH})$. First, we set β to ensure that $\frac{\lambda_2}{\beta\lambda_1}$ approximates the abscissa of the lowest point of $\lambda_{\max}(I - \gamma\mathcal{LH})^T(I - \gamma\mathcal{LH})$. Using β, the feasible range length of γ_1 is enlarged from γ^* to $\overline{\gamma}^*$, which ensures the convergence of the proposed scheme during the early stages of the learning process. We then choose the value of the parameter α to ensure that the value of $\overline{q}(\gamma)$ is less than that of $\lambda_{\max}(I - \gamma\mathcal{LH})^T(I - \gamma\mathcal{LH})$ for all values of γ. In this manner, the problems related to the large value of K_i are eliminated, and thus, the performance of the original MSL mechanism (5.28)–(5.30) is significantly improved.

The MSL mechanism (defined by (5.28)–(5.30)) is revised accordingly:

$$\overline{E_1} = \mathbb{E}[\|\Delta\boldsymbol{u}_1\|^2], \tag{5.37}$$

$$K_i = \min\left\{k \in \mathbf{N} \,\middle|\, \overline{q}(\gamma_i)^{k+1}\overline{E}_i \le \frac{{\gamma_i}^2}{1 - \overline{q}(\gamma_i)}\sigma_\xi^2\right\}, \tag{5.38}$$

$$\overline{E_{i+1}} = 2\overline{q}(\gamma_i)^{K_i}\overline{E}_i, \tag{5.39}$$

where $\gamma_i = \gamma_1\theta^{i-1}$, $\gamma_1 \in (0, \overline{\gamma}^*)$, and $\theta \in (0, 1)$. The learning gain sequence is then given by

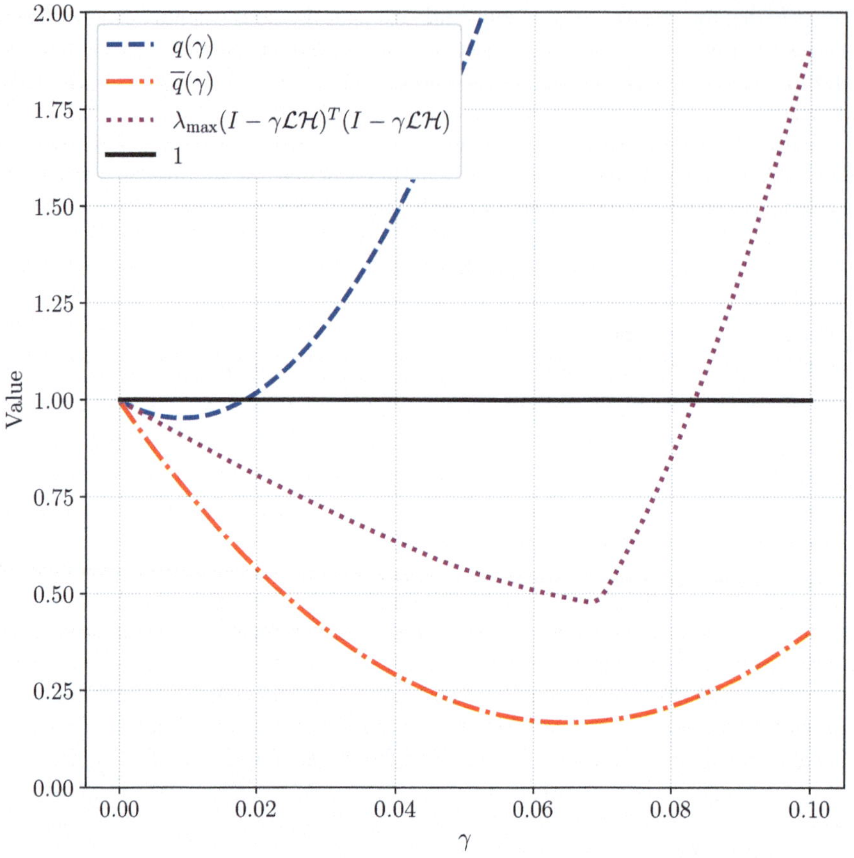

Fig. 5.2 The profiles of $q(\gamma)$, $\overline{q}(\gamma_i)$, and $\lambda_{\max}(I - \gamma\mathcal{LH})^T(I - \gamma\mathcal{LH})$

$$\left\{\gamma_k \middle| \gamma_k = \gamma_1\theta^{i-1}, \sum_{i=1}^{i-1} K_i < k \leq \sum_{i=1}^{i} K_i, i, k \in \mathbf{N}\right\}. \tag{5.40}$$

We find that (5.37)–(5.39) give an alternative MSL mechanism that generates the learning gain sequence. The proposed methodology benefits from this alternative mechanism, which enhances convergence performance. Specifically, the adjustment of β offers greater flexibility in selecting the initial gain, while the adjustment of α ensures that $\overline{q}(\gamma)$ closely approximates the actual contraction situation. Moreover, by dividing the learning process into multiple event-triggered stages with constant learning gains, the convergence speed is accelerated. Through transition between these stages, asymptotic elimination of random noise influence is achieved.

5.1.4 Convergence Analysis

In this subsection, we analyze the convergence of (5.9), where the gain sequence is generated by (5.37)–(5.39). This analysis is conducted in two steps. Firstly, we establish the convergence of (5.9) under the general condition of the learning gain sequence given in (5.10) (Theorem 5.1). Subsequently, we demonstrate that the gain sequence generated by (5.37)–(5.39) satisfies the general condition (Lemma 5.4). By combining these two steps, we derive our main result (Theorem 5.2).

Using Lemma 2.2, we obtain the convergence of (5.9) under the general condition given in (5.10), as shown below.

Theorem 5.1 *Consider the discrete-time stochastic system* (5.1) *that satisfies Assumptions 5.1–5.3. Apply the scheme* (5.9) *and assume that all the eigenvalues of $L_t C_{t+1} B_t$ have positive real parts as result of the selection of L_t. Then, the following statements are valid.*

(a) If the variable gain, γ_k, satisfies that $\gamma_k > 0$, $\gamma_k \longrightarrow 0$, and $\sum_{k=1}^{\infty} \gamma_k = \infty$, the input error $\Delta \boldsymbol{u}_k$ must converge to zero in the mean square sense, i.e., $\lim_{k\to\infty} \mathbb{E}[\|\Delta \boldsymbol{u}_k\|^2] = 0$.
(b) If γ_k additionally satisfies $\sum_{k=1}^{\infty} \gamma_k^2 < \infty$, the input error $\Delta \boldsymbol{u}_k$ converges to zero in the almost-sure sense, i.e., $\mathbb{P}(\lim_{k\to\infty} \|\Delta \boldsymbol{u}_k\|^2 = 0) = 1$.

Proof. We first prove (a). Considering (5.22) and replacing γ with γ_k, we obtain

$$\mathbb{E}[\|\Delta \boldsymbol{u}_{k+1}\|^2] \le q(\gamma_k)\mathbb{E}[\|\Delta \boldsymbol{u}_k\|^2] + {\gamma_k}^2\sigma_\xi^2. \tag{5.41}$$

Denoting $q_{i,j} = \prod_{l=j}^{i} q(\gamma_l)$ for $i \ge j$ and $q_{i,i+1} = 1$, we find that

$$\mathbb{E}[\|\Delta \boldsymbol{u}_{k+1}\|^2] \le q_{k,1}\mathbb{E}[\|\Delta \boldsymbol{u}_1\|^2] + \sum_{i=1}^{k} q_{k,i+1}{\gamma_i}^2\sigma_\xi^2. \tag{5.42}$$

We note that $q_{k,1} = \exp\left(\sum_{i=1}^{k} \ln(1 - 2\lambda_2\gamma_i + \lambda_1\gamma_i^2)\right)$. Via the power series expansion of $ln(1-x)$ and the expression $\sum_{k=1}^{\infty} \gamma_k = \infty$, we obtain $\lim_{k\to\infty} q_{k,1} = 0$. Thus, the first term of (5.42) converges to zero. Regarding the second term in (5.42),

$$\sum_{i=1}^{k} q_{k,i+1}{\gamma_i}^2\sigma_\xi^2 = \sigma_\xi^2 \sum_{i=1}^{k} q_{k,i+1}(1 - q(\gamma_i))\frac{{\gamma_i}^2}{(1 - q(\gamma_i))}. \tag{5.43}$$

By direct calculation we find that

$$\sum_{i=1}^{k} q_{k,i+1}(1 - q(\gamma_i)) = 1 - \prod_{i=1}^{k} q(\gamma_i) = 1 - q_{k,i}.$$

Then, $\lim_{k\to\infty}\sum_{i=1}^{k} q_{k,i+1}(1-q(\gamma_i)) = 1$. In addition, we have $\lim_{k\to\infty}\frac{\gamma_k{}^2}{(1-q(\gamma_k))} = \lim_{k\to\infty}\frac{\gamma_k}{2\lambda_2-\lambda_1\gamma_k} = 0$.

Considering the Toeplitz Lemma, we obtain that

$$\begin{aligned}&\lim_{k\to\infty}\sum_{i=1}^{k} q_{k,i+1}(1-q(\gamma_i))\frac{\gamma_i{}^2}{(1-q(\gamma_i))}\\ &=\lim_{k\to\infty}\frac{\gamma_k{}^2}{(1-q(\gamma_k))}=0.\end{aligned}$$

Thus, the right-hand side of (5.42) converges to zero. In other words, $\lim_{k\to\infty}\mathbb{E}[\|\Delta \boldsymbol{u}_k\|^2] = 0$.

We then proceed to prove (b). Similar to the case of (5.21), replacing γ with γ_k yields

$$\begin{aligned}&\mathbb{E}[\|\Delta \boldsymbol{u}_{k+1}\|^2|\mathcal{F}_k]\\ &\le \|\Delta \boldsymbol{u}_k\|^2+\gamma_k{}^2\lambda_1\|\Delta \boldsymbol{u}_k\|^2\\ &\quad -2\gamma_k\lambda_2\|\Delta \boldsymbol{u}_k\|^2+\gamma_k{}^2\sigma_\xi^2\\ &\le \|\Delta \boldsymbol{u}_k\|^2+\gamma_k{}^2(\lambda_1\|\Delta \boldsymbol{u}_k\|^2+\sigma_\xi^2).\end{aligned} \tag{5.44}$$

Comparing (5.44) and equations in Lemma 2.2, we notice that $\gamma_k{}^2(\lambda_1\|\Delta \boldsymbol{u}_k\|^2 +\sigma_\xi^2)$ and $\|\Delta \boldsymbol{u}_k\|^2$ correspond to τ_k and r_k, respectively, while $\kappa_k = s_k = 0$. To complete the proof, it is sufficient to verify $\sum_{k=1}^{\infty}\gamma_k{}^2\lambda_1\|\Delta \boldsymbol{u}_{k+1}\|^2 < \infty$.

We have proved in (a) that $\mathbb{E}[\|\Delta \boldsymbol{u}_{k+1}\|^2]$ converges to zero. Thus, $\mathbb{E}[\|\Delta \boldsymbol{u}_{k+1}\|^2]$ is bounded. By the condition of γ_k in (b), we obtain $\sum_{k=1}^{\infty}\gamma_k^2\lambda_1\mathbb{E}[\|\Delta \boldsymbol{u}_{k+1}\|^2] < \infty$ and $\|\Delta \boldsymbol{u}_{k+1}\|^2$ is a nonnegative stochastic process. By the monotone convergence theorem, $\sum_{k=1}^{\infty}\gamma_k{}^2\lambda_1\|\Delta \boldsymbol{u}_{k+1}\|^2 < \infty$ almost surely. Thus, we obtain that $\sum_{k=1}^{\infty}\tau_k < \infty$. Hence, by Lemma 2.2 and $\lim_{k\to\infty}\mathbb{E}[\|\Delta \boldsymbol{u}_k\|^2] = 0$, we obtain that $\|\Delta \boldsymbol{u}_{k+1}\|^2$ converges to zero almost surely. The proof is completed.

Theorem 5.1 implies that the convergence to zero of the input error is guaranteed by the general conditions stated in (5.10). Therefore, to establish the convergence of a new scheme, it suffices to verify the validity of these conditions for the newly generated learning gain sequence.

To prove the asymptotic convergence of the input sequence, we demonstrate that $\{\gamma_k\}$, which is given by (5.37)–(5.39), satisfies the general conditions given in Theorem 5.1. This result is demonstrated as follows.

Lemma 5.4 *Consider the variable gain sequence* γ_k *given by* (5.37)–(5.39). *We see that*

$$\sum_{k=0}^{\infty}\gamma_k=\infty,\quad \sum_{k=0}^{\infty}\gamma_k{}^2\le\infty. \tag{5.45}$$

Proof. From the definition of γ_k, we see that

$$\sum_{k=1}^{\infty}\gamma_k = \gamma_1\sum_{i=1}^{\infty}K_i\theta^{i-1},\ \sum_{k=1}^{\infty}\gamma_k^2 = \gamma_1^2\sum_{i=1}^{\infty}K_i\theta^{2i-2}. \tag{5.46}$$

According to (5.37)–(5.39), the trigger for the transition between stages can be expressed by

$$K_i = \min\left\{K \in \mathbf{N}\,\middle|\, \overline{q}(\gamma_i)^{k+1}2^i\prod_{j=1}^{i-1}\overline{q}(\gamma_j)^{K_j}\overline{E}_1 \le \frac{{\gamma_i}^2}{1-\overline{q}(\gamma_i)}\sigma_\xi^2\right\}. \tag{5.47}$$

Then, we know that the following inequalities hold:

$$\overline{q}(\gamma_i)^{K_i}2^i\prod_{j=1}^{i-1}\overline{q}(\gamma_j)^{K_j}\overline{E}_1 > \frac{\gamma_i^2}{1-\overline{q}(\gamma_i)}\sigma_\xi^2, \tag{5.48}$$

$$\overline{q}(\gamma_i)^{K_i+1}2^i\prod_{j=1}^{i-1}\overline{q}(\gamma_j)^{K_j}\overline{E}_1 \le \frac{\gamma_i^2}{1-\overline{q}(\gamma_i)}\sigma_\xi^2, \tag{5.49}$$

$$2^{i-1}\prod_{j=1}^{i-1}\overline{q}(\gamma_j)^{K_j}\overline{E}_1 > \frac{\gamma_{i-1}^2}{1-\overline{q}(\gamma_{i-1})}\sigma_\xi^2, \tag{5.50}$$

$$\overline{q}(\gamma_{i-1})2^{i-1}\prod_{j=1}^{i-1}\overline{q}(\gamma_j)^{K_j}\overline{E}_1 \le \frac{\gamma_{i-1}^2}{1-\overline{q}(\gamma_{i-1})}\sigma_\xi^2. \tag{5.51}$$

From (5.50) and (5.51), we obtain

$$\prod_{j=1}^{i-1}\overline{q}(\gamma_j)^{K_j} > \frac{\gamma_{i-1}^2}{2^{i-1}\overline{E}_1(1-\overline{q}(\gamma_{i-1}))}\sigma_\xi^2, \tag{5.52}$$

$$\prod_{j=1}^{i-1}\overline{q}(\gamma_j)^{K_j} \le \frac{\gamma_{i-1}^2}{\overline{q}(\gamma_{i-1})2^{i-1}\overline{E}_1(1-\overline{q}(\gamma_{i-1}))}\sigma_\xi^2. \tag{5.53}$$

Substituting (5.52) and (5.53) into (5.48), and considering (5.49), we see that

$$\overline{q}(\gamma_i)^{K_i} > \frac{\overline{q}(\gamma_{i-1})\gamma_i^2(1-\overline{q}(\gamma_{i-1}))}{2\gamma_{i-1}^2(1-\overline{q}(\gamma_i))}, \tag{5.54}$$

$$\overline{q}(\gamma_i)^{K_i+1} \le \frac{\gamma_i^2(1-\overline{q}(\gamma_{i-1}))}{2\gamma_{i-1}^2(1-\overline{q}(\gamma_i))}. \tag{5.55}$$

This leads to a bound for $\overline{q}(\gamma_i)^{K_i}$:

$$\frac{\theta^2\overline{q}(\gamma_{i-1})(1-\overline{q}(\gamma_{i-1}))}{2(1-\overline{q}(\gamma_i))} < \overline{q}(\gamma_i)^{K_i} \leq \frac{\theta^2(1-\overline{q}(\gamma_{i-1}))}{2\overline{q}(\gamma_i)(1-\overline{q}(\gamma_i))}. \tag{5.56}$$

By direct calculation, we obtain that for $i \geq 2$,

$$\begin{aligned}\frac{\theta(1-\overline{q}(\gamma_{i-1}))}{1-\overline{q}(\gamma_i)} &= \frac{2\alpha\beta\lambda_2\gamma_1\theta^{i-1}-\alpha\beta^2\lambda_1\gamma_1^2\theta^{2i-3}}{2\alpha\beta\lambda_2\gamma_1\theta^{i-1}-\alpha\beta^2\lambda_1\gamma_1^2\theta^{2i-2}} \\ &= 1-\frac{\alpha\beta^2\lambda_1\gamma_1^2(1-\theta)\theta^{i-2}}{2\alpha\beta\lambda_2\gamma_1-\alpha\beta^2\lambda_1\gamma_1^2\theta^{i-1}}.\end{aligned} \tag{5.57}$$

For the right-hand side of the above expression, a minimum is achieved at $i = 2$. Noting that $2\alpha\beta\lambda_2\gamma_1 - \alpha\beta^2\lambda_1\gamma_1^2\theta^{i-1} > 0$, we have

$$1-\frac{\alpha\beta^2\lambda_1\gamma_1^2(1-\theta)}{2\alpha\beta\lambda_2\gamma_1-\alpha\beta^2\lambda_1\gamma_1^2\theta} < \frac{\theta(1-\overline{q}(\gamma_{i-1}))}{1-\overline{q}(\gamma_i)} < 1. \tag{5.58}$$

Furthermore, as i increases, $\overline{q}(\gamma_i) \in (0, 1)$ can be seen to increase to one. Supposing that $\frac{\theta}{2q(\gamma_i)} < 1$ holds for all $i \geq N_1$, where N_1 is a sufficiently large number, we obtain

$$\frac{\theta\overline{q}(\gamma_1)}{2} < \frac{\theta\overline{q}(\gamma_{i-1})}{2} < 1, \quad \frac{\theta}{2} < \frac{\theta}{2q(\gamma_i)} < 1, i \geq N_1 \tag{5.59}$$

By substituting relations (5.58) and (5.59) into (5.56), we obtain

$$c_1 < \overline{q}(\gamma_i)^{K_i} \leq c_2, i \geq N_1, \tag{5.60}$$

where $0 < c_1 < c_2 < 1$ are some constants.

We take the logarithm of (5.60). Noting that $\overline{q}(\gamma_i) \in (0, 1)$, we see that

$$\frac{-\ln c_2}{-\ln(1-(1-\overline{q}(\gamma_i)))} \leq K_i < \frac{-\ln c_1}{-\ln(1-(1-\overline{q}(\gamma_i)))}, \tag{5.61}$$

for $i \geq N_1$. Considering the inequality $x < -\ln(1-x) < \frac{x}{1-x}$ for $x \in (0, 1)$, it follows that

$$\frac{-\overline{q}(\gamma_i)\ln c_2}{1-\overline{q}(\gamma_i)} \leq K_i < \frac{-\ln c_1}{1-\overline{q}(\gamma_i)}, \quad \forall i \geq N_1. \tag{5.62}$$

Given that $1-\overline{q}(\gamma_i) = 2\alpha\beta\lambda_2\gamma_i - \alpha\beta^2\lambda_1{\gamma_i}^2 = \alpha\beta\lambda_2\gamma_i\,(2-\frac{\beta\lambda_1}{\lambda_2}\gamma_i)$ and $0 < \gamma_i < \frac{\lambda_2}{\beta\lambda_1}$, we obtain $\alpha\beta\lambda_2\gamma_1\theta^{i-1} < 1-\overline{q}(\gamma_i) < 2\alpha\beta\lambda_2\gamma_1\theta^{i-1}$. Noting $\overline{q}(\gamma_i) \in [\overline{q}(\gamma_1), 1)$, we obtain

$$\frac{c_3}{\theta^{i-1}} \le K_i < \frac{c_4}{\theta^{i-1}}, \quad \forall i \ge N_1, \tag{5.63}$$

where $c_3 = \frac{-\overline{q}(\gamma_1)\ln c_2}{2\alpha\beta\lambda_2\gamma_1}$ and $c_4 = \frac{-\ln c_1}{\alpha\beta\lambda_2\gamma_1}$.

Then, substituting (5.63) into (5.46), we find that

$$\sum_{k=1}^{\infty}\gamma_k \ge \gamma_1 \sum_{i=N_1}^{\infty}\frac{c_3}{\theta^{i-1}}\theta^{i-1} = \infty, \tag{5.64}$$

$$\sum_{k=1}^{\infty}\gamma_k^2 < \gamma_1^2 \sum_{i=1}^{N_1-1} K_i\theta^{2i-2} + \gamma_1^2 \sum_{i=N_1}^{\infty}\frac{c_4}{\theta^{i-1}}\theta^{2i-2} < \infty. \tag{5.65}$$

The proof is thus completed.

Lemma 5.4 demonstrates that the general conditions of the learning gain given by (5.10) is satisfied by the gain sequence generated by the MSL mechanism defined by (5.37)–(5.39). Combining Theorem 5.1 and Lemma 5.4, we obtain the following theorem.

Theorem 5.2 *Consider the discrete-time stochastic system* (5.1) *that satisfies Assumptions 5.1-5.3. The DRM L_t is such that the real parts of all the eigenvalues of $L_t C_{t+1} B_t$ are positive. If we consider the scheme defined in* (5.9), *where the gain is generated by the MSL mechanism given by* (5.37)–(5.39), *the input error converges to zero in both mean square and almost-sure senses, i.e.,* $\lim_{k\to\infty}\mathbb{E}[\|\Delta \boldsymbol{u}_k\|^2] = 0$ *and* $\mathbb{P}(\lim_{k\to\infty}\|\Delta \boldsymbol{u}_k\|^2 = 0) = 1$.

Theorem 5.2 establishes the asymptotic convergence of the learning control scheme, where the gain sequence is generated by an MSL mechanism. It is noteworthy that the new gain sequence enhances the transient convergence speed rather than altering the asymptotic convergence rate. In other words, it improves the early iteration performance of convergence. The estimation of the asymptotic convergence rate remains a subject for further research.

5.1.5 Illustrative Simulations

To illustrate the effectiveness of our proposed method, we simulate the proposed schemes on an industrial robot model of the closed-loop joint given in Zhang et al. (2014). The transfer function of the model is given by

$$G(s) = \frac{948}{s^2 + 42s + 948}.$$

The discrete-time state-space formulation of the model with (A, B, C) is given by

$$A = \begin{bmatrix} 0.6213 & -7.6196 \\ 0.008 & 0.9589 \end{bmatrix}, \quad B = \begin{bmatrix} 0.008 \\ 0 \end{bmatrix}, \quad C = \begin{bmatrix} 948 & 0 \end{bmatrix}.$$

The initial state and initial input are given by $x_k(0) = x_d(0) = 0$ and $u_0(t) = 0$, $t \in [0, N]$, respectively. Noise that satisfies $N(0, \sigma^2)$ with $\sigma = 0.1$ is added to the output simulating the effect of random noise. The desired reference is given by $y_d(t) = \sin(1.6t/5) + \sin(0.8t/5)$. The time length of an iteration is set to $N = 50$.

The parameters used to evaluate the proposed schemes are given as follows: The DRM is set to $L_t = 1.1$. It is clear that all eigenvalues of $L_t C_{t+1} B_t$ have positive real parts. The parameters used in (5.37)–(5.39) are set to $\gamma_1 = \overline{\gamma}^*$, $\theta = 0.9$, $\alpha = 15$, and $\beta = 0.14$. For comparison, we also simulate the scheme (5.8) using a constant gain of $\gamma = \overline{\gamma}^*$ and the scheme (5.9) using a common decreasing gain $\gamma_k = \overline{\gamma}^*/k$.

5.1.5.1 Performance of the Proposed Schemes

Figure 5.3 shows the actual outputs and inputs at the 40th and 80th iterations, as well as the desired reference. In Fig. 5.3, the two subplots correspond to the output and

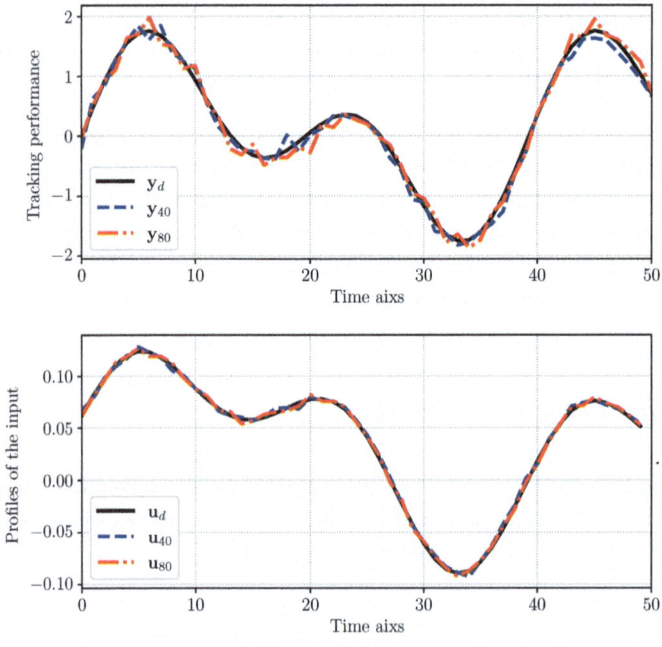

Fig. 5.3 Profiles for the proposed scheme with multi-stage adaptive gain implemented at the 40th and 80th iteration:(top) the output and (bottom) the input.eps

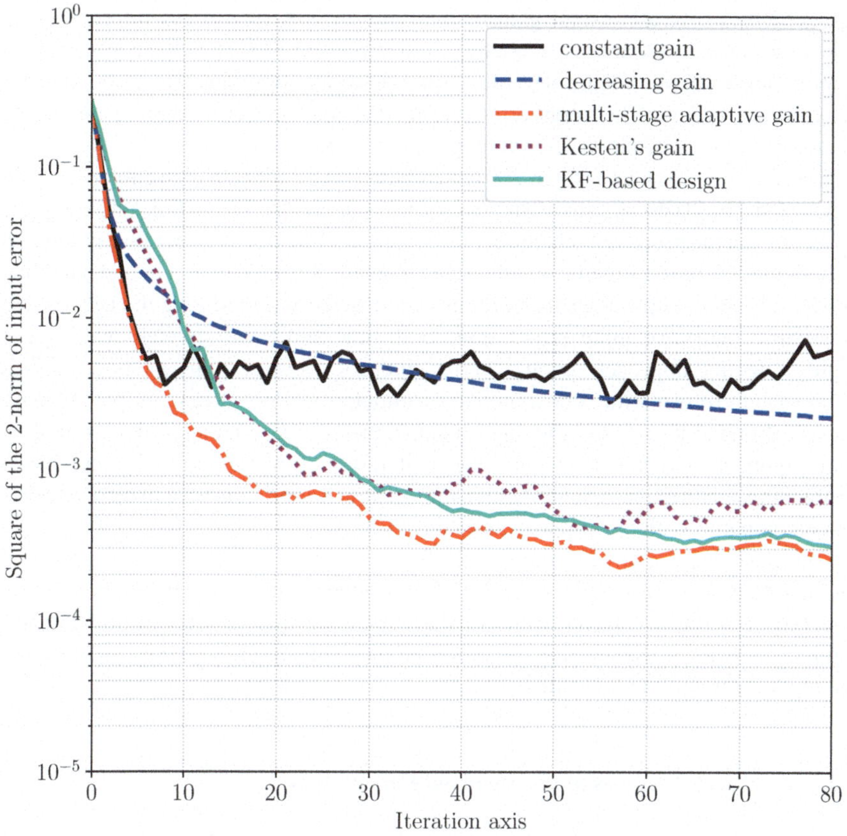

Fig. 5.4 Input error norm profiles for five schemes: constant gain, decreasing gain, multi-stage adaptive gain, Kesten's gain, and the KF-based scheme

input profiles at the 40th and 80th iterations. It is clear that the tracking performance is acceptable after a few iterations. The fluctuations in the output profiles are due to the random noise present in the system. The overlap of the actual input profiles and the desired input profile illustrates the effectiveness of the proposed scheme.

To demonstrate the asymptotic convergence of the proposed methodology and compare it with other existing schemes, Fig. 5.4 presents the norm profiles of input errors, defined as $\|\Delta \boldsymbol{u}_k\|^2$, for five different methodologies. The first three schemes are P-type learning control schemes with constant, decreasing, and multi-stage adaptive gains. It is observed that when a constant learning gain is used, the input error profile stops decreasing after a small number of iterations and begins to fluctuate around a constant value. This occurs because a constant gain fails to suppress the noise effect; however, the decreasing gain methodology can effectively mitigate this noise effect. Consequently, the input error profile in the case of decreasing gain exhibits a declining trend; albeit at a slower convergence speed compared to that achieved by using a constant gain method due to lack of process information in selecting the gain sequence. In contrast, our proposed MSL mechanism generates a multi-stage adaptive learning gain sequence which significantly accelerates the convergence speed of the input error profile. This performance comparison clearly demonstrates an advantage offered by our proposed mechanism.

The KF-based scheme (Saab 2001a) and the P-type scheme using Kesten's gain (Shen and Xu 2020) are also simulated for comparison. From Fig. 5.4, it can be observed that the proposed scheme achieves superior tracking performance compared to both the KF-based scheme and the scheme using Kesten's gain. On one hand, the KF-based scheme optimizes its learning gain matrix by minimizing the trace of the covariance matrix of input errors along both time and iteration axes, making it a highly effective approach among P-type schemes. However, this method requires more system information and significant computational resources compared to our proposed methodology. On the other hand, Kensten's gain method, determined by multiplying error vectors from two adjacent iterations, may randomly decrease based on tracking performance. Nevertheless, this method is more sensitive to random noise when compared with our proposed scheme.

The norm profiles of the input error for three noise levels using the proposed methodology are presented in Fig. 5.5. We simulate three cases parameterized by $\sigma = 0.05, 0.1$, and 0.2. It can be observed that the input error profiles are significantly influenced by the noise level, as our proposed mechanism relies on the ratio between the input error and the noise drift.

5.1.5.2 Effect Analysis of the Involved Parameters

We first illustrate the necessity of the stretching parameters α and β. These parameters solve two practical problems: (1) the interval bound γ^* might be very small, which results in a small initial learning gain γ_1 and (2) there is a large iteration-delay in the gain switching. In Table 5.1, we list three groups of values for $\mathcal{L}$, γ^*, $\gamma^*\mathcal{L}$, $\overline{\gamma}^*$, and $\overline{\gamma}^*\mathcal{L}$, where $\gamma^* = \frac{\lambda_2}{\lambda_1}$, $\overline{\gamma}^* = \frac{\lambda_2}{\beta\lambda_1}$, $\lambda_1 = \lambda_{\max}(\mathcal{H}^T\mathcal{L}^T\mathcal{L}\mathcal{H})$, and $\lambda_2 = \lambda_{\min}(\frac{\mathcal{H}^T\mathcal{L}^T+\mathcal{L}\mathcal{H}}{2})$. The table reflects the expansion of the interval bound of γ^* via the use of the transverse stretch parameter β. It can be seen from the second row that the value of γ_1 changes from 0.0091 to 0.0653. The improvement of the transient performance can be seen from the blue line in Fig. 5.6. Furthermore, we study the effect of the DRM on γ^* at different scales specified in Table 5.1. It can be seen from the Table 5.1 that different DRM affects the value of γ^*, but the product $\gamma^*\mathcal{L}$ remains unchanged. Figure 5.6 shows that the influence of different DRMs on the transient performance is not significant. This demonstrates that the selection of the DRM has little influence on the convergence performance.

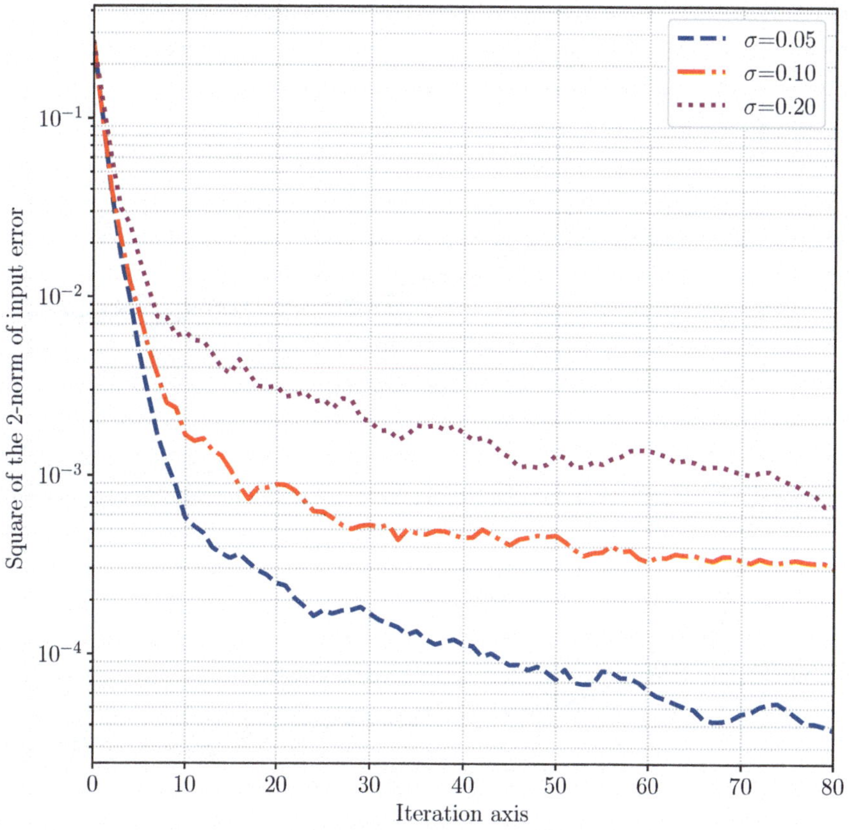

Fig. 5.5 Input error norm profiles for three magnitudes of noise: $\sigma = 0.05, 0.1$, and 0.2

Table 5.1 Parameter values

$\mathcal{L}$	γ^*	$\gamma^*\mathcal{L}$	$\overline{\gamma}^*$	$\overline{\gamma}^*\mathcal{L}$
2.2I	0.0046	0.0101I	0.0327	0.0718I
1.1I	0.0091	0.0101I	0.0653	0.0718I
0.55I	0.0182	0.0101I	0.1306	0.0718I

Investigating the influence of the longitudinal stretch parameter, α, and transverse stretch parameter, β, is of great interest. To clearly observe the impact of these two parameters, we independently adjust the selection of α and β. The results for varying α and β separately are presented in Figs. 5.7 and 5.8, respectively. Both figures consist of upper subplots displaying input error norm profiles and lower subplots showing the learning gain sequence. In Fig. 5.7, we examine the outcomes for a fixed value of $\beta = 0.15$ while considering three different values for α: 1, 8, and 15. It can be observed that smaller values of α result in longer stages, particularly noticeable when

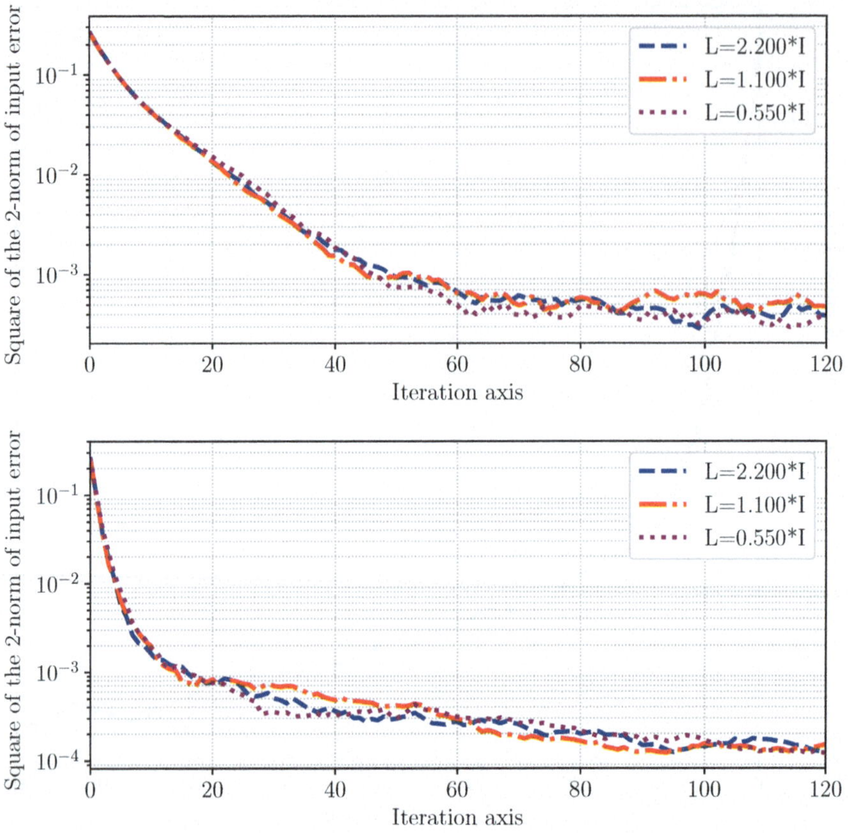

Fig. 5.6 Input error norm profiles with three selections of the DRM $\mathcal{L}$. Top: the MSL mechanism defined by (5.28)–(5.30); bottom: the improved mechanism described by (5.37)–(5.39)

set to 1. This leads to an iteration-delay before reducing the learning gain, ultimately slowing down convergence speed. Figure 5.8 illustrates the results obtained with a fixed value of $\alpha = 15$ while varying three different values for β: 0.15, 0.5, and 1.0. Notably, smaller values of β provide a wider range for initial gain; hence, there are differences in initial gains among these three chosen values as we select it to be equal to $\overline{\gamma}^* = \frac{\lambda_2}{\beta\lambda_1}$. The convergence speed during early stages is seen to increase significantly. To summarize, parameter adjustment reveals that convergence speed during early learning stage is related to parameter β, whereas appropriate reduction timing in this stage depends on adjusting parameter α.

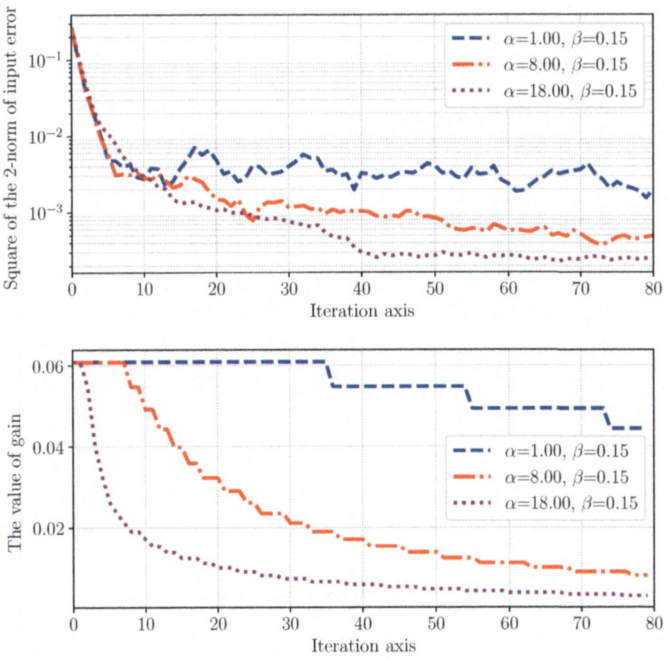

Fig. 5.7 The effect of different values of α for a fixed value of β. Top: input error norm profiles; bottom: gain sequence

5.2 Accelerated Multi-stage Event-Triggering Gain Design

5.2.1 *Motivation of Accelerated Multi-stage Event-Triggering Gain*

The constant gain is typically replaced with various gain designs to ensure the perfect convergence of the P-type scheme for stochastic systems. A promising approach involves applying a priori decreasing gain sequence (Shen and Qu 2020; Shen and Yu 2022; Qu et al. 2023), drawing inspiration from stochastic approximation theory (Borkar 2009). However, if not correctly selected, this independent sequence may lead to unacceptable transient performance in early iterations. Additionally, premature decrease in the learning gain slows down the convergence speed. Therefore, a recent investigation has focused on a process-dependent design of the learning gain, such as adaptive gain (Shen and Xu 2020). This investigation is still at an initial stage and has yet to resolve the conflict between transient convergence speed and final tracking precision.

To achieve a desirable trade-off between convergence speed and tracking precision, an intuitive idea emerges: can we design an adaptive performance-dependent decreasing gain sequence that caters to practical processes? The answer is affirmative.

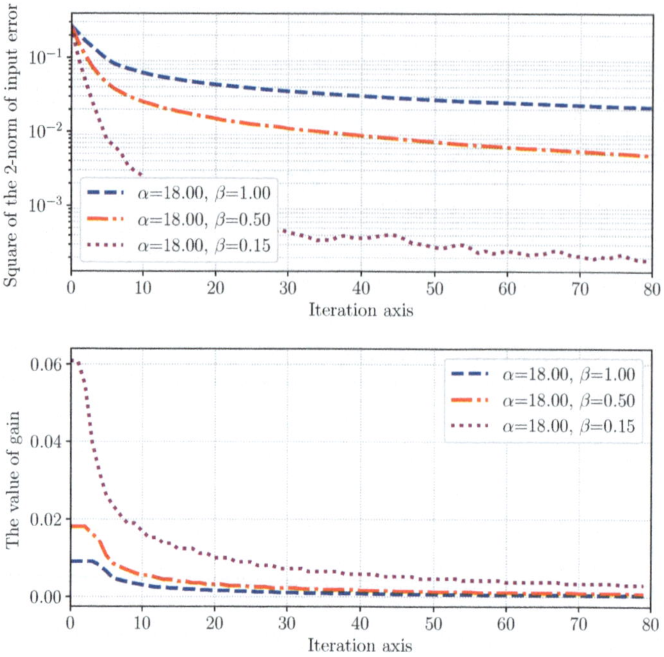

Fig. 5.8 The effect of different values of β for a fixed value of α. Top: input error norm profiles; bottom: gain sequence

In Sect. 5.1, the author presents a novel multistage learning control strategy (MS-LCS), wherein the iteration process is divided into successive stages, each consisting of multiple iterations (Cheng et al. 2023). Within each stage, the learning gain remains constant while decreasing from one stage to another. This approach facilitates faster convergence by employing a consistent gain within each stage and enhances final tracking performance through gradual reduction of the learning gain. Consequently, determining the transition iteration from one stage to the next becomes crucial for this strategy's effectiveness. However, in Sect. 5.1, this determination relies on comparing the contracted initial input error with accumulated noise drift influence alone, which fails to adequately reflect the rate at which tracking performance descends. Moreover, generating only contracted initial input error and accumulated noise drift influence does not fully capture the actual state of the system, leaving room for improvement in transient learning processes.

We propose an innovative strategy, called the accelerated MS-LCS, to effectively address the deficiencies of Sect. 5.1. Our approach focuses on a stochastic linear system and begins by iteratively calculating the mean-square error of the input under a P-type scheme using a general decreasing gain sequence. This mathematical analysis demonstrates how decreasing gain impacts the learning process of system input. Subsequently, we introduce three progressive schemes for generating gain sequences based on virtual performance metrics that measure both the descending speed of mean-squared input error and its estimated upper bound.

The first scheme is the ideal edition, where the iterative formula of the mean-square input error is utilized to calculate the descent rate for both the current and next stage's gain. By comparing these rates, one can determine whether to decrease the gain sequence and initiate a new stage. However, we have observed that calculating the mean-square input error incurs a significant computational burden due to various matrix operations. Therefore, we propose a second scheme called the practical approach, in which we replace the mean-square input error with its upper bound. This upper bound can be easily calculated iteratively. With this approximation, we can obtain approximate descent rates to determine when to switch stages. Nevertheless, our further observation reveals that these approximate descent rates may deviate significantly from those generated by the first scheme. Consequently, while using this second scheme improves transient performance compared to not employing any switching mechanism at all, it might still fall short of achieving performance similar to that obtained by utilizing only the first scheme alone. To address this limitation, we introduce two additional stretching parameters and propose an improved third scheme.

5.2.2 P-Type Scheme and Stage-Based Gain Design

In this subsection, we initially revisit the fundamental P-type LCS. Subsequently, we elucidate the motivation and principle behind stage-based gain design based on transient performance analysis for a regular diminishing-type gain. Furthermore, we present a comprehensive framework of MS-LCS along with a sequential arrangement of stage-based gains. Lastly, we explicate several prerequisites for the design and analysis.

5.2.2.1 System Formulation

To this end, we employ the system model and formulation in Sect. 5.1. For convenience, we copy the main formulations below. Consider the following discrete-time stochastic system:

$$\begin{cases} x_k(t+1) = A_t x_k(t) + B_t u_k(t) + \omega_k(t), \\ y_k(t) = C_t x_k(t) + \nu_k(t), \end{cases} \tag{5.66}$$

where t and k denote the time index and iteration number, respectively. Here, $t = 0, 1, \cdots, N$, where N is the iteration length, and $k = 1, 2, \cdots$. Vectors $x_k(t) \in \mathbf{R}^n$, $u_k(t) \in \mathbf{R}^p$, and $y_k(t) \in \mathbf{R}^q$ denote the system state, input, and output, respectively. The vectors $\omega_k(t)$ and $\nu_k(t)$ denote random noises, independent of different time indices and iteration numbers. Besides, $\omega_k(t)$ and $\nu_k(t)$ satisfy $\mathbb{E}[\omega_k(t)] = \mathbb{E}[\nu_k(t)] = 0$, $\mathbb{E}[\|\omega_k(t)\|^2] \le \sigma_\omega^2$, and $\mathbb{E}[\|\nu_k(t)\|^2] \le \sigma_\nu^2$. Time-varying matrices A_t,

B_t, and C_t are with appropriate dimensions indicating the system's information. Assume that the matrix $C_{t+1}B_t$ is of full column rank.

By the lifting technique, we obtain

$$\boldsymbol{y}_k = \mathcal{H}\boldsymbol{u}_k + \mathcal{G}\boldsymbol{\omega}_k + \boldsymbol{v}_k + \mathcal{M}x_k(0), \tag{5.67}$$

where $\boldsymbol{u}_k = [u_k(0)^T, u_k(1)^T, \cdots, u_k(N-1)^T]^T$, $\boldsymbol{y}_k = [y_k(1)^T, y_k(2)^T, \cdots, y_k(N)^T]^T$, $\boldsymbol{\omega}_k = [\omega_k(0)^T, \cdots, \omega_k(N-1)^T]^T$, $\boldsymbol{v}_k = [v_k(1)^T, \cdots, v_k(N)^T]^T$, and

$$\mathcal{H} = \begin{bmatrix} C_1A_{0,1}B_0 & 0 & 0 & \cdots & 0 \\ C_2A_{1,1}B_0 & C_2A_{1,2}B_1 & 0 & \cdots & 0 \\ \vdots & \vdots & \vdots & \ddots & \vdots \\ C_NA_{N-1,1}B_0 & C_NA_{N-1,2}B_1 & \cdots & \cdots & C_NA_{N-1,N}B_{N-1} \end{bmatrix},$$

$$\mathcal{G} = \begin{bmatrix} C_1A_{0,1} & 0 & 0 & \cdots & 0 \\ C_2A_{1,1} & C_2A_{1,2} & 0 & \cdots & 0 \\ \vdots & \vdots & \vdots & \ddots & \vdots \\ C_NA_{N-1,1} & C_NA_{N-1,2} & \cdots & \cdots & C_NA_{N-1,N} \end{bmatrix},$$

$$\mathcal{M} = [(C_1A_0)^T, (C_2A_1A_0)^T, \cdots, (C_NA_{N-1,1})^T]^T.$$

Let $y_d(t)$ be the desired tracking reference and assume it is realizable. Assume that $x_k(0) = x_d(0)$, $\forall k$. Denote the tracking error by $e_k(t) \triangleq y_d(t) - y_k(t)$, and its lifting vector is given by

$$\boldsymbol{e}_k \triangleq \boldsymbol{y}_d - \boldsymbol{y}_k = \mathcal{H}\Delta\boldsymbol{u}_k - \mathcal{G}\boldsymbol{\omega}_k - \boldsymbol{v}_k, \tag{5.68}$$

where $\Delta\boldsymbol{u}_k \triangleq \boldsymbol{u}_d - \boldsymbol{u}_k$ denotes the input error.

5.2.2.2 P-Type Learning Control Scheme and Performance Analysis

If noises are absent in the system (5.66), i.e., $\omega_k(t) = v_k(t) = 0$, the precise tracking is easy to achieve by employing the following P-type LCS:

$$\boldsymbol{u}_{k+1} = \boldsymbol{u}_k + \eta\mathcal{L}\boldsymbol{e}_k, \tag{5.69}$$

where η is a constant learning gain and $\mathcal{L}$ is the direction regulation matrix (DRM). Here, $\mathcal{L} = \text{diag}\{L_0, L_1, \cdots, L_{N-1}\}$, where $L_i \in \mathbf{R}^{p\times q}$ $(0 \le i \le N-1)$ satisfies that all eigenvalues of $L_tC_{t+1}B_t$ have positive real parts (Shen and Qu 2020; Shen and Yu 2022).

While noises ω_k and ν_k are present, (5.69) cannot ensure the convergence of $\boldsymbol{u}_k$ to $\boldsymbol{u}_d$. The scheme is usually revised using a decreasing gain sequence as follows:

$$\boldsymbol{u}_{k+1} = \boldsymbol{u}_k + \eta_k \mathcal{L} \boldsymbol{e}_k, \tag{5.70}$$

where η_k satisfies

$$\eta_k > 0, \quad \eta_k \longrightarrow 0, \quad \sum_{k=1}^{\infty} \eta_k = \infty, \quad \sum_{k=1}^{\infty} \eta_k^2 < \infty. \tag{5.71}$$

The scheme described in (5.70) along with (5.71) is widely utilized in stochastic ILC. However, the predetermined nature of this gain sequence often leads to unsatisfactory transient convergence performance for practical applications. This can be attributed to inappropriate selection of the candidate learning gain during simulations, resulting in either an initial explosion of tracking error or a persistently large error even after multiple learning iterations.

To comprehend this issue, we commence with analyzing the transient performance of the scheme (5.70) by considering $\mathbb{E}[\|\Delta \boldsymbol{u}_k\|^2]$. The evolution of $\mathbb{E}[\|\Delta \boldsymbol{u}_k\|^2]$ is initially illustrated through the following lemma.

Lemma 5.5 *Consider the system* (5.66). *Assume that* $\mathcal{L} = diag\{L_0, L_1, \cdots, L_{N-1}\}$ *satisfies that all eigenvalues of* $L_t C_{t+1} B_t$ *have positive real parts. For* $\boldsymbol{u}_k$ *generated by the scheme* (5.70), $\mathbb{E}[\|\Delta \boldsymbol{u}_k\|^2]$ *is calculated as follows:*

$$\mathbb{E}[\|\Delta \boldsymbol{u}_k\|^2] = trace(\mathbb{E}[\Delta \boldsymbol{u}_k \Delta \boldsymbol{u}_k^T]), \tag{5.72}$$

$$\begin{aligned} &\mathbb{E}[\Delta \boldsymbol{u}_{k+1} \Delta \boldsymbol{u}_{k+1}^T] \\ &\quad = (I - \eta_k \mathcal{L}\mathcal{H}) \mathbb{E}[\Delta \boldsymbol{u}_k \Delta \boldsymbol{u}_k^T](I - \eta_k \mathcal{L}\mathcal{H})^T \\ &\qquad + \eta_k^2 \mathcal{L}(\sigma_\omega^2 \mathcal{G}\mathcal{G}^T + \sigma_\nu^2 I)\mathcal{L}^T, \end{aligned} \tag{5.73}$$

where $trace(\cdot)$ *denotes the trace of a square matrix.*

Proof. Equation (5.72) is evident. We now prove the validity of (5.73). Subtracting both sides of (5.69) from the desired input $\boldsymbol{u}_d$ leads to

$$\Delta \boldsymbol{u}_{k+1} = \Delta \boldsymbol{u}_k - \eta_k \mathcal{L} \boldsymbol{e}_k. \tag{5.74}$$

Substituting (5.68) into (5.74) leads to

$$\begin{aligned} \Delta \boldsymbol{u}_{k+1} &= \Delta \boldsymbol{u}_k - \eta_k \mathcal{L}[\mathcal{H} \Delta \boldsymbol{u}_k - \xi_k] \\ &= (I - \eta_k \mathcal{L}\mathcal{H}) \Delta \boldsymbol{u}_k + \eta_k \mathcal{L} \xi_k, \end{aligned}$$

where $\xi_k = \mathcal{G}\boldsymbol{w}_k + \boldsymbol{v}_k$. Then, we obtain

$$\begin{aligned}\Delta\boldsymbol{u}_{k+1}\Delta\boldsymbol{u}_{k+1}^T &= (I-\eta_k\mathcal{L}\mathcal{H})\Delta\boldsymbol{u}_k\Delta\boldsymbol{u}_k^T(I-\eta_k\mathcal{L}\mathcal{H})^T\\ &\quad+\eta_k(I-\eta_k\mathcal{L}\mathcal{H})\Delta\boldsymbol{u}_k\xi_k^T\mathcal{L}^T\\ &\quad+\eta_k\mathcal{L}\xi_k\Delta\boldsymbol{u}_k^T(I-\eta_k\mathcal{L}\mathcal{H})^T\\ &\quad+\eta_k^2\mathcal{L}\xi_k\xi_k^T\mathcal{L}^T. \end{aligned} \tag{5.75}$$

Denote $\mathcal{F}_k$ as a σ-algebra indicating the possible history of all involved random factors up to iteration k, i.e., $\mathcal{F}_k \triangleq \sigma\{x_d(0), u_1(t), \omega_i(t-1), v_i(t), 1 \le i \le k, 1 \le t \le N\}$. Then, because ξ_k is independent of $\mathcal{F}_{k-1}$, we obtain $\mathbb{E}[\xi_k|\mathcal{F}_{k-1}] = \mathbf{0}$. Besides, $\Delta\boldsymbol{u}_k$ depends merely on the information of the $(k-1)$th iteration; thus, $\Delta\boldsymbol{u}_k \in \mathcal{F}_{k-1}$. Hence, considering the conditional expectation of (5.75) with respect to $\mathcal{F}_{k-1}$, we obtain

$$\begin{aligned}&\mathbb{E}[\Delta\boldsymbol{u}_{k+1}\Delta\boldsymbol{u}_{k+1}^T|\mathcal{F}_{k-1}]\\ &\quad=(I-\eta_k\mathcal{L}\mathcal{H})\mathbb{E}[\Delta\boldsymbol{u}_k\Delta\boldsymbol{u}_k^T](I-\eta_k\mathcal{L}\mathcal{H})^T\\ &\qquad+\eta_k^2\mathcal{L}\mathbb{E}[\xi_k\xi_k^T]\mathcal{L}^T\\ &\quad=(I-\eta_k\mathcal{L}\mathcal{H})\mathbb{E}[\Delta\boldsymbol{u}_k\Delta\boldsymbol{u}_k^T](I-\eta_k\mathcal{L}\mathcal{H})^T\\ &\qquad+\eta_k^2\mathcal{L}(\sigma_\omega^2\mathcal{G}\mathcal{G}^T+\sigma_v^2 I)\mathcal{L}^T.\end{aligned}$$

Taking the expectation of the above equality leads to

$$\begin{aligned}\mathbb{E}[\Delta\boldsymbol{u}_{k+1}\Delta\boldsymbol{u}_{k+1}^T] &= (I-\eta_k\mathcal{L}\mathcal{H})\mathbb{E}[\Delta\boldsymbol{u}_k\Delta\boldsymbol{u}_k^T](I-\eta_k\mathcal{L}\mathcal{H})^T\\ &\quad+\eta_k^2\mathcal{L}(\sigma_\omega^2\mathcal{G}\mathcal{G}^T+\sigma_v^2 I)\mathcal{L}^T.\end{aligned}$$

Therefore, (5.73) is obtained, completing the proof.

Lemma 5.5 establishes a recursive calculation of the mean-squared input error $\mathbb{E}(||\Delta\boldsymbol{u}_{k+1}||^2)$, which quantifies the impact of noise on the update process of (5.70). The last term in (5.73) converges to zero as the gain η_k approaches zero, indicating that reducing η_k mitigates the effect of noise. As for the first term on the right-hand side of (5.73), when the gain η_k converges to zero, the matrix $I-\eta_k\mathcal{L}\mathcal{H}$ approaches identity matrix I. In other words, decreasing η_k slows down the update rate of $\Delta\boldsymbol{u}_k$. Therefore, there exists a trade-off between noise suppression and input refinement. This observation motivates us to employ an MS-LCS proposed in Sect. 5.1.

5.2.2.3 Multistage Learning Mechanism and Stage-Based Gain

To reconcile the conflicting demands of noise suppression and input refinement in (5.70), we propose the integration of an MS-LCS and a stage-based gain design. In this mechanism, each stage consists of a series of consecutive iterations. Let

k_l denote a constant representing the length of the lth stage, which is determined based on specific schemes. The learning gain remains constant within each stage and decreases by a predetermined ratio as we transition from one stage to another.

Define a decreasing gain sequence $\{\gamma_l = \gamma_1\theta^{l-1}\}$, where $0 < \theta < 1$ is a constant. We obtain a stage-based gain sequence:

$$\left\{\eta_k \middle| \eta_k = \gamma_l, \sum_{i=1}^{l-1} k_i < k \leq \sum_{i=1}^{l} k_i, i, l, k \in \mathbf{N}\right\}. \tag{5.76}$$

The fundamental concept of the MS-LCS is twofold. Firstly, we aim to maintain a consistent learning gain in each stage while maximizing the update of $\boldsymbol{u}_k$, ensuring that random noise does not hinder the process. Secondly, by mitigating the overall decrease in the learning gain, we can ultimately eliminate the impact of noise and guarantee a stable convergence of (5.70). To provide further clarity on the MS-LCS, it is crucial to determine the switching iteration that divides these stages.

Throughout this section, we refer to the LCS (5.70) with a well-defined stage-based gain sequence (5.76) as MS-LCS. The key aspect of MS-LCS lies in establishing an effective method for determining the values of γ_1, θ, and k_l. We present three variations of MS-LCS for generating the gain sequence. The first scheme is an ideal version where the performance index is directly calculated. However, this approach incurs a significant computational burden. The second scheme offers an alternative by replacing the original performance index with its upper bound, resulting in a substantial reduction in computational complexity but sacrificing performance evaluation accuracy. Finally, we propose a third scheme that further enhances performance evaluation by introducing stretching parameters.

To facilitate subsequent analysis, we impose additional assumptions on $\mathcal{L}$. Apart from the requirement that all eigenvalues of $L_t C_{t+1} B_t$ have positive real parts, it is necessary to design $\mathcal{L}$ in such a way that $\mathcal{L}\mathcal{H}$ becomes a quasi-positive-definite matrix, i.e., $\mathbf{x}^T\mathcal{L}\mathcal{H}\mathbf{x} > 0$, for all non-zero vectors $\mathbf{x}$.

The selection of $\mathcal{L}$ is feasible. Given that all eigenvalues of $L_t C_{t+1} B_t$ have positive real parts and we define $\mathcal{L}_0 = \text{diag}\{L_0, L_1, \ldots, L_{N-1}\}$, for any positive-definite matrix $\mathcal{S}$, the equation $(\mathcal{L}_0\mathcal{H})^T\Phi + \Phi(\mathcal{L}_0\mathcal{H}) = \mathcal{S}$ has a unique solution of Φ as a positive-definite matrix. Let $\mathcal{L} = \Phi\mathcal{L}_0$. Then, for any $\mathbf{x} \neq 0$, $x \in \mathbf{R}^{p\times N}$, we obtain

$$\begin{aligned}
\mathbf{x}^T\mathcal{L}\mathcal{H}\mathbf{x} &= \mathbf{x}^T\left(\frac{(\mathcal{L}\mathcal{H})^T + (\mathcal{L}\mathcal{H})}{2}\right)\mathbf{x} \\
&= \mathbf{x}^T\left(\frac{(\mathcal{L}_0\mathcal{H})^T\Phi + \Phi(\mathcal{L}_0\mathcal{H})}{2}\right)\mathbf{x} \\
&= \frac{\mathbf{x}^T\mathcal{S}\mathbf{x}}{2} > 0.
\end{aligned}$$

For three MS-LCSs, the requirements of γ_1 are respectively given as follows:

(a) γ_1 satisfies

$$\begin{aligned}
&\gamma_1 \mathcal{L}\mathcal{H}\mathbb{E}[\Delta \boldsymbol{u}_1 \Delta \boldsymbol{u}_1^T] + \gamma_1 \mathbb{E}[\Delta \boldsymbol{u}_1 \Delta \boldsymbol{u}_1^T](\mathcal{L}\mathcal{H})^T \\
&\quad - \gamma_1^2 \mathcal{L}\mathcal{H}\mathbb{E}[\Delta \boldsymbol{u}_1 \Delta \boldsymbol{u}_1^T](\mathcal{L}\mathcal{H})^T \\
&\quad - \gamma_1^2 \mathcal{L}(\sigma_\omega^2 \mathcal{G}\mathcal{G}^T + \sigma_v^2 I)\mathcal{L}^T \geq 0, \qquad (5.77)\\
&\lambda_{\max}(I - \gamma_1 \mathcal{L}\mathcal{H})(I - \gamma_1 \mathcal{L}\mathcal{H})^T < 1. \qquad (5.78)
\end{aligned}$$

(b) γ_1 satisfies $\gamma_1 \in (0, \gamma^*]$, where $\gamma^* = \lambda_2/\lambda_1$, $\lambda_1 = \lambda_{\max}(\mathcal{H}^T \mathcal{L}^T \mathcal{L}\mathcal{H})$, and $\lambda_2 = \lambda_{\min}(\frac{\mathcal{H}^T \mathcal{L}^T + \mathcal{L}\mathcal{H}}{2})$.

(c) γ_1 satisfies $\gamma_1 \in (0, \overline{\gamma}^*]$, where $\overline{\gamma}^* = \beta\lambda_2/\lambda_1$ and $\beta > 1$ is a constant.

The condition stated in (5.77) is a technical requirement for subsequent derivations. It imposes a constraint on the initial gain γ_1 to ensure a smooth learning process during the early stages. Otherwise, if γ_1 is set too high, it may lead to rapid increase in input error at the initial iterations and disrupt the learning process. In this inequality, because $\mathcal{L}\mathcal{H}$ is a quasi-positive-definite matrix, $\mathcal{L}\mathcal{H}\mathbb{E}[\Delta \boldsymbol{u}_1^l \Delta \boldsymbol{u}_1^{l\,T}] + \mathbb{E}[\Delta \boldsymbol{u}_1^l \Delta \boldsymbol{u}_1^{l\,T}](\mathcal{L}\mathcal{H})^T$ is positive-definite matrix. Thus, there exists γ_1 satisfying (5.77). Besides, through direction calculation, we obtain

$$\begin{aligned}
&\lambda_{\max}(I - \gamma_1 \mathcal{L}\mathcal{H})(I - \gamma_1 \mathcal{L}\mathcal{H})^T \\
&= 1 - \gamma_1 \lambda_{\min}(\mathcal{L}\mathcal{H} + \mathcal{H}^T \mathcal{L}^T - \gamma_1 \mathcal{L}\mathcal{H}\mathcal{H}^T \mathcal{L}^T]. \qquad (5.79)
\end{aligned}$$

Thus, there exists γ_1 satisfying (5.78). Furthermore, according to (5.79), if $\lambda_{\max}(I - \gamma_1 \mathcal{L}\mathcal{H})(I - \gamma_1 \mathcal{L}\mathcal{H})^T < 1$, then $\lambda_{\max}(I - \gamma_l \mathcal{L}\mathcal{H})(I - \gamma_l \mathcal{L}\mathcal{H})^T < 1$ for all l.

For conditions (b) and (c), $\mathcal{L}\mathcal{H}$ should be a quasi-positive-definite matrix to ensure $\lambda_2 > 0$.

5.2.3 *Ideal MS-LCS*

In this section, we give the ideal MS-LCS to determine the value of k_l in each stage, where the transient performance is evaluated regardless of computation costs.

To facilitate the elaboration of designing k_l, we rewrite (5.70) with the stage-based gain (5.76). Denote $\boldsymbol{u}_k^l$ as the system input, where the superscript is the lth stage and the subscript is the iteration number in this stage. Denote $\boldsymbol{e}_k^l$ as the tracking error generated by $\boldsymbol{u}_k^l$. We can rewrite (5.70) as follows:

$$\begin{aligned}
\boldsymbol{u}_{k+1}^l &= \boldsymbol{u}_k^l + \gamma_l \boldsymbol{e}_k^l, \quad k = 1, \ldots, k_l, \\
\boldsymbol{u}_1^{l+1} &= \boldsymbol{u}_{k_l+1}^l.
\end{aligned}$$

To generate a stage-based gain sequence, according to Lemma 5.5, we iteratively calculate $\mathbb{E}[\|\Delta \boldsymbol{u}_k^l\|^2]$ as follows:

$$\mathbb{E}[\|\Delta \boldsymbol{u}_k^l\|^2] = \text{trace}(\mathbb{E}[\Delta \boldsymbol{u}_k^l \Delta \boldsymbol{u}_k^{l\,T}]), \tag{5.80}$$

$$\begin{aligned}
&\mathbb{E}[\Delta \boldsymbol{u}_{k+1}^l \Delta \boldsymbol{u}_{k+1}^{l\;T}] \\
&\quad = (I - \gamma_l \mathcal{L}\mathcal{H})\mathbb{E}[\Delta \boldsymbol{u}_k^l \Delta \boldsymbol{u}_k^{l\,T}](I - \gamma_l \mathcal{L}\mathcal{H})^T \\
&\qquad + \gamma_l^2 \mathcal{L}(\sigma_\omega^2 \mathcal{G}\mathcal{G}^T + \sigma_\nu^2 I)\mathcal{L}^T, \; k = 1, \ldots, k_l + 1,
\end{aligned} \tag{5.81}$$

$$\mathbb{E}[\Delta \boldsymbol{u}_1^{l+1} \Delta \boldsymbol{u}_1^{l+1\,T}] = \mathbb{E}[\Delta \boldsymbol{u}_{k_l+1}^l \Delta \boldsymbol{u}_{k_l+1}^{l\;T}]. \tag{5.82}$$

Then, we define the descending speed of $\mathbb{E}[\|\Delta \boldsymbol{u}_k^l\|^2]$ by

$$\begin{aligned}
s_k^l &= \mathbb{E}[\|\Delta \boldsymbol{u}_k^l\|^2] - \mathbb{E}[\|\Delta \boldsymbol{u}_{k+1}^l\|^2], \\
&\qquad k = 1, \ldots, k_l + 1, \\
s_1^{l+1} &= \mathbb{E}[\|\Delta \boldsymbol{u}_1^{l+1}\|^2] - \mathbb{E}[\|\Delta \boldsymbol{u}_2^{l+1}\|^2].
\end{aligned}$$

Note that we denote the number of iterations for each stage by $k_l + 1$ because we must compare $s_{k_l+1}^l$ and s_1^{l+1}. We have the following lemma regarding the descending index s_k^l.

Lemma 5.6 *Consider the stochastic system* (5.66). *Let* $\mathcal{L}$=*diag*$\{L_0, L_1, \cdots, L_{N-1}\}$ *satisfy that all eigenvalues of* $L_t C_{t+1} B_t$ *have positive real parts and* $\mathcal{L}\mathcal{H}$ *be a positive-definite matrix. Moreover,* γ_1 *satisfies* (5.77) *and* (5.78). *For the recursive calculations* (5.80)–(5.82), *we have*

(a) For the matrix $\mathbb{E}[\Delta \boldsymbol{u}_k^l \Delta \boldsymbol{u}_k^{l\,T}] - \mathbb{E}[\Delta \boldsymbol{u}_{k+1}^l \Delta \boldsymbol{u}_{k+1}^{l\;T}]$ *that determines* s_k^l, *we have*

$$\begin{aligned}
&\mathbb{E}[\Delta \boldsymbol{u}_{k+1}^l \Delta \boldsymbol{u}_{k+1}^{l\;T}] - \mathbb{E}[\Delta \boldsymbol{u}_{k+2}^l \Delta \boldsymbol{u}_{k+2}^{l\;T}] \\
&\quad = (I - \gamma_l \mathcal{L}\mathcal{H})\mathbb{E}[\Delta \boldsymbol{u}_k^l \Delta \boldsymbol{u}_k^{l\,T}](I - \gamma_l \mathcal{L}\mathcal{H})^T \\
&\qquad - (I - \gamma_l \mathcal{L}\mathcal{H})\mathbb{E}[\Delta \boldsymbol{u}_{k+1}^l \Delta \boldsymbol{u}_{k+1}^{l\;T}](I - \gamma_l \mathcal{L}\mathcal{H})^T,
\end{aligned} \tag{5.83}$$

$$\begin{aligned}
&\mathbb{E}[\Delta \boldsymbol{u}_1^{l+1} \Delta \boldsymbol{u}_1^{l+1\,T}] - \mathbb{E}[\Delta \boldsymbol{u}_2^{l+1} \Delta \boldsymbol{u}_2^{l+1\,T}] \\
&\quad = \frac{\gamma_{l+1}}{\gamma_l}\left(\mathbb{E}[\Delta \boldsymbol{u}_{k_l+1}^l \Delta \boldsymbol{u}_{k_l+1}^{l\;T}] - \mathbb{E}[\Delta \boldsymbol{u}_{k_l+2}^l \Delta \boldsymbol{u}_{k_l+2}^{l\;T}]\right) \\
&\qquad + (\gamma_l - \gamma_{l+1})\gamma_{l+1}\{\mathcal{L}(\sigma_\omega^2 \mathcal{G}\mathcal{G}^T + \sigma_\nu^2 I)\mathcal{L}^T \\
&\qquad + \mathcal{L}\mathcal{H}\mathbb{E}[\Delta \boldsymbol{u}_{k_l+1}^l \Delta \boldsymbol{u}_{k_l+1}^{l\;T}](\mathcal{L}\mathcal{H})^T\}.
\end{aligned} \tag{5.84}$$

(b) The matrix $\mathbb{E}[\Delta \boldsymbol{u}_k^l \Delta \boldsymbol{u}_k^{l\,T}] - \mathbb{E}[\Delta \boldsymbol{u}_{k+1}^l \Delta \boldsymbol{u}_{k+1}^{l\;T}]$ *is semi-positive definite, i.e.,*

$$\begin{aligned}
&\mathbb{E}[\Delta \boldsymbol{u}_k^l \Delta \boldsymbol{u}_k^{l\,T}] - \mathbb{E}[\Delta \boldsymbol{u}_{k+1}^l \Delta \boldsymbol{u}_{k+1}^{l\;T}] \geq 0, \\
&\forall l, \quad k = 1, \cdots, k_l + 1.
\end{aligned} \tag{5.85}$$

(c) During the lth stage, $s_k^l > s_{k+1}^l$.
(d) As the gain decreases from γ_l *to* γ_{l+1}*, we have*

$$\begin{aligned}s_1^{l+1} &= \theta s_{k_l+1}^l \\ &\quad + (\gamma_l - \gamma_{l+1})\gamma_{l+1} trace\Big\{\mathcal{L}(\sigma_\omega^2 \mathcal{G}\mathcal{G}^T + \sigma_v^2 I)\mathcal{L}^T \\ &\quad + \mathcal{L}\mathcal{H}\mathbb{E}[\Delta \boldsymbol{u}_{k_l+1}^l \Delta \boldsymbol{u}_{k_l+1}^{l}{}^T](\mathcal{L}\mathcal{H})^T\Big\}. \end{aligned} \tag{5.86}$$

Proof. According to (5.81), we have

$$\begin{aligned}&\mathbb{E}[\Delta \boldsymbol{u}_{k+1}^l \Delta \boldsymbol{u}_{k+1}^{l}{}^T] - \mathbb{E}[\Delta \boldsymbol{u}_{k+2}^l \Delta \boldsymbol{u}_{k+2}^{l}{}^T] \\ &= (I - \gamma_l \mathcal{L}\mathcal{H})\mathbb{E}[\Delta \boldsymbol{u}_k^l \Delta \boldsymbol{u}_k^{l}{}^T](I - \gamma_l \mathcal{L}\mathcal{H})^T \\ &\quad + \gamma_l^2 \mathcal{L}(\sigma_\omega^2 \mathcal{G}\mathcal{G}^T + \sigma_v^2 I)\mathcal{L}^T \\ &\quad - (I - \gamma_l \mathcal{L}\mathcal{H})\mathbb{E}[\Delta \boldsymbol{u}_{k+1}^l \Delta \boldsymbol{u}_{k+1}^{l}{}^T](I - \gamma_l \mathcal{L}\mathcal{H})^T \\ &\quad - \gamma_l^2 \mathcal{L}(\sigma_\omega^2 \mathcal{G}\mathcal{G}^T + \sigma_v^2 I)\mathcal{L}^T \\ &= (I - \gamma_l \mathcal{L}\mathcal{H})\mathbb{E}[\Delta \boldsymbol{u}_k^l \Delta \boldsymbol{u}_k^{l}{}^T](I - \gamma_l \mathcal{L}\mathcal{H})^T \\ &\quad - (I - \gamma_l \mathcal{L}\mathcal{H})\mathbb{E}[\Delta \boldsymbol{u}_{k+1}^l \Delta \boldsymbol{u}_{k+1}^{l}{}^T](I - \gamma_l \mathcal{L}\mathcal{H})^T. \end{aligned}$$

Then, we prove (5.84). According to (5.81) and (5.82), we have

$$\begin{aligned}&\mathbb{E}[\Delta \boldsymbol{u}_1^{l+1} \Delta \boldsymbol{u}_1^{l+1}{}^T] - \mathbb{E}[\Delta \boldsymbol{u}_2^{l+1} \Delta \boldsymbol{u}_2^{l+1}{}^T] \\ &= (I - \gamma_l \mathcal{L}\mathcal{H})\mathbb{E}[\Delta \boldsymbol{u}_{k_l}^l \Delta \boldsymbol{u}_{k_l}^{l}{}^T](I - \gamma_l \mathcal{L}\mathcal{H})^T \\ &\quad - (I - \gamma_{l+1} \mathcal{L}\mathcal{H})\mathbb{E}[\Delta \boldsymbol{u}_{k_l+1}^l \Delta \boldsymbol{u}_{k_l+1}^{l}{}^T](I - \gamma_{l+1} \mathcal{L}\mathcal{H})^T \\ &\quad + (\gamma_l^2 - \gamma_{l+1}^2)\mathcal{L}(\sigma_\omega^2 \mathcal{G}\mathcal{G}^T + \sigma_v^2 I)\mathcal{L}^T \\ &= (I - \gamma_l \mathcal{L}\mathcal{H})\mathbb{E}[\Delta \boldsymbol{u}_{k_l}^l \Delta \boldsymbol{u}_{k_l}^{l}{}^T](I - \gamma_l \mathcal{L}\mathcal{H})^T \\ &\quad - (I - \gamma_l \mathcal{L}\mathcal{H} + \gamma_l \mathcal{L}\mathcal{H} - \gamma_{l+1}\mathcal{L}\mathcal{H}) \\ &\qquad \times \mathbb{E}[\Delta \boldsymbol{u}_{k_l+1}^l \Delta \boldsymbol{u}_{k_l+1}^{l}{}^T](I - \gamma_l \mathcal{L}\mathcal{H} + \gamma_l \mathcal{L}\mathcal{H} - \gamma_{l+1}\mathcal{L}\mathcal{H})^T \\ &\quad + (\gamma_l^2 - \gamma_{l+1}^2)\mathcal{L}(\sigma_\omega^2 \mathcal{G}\mathcal{G}^T + \sigma_v^2 I)\mathcal{L}^T\} \\ &= (I - \gamma_l \mathcal{L}\mathcal{H})\mathbb{E}[\Delta \boldsymbol{u}_{k_l}^l \Delta \boldsymbol{u}_{k_l}^{l}{}^T](I - \gamma_l \mathcal{L}\mathcal{H})^T \\ &\quad - (I - \gamma_l \mathcal{L}\mathcal{H})\mathbb{E}[\Delta \boldsymbol{u}_{k_l+1}^l \Delta \boldsymbol{u}_{k_l+1}^{l}{}^T](I - \gamma_l \mathcal{L}\mathcal{H})^T \\ &\quad - (I - \gamma_l \mathcal{L}\mathcal{H})\mathbb{E}[\Delta \boldsymbol{u}_{k_l+1}^l \Delta \boldsymbol{u}_{k_l+1}^{l}{}^T](\gamma_l - \gamma_{l+1})(\mathcal{L}\mathcal{H})^T \\ &\quad - (\gamma_l - \gamma_{l+1})\mathcal{L}\mathcal{H}\mathbb{E}[\Delta \boldsymbol{u}_{k_l+1}^l \Delta \boldsymbol{u}_{k_l+1}^{l}{}^T](I - \gamma_l \mathcal{L}\mathcal{H})^T \\ &\quad - (\gamma_l - \gamma_{l+1})^2 \mathcal{L}\mathcal{H}\mathbb{E}[\Delta \boldsymbol{u}_{k_l+1}^l \Delta \boldsymbol{u}_{k_l+1}^{l}{}^T](\mathcal{L}\mathcal{H})^T \end{aligned}$$

$$
\begin{aligned}
&+ (\gamma_l^2 - \gamma_{l+1}^2)\mathcal{L}(\sigma_\omega^2 \mathcal{G}\mathcal{G}^T + \sigma_v^2 I)\mathcal{L}^T \\
&= \mathbb{E}[\Delta \boldsymbol{u}^l_{k_l+1} \Delta \boldsymbol{u}^{l}_{k_l+1}{}^T] - \mathbb{E}[\Delta \boldsymbol{u}^l_{k_l+2} \Delta \boldsymbol{u}^{l}_{k_l+2}{}^T] \\
&- (\gamma_l - \gamma_{l+1})\mathbb{E}[\Delta \boldsymbol{u}^l_{k_l+1} \Delta \boldsymbol{u}^{l}_{k_l+1}{}^T](\mathcal{L}\mathcal{H})^T \\
&- (\gamma_l - \gamma_{l+1})\mathcal{L}\mathcal{H}\mathbb{E}[\Delta \boldsymbol{u}^l_{k_l+1} \Delta \boldsymbol{u}^{l}_{k_l+1}{}^T] \\
&+ (\gamma_l^2 - \gamma_{l+1}^2)\mathcal{L}\mathcal{H}\mathbb{E}[\Delta \boldsymbol{u}^l_{k_l+1} \Delta \boldsymbol{u}^{l}_{k_l+1}{}^T](\mathcal{L}\mathcal{H})^T \\
&+ (\gamma_l^2 - \gamma_{l+1}^2)\mathcal{L}(\sigma_\omega^2 \mathcal{G}\mathcal{G}^T + \sigma_v^2 I)\mathcal{L}^T \\
&= \mathbb{E}[\Delta \boldsymbol{u}^l_{k_l+1} \Delta \boldsymbol{u}^{l}_{k_l+1}{}^T] - \mathbb{E}[\Delta \boldsymbol{u}^l_{k_l+2} \Delta \boldsymbol{u}^{l}_{k_l+2}{}^T] \\
&+ \frac{\gamma_l - \gamma_{l+1}}{\gamma_l}\Big\{ - \gamma_l \mathbb{E}[\Delta \boldsymbol{u}^l_{k_l+1} \Delta \boldsymbol{u}^{l}_{k_l+1}{}^T](\mathcal{L}\mathcal{H})^T \\
&- \gamma_l \mathcal{L}\mathcal{H}\mathbb{E}[\Delta \boldsymbol{u}^l_{k_l+1} \Delta \boldsymbol{u}^{l}_{k_l+1}{}^T] \\
&+ (\gamma_l^2 + \gamma_l\gamma_{l+1})\mathcal{L}\mathcal{H}\mathbb{E}[\Delta \boldsymbol{u}^l_{k_l+1} \Delta \boldsymbol{u}^{l}_{k_l+1}{}^T](\mathcal{L}\mathcal{H})^T \\
&+ (\gamma_l^2 + \gamma_l\gamma_{l+1})\mathcal{L}(\sigma_\omega^2 \mathcal{G}\mathcal{G}^T + \sigma_v^2 I)\mathcal{L}^T \Big\}.
\end{aligned}
\tag{5.87}
$$

By (5.81), we have

$$
\begin{aligned}
&\mathbb{E}[\Delta \boldsymbol{u}^l_{k_l+2} \Delta \boldsymbol{u}^{l}_{k_l+2}{}^T] - \mathbb{E}[\Delta \boldsymbol{u}^l_{k_l+1} \Delta \boldsymbol{u}^{l}_{k_l+1}{}^T] \\
&= -\gamma_l \mathcal{L}\mathcal{H}\mathbb{E}[\Delta \boldsymbol{u}^l_{k_l+1} \Delta \boldsymbol{u}^{l}_{k_l+1}{}^T] \\
&\quad - \gamma_l \mathbb{E}[\Delta \boldsymbol{u}^l_{k_l+1} \Delta \boldsymbol{u}^{l}_{k_l+1}{}^T](\mathcal{L}\mathcal{H})^T \\
&\quad + \gamma_l^2 \mathcal{L}\mathcal{H}\mathbb{E}[\Delta \boldsymbol{u}^l_{k_l+1} \Delta \boldsymbol{u}^{l}_{k_l+1}{}^T](\mathcal{L}\mathcal{H})^T \\
&\quad + \gamma_l^2 \mathcal{L}(\sigma_\omega^2 \mathcal{G}\mathcal{G}^T + \sigma_v^2 I)\mathcal{L}^T.
\end{aligned}
\tag{5.88}
$$

Substituting (5.88) into (5.87) leads to

$$
\begin{aligned}
&\mathbb{E}[\Delta \boldsymbol{u}^{l+1}_1 \Delta \boldsymbol{u}^{l+1}_1{}^T] - \mathbb{E}[\Delta \boldsymbol{u}^{l+1}_2 \Delta \boldsymbol{u}^{l+1}_2{}^T] \\
&= \frac{\gamma_{l+1}}{\gamma_l}\left(\mathbb{E}[\Delta \boldsymbol{u}^l_{k_l+1} \Delta \boldsymbol{u}^{l}_{k_l+1}{}^T] - \mathbb{E}[\Delta \boldsymbol{u}^l_{k_l+2} \Delta \boldsymbol{u}^{l}_{k_l+2}{}^T]\right) \\
&\quad + (\gamma_l - \gamma_{l+1})\gamma_{l+1}\big\{\mathcal{L}(\sigma_\omega^2 \mathcal{G}\mathcal{G}^T + \sigma_v^2 I)\mathcal{L}^T \\
&\quad + \mathcal{L}\mathcal{H}\mathbb{E}[\Delta \boldsymbol{u}^l_{k_l+1} \Delta \boldsymbol{u}^{l}_{k_l+1}{}^T](\mathcal{L}\mathcal{H})^T\big\}.
\end{aligned}
$$

Therefore, item (a) has been proven.

Next, we proceed to item (b). To this end, we first assume

$$
\mathbb{E}[\Delta \boldsymbol{u}^l_1 \Delta \boldsymbol{u}^{l}_1{}^T] - \mathbb{E}[\Delta \boldsymbol{u}^l_2 \Delta \boldsymbol{u}^{l}_2{}^T] \geq 0, \quad \forall l. \tag{5.89}
$$

Then, by (5.83), we have

$$\begin{aligned}&\mathbb{E}[\Delta \boldsymbol{u}_{k+1}^{l}\Delta \boldsymbol{u}_{k+1}^{l}{}^{T}]-\mathbb{E}[\Delta \boldsymbol{u}_{k+2}^{l}\Delta \boldsymbol{u}_{k+2}^{l}{}^{T}]\\&=(I-\gamma_{l}\mathcal{L}\mathcal{H})(\mathbb{E}[\Delta \boldsymbol{u}_{k}^{l}\Delta \boldsymbol{u}_{k}^{l}{}^{T}]\\&\quad-\mathbb{E}[\Delta \boldsymbol{u}_{k+1}^{l}\Delta \boldsymbol{u}_{k+1}^{l}{}^{T}])(I-\gamma_{l}\mathcal{L}\mathcal{H})^{T}.\end{aligned}\tag{5.90}$$

From (5.90), if $\mathbb{E}[\Delta \boldsymbol{u}_{k}^{l}\Delta \boldsymbol{u}_{k}^{l}{}^{T}]-\mathbb{E}[\Delta \boldsymbol{u}_{k+1}^{l}\Delta \boldsymbol{u}_{k+1}^{l}{}^{T}]\geq 0$ is valid, then $\mathbb{E}[\Delta \boldsymbol{u}_{k+1}^{l}\Delta \boldsymbol{u}_{k+1}^{l}{}^{T}]-\mathbb{E}[\Delta \boldsymbol{u}_{k+2}^{l}\Delta \boldsymbol{u}_{k+2}^{l}{}^{T}]\geq 0$ holds. Thus, (5.85) holds for all iterations in the lth stage under the premise of (5.89). To prove item (b), it is sufficient to verify (5.89) for all stages l. We complete this step using the induction principle.

Initial step: $l=1$. By (5.81) and (5.77), we have

$$\begin{aligned}&\mathbb{E}[\Delta \boldsymbol{u}_{1}^{1}\Delta \boldsymbol{u}_{1}^{1}{}^{T}]-\mathbb{E}[\Delta \boldsymbol{u}_{2}^{1}\Delta \boldsymbol{u}_{2}^{1}{}^{T}]\\&=\gamma_{1}\mathcal{L}\mathcal{H}\mathbb{E}[\Delta \boldsymbol{u}_{1}^{1}\Delta \boldsymbol{u}_{1}^{1}{}^{T}]+\gamma_{1}\mathbb{E}[\Delta \boldsymbol{u}_{1}^{l}\Delta \boldsymbol{u}_{1}^{1}{}^{T}](\mathcal{L}\mathcal{H})^{T}\\&\quad-\gamma_{1}^{2}\mathcal{L}\mathcal{H}\mathbb{E}[\Delta \boldsymbol{u}_{1}^{1}\Delta \boldsymbol{u}_{1}^{1}{}^{T}](\mathcal{L}\mathcal{H})^{T}-\gamma_{1}^{2}\mathcal{L}(\sigma_{\omega}^{2}\mathcal{G}\mathcal{G}^{T}+\sigma_{\upsilon}^{2}I)\mathcal{L}^{T}\\&\geq 0.\end{aligned}$$

Inductive step: Suppose (5.89) holds for index l. Then, we obtain that $\mathbb{E}[\Delta \boldsymbol{u}_{k_l+1}^{l}\Delta \boldsymbol{u}_{k_l+1}^{l}{}^{T}]-\mathbb{E}[\Delta \boldsymbol{u}_{k_l+2}^{l}\Delta \boldsymbol{u}_{k_l+2}^{l}{}^{T}]\geq 0$. By (5.84), $\mathbb{E}[\Delta \boldsymbol{u}_{1}^{l+1}\Delta \boldsymbol{u}_{1}^{l+1}{}^{T}]-\mathbb{E}[\Delta \boldsymbol{u}_{2}^{l+1}\Delta \boldsymbol{u}_{2}^{l+1}{}^{T}]$ ≥ 0 holds for index $l+1$. According to the induction principle, the conclusion is valid. Therefore, item (b) has been proven.

Now, we come to item (c). We expand s_{k+1}^{l} as follows:

$$\begin{aligned}s_{k+1}^{l}&=\operatorname{trace}\{\mathbb{E}[\Delta \boldsymbol{u}_{k+1}^{l}\Delta \boldsymbol{u}_{k+1}^{l}{}^{T}]-\mathbb{E}[\Delta \boldsymbol{u}_{k+2}^{l}\Delta \boldsymbol{u}_{k+2}^{l}{}^{T}]\}\\&=\operatorname{trace}\{(I-\gamma_{l}\mathcal{L}\mathcal{H})\mathbb{E}[\Delta \boldsymbol{u}_{k}^{l}\Delta \boldsymbol{u}_{k}^{l}{}^{T}](I-\gamma_{l}\mathcal{L}\mathcal{H})^{T}\\&\quad-(I-\gamma_{l}\mathcal{L}\mathcal{H})\mathbb{E}[\Delta \boldsymbol{u}_{k+1}^{l}\Delta \boldsymbol{u}_{k+1}^{l}{}^{T}](I-\gamma_{l}\mathcal{L}\mathcal{H})^{T}\}\\&=\operatorname{trace}\{(I-\gamma_{l}\mathcal{L}\mathcal{H})(\mathbb{E}[\Delta \boldsymbol{u}_{k}^{l}\Delta \boldsymbol{u}_{k}^{l}{}^{T}]\\&\quad-\mathbb{E}[\Delta \boldsymbol{u}_{k+1}^{l}\Delta \boldsymbol{u}_{k+1}^{l}{}^{T}])(I-\gamma_{l}\mathcal{L}\mathcal{H})^{T}\}\\&=\operatorname{trace}\{(I-\gamma_{l}\mathcal{L}\mathcal{H})(I-\gamma_{l}\mathcal{L}\mathcal{H})^{T}(\mathbb{E}[\Delta \boldsymbol{u}_{k}^{l}\Delta \boldsymbol{u}_{k}^{l}{}^{T}]\\&\quad-\mathbb{E}[\Delta \boldsymbol{u}_{k+1}^{l}\Delta \boldsymbol{u}_{k+1}^{l}{}^{T}])\}\\&\leq\lambda_{\max}\left((I-\gamma_{l}\mathcal{L}\mathcal{H})(I-\gamma_{l}\mathcal{L}\mathcal{H})^{T}\right)\\&\quad\times\operatorname{trace}\{(\mathbb{E}[\Delta \boldsymbol{u}_{k}^{l}\Delta \boldsymbol{u}_{k}^{l}{}^{T}]-\mathbb{E}[\Delta \boldsymbol{u}_{k+1}^{l}\Delta \boldsymbol{u}_{k+1}^{l}{}^{T}])\}\\&=\lambda_{\max}\left((I-\gamma_{l}\mathcal{L}\mathcal{H})(I-\gamma_{l}\mathcal{L}\mathcal{H})^{T}\right)s_{k}^{l}\\&<s_{k}^{l}.\end{aligned}\tag{5.91}$$

The first inequality in (5.91) holds because $\text{trace}(AB) < \lambda_{\max}(A)\text{trace}(B)$, where A and B are semi-positive-definite square matrices. The last inequality holds due to the fact that $\lambda_{\max}\left((I-\gamma_l\mathcal{LH})(I-\gamma_l\mathcal{LH})^T\right) < 1$. Hence, the proof of item (c) is concluded.

We proceed to item (d). By (5.80) and (5.84), we have

$$\begin{aligned}
s_1^{l+1} &= \text{trace}\left\{\mathbb{E}[\Delta \boldsymbol{u}_1^{l+1}\Delta \boldsymbol{u}_1^{l+1^T}] - \mathbb{E}[\Delta \boldsymbol{u}_2^{l+1}\Delta \boldsymbol{u}_2^{l+1^T}]\right\}\\
&= \text{trace}\left\{\frac{\gamma_{l+1}}{\gamma_l}(\mathbb{E}[\Delta \boldsymbol{u}_{k_l+1}^{l}\Delta \boldsymbol{u}_{k_l+1}^{l}{}^T]\right.\\
&\quad \left. - \mathbb{E}[\Delta \boldsymbol{u}_{k_l+2}^{l}\Delta \boldsymbol{u}_{k_l+2}^{l}{}^T])\right\}\\
&\quad + (\gamma_l - \gamma_{l+1})\gamma_{l+1}\text{trace}\{\mathcal{L}(\sigma_\omega^2\mathcal{G}\mathcal{G}^T + \sigma_v^2 I)\mathcal{L}^T\\
&\quad + \mathcal{LH}\mathbb{E}[\Delta \boldsymbol{u}_{k_l+1}^{l}\Delta \boldsymbol{u}_{k_l+1}^{l}{}^T](\mathcal{LH})^T\}\\
&=\theta s_{k_l+1}^l + (\gamma_l - \gamma_{l+1})\gamma_{l+1}\text{trace}\{\mathcal{L}(\sigma_\omega^2\mathcal{G}\mathcal{G}^T + \sigma_v^2 I)\mathcal{L}^T\\
&\quad + \mathcal{LH}\mathbb{E}[\Delta \boldsymbol{u}_{k_l+1}^{l}\Delta \boldsymbol{u}_{k_l+1}^{l}{}^T](\mathcal{LH})^T\}.
\end{aligned}$$

The proof is completed.

Algorithm 5.1 Gain adaptation for the ideal MS-LCS

Initialization: $\theta > 0$, $\mathbb{E}[\Delta \boldsymbol{u}_1^1\Delta \boldsymbol{u}_1^{1^T}]$, γ_1 satisfied (5.77) and (5.78), $l = 1$.
For $l = 1, 2, \cdots$ do
  a) calculate K_l by event trigger
    $K_l = \min\{k_l \in \mathbf{N} | s_{k_l+1}^l \le s_{l+1}^1\}$;
  b) $\left\{\tau_k \middle| \tau_k = \gamma_l, \sum_{l=1}^{l-1} K_l < k \le \sum_{l=1}^{l} K_l\right\}$;
  c) calculate $\mathbb{E}[\Delta \boldsymbol{u}_{K_l+1}^l\Delta \boldsymbol{u}_{K_l+1}^l{}^T]$ by (5.81);
  d) calculate $\mathbb{E}[\Delta \boldsymbol{u}_1^{l+1}\Delta \boldsymbol{u}_1^{l+1^T}]$ by (5.82);
  e) $l = l + 1$.
End for

The primary idea of designing k_l is to reduce the gain while maintaining a rapid descending speed of $\mathbb{E}[\|\Delta \boldsymbol{u}_k^l\|^2]$. That is, the stage length k_l should satisfy

$$s_k^l > s_1^{l+1}, k = 1, 2\cdots, k_l, \quad s_{k_l+1}^l < s_1^{l+1}. \tag{5.92}$$

From item (d) of Lemma 5.6, a finite stage length k_l exists such that $s_{k_l+1}^l < s_1^{l+1}$ holds, $\forall l$. To see this point, checking item (b) of Lemma 5.6, we find that $s_k^l > 0$, $k = 1, 2, \cdots, k_l + 1, \forall l$. By item (c) of Lemma 5.6, $s_{k_l+1}^l$ is constantly decreasing in this stage as k_l increases. That is, $s_{k_l+1}^l$ decreases monotonically to 0 as k_l increases to infinity. However, in (5.86), the term $(\gamma_l - \gamma_{l+1})\gamma_{l+1}\text{trace}\{\mathcal{L}(\sigma_\omega^2\mathcal{G}\mathcal{G}^T + \sigma_v^2 I)\mathcal{L}^T +$

$\mathcal{LH}\mathbb{E}[\Delta \boldsymbol{u}^l_{k_l+1}\Delta \boldsymbol{u}^{l}_{k_l+1}{}^T](\mathcal{LH})^T\}$ is a positive constant away from zero. Therefore, there must be a finite k_l satisfying (5.92).

To generate k_l, we introduce an event trigger as follows:

$$K_l = \min\{k_l \in \mathbf{N} | s^l_{k_l+1} \le s^{l+1}_1\}. \tag{5.93}$$

Through the recursive formula (5.80)–(5.82) and the event trigger (5.93), we present the ideal MS-LCS, where the stage-based gain sequence is defined by

$$\left\{\tau_k \middle| \tau_k = \gamma_l, \sum_{i=1}^{l-1} K_i < k \le \sum_{i=1}^{l} K_i, i, l, k \in \mathbf{N}\right\}. \tag{5.94}$$

The pseudo-code for generating τ_k is presented in Algorithm 5.1. This algorithm illustrates the theoretical principle of the proposed ideal MS-LCS, where the length of each stage is computed to determine the stage switching. However, in practical implementation, there is no need to pre-compute the stage length and instead, the gain sequence can be generated iteratively. Hence, we provide an alternative pseudo-code for generating the stage-dependent gain τ_k in Algorithm 5.2, which offers a more practical execution process. The same approach applies to both practical and improved MS-LCSs discussed in Sect. 5.2.4.

Algorithm 5.2 Gain adaptation for the ideal MS-LCS (Iterative)

Initialization: $\theta > 0$, $\mathbb{E}[\Delta \boldsymbol{u}^1_1 \Delta \boldsymbol{u}^{1}_1{}^T]$, γ_1 satisfied (5.77) and (5.78), $l = 1$, $k_1 = 0$, $k = 0$.
For $k = 0, 1, 2, \cdots$ do
 a) $k_l = k_l + 1$;
 b) $k = k + 1$;
 c) calculate $\mathbb{E}[\|\Delta \boldsymbol{u}^l_{k_l+1}\|^2]$, by (5.81);
 d) calculate $s^l_{k_l} = \mathbb{E}[\|\Delta \boldsymbol{u}^l_{k_l}\|^2] - \mathbb{E}[\|\Delta \boldsymbol{u}^l_{k_l+1}\|^2]$;
 e) calculate s^{l+1}_1 by (5.86).
 If $s^l_{k_l} < s^{l+1}_1$ then
  i) $k_l = k_l - 1$;
  ii) $k = k - 1$;
  iii) $\mathbb{E}[\|\Delta \boldsymbol{u}^{l+1}_1\|^2] = \mathbb{E}[\|\Delta \boldsymbol{u}^l_{k_l+1}\|^2]$;
  iv) $l = l + 1$;
  v) $k_l = 0$.
 Else
  i) $\tau_k = \gamma_l$.
 End if
End for

However, it is challenging to analyze whether the gain sequence $\{\tau_k\}$ defined by (5.93) and (5.94) satisfies the convergence condition (5.71). Furthermore, calculating $\mathbb{E}[\|\Delta \boldsymbol{u}^l_{K_l+1}\|^2]$ using (5.80)–(5.82) is troublesome, especially when the dimension of u^l_k is very high. Therefore, we propose a practical MS-LCS and its improved version by introducing an upper bound of $\mathbb{E}[\|\Delta \boldsymbol{u}^l_{K_l+1}\|^2]$, which is easier to implement in applications given in the following section.

5.2.4 Practical and Improved MS-LCSs

This section introduces the practical and improved MS-LCSs to facilitate practical applications. Before the specific MS-LCSs, we first introduce an upper bound description of $\mathbb{E}[\|\Delta \boldsymbol{u}^l_k\|^2]$.

5.2.4.1 Upper Bound Model

We give the following lemma to describe an upper bound of $\mathbb{E}[\|\Delta \boldsymbol{u}^l_k\|^2]$.

Lemma 5.7 *Consider the stochastic system* (5.66). *Let* $\mathcal{L}$=*diag*$\{L_0, L_1, \cdots, L_{N-1}\}$ *satisfy that all eigenvalues of* $L_t C_{t+1} B_t$ *have positive real parts and* $\mathcal{LH}$ *be a quasi-positive-definite matrix. For the recursive formula* (5.80)–(5.82), *we have*

(a) Define $\lambda_1 = \lambda_{max}(\mathcal{H}^T\mathcal{L}^T\mathcal{LH})$ *and* $\lambda_2 = \lambda_{\min}(\frac{\mathcal{H}^T\mathcal{L}^T+\mathcal{LH}}{2})$. *We have*

$$\mathbb{E}[\|\Delta \boldsymbol{u}^l_{k+1}\|^2] \le q(\gamma_l)\mathbb{E}[\|\Delta \boldsymbol{u}^l_k\|^2] + \gamma_l^2\sigma_\xi^2,$$

$$\mathbb{E}[\|\Delta \boldsymbol{u}^l_{k+1}\|^2] \le q(\gamma_l)^k\mathbb{E}[\|\Delta \boldsymbol{u}^l_1\|^2] + \gamma_l^2\sigma_\xi^2 \sum_{i=0}^{i=k-1} q(\gamma_l)^i,$$

where $q(\gamma) = 1 - 2\lambda_2\gamma + \lambda_1\gamma^2$, $\sigma_\xi^2 = \lambda_{max}(\mathcal{L}^T\mathcal{L})qN\sigma_\nu^2 + \lambda_{max}(\mathcal{G}^T\mathcal{L}^T\mathcal{LG})p\, N\sigma_\omega^2$.

(b) We have an upper bound $\overline{\boldsymbol{u}}^l_k$ *satisfying that*

$$\mathbb{E}[\|\Delta \boldsymbol{u}^l_{k+1}\|^2] \le \overline{\boldsymbol{u}}^l_{k+1}, \quad \forall l, \quad k = 1, 2, \cdots, k_l,$$

where $\overline{\boldsymbol{u}}^l_{k+1}$ *is generated by the following recursive formula:*

$$\overline{\boldsymbol{u}}^1_1 = \mathbb{E}[\|\Delta \boldsymbol{u}^1_1\|^2], \tag{5.95}$$

$$\overline{\boldsymbol{u}}^l_{k+1} = q(\gamma_l)^k\overline{\boldsymbol{u}}^l_1 + \gamma_l^2\sigma_\xi^2 \sum_{i=0}^{i=k-1} q(\gamma_l)^i, \tag{5.96}$$

$$\overline{\boldsymbol{u}}^{l+1}_1 = \overline{\boldsymbol{u}}^l_{k_l+1}. \tag{5.97}$$

Proof. We first prove item (a). From (5.81), we have

$$\begin{aligned}
&\text{trace}\left(\mathbb{E}[\Delta \boldsymbol{u}_{k+1}^{l} \Delta \boldsymbol{u}_{k+1}^{l}{}^{T}]\right) \\
&= \text{trace}((I - \gamma_l \mathcal{L}\mathcal{H})\mathbb{E}[\Delta \boldsymbol{u}_k^l \Delta \boldsymbol{u}_k^l{}^T](I - \gamma_l \mathcal{L}\mathcal{H})^T) \\
&\quad + \text{trace}(\gamma_l^2 \mathcal{L}(\sigma_\omega^2 \mathcal{G}\mathcal{G}^T + \sigma_v^2 I)\mathcal{L}^T) \\
&= \mathbb{E}[\text{trace}((I - \gamma_l \mathcal{L}\mathcal{H})\Delta \boldsymbol{u}_k^l \Delta \boldsymbol{u}_k^l{}^T (I - \gamma_l \mathcal{L}\mathcal{H})^T)] \\
&\quad + \gamma_l^2[\sigma_\omega^2 \text{trace}(\mathcal{L}\mathcal{G}\mathcal{G}^T \mathcal{L}^T) + \sigma_v^2 \text{trace}(\mathcal{L}\mathcal{L}^T)] \\
&= \mathbb{E}[\text{trace}(\Delta \boldsymbol{u}_k^l{}^T (I - \gamma_l \mathcal{L}\mathcal{H})^T (I - \gamma_l \mathcal{L}\mathcal{H})\Delta \boldsymbol{u}_k^l)] \\
&\quad + \gamma_l^2[\sigma_\omega^2 \text{trace}(\mathcal{G}^T \mathcal{L}^T \mathcal{L}\mathcal{G}) + \sigma_v^2 \text{trace}(\mathcal{L}^T \mathcal{L})].
\end{aligned} \tag{5.98}$$

Thus, according to (5.80) and (5.98), we obtain

$$\begin{aligned}
&\mathbb{E}[\|\Delta \boldsymbol{u}_{k+1}^l\|^2] \\
&= \mathbb{E}[\Delta \boldsymbol{u}_k^l{}^T (I - \gamma_l \mathcal{L}\mathcal{H})^T (I - \gamma_l \mathcal{L}\mathcal{H})\Delta \boldsymbol{u}_k^l] \\
&\quad + \gamma_l^2[\sigma_\omega^2 \text{trace}(\mathcal{G}^T \mathcal{L}^T \mathcal{L}\mathcal{G}) + \sigma_v^2 \text{trace}(\mathcal{L}^T \mathcal{L})].
\end{aligned} \tag{5.99}$$

For the first term on the right-hand side (5.99), we have

$$\begin{aligned}
&\mathbb{E}[\Delta \boldsymbol{u}_k^l{}^T (I - \gamma_l \mathcal{L}\mathcal{H})^T (I - \gamma_l \mathcal{L}\mathcal{H})\Delta \boldsymbol{u}_k^l] \\
&= \mathbb{E}[\Delta \boldsymbol{u}_k^l{}^T \Delta \boldsymbol{u}_k^l + \gamma_l^2 \Delta \boldsymbol{u}_k^l{}^T \mathcal{H}^T \mathcal{L}^T \mathcal{L}\mathcal{H}\Delta \boldsymbol{u}_k \\
&\quad - \gamma_l \Delta \boldsymbol{u}_k^l{}^T (\mathcal{H}^T \mathcal{L}^T + \mathcal{L}\mathcal{H})\Delta \boldsymbol{u}_k^l] \\
&\leq (1 - 2\lambda_2 \gamma_l + \lambda_1 \gamma_l^2)\mathbb{E}[||\Delta \boldsymbol{u}_k^l||^2].
\end{aligned} \tag{5.100}$$

For the last term of (5.99), we have

$$\begin{aligned}
&\sigma_\omega^2 \text{trace}(\mathcal{G}^T \mathcal{L}^T \mathcal{L}\mathcal{G}^T) + \sigma_v^2 \text{trace}(\mathcal{L}^T \mathcal{L}) \\
&\leq \lambda_{max}(\mathcal{L}^T \mathcal{L}) q N \sigma_v^2 + \lambda_{max}(\mathcal{G}^T \mathcal{L}^T \mathcal{L}\mathcal{G}) p N \sigma_\omega^2 \\
&= \sigma_\xi^2.
\end{aligned} \tag{5.101}$$

Substituting (5.100) and (5.101) into (5.99) yields

$$\mathbb{E}[||\Delta \boldsymbol{u}_{k+1}^l||^2] \leq q(\gamma_l)\mathbb{E}[||\Delta \boldsymbol{u}_k^l||^2] + \gamma_l^2 \sigma_\xi^2.$$

This further leads to

$$\mathbb{E}[\|\Delta \boldsymbol{u}_{k+1}^l\|^2] \leq q(\gamma_l)^k \mathbb{E}[\|\Delta \boldsymbol{u}_1^l\|^2] + \gamma_l^2 \sigma_\xi^2 \sum_{i=0}^{i=k-1} q(\gamma_l)^i.$$

We proceed to item (b). First, suppose $\mathbb{E}[\|\Delta \boldsymbol{u}_1^l\|^2] \leq \overline{\boldsymbol{u}}_1^l, \forall l$. By item (a), we have

$$\mathbb{E}[\|\Delta \boldsymbol{u}_{k+1}^l\|^2] \leq \overline{\boldsymbol{u}}_{k+1}^l, \quad k = 2, 3, \ldots, k_l, \ \forall l. \tag{5.102}$$

Thus, it is sufficient to prove the following inequality using the induction principle:

$$\mathbb{E}[\|\Delta \boldsymbol{u}_1^l\|^2] \leq \overline{\boldsymbol{u}}_1^l, \quad \forall l. \tag{5.103}$$

By (5.95), (5.103) holds for the initial case $l = 1$. Suppose (5.103) holds for the index l. According to (5.102), we have $\mathbb{E}[\|\Delta \boldsymbol{u}_{k_l+1}^l\|^2] \leq \overline{\boldsymbol{u}}_{k_l+1}^l$. Using (5.82) and (5.97), we conclude that (5.103) holds for the index $l + 1$. Therefore, by combining (5.102) and (5.103), the proof is completed.

By applying Lemma 5.7, we introduce an upper bound for $\mathbb{E}[\|\Delta \boldsymbol{u}_k^l\|^2]$ using the recursive formulas (5.95)–(5.97). This upper bound serves as an approximation of $\mathbb{E}[\|\Delta \boldsymbol{u}_k^l\|^2]$. It is worth noting that these recursive formulas do not involve any matrix operations, resulting in significantly reduced computational burden compared to(5.80)–(5.82). Consequently, this upper bound can be utilized to generate a stage-based gain sequence, which will be further elaborated in the subsequent subsection.

To facilitate the subsequent analysis, we provide some properties of $q(\gamma) = 1 - 2\lambda_2\gamma + \lambda_1\gamma^2$ in the following lemma.

Lemma 5.8 *Define* $q(\gamma) = 1 - 2\lambda_2\gamma + \lambda_1\gamma^2$, $\gamma \in (0, \gamma^*]$, *where* $\gamma^* = \frac{\lambda_2}{\lambda_1}$. *Moreover, define two auxiliary functions:*

$$f(\gamma) = \frac{1 - q(\theta\gamma)}{[1 - q(\gamma)][q(\theta\gamma) - q(\gamma)]} - \frac{\theta^2}{q(\theta\gamma) - q(\gamma)},$$
$$g(\gamma) = \frac{1}{1 - q(\gamma)} - \frac{\theta^2}{1 - q(\theta\gamma)}.$$

Then, we have the following conclusions.

(a) The range of $q(\gamma)$ *is given by* $0 < q(\gamma) < 1$, $\forall\gamma \in (0, \gamma^*]$.
(b) If $0 < \theta < q(\gamma^*)$, *then* $f(\gamma) > 0$, $q(\gamma)f(\gamma) - \theta^2 f(\theta\gamma) > 0$, $\forall\gamma \in (0, \gamma^*]$.
(c) If $0 < \theta < 1$, *there is a constant* c_1 *such that* $g(\gamma) > c_1$, $\forall\gamma \in (0, \gamma^*]$.
(d) If $0 < \theta < q(\gamma^*)$, *there are two constants* $0 < c_2 < c_3 < 1$ *such that* $c_2 < \frac{\theta^2 f(\theta\gamma)}{q(\gamma)f(\gamma)+g(\gamma)} < c_3$, $\forall\gamma \in (0, \gamma^*]$.

Proof. We first prove item (a). Note that $q(\gamma)$ is a quadratic function with a lowest point $(\frac{\lambda_2}{\lambda_1}, 1 - \frac{\lambda_2^2}{\lambda_1})$, and $q(0) = 1$. Thus, we only need to prove $0 < \frac{\lambda_2^2}{\lambda_1} < 1$.

Because the maximum singular value of a matrix is not smaller than its spectral radius, we obtain $\lambda_1^{\frac{1}{2}} = \lambda_{\max}^{\frac{1}{2}}(\mathcal{H}^T\mathcal{L}^T\mathcal{L}\mathcal{H}) \geq \lambda_{\max}(\mathcal{L}\mathcal{H})$.

We denote the smallest eigenvalue of $\frac{\mathcal{L}^T\mathcal{H}^T + \mathcal{L}\mathcal{H}}{2}$ by λ_3 and its corresponding eigenvector by $\mathbf{z}_3$. For any $\mathbf{z} \in \mathbf{C}^n$, we obtain

$$\mathrm{Re}(\lambda_3) = \frac{\mathbf{z}_3^H(\mathcal{L}^T\mathcal{H}^T + \mathcal{L}\mathcal{H})\mathbf{z}_3}{2\mathbf{z}_3^H\mathbf{z}_3} \leq \frac{\mathbf{z}^H(\mathcal{L}^T\mathcal{H}^T + \mathcal{L}\mathcal{H})\mathbf{z}}{2\mathbf{z}^H\mathbf{z}}, \tag{5.104}$$

where $\mathbf{z}^H$ represents the conjugate transpose of $\mathbf{z}$, and $Re(\lambda_3)$ represents the real part of λ_3.

Suppose that λ_4 is the eigenvalue with the smallest modulus among the eigenvalues of $\mathcal{L}\mathcal{H}$ and denote its corresponding eigenvector by $\mathbf{z}_4$. Then, according to (5.104), we obtain

$$\frac{\mathbf{z}_3^H(\mathcal{L}^T\mathcal{H}^T + \mathcal{L}\mathcal{H})\mathbf{z}_3}{2\mathbf{z}_3^H\mathbf{z}_3} \leq \frac{\mathbf{z}_4^H(\mathcal{L}^T\mathcal{H}^T + \mathcal{L}\mathcal{H})\mathbf{z}_4}{2\mathbf{z}_4^H\mathbf{z}_4} = \mathrm{Re}(\lambda_4),$$

Note that $\lambda_2 = \|\lambda_3\| = \mathrm{Re}(\lambda_3)$ and $\mathrm{Re}(\lambda_4) < \lambda_{\min}(\mathcal{L}\mathcal{H})$. Thus, we obtain $\lambda_1^{\frac{1}{2}} \geq \lambda_{\max}(\mathcal{L}\mathcal{H}) > \lambda_{\min}(\mathcal{L}\mathcal{H}) \geq \lambda_2 > 0$. Therefore, $0 < \frac{\lambda_2^2}{\lambda_1} < 1$ holds, proving item (a).

We now prove item (b). We have

$$\begin{aligned} f(\gamma) &= \frac{1 - q(\theta\gamma)}{[1 - q(\gamma)][q(\theta\gamma) - q(\gamma)]} - \frac{\theta^2}{q(\theta\gamma) - q(\gamma)} \\ &= \frac{2\lambda_2(1-\theta)\theta\gamma}{(2\lambda_2\gamma - \lambda_1\gamma^2)(2\lambda_2\gamma - 2\lambda_2\theta\gamma + \lambda_1\theta^2\gamma^2 - \lambda_1\gamma^2)}, \end{aligned} \tag{5.105}$$

$$\begin{aligned} f(\theta\gamma) &= \frac{2\lambda_2(1-\theta)\theta^2\gamma}{(2\lambda_2\theta\gamma - \lambda_1\theta^2\gamma^2)(2\lambda_2\theta\gamma - 2\lambda_2\theta^2\gamma + \lambda_1\theta^4\gamma^2 - \lambda_1\theta^2\gamma^2)} \\ &< \frac{2\lambda_2(1-\theta)\theta^2\gamma}{(2\lambda_2\theta\gamma - \lambda_1\theta\gamma^2)(2\lambda_2\theta\gamma - 2\lambda_2\theta^2\gamma + \lambda_1\theta^3\gamma^2 - \lambda_1\theta\gamma^2)} \\ &= \frac{2\lambda_2(1-\theta)\gamma}{(2\lambda_2\gamma - \lambda_1\gamma^2)(2\lambda_2\gamma - 2\lambda_2\theta\gamma + \lambda_1\theta^2\gamma^2 - \lambda_1\gamma^2)}. \end{aligned} \tag{5.106}$$

In (5.106), the inequality holds because $2\lambda_2\theta\gamma - \lambda_1\theta^2\gamma^2 > 2\lambda_2\theta\gamma - \lambda_1\theta\gamma^2 > 0$ and $2\lambda_2\theta\gamma - 2\lambda\theta^2\gamma + \lambda_1\theta^4\gamma^2 - \lambda_1\theta^2\gamma^2 > 2\lambda_2\theta\gamma - 2\lambda\theta^2\gamma + \lambda_1\theta^3\gamma^2 - \lambda_1\theta\gamma^2 > 0, \forall\gamma \in (0, \gamma^*]$. Then, we have $q(\gamma)f(\gamma) - \theta^2 f(\theta\gamma) > \frac{2\lambda_2(1-\theta)(q(\gamma)-\theta)\theta\gamma}{(2\lambda_2\gamma-\lambda_1\gamma^2)(2\lambda_2\gamma-2\lambda\theta\gamma+\lambda_1\theta^2\gamma^2-\lambda_1\gamma^2)}$.

Note that $q(\gamma)$ is a monotonically decreasing function on $(0, \gamma^*]$; thus, $q(\gamma) \geq q(\gamma^*)$. If $0 < \theta < q(\gamma^*)$, we have $q(\gamma)f(\gamma) - \theta^2 f(\theta\gamma) > 0, \forall\gamma \in (0, \gamma^*]$. Furthermore, (5.105) tells us $f(\gamma) > 0, \forall\gamma \in (0, \gamma^*]$. Thus, item (b) is valid.

Next, we prove item (c). By the definition of $g(\gamma)$, we have

$$g(\gamma) = \frac{2\lambda_2\gamma(1-\theta)\theta}{(2\lambda_2\gamma - \lambda_1\gamma^2)(2\lambda_2\theta\gamma - \lambda_1\theta^2\gamma^2)} > 0, \forall\gamma \in (0, \gamma^*].$$

Calculating the limit value of $g(\gamma)$ at 0, we obtain

$$\lim_{\gamma\to0^+} g(\gamma) = \lim_{\gamma\to0^+} \frac{2\lambda_2\gamma(1-\theta)\theta}{(2\lambda_2\gamma - \lambda_1\gamma^2)(2\lambda_2\theta\gamma - \lambda_1\theta^2\gamma^2)} = +\infty.$$

Thus, there is a constant c_1 satisfying (c).

Lastly, we prove item (d). It is evident that

$$\begin{aligned}
&\lim_{\gamma\to 0^+}\frac{\theta^2 f(\theta\gamma)}{q(\gamma)f(\gamma)+g(\gamma)}\\
=&\lim_{\gamma\to 0^+}\frac{\theta^2\frac{2\lambda_2(1-\theta)\theta^2\gamma}{2\lambda_2\theta\gamma(2\lambda_2\theta\gamma-2\lambda_2\theta^2\gamma)}}{q(\gamma)\frac{2\lambda_2(1-\theta)\theta\gamma}{2\lambda_2\gamma(2\lambda_2\gamma-2\lambda_2\theta\gamma)}+\frac{2\lambda_2(1-\theta)\theta\gamma}{(2\lambda_2\gamma)(2\lambda_2\theta\gamma)}}\\
=&\theta^2<1.
\end{aligned}$$

From item (a) and item (b), we obtain $\frac{\theta^2 f(\theta\gamma)}{q(\gamma)f(\gamma)+g(\gamma)}<1, \forall\gamma\in(0,\gamma^*]$. Thus, two constants $0<c_2<c_3<1$ exist such that (d) hold. Thus, the proof is completed.

5.2.4.2 Practical MS-LCS

In Lemma 5.7, we define an upper bound of $\mathbb{E}[\|\Delta\boldsymbol{u}_k^l\|^2]$ using (5.95)–(5.97). Using this upper bound, we estimate s_k^l to trigger the generation of a novel stage-based gain sequence. Assume that $\gamma_l\in(0,\gamma^*]$. In the following, we use $q_l\triangleq q(\gamma_l)$ to simplify the expressions.

We define $\hat{s}_k^l=\overline{\boldsymbol{u}}_k^l-\overline{\boldsymbol{u}}_{k+1}^l$ as the descending speed of $\overline{\boldsymbol{u}}_k^l$. According to (5.95), (5.96), and (5.97), we obtain

$$\hat{s}_1^1=\overline{\boldsymbol{u}}_1^1-\overline{\boldsymbol{u}}_2^1=(1-q_1)\mathbb{E}[\|\Delta\boldsymbol{u}_1^1\|^2]-\gamma_1^2{\sigma_\xi}^2, \tag{5.107}$$

$$\hat{s}_k^l=\overline{\boldsymbol{u}}_k^l-\overline{\boldsymbol{u}}_{k+1}^l={q_l}^{k-1}(1-q_l)\overline{\boldsymbol{u}}_1^l-{q_l}^{k-1}\gamma_l^2{\sigma_\xi}^2, \tag{5.108}$$

$$\hat{s}_1^{l+1}=\overline{\boldsymbol{u}}_1^{l+1}-\overline{\boldsymbol{u}}_2^{l+1}=(1-q_{l+1})\overline{\boldsymbol{u}}_1^{l+1}-\gamma_{l+1}^2{\sigma_\xi}^2. \tag{5.109}$$

The corresponding event trigger becomes:

$$K_l=\min\{k_l\in\mathbf{N}\mid\hat{s}_{k_l+1}^l\le\hat{s}_1^{l+1}\}. \tag{5.110}$$

We demonstrate that the event trigger (5.110) is well defined. Therefore, a finite k_l exists satisfying $\hat{s}_k^l>\hat{s}_1^{l+1}, k=1,2\cdots,k_l$ and $\hat{s}_{k_l+1}^l\le\hat{s}_1^{l+1}, \forall l$. The event trigger (5.110) can be rewritten as follows:

$$\begin{aligned}
K_l=\min\{k_l\in\mathbf{N}\mid {q_l}^{k_l}(1-q_l)\overline{\boldsymbol{u}}_1^l-{q_l}^{k_l}\gamma_l^2{\sigma_\xi}^2\\
\le(1-q_{l+1})\overline{\boldsymbol{u}}_1^{l+1}-\gamma_{l+1}^2{\sigma_\xi}^2\}.
\end{aligned} \tag{5.111}$$

Substituting (5.97) into (5.111) yields

$$K_l = \min\{k_l \in \mathbf{N} \mid {q_l}^{k_l}(1-q_l)\overline{\boldsymbol{u}}_1^l - {q_l}^{k_l}\gamma_l^2{\sigma_\xi}^2 \leq (1-q_{l+1})\left({q_l}^{k_l}\overline{\boldsymbol{u}}_1^l + \gamma_l^2\sigma_\xi^2\sum_{i=0}^{i=k_l-1}{q_l}^i\right) - \gamma_{l+1}^2{\sigma_\xi}^2\},$$

which is equivalent to

$$K_l = \min\left\{k_l \in \mathbf{N} \mid {q_l}^{k_l}(q_{l+1}-q_l)\overline{\boldsymbol{u}}_1^l \leq \gamma_l^2\sigma_\xi^2\sum_{i=0}^{i=k_l}{q_l}^i - q_{l+1}\gamma_l^2\sigma_\xi^2\sum_{i=0}^{i=k_l-1}{q_l}^i - \gamma_{l+1}^2{\sigma_\xi}^2\right\}.$$

By the summation of a geometric sequence, we obtain

$$K_l = \min\left\{k_l \in \mathbf{N} \mid {q_l}^{k_l}(q_{l+1}-q_l)\overline{\boldsymbol{u}}_1^l \leq \frac{1-{q_l}^{k_l+1}}{1-q_l}\gamma_l^2\sigma_\xi^2 - \frac{1-{q_l}^{k_l}}{1-q_l}q_{l+1}\gamma_l^2\sigma_\xi^2 - \gamma_{l+1}^2{\sigma_\xi}^2\right\}.$$

Direct calculation yields

$$K_l = \min\left\{k_l \in \mathbf{N}|{q_l}^{k_l}(\overline{\boldsymbol{u}}_1^l - \frac{\gamma_l^2\sigma_\xi^2}{1-q_l}) \leq \frac{(1-q_{l+1})\gamma_l^2\sigma_\xi^2}{(1-q_l)(q_{l+1}-q_l)} - \frac{\gamma_{l+1}^2{\sigma_\xi}^2}{q_{l+1}-q_l}\right\}. \tag{5.112}$$

The event trigger (5.112) is equivalent to the event trigger (5.110). The former puts the parameter k_l on the left-hand side, facilitating the subsequent analysis. According to (5.112), k_l indicates that

$${q_l}^{K_l-1}\left(\overline{\boldsymbol{u}}_1^l - \frac{\gamma_l^2\sigma_\xi^2}{1-q_l}\right) > \frac{(1-q_{l+1})\gamma_l^2\sigma_\xi^2}{(1-q_l)(q_{l+1}-q_l)} - \frac{\gamma_{l+1}^2{\sigma_\xi}^2}{q_{l+1}-q_l}, \tag{5.113}$$

$${q_l}^{K_l}\left(\overline{\boldsymbol{u}}_1^l - \frac{\gamma_l^2\sigma_\xi^2}{1-q_l}\right) \leq \frac{(1-q_{l+1})\gamma_l^2\sigma_\xi^2}{(1-q_l)(q_{l+1}-q_l)} - \frac{\gamma_{l+1}^2{\sigma_\xi}^2}{q_{l+1}-q_l}. \tag{5.114}$$

These two inequalities will help analyze the convergence.

We start from (5.112) to prove that the event trigger (5.110) is well defined, as summarized in the following lemma.

Lemma 5.9 *Given the values of $\overline{\boldsymbol{u}}_1^1$ and γ_1, if $0 < \theta < q_1$, the event trigger* (5.110) *is well defined for all l.*

Proof. For any $l \in \mathbf{N}$, by Lemma 5.8, we have

$$\frac{(1-q_{l+1})\gamma_l^2\sigma_\xi^2}{(1-q_l)(q_{l+1}-q_l)} - \frac{\gamma_{l+1}^2{\sigma_\xi}^2}{q_{l+1}-q_l} = f(\gamma_l)\gamma_l^2\sigma_\xi^2 > 0.$$

Because $0 < q_l < 1$, there must exist a finite k_l such that $q_l{}^{k_l}(\overline{\boldsymbol{u}}_1^l - \frac{\gamma_l^2\sigma_\xi^2}{1-q_l}) \le \frac{(1-q_{l+1})\gamma_l^2\sigma_\xi^2}{(1-q_l)(q_{l+1}-q_l)} - \frac{\gamma_{l+1}^2{\sigma_\xi}^2}{q_{l+1}-q_l}$. Moreover, we still need to prove that

$$\overline{\boldsymbol{u}}_1^l - \frac{\gamma_l^2\sigma_\xi^2}{1-q_l} > \frac{(1-q_{l+1})\gamma_l^2\sigma_\xi^2}{(1-q_l)(q_{l+1}-q_l)} - \frac{\gamma_{l+1}^2\sigma_\xi^2}{q_{l+1}-q_l}, \ \forall l. \tag{5.115}$$

This is done using the induction principle. For the initial step, i.e., $l = 1$, without loss of generality, assume that $\overline{\boldsymbol{u}}_1^1$ is larger than σ_ξ^2. Otherwise, the initial learning gain γ_1 can be shrunk.

We come to the inductive step. Suppose (5.115) holds for the index l. That is, k_l exists and (5.113) holds. Then, by (5.96) and (5.97), we have

$$\begin{aligned}
&\overline{\boldsymbol{u}}_1^{l+1} - \frac{\gamma_{l+1}^2\sigma_\xi^2}{1-q_{l+1}} \\
&= q_l q_l^{k_l-1}\left(\overline{\boldsymbol{u}}_1^l - \frac{\gamma_l{}^2\sigma_\xi^2}{1-q_l}\right) + \frac{\gamma_l{}^2\sigma_\xi^2}{1-q_l} - \frac{\gamma_{l+1}^2\sigma_\xi^2}{1-q_{l+1}} \\
&> \frac{q_l(1-q_{l+1})\gamma_l^2\sigma_\xi^2}{(1-q_l)(q_{l+1}-q_l)} - \frac{q_l\gamma_{l+1}^2\sigma_\xi^2}{q_{l+1}-q_l} + \frac{\gamma_l{}^2\sigma_\xi^2}{1-q_l} - \frac{\gamma_{l+1}^2\sigma_\xi^2}{1-q_{l+1}}.
\end{aligned}$$

Thus, the validity of (5.115) for $l + 1$ is equivalent to

$$\begin{aligned}
&\frac{q_l(1-q_{l+1})\gamma_l^2\sigma_\xi^2}{(1-q_l)(q_{l+1}-q_l)} - \frac{q_l\gamma_{l+1}^2\sigma_\xi^2}{q_{l+1}-q_l} + \frac{\gamma_l{}^2\sigma_\xi^2}{1-q_l} - \frac{\gamma_{l+1}^2\sigma_\xi^2}{1-q_{l+1}} \\
&> \frac{(1-q_{l+2})\gamma_{l+1}^2\sigma_\xi^2}{(1-q_{l+1})(q_{l+2}-q_{l+1})} - \frac{\gamma_{l+2}{}^2\sigma_\xi^2}{q_{l+2}-q_{l+1}}.
\end{aligned} \tag{5.116}$$

We can rewrite (5.116) as $[q(\gamma_l)f(\gamma_l) - \theta^2 f(\theta\gamma_l) + g(\gamma_l)]\gamma_l^2\sigma_\xi^2 > 0$. By Lemma 5.8, this inequality holds. This proof is completed.

Through the recursive formula (5.95)–(5.97) and the event trigger (5.112), we present the practical MS-LCS, where the stage-based gain is defined by

$$\left\{\overline{\tau}_k \middle| \overline{\tau}_k = \gamma_l, \sum_{i=1}^{l-1} K_i < k \le \sum_{i=1}^{l} K_i, i, l, k \in \mathbf{N}\right\}. \tag{5.117}$$

Algorithm 5.3 Gain adaptation for the practical MS-LCS

Initialization: $\gamma_1 \leq \gamma^*, \theta < q_1, \overline{\boldsymbol{u}}_1^1, \sigma_\xi^2, l = 1$.
Step 1:
a) calculate K_l by event trigger

$$K_l = \min\{k_l \in \mathbf{N} \Big| q_l{}^{k_l}(\overline{\boldsymbol{u}}_1^l - \frac{\gamma_l^2\sigma_\xi^2}{1-q_l}) \geq \frac{(1-q_{l+1})\gamma_l^2\sigma_\xi^2}{(1-q_l)(q_{l+1}-q_l)} - \frac{\gamma_{l+1}^2\sigma_\xi{}^2}{q_{l+1}-q_l}\};$$

b) $\left\{\overline{\tau}_k \Big| \overline{\tau}_k = \gamma_l, \sum_{l=1}^{l-1} K_l < k \leq \sum_{l=1}^{l} K_l\right\}$;
c) calculate $\overline{\boldsymbol{u}}_{K_l+1}^l$ by $\overline{\boldsymbol{u}}_{k+1}^l = q_l{}^k\overline{\boldsymbol{u}}_1^l + \gamma_l{}^2\sigma_\xi^2\sum_{i=0}^{i=k-1} q_l{}^i$;
d) calculate $\overline{\boldsymbol{u}}_1^{l+1}$ by $\overline{\boldsymbol{u}}_1^{l+1} = \overline{\boldsymbol{u}}_{K_l+1}^l$;
Step 2:
e) $l = l + 1$, repeat whole process.

The pseudo-code for generating $\overline{\tau}_k$ is shown in Algorithm 5.3.

The practical MS-LCS provides an alternative approach for generating the stage-based gain sequence, which is more suitable for practical implementation. In comparison with (5.70), enhancing the practical MS-LCS can be explained as follows. Firstly, improving the upper bound $\overline{\boldsymbol{u}}_{k+1}^l$ can enhance transient performance. Additionally, calculating $\overline{\boldsymbol{u}}_k^l$ can be considered as an approximation of $\mathbb{E}[\|\Delta\boldsymbol{u}_k^l\|^2]$. Although there may be significant differences between the values of $\overline{\boldsymbol{u}}_k^l$ and $\mathbb{E}[\|\Delta\boldsymbol{u}_k^l\|^2]$, the descent rates of $\hat{s}_k^l$ and s_k^l can still be similar. Therefore, a trigger event generated by $\overline{\boldsymbol{u}}_k^l$ (Algorithm 5.3) is comparable to that generated by $\mathbb{E}[\|\Delta\boldsymbol{u}_k^l\|^2]$ (Algorithm 5.1). However, this observation motivates us to further improve the practical MS-LCS by relaxing $q(\gamma)$ in order to achieve a better approximation of $\mathbb{E}[\|\Delta\boldsymbol{u}_k^l\|^2]$.

5.2.4.3 Improved MS-LCS

We improve the practical MS-LCS by generating the upper bound $\overline{\boldsymbol{u}}_k^l$ to better approximate $\mathbb{E}[\|\Delta\boldsymbol{u}_k^l\|^2]$. We observe that $q(\gamma_l)$ plays a critical role in the generation process of $\overline{\boldsymbol{u}}_k^l$. Note that $q(\gamma_l)$ is derived from scaling (5.100). The inappropriate scaling of $q(\gamma_l)$ might differ significantly from the actual contraction effect of $\mathbb{E}[\|\Delta\boldsymbol{u}_k^l\|^2]$. Besides, this scaling can considerably reduce the value of γ^*, leading to an excessively small value of γ_1.

Therefore, we adjust the expression of $q(\gamma)$ by introducing two stretching parameters $\beta > 0$ and $\alpha > 0$. The new scaling function is denoted by $\tilde{q}(\gamma)$:

$$\tilde{q}(\gamma) = 1 - 2\alpha\beta\lambda_2\gamma + \alpha\beta^2\lambda_1\gamma^2.$$

where $1 \leq \alpha < \frac{\lambda_1}{\lambda_2^2}, 0 < \beta \leq 1$.

For the function $\tilde{q}(\gamma)$ with $\gamma \in (0, \tilde{\gamma}^*]$, where $\tilde{\gamma}^* = \frac{\lambda_2}{\beta\lambda_1}$, define two auxiliary functions $\tilde{f}(\gamma) = \frac{1-\tilde{q}(\theta\gamma)}{[1-\tilde{q}(\gamma)][\tilde{q}(\theta\gamma)-\tilde{q}(\gamma)]} - \frac{\theta^2}{\tilde{q}(\theta\gamma)-\tilde{q}(\gamma)}$ and $\tilde{g}(\gamma) = \frac{1}{1-\tilde{q}(\gamma)} - \frac{\theta^2}{1-\tilde{q}(\theta\gamma)}$. Then, similar results of Lemma 5.8 are still valid.

We use $\tilde{q}(\gamma)$ as the multistage generation function. The longitudinal and transverse stretching parameters are parameters α and β, respectively. Note that $\tilde{q}(\gamma)$ is a quadratic function with the lowest point $\left(\frac{\lambda_2}{\beta\lambda_1}, 1-\frac{\alpha\lambda_2^2}{\lambda_1}\right)$. The range of α ensures $0 < 1-\alpha\lambda_2^2/\lambda_1 < 1$. The initial gain γ_1 should be restricted within the interval $(0, \tilde{\gamma}^*]$, where $\tilde{\gamma}^* = \frac{\lambda_2}{\beta\lambda_1}$. With this relaxation, one can decrease β to enlarge the interval bound $\tilde{\gamma}^*$ such that the learning gain's possible range is enlarged. One can adjust α such that generating $\overline{\boldsymbol{u}}_k^l$ better approximates to that of $\mathbb{E}[\|\Delta\boldsymbol{u}_k^l\|^2]$. It is a straightforward method to adjust the expression of $q(\gamma)$.

Remark 5.3 We provide several guidelines for parameter selection, specifically α and β, by investigating their impact on learning performance. Firstly, the introduction of parameter β aims to expand the range of initial gain γ_1. Inadequate design of function $\mathcal{L}$ may result in a small value of optimal gain γ^*, thereby limiting the range of γ_1 and slowing down iterative updates during the early stage. Selecting a sufficiently small value for parameter β allows for a larger optimal gain ($\tilde{\gamma}^*$) and facilitates a greater initial gain (γ_1). Subsequently, we introduce parameter α to minimize the gap between approximation function $q(\gamma)$ and actual contraction effect represented by $\mathbb{E}[\|\Delta\boldsymbol{u}_k^l\|^2]$. If we were to remove parameter α, $q(\gamma)$ could closely approach one under sufficiently small values of β. This scenario would lead to an extended duration for the initial stage (K_l), resulting in significant delays in stage switching. Therefore, it is crucial to set parameter α appropriately large so as to accurately approximate the actual contraction effect. However, determining an optimal selection for both parameters remains unresolved. Future research should focus on developing well-founded designs for selecting appropriate values of both parameters or exploring alternative variants of $q(\gamma)$ that enhance approximation performance. Without explicit expressions for these parameters, proving superior performance over practical MS-LCS becomes challenging.

Based on the new function $\tilde{q}(\gamma)$, the recursive formula (5.95)–(5.97), (5.107)–(5.109), and event trigger (5.112) can be reformulated as

$$\overline{\boldsymbol{u}}_1^1 = \mathbb{E}[\|\Delta\boldsymbol{u}_1^1\|^2], \tag{5.118}$$

$$\overline{\boldsymbol{u}}_{k+1}^l = \tilde{q}(\gamma_l)^k\overline{\boldsymbol{u}}_1^l + \gamma_l^2\sigma_\xi^2\sum_{i=0}^{i=k-1}\tilde{q}(\gamma_l)^i, \tag{5.119}$$

$$\overline{\boldsymbol{u}}_1^{l+1} = \overline{\boldsymbol{u}}_{K_l+1}^l, \tag{5.120}$$

$$\hat{s}_1^1 = \overline{\boldsymbol{u}}_1^1 - \overline{\boldsymbol{u}}_2^1 = (1-\tilde{q}_1)\mathbb{E}[\|\Delta\boldsymbol{u}_1^1\|^2] - \gamma_1^2{\sigma_\xi}^2, \tag{5.121}$$

$$\hat{s}_k^l = \overline{\boldsymbol{u}}_k^l - \overline{\boldsymbol{u}}_{k+1}^l = \tilde{q}_l^{k-1}(1-\tilde{q}_l)\overline{\boldsymbol{u}}_1^l - {q_l}^{k-1}\gamma_l^2{\sigma_\xi}^2, \tag{5.122}$$

$$\hat{s}_1^{l+1} = \overline{\boldsymbol{u}}_1^{l+1} - \overline{\boldsymbol{u}}_2^{l+1} = (1-\tilde{q}_{l+1})\overline{\boldsymbol{u}}_1^{l+1} - \gamma_{l+1}^2{\sigma_\xi}^2, \tag{5.123}$$

Algorithm 5.4 Gain adaptation of the improved MS-LCS

Initialization: $1 \le \alpha < \frac{\lambda_1}{\lambda_2^2}, 0 < \beta \le 1, \gamma_1 \le \tilde{\gamma}^*, \theta \le \tilde{q}_1, \overline{\boldsymbol{u}}_1^1, \sigma_\xi^2$.
Step 1:
a) calculate K_l by event trigger
$K_l = \min\{K \in \mathbf{N}\Big|$
$\tilde{q}_l^K(\overline{\boldsymbol{u}}_1^l - \frac{\gamma_l^2\sigma_\xi^2}{1-\tilde{q}_l}) < \frac{(1-\tilde{q}_{l+1})\gamma_l^2\sigma_\xi^2}{(1-\tilde{q}_l)(\tilde{q}_{l+1}-\tilde{q}_l)} - \frac{\gamma_{l+1}^2{\sigma_\xi}^2}{\tilde{q}_{l+1}-\tilde{q}_l}\}$;
b) $\left\{\tilde{\tau}_k \middle| \tilde{\tau}_k = \gamma_l, \sum_{l=1}^{l-1} K_l < k \le \sum_{l=1}^{l} K_l\right\}$;
c) calculate $\overline{\boldsymbol{u}}_{K_l+1}^l$ by $\overline{\boldsymbol{u}}_{k+1}^l = \tilde{q}_l^k \overline{\boldsymbol{u}}_1^l + \gamma_l^2\sigma_\xi^2 \sum_{i=0}^{i=k-1} \tilde{q}_l^i$;
d) calculate $\overline{\boldsymbol{u}}_1^{l+1}$ by $\overline{\boldsymbol{u}}_1^{l+1} = \overline{\boldsymbol{u}}_{K_l+1}^l$;
Step 2:
a) $l = l + 1$, repeat whole process.

$$K_l = \min\{K \in \mathbf{N}|\tilde{q}_l^K(\overline{\boldsymbol{u}}_1^l - \frac{\gamma_l^2\sigma_\xi^2}{1-\tilde{q}_l}) \le \frac{(1-\tilde{q}_{l+1})\gamma_l^2\sigma_\xi^2}{(1-\tilde{q}_l)(\tilde{q}_{l+1}-\tilde{q}_l)} - \frac{\gamma_{l+1}^2{\sigma_\xi}^2}{\tilde{q}_{l+1}-\tilde{q}_l}\}, \tag{5.124}$$

where $\tilde{q}_l = \tilde{q}(\gamma_l)$.

Through the recursive formula (5.118)–(5.120) and the event trigger (5.124), we establish the improved MS-LCS, where the stage-based gain is defined by

$$\left\{\tilde{\tau}_k \middle| \tilde{\tau}_k = \gamma_l, \sum_{i=1}^{l-1} K_i < k \le \sum_{i=1}^{l} K_i, i, l, k \in \mathbf{N}\right\}. \tag{5.125}$$

The pseudo-code for generating $\tilde{\tau}_k$ is shown in Algorithm 5.4.

In summary, the enhanced MS-LCS offers a practical alternative to the existing MS-LCS by significantly improving convergence performance (see simulations in Sect. 5.2.6). The computational costs of both the enhanced and practical MS-LCSs remain identical as the variation of $q(\gamma)$ does not introduce any additional computation. Furthermore, compared to the ideal MS-LCS, the enhanced version requires fewer computational costs due to the elimination of matrix operations (see Lemma 5.7). The convergence analyses for both practical and enhanced MS-LCSs are presented in the subsequent subsection.

We have derived three accelerated MS-LCSs based on the descending rate of the mean-squared input error and its estimated upper bound. Unlike Sect. 5.1, where stage switching is determined by comparing the effect of input error and random noise, these accelerated MS-LCSs offer a more intuitive approach to enhance transient performance. Moreover, the slight difference between the upper bound values $\overline{\boldsymbol{u}}_k^l$ and $\mathbb{E}[\|\Delta \boldsymbol{u}_k^l\|^2]$ has negligible influence on determining stage-switching using descending speed as the decision variable. This implies that generating the upper bound should be similar to that of $\mathbb{E}[\|\Delta \boldsymbol{u}_k^l\|^2]$.

5.2.5 Convergence Analysis

In this section, we prove the convergence of (5.70) with stage-based gain sequences defined by (5.117) and (5.125), where the former is based on (5.107)–(5.109), and the latter is based on (5.118)–(5.120). Noticing that (5.117) is a special case of (5.125) with $\alpha = 1$ and $\beta = 1$, it is sufficient to show the convergence of (5.125). To this end, we use Lemma 2.2.

The convergence of (5.70) with the stage-based gain sequence (5.125) is presented in the following theorem.

Theorem 5.3 *Consider the stochastic system* (5.66). *Let $\mathcal{L} = diag\{L_0, L_1, \cdots, L_{N-1}\}$ satisfy that all eigenvalues of $L_t C_{t+1} B_t$ have positive real parts and $\mathcal{LH}$ be a quasi-positive-definite matrix. For the improved MS-LCS, defined by* (5.70), (5.125), *and* (5.118)–(5.120), *we have the following conclusions:*

(a) The gain sequence $\tilde{\tau}_k$ satisfies $\sum_{k=0}^{\infty} \tilde{\tau}_k = \infty$ and $\sum_{k=0}^{\infty} \tilde{\tau}_k^2 \leq \infty$.
(b) The input error $\Delta \boldsymbol{u}_k$ converges to zero in the mean-square sense, i.e.,

$$\lim_{k\to\infty} \mathbb{E}[\|\Delta \boldsymbol{u}_k\|^2] = 0.$$

(c) The input error $\Delta \boldsymbol{u}_k$ converges to zero in the almost sure sense, i.e.,

$$\mathbb{P}(\lim_{k\to\infty} \|\Delta \boldsymbol{u}_k\|^2 = 0) = 1.$$

Proof. By the definition of γ_l, we have

$$\sum_{k=1}^{\infty} \tilde{\tau}_k = \gamma_1 \sum_{i=1}^{\infty} K_j \theta^{i-1}, \quad \sum_{k=1}^{\infty} \tilde{\tau}_k^2 = \gamma_1^2 \sum_{i=1}^{\infty} K_i \theta^{2i-2}. \tag{5.126}$$

According to (5.113) and (5.114), K_{l-1} and K_l satisfy

$$\tilde{q}_l^{K_l}(\overline{\boldsymbol{u}}_1^l - \frac{\gamma_l^2 \sigma_\xi^2}{1-\tilde{q}_l}) \leq \frac{(1-\tilde{q}_{l+1})\gamma_l^2\sigma_\xi^2}{(1-\tilde{q}_l)(\tilde{q}_{l+1}-\tilde{q}_l)} - \frac{\gamma_{l+1}^2 {\sigma_\xi}^2}{\tilde{q}_{l+1}-\tilde{q}_l}, \tag{5.127}$$

$$\tilde{q}_l^{K_l-1}(\overline{\boldsymbol{u}}_1^l - \frac{\gamma_l^2 \sigma_\xi^2}{1-\tilde{q}_l}) > \frac{(1-\tilde{q}_{l+1})\gamma_l^2\sigma_\xi^2}{(1-\tilde{q}_l)(\tilde{q}_{l+1}-\tilde{q}_l)} - \frac{\gamma_{l+1}^2 {\sigma_\xi}^2}{\tilde{q}_{l+1}-\tilde{q}_l}, \tag{5.128}$$

$$\tilde{q}_{l-1}^{K_{l-1}}(\overline{\boldsymbol{u}}_1^{l-1} - \frac{\gamma_{l-1}^2 \sigma_\xi^2}{1-\tilde{q}_{l-1}}) \leq \frac{(1-\tilde{q}_l)\gamma_{l-1}^2\sigma_\xi^2}{(1-\tilde{q}_{l-1})(\tilde{q}_l-\tilde{q}_{l-1})} - \frac{{\gamma_l}^2 {\sigma_\xi}^2}{\tilde{q}_l-\tilde{q}_{l-1}}, \tag{5.129}$$

$$\tilde{q}_{l-1}^{K_{l-1}-1}(\overline{\boldsymbol{u}}_1^{l-1} - \frac{\gamma_{l-1}^2 \sigma_\xi^2}{1-\tilde{q}_{l-1}}) > \frac{(1-\tilde{q}_l)\gamma_{l-1}^2\sigma_\xi^2}{(1-\tilde{q}_{l-1})(\tilde{q}_l-\tilde{q}_{l-1})} - \frac{{\gamma_l}^2 {\sigma_\xi}^2}{\tilde{q}_l-\tilde{q}_{l-1}}. \tag{5.130}$$

From (5.127) (5.128), we obtain

$$\tilde{q}_l^{K_l}[\tilde{q}_{l-1}^{K_{l-1}}(\overline{\boldsymbol{u}}_1^{l-1} - \frac{\gamma_{l-1}^2\sigma_\xi^2}{1-\tilde{q}_{l-1}}) + \frac{\gamma_{l-1}^2\sigma_\xi^2}{1-\tilde{q}_{l-1}} - \frac{\gamma_l^2\sigma_\xi^2}{1-\tilde{q}_l}]$$
$$\leq \frac{(1-\tilde{q}_{l+1})\gamma_l^2\sigma_\xi^2}{(1-\tilde{q}_l)(\tilde{q}_{l+1}-\tilde{q}_l)} - \frac{\gamma_{l+1}^2{\sigma_\xi}^2}{\tilde{q}_{l+1}-\tilde{q}_l}, \tag{5.131}$$

$$\tilde{q}_l^{K_l-1}[\tilde{q}_{l-1}^{K_{l-1}}(\overline{\boldsymbol{u}}_1^{l-1} - \frac{\gamma_{l-1}^2\sigma_\xi^2}{1-\tilde{q}_{l-1}}) + \frac{\gamma_{l-1}^2\sigma_\xi^2}{1-\tilde{q}_{l-1}} - \frac{\gamma_l^2\sigma_\xi^2}{1-\tilde{q}_l}]$$
$$> \frac{(1-\tilde{q}_{l+1})\gamma_l^2\sigma_\xi^2}{(1-\tilde{q}_l)(\tilde{q}_{l+1}-\tilde{q}_l)} - \frac{\gamma_{l+1}^2{\sigma_\xi}^2}{\tilde{q}_{l+1}-\tilde{q}_l}. \tag{5.132}$$

To obtain a bound of $\tilde{q}_l^{K_l}$, substituting (5.129) and (5.130) into (5.131) and (5.132) yields

$$\frac{\tilde{q}_l\left(\frac{(1-\tilde{q}_{l+1})\gamma_l^2\sigma_\xi^2}{(1-\tilde{q}_l)(\tilde{q}_{l+1}-\tilde{q}_l)} - \frac{\gamma_{l+1}^2\sigma_\xi^2}{\tilde{q}_{l+1}-\tilde{q}_l}\right)}{\frac{(1-\tilde{q}_l)\gamma_{l-1}^2\sigma_\xi^2}{(1-\tilde{q}_{l-1})(\tilde{q}_l-\tilde{q}_{l-1})} - \frac{\gamma_l{}^2\sigma_\xi^2}{\tilde{q}_l-\tilde{q}_{l-1}} + \frac{\gamma_{l-1}^2\sigma_\xi^2}{1-\tilde{q}_{l-1}} - \frac{\gamma_l^2\sigma_\xi^2}{1-\tilde{q}_l}} < \tilde{q}_l^{K_l}$$
$$< \frac{\frac{(1-\tilde{q}_{l+1})\gamma_l^2\sigma_\xi^2}{(1-\tilde{q}_l)(\tilde{q}_{l+1}-\tilde{q}_l)} - \frac{\gamma_{l+1}^2\sigma_\xi^2}{\tilde{q}_{l+1}-\tilde{q}_l}}{\tilde{q}_{l-1}\left(\frac{(1-\tilde{q}_l)\gamma_{l-1}^2\sigma_\xi^2}{(1-\tilde{q}_{l-1})(\tilde{q}_l-\tilde{q}_{l-1})} - \frac{\gamma_l{}^2\sigma_\xi^2}{\tilde{q}_l-\tilde{q}_{l-1}}\right) + \frac{\gamma_{l-1}^2\sigma_\xi^2}{1-\tilde{q}_{l-1}} - \frac{\gamma_l^2\sigma_\xi^2}{1-\tilde{q}_l}}. \tag{5.133}$$

From the results of Lemma 5.8, where the function $q(\gamma) = 1 - 2\lambda_2\gamma + \lambda_1\gamma^2$ is replaced by $\tilde{q}(\gamma) = 1 - 2\alpha\beta\lambda_2\gamma + \alpha\beta^2\lambda_1\gamma^2$, (5.133) is transformed into

$$\tilde{q}_l^{K_l} < \frac{\tilde{f}(\theta\gamma_{l-1})\theta^2}{\tilde{q}(\gamma_{l-1})\tilde{f}(\gamma_{l-1}) + \tilde{g}(\gamma_{l-1})}, \tag{5.134}$$
$$\tilde{q}_l^{K_l} > \frac{\tilde{q}(\theta\gamma_{l-1})\tilde{f}(\theta\gamma_{l-1})\theta^2}{\tilde{f}(\gamma_{l-1}) + \tilde{g}(\gamma_{l-1})}. \tag{5.135}$$

From (5.135), we further have $\tilde{q}_l^{K_l} > \frac{\tilde{q}(\theta\gamma_{l-1})\tilde{q}(\gamma_{l-1})\tilde{f}(\theta\gamma_{l-1})\theta^2}{\tilde{q}(\gamma_{l-1})\tilde{f}(\gamma_{l-1})+\tilde{g}(\gamma_{l-1})}$. Then, we obtain

$$c_4 < \tilde{q}(\theta\gamma_{l-1})\tilde{q}(\gamma_{l-1})c_2 < \tilde{q}_l^{K_l} < c_3, \tag{5.136}$$

where $c_4 = \tilde{q}_1^2 c_2$.

From the logarithm of (5.136) and noticing $\tilde{q}(\gamma_l) \in (0, 1)$, we have $\frac{-\ln c_3}{-\ln(1-(1-\tilde{q}_l))} < K_l < \frac{-\ln c_4}{-\ln(1-(1-\tilde{q}_l))}$. By the basic inequality $x < -\ln(1-x) < \frac{x}{1-x}$, $\forall x \in (0, 1)$, it follows that $\frac{-\tilde{q}_l \ln c_3}{1-\tilde{q}_l} < K_t < \frac{-\ln c_4}{1-\tilde{q}_l}$.

Because $1 - \tilde{q}_l = 2\alpha\beta\lambda_2\gamma_l - \alpha\beta^2\lambda_1\gamma_l^2 = \alpha\beta\lambda_2\gamma_l(2 - \frac{\beta\lambda_1}{\lambda_2}\gamma_l)$ and $0 < \gamma_l < \frac{\lambda_2}{\beta\lambda_1}$, we obtain $\alpha\beta\lambda_2\gamma_1\theta^{l-1} < 1 - \tilde{q}(\gamma_l) < 2\alpha\beta\lambda_2\gamma_1\theta^{l-1}$. Then, we have

$$\frac{c_5}{\theta^{l-1}} < K_l < \frac{c_6}{\theta^{l-1}}, \tag{5.137}$$

where $c_5 = \frac{-\tilde{q}_1 \ln c_3}{2\alpha\beta\lambda_2\gamma_1}$ and $c_6 = \frac{-\ln c_4}{\alpha\beta\lambda_2\gamma_1}$

Substituting (5.137) into (5.126) leads to

$$\sum_{k=1}^{\infty} \tilde{\tau}_k > \gamma_1 \sum_{l=1}^{\infty} \frac{c_3}{\theta^{l-1}} \theta^{l-1} = \infty,$$
$$\sum_{k=1}^{\infty} \tilde{\tau}_k^2 < \gamma_1^2 \sum_{l=1}^{\infty} \frac{c_4}{\theta^{l-1}} \theta^{2l-2} < \infty.$$

Thus, item (a) has been proved.

Now, we proceed to item (b). By (5.75), replacing η_k with $\tilde{\tau}_k$ and taking trace leads to

$$\begin{aligned}
&\operatorname{trace}(\mathbb{E}[\Delta \boldsymbol{u}_{k+1} \Delta \boldsymbol{u}_{k+1}^T | \mathcal{F}_{k-1}]) \\
&= \operatorname{trace}[(I - \tilde{\tau}_k \mathcal{L}\mathcal{H})\mathbb{E}[\Delta \boldsymbol{u}_k \Delta \boldsymbol{u}_k^T](I - \tilde{\tau}_k \mathcal{L}\mathcal{H})^T] \\
&\quad + \operatorname{trace}[\tilde{\tau}_k^2 \mathcal{L}(\sigma_\omega^2 \mathcal{G}\mathcal{G}^T + \sigma_v^2 I)\mathcal{L}^T].
\end{aligned}$$

Then, by direct calculation, we obtain

$$\begin{aligned}
&\mathbb{E}[\|\Delta \boldsymbol{u}_{k+1}\|^2 | \mathcal{F}_{k-1}] \\
&= \mathbb{E}[\Delta \boldsymbol{u}_k{}^T \Delta \boldsymbol{u}_k + \tilde{\tau}_k^2 \Delta \boldsymbol{u}_k^T \mathcal{H}^T \mathcal{L}^T \mathcal{L}\mathcal{H} \Delta \boldsymbol{u}_k \\
&\quad - \tilde{\tau}_k \Delta \boldsymbol{u}_k^T (\mathcal{H}^T \mathcal{L}^T + \mathcal{L}\mathcal{H}) \Delta \boldsymbol{u}_k] \\
&\quad + \tilde{\tau}_k^2 [\sigma_\omega^2 \operatorname{trace}(\mathcal{G}^T \mathcal{L}^T \mathcal{L}\mathcal{G}^T) + \sigma_v^2 \operatorname{trace}(\mathcal{L}^T \mathcal{L})] \\
&\le q(\tilde{\tau}_k)\mathbb{E}[\|\Delta \boldsymbol{u}_k\|^2] + \tilde{\tau}_k^2 \sigma_\xi^2.
\end{aligned} \tag{5.138}$$

Taking the expectation of the above equality yields

$$\mathbb{E}[\|\Delta \boldsymbol{u}_{k+1}\|^2] \le q(\tilde{\tau}_k)\mathbb{E}[\|\Delta \boldsymbol{u}_k\|^2] + \tilde{\tau}_k^2 \sigma_\xi^2.$$

Denote $q_{i,j} = \prod_{l=j}^{i} q(\tilde{\tau}_k)$ for $i \ge j$ and $q_{i,i+1} = 1$. Then,

$$\mathbb{E}[\|\Delta \boldsymbol{u}_{k+1}\|^2] \le q_{k,1}\mathbb{E}[\|\Delta \boldsymbol{u}_1\|^2] + \sum_{i=1}^{k} q_{k,i+1} \tilde{\tau}_i^2 \sigma_\xi^2. \tag{5.139}$$

Notice that $q_{k,1} = \exp\left(\sum_{i=1}^{k} \ln(1 - 2\lambda_2 \tilde{\tau}_i + \lambda_1 \tilde{\tau}_i^2)\right)$. By the power series expansion of $\ln(1-x)$ and $\sum_{k=1}^{\infty} \tilde{\tau}_k = \infty$, we obtain $\lim_{k\to\infty} q_{k,1} = 0$. Thus, the first term on the right-hand side of (5.139) converges to zero. For the second term,

$$\sum_{i=1}^{k} q_{k,i+1} \tilde{\tau}_i^2 \sigma_\xi^2 = \sigma_\xi^2 \sum_{i=1}^{k} q_{k,i+1}(1 - q(\tilde{\tau}_i)) \frac{\tilde{\tau}_i^2}{(1 - q(\tilde{\tau}_i))}.$$

By direct calculation, we have

$$\sum_{i=1}^{k} q_{k,i+1}(1-q(\tilde{\tau}_i)) = 1 - \prod_{i=1}^{k} q(\tilde{\tau}_i) = 1 - q_{k,i}.$$

This yields $\lim_{k\to\infty}\sum_{i=1}^{k} q_{k,i+1}(1-q(\tilde{\tau}_i)) = 1$. Furtheremore, $\lim_{k\to\infty}\frac{\tilde{\tau}_k^2}{(1-q(\tilde{\tau}_k))} = \lim_{k\to\infty}\frac{\tilde{\tau}_k}{2\lambda_2-\lambda_1\tilde{\tau}_k} = 0$. By the Toeplitz lemma, we obtain $\lim_{k\to\infty}\sum_{i=1}^{k} q_{k,i+1}(1-q(\tilde{\tau}_i))\frac{\tilde{\tau}_i^2}{(1-q(\tilde{\tau}_i))} = \lim_{k\to\infty}\frac{\tilde{\tau}_k^2}{(1-q(\tilde{\tau}_k))} = 0$. Summarizing these estimates, the right-hand side of (5.139) converges to zero. Therefore, $\lim_{k\to\infty}\mathbb{E}[\|\Delta \boldsymbol{u}_k\|^2] = 0$.

Lastly, we prove item (c). According to (5.138), we obtain

$$\begin{aligned}\mathbb{E}[\|\Delta \boldsymbol{u}_{k+1}\|^2|\mathcal{F}_{k-1}] &\le \|\Delta \boldsymbol{u}_k\|^2 + \tilde{\tau}_k^2\lambda_1\|\Delta \boldsymbol{u}_k\|^2 \\ &\quad - 2\tilde{\tau}_k\lambda_2\|\Delta \boldsymbol{u}_k\|^2 + \tilde{\tau}_k^2\sigma_\xi^2 \\ &\le \|\Delta \boldsymbol{u}_k\|^2 + \tilde{\tau}_k^2(\lambda_1\|\Delta \boldsymbol{u}_k\|^2 + \sigma_\xi^2).\end{aligned}$$

To use Lemma 2.2, we take $\kappa_k = s_k = 0, r_k = \|\Delta \boldsymbol{u}_k\|^2, \varepsilon_k = \tilde{\tau}_k^2(\lambda_1\|\Delta \boldsymbol{u}_k\|^2 + \sigma_\xi^2)$ and $\mathcal{G}_k = \mathcal{F}_{k-1}$, respectively. According to Lemma 2.2, if $\sum_{k=1}^{\infty}\varepsilon_k < \infty$, $\|\Delta \boldsymbol{u}_k\|^2$ converges almost surely. Therefore, we must only verify $\sum_{k=1}^{\infty}\varepsilon_k < \infty$.

Because $\lim_{k\to\infty}\mathbb{E}[\|\Delta \boldsymbol{u}_k\|^2] = 0$, there is a constant c such that $\mathbb{E}[\|\Delta \boldsymbol{u}_k\|^2] < c$ for all k. Because $\sum_{k=1}^{\infty}\tilde{\tau}_k^2 < \infty$, we have $\sum_{k=1}^{\infty}\tilde{\tau}_k^2\lambda_1\mathbb{E}[\|\Delta \boldsymbol{u}_{k+1}\|^2] < c\lambda_1\sum_{k=1}^{\infty}\tilde{\tau}_k^2 < \infty$. Note that $\|\Delta \boldsymbol{u}_{k+1}\|^2$ is a nonnegative stochastic process. Thus, by the monotone convergence theorem, we obtain $\sum_{k=1}^{\infty}\tilde{\tau}_k^2\lambda_1\|\Delta \boldsymbol{u}_{k+1}\|^2 < \infty$ almost surely. Therefore, $\sum_{k=1}^{\infty}\varepsilon_k = \sum_{k=1}^{\infty}\tilde{\tau}_k^2(\lambda_1\|\Delta \boldsymbol{u}_k\|^2 + \sigma_\xi^2) < \infty$. The proof is completed.

Theorem 5.3 characterizes the asymptotic convergence of the practical and improved MS-LCSs, while the convergence of the ideal MS-LCS remains an intricate open problem. It is worth emphasizing that stage-based gain sequences can expedite transient convergence speed, thereby enhancing early iteration convergence rates. Estimating the asymptotic convergence rate as iterations tend towards infinity remains an area for further research.

5.2.6 Illustrative Simulations

We illustrate the effectiveness of the three MS-LCSs with a linear time-varying system, of which the system matrices are given by

$$A_t = \begin{pmatrix} 0.05\sin(\frac{t}{5}) & -0.2 & 0.02t \\ 0.1 & -0.01t & -0.02\cos(\frac{t}{2}) \\ 0.1 & 0.1 & 0.2 + 0.05\cos(\frac{t}{5}) \end{pmatrix},$$

$$B_t = \begin{pmatrix} 4.2 - 0.84\sin^2(\frac{\pi t}{2}) & 0 \\ 0.042t & 0.042t \\ 0 & 4.2 + 0.42\sin(\frac{\pi t}{2}) \end{pmatrix},$$

$$C_t = \begin{pmatrix} 0.84 + 0.42\sin^2(\frac{\pi t}{2}) & 0.42 & -0.42 \\ 0 & 0.42 & 0.84 - 0.4\sin(\frac{\pi t}{2}) \end{pmatrix}.$$

The initial state and input are $x_k(0) = x_d(0) = 0$, $\forall k$ and $u_0(t) = [0, 0]^T$, $\forall t \in [0, N]$, respectively. The stochastic noises are $\omega_k(t) \sim N(0, 0.1^2)$ and $v_k(t) \sim N(0, 0.1^2)$.

The desired reference is given by $y_d(t) = [1.2\sin(\pi t/6) + 1.1(1 - \cos(\pi t/8)),$ $-1.2\sin(\pi t/6) - 1.1(1 - \cos(\pi t/8))]^T$. The iteration length is set to $N = 50$.

We used the following parameters for these MS-LCSs. We set the total iteration number to 80. The DRM is $\mathcal{L} = 0.13 \times I$. This DRM ensures that all eigenvalues of $L_t C_{t+1} B_t$ have positive real parts, $\forall t$. The parameters for the proposed schemes are listed as follows: for the ideal MS-LCS, $\gamma_1 = \tilde{\gamma}^*$, $\theta = 0.8$, for the practical MS-LCS, $\gamma_1 = \gamma^*$, $\theta = 0.8$, and for the improved MS-LCS, $\gamma_1 = \tilde{\gamma}^*$, $\theta = 0.95$, $\alpha = 15$, $\beta = 0.14$. Note that θ could be larger than $\tilde{q}(\gamma_1)$ in practice. Furthermore, we simulate (5.69) with a constant gain $\eta = 1$ and (5.70) with a common decreasing gain $\eta_k = 2/k$ for comparison.

We demonstrate the enhanced transient performance of the ideal and improved MS-LCSs, while excluding the practical MS-LCS as it represents a special case of the improved MS-LCS. Figure 5.9 illustrates the variation of $\|\Delta \boldsymbol{u}_k\|^2$ for five methodologies. It is observed that (5.69), with a constant learning gain, can only ensure bounded convergence of input errors due to its inability to effectively eliminate noise effects. Furthermore, (5.70), employing a decreasing gain $\eta_k = 2/k$, guarantees asymptotic convergence but fails to improve transient performance due to empirical selection without considering practical processes. In contrast, our proposed MS-LCSs utilize successful event triggers to capture process information and generate a stage-based gain sequence, thereby accelerating convergence rate and enhancing transient performance as evidenced by the rapid decrease in trends shown in Fig. 5.9. Moreover, although slightly superior in performance compared to the improved MS-LCSs, this result is reasonable since the ideal MS-LCS incorporates comprehensive information without considering computational costs. This finding highlights the distinct advantage of our improved MS-LCS approach which achieves comparable performance with significant savings in computation resources.

The KF-based scheme (Saab 2001a) and P-type schemes with a multistage adaptive gain (Cheng et al. 2023) are simulated for comparison (Fig. 5.9). The tracking performance of the ideal and improved MS-LCSs surpasses that of the P-type scheme with a multistage adaptive gain, thereby demonstrating the accelerated nature of the proposed schemes. Furthermore, while achieving satisfactory tracking performance with limited information and resources, it is worth noting that the KF-based scheme

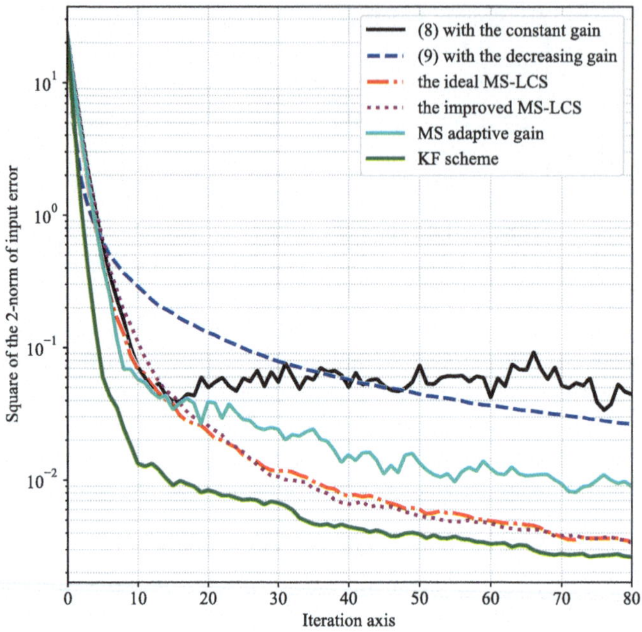

Fig. 5.9 Input error profiles for the P-type schemes (5.69), (5.70), the ideal MS-LCS, the improved MS-LCS, the MS adaptive gain, and the KF scheme

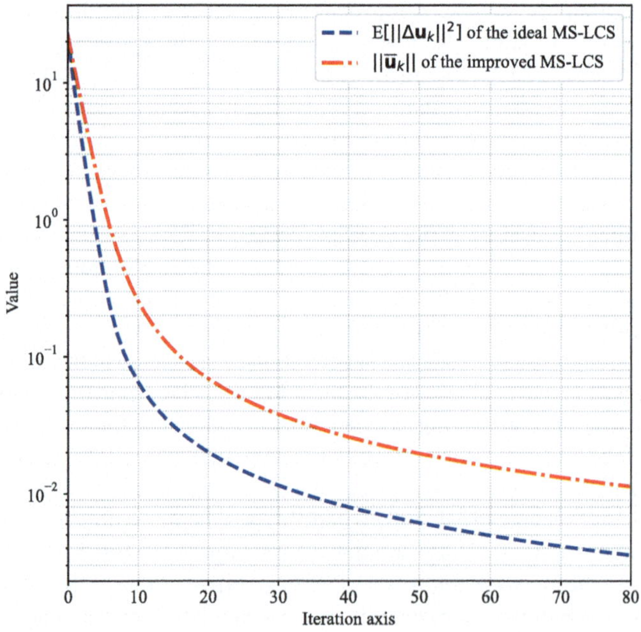

Fig. 5.10 The value profiles of $\mathbb{E}[\|\Delta \boldsymbol{u}_k\|^2]$ and $\|\overline{\boldsymbol{u}}_k\|^2$ for the ideal and improved MS-LCSs, respectively

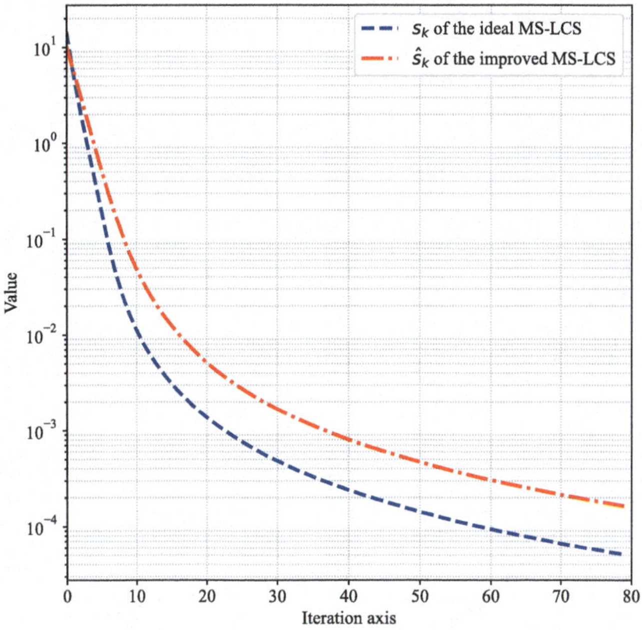

Fig. 5.11 The value profiles of s_k and $\hat{s}_k$ corresponding to the ideal and improved MS-LCSs, respectively

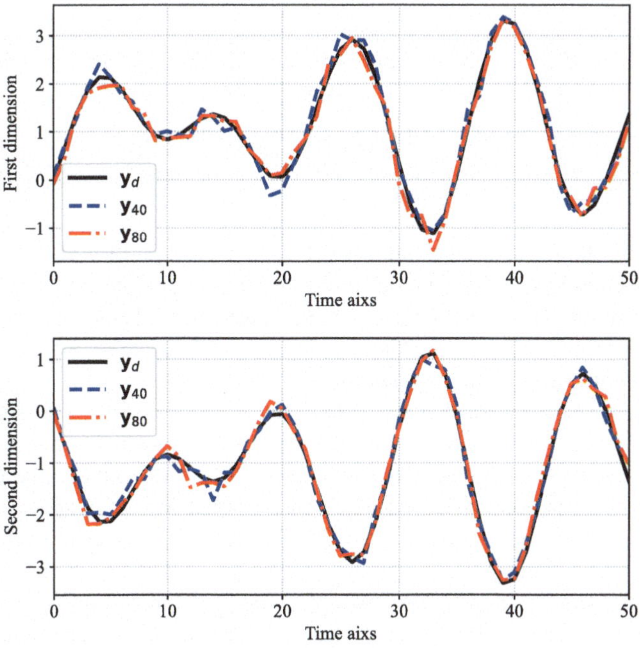

Fig. 5.12 Tracking performance of the improved MS-LCS

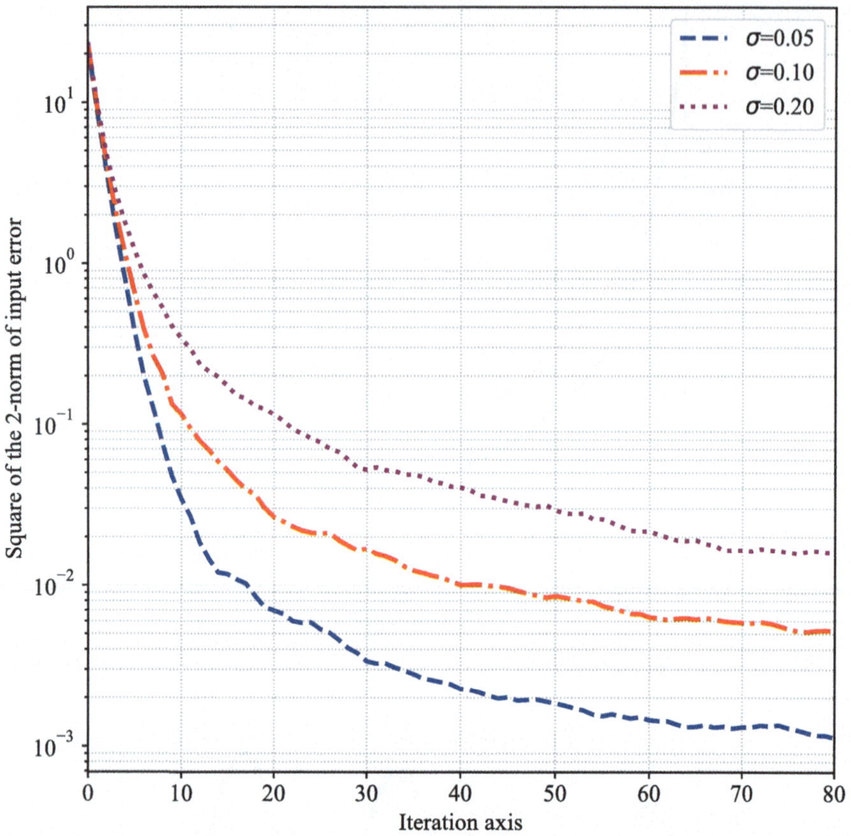

Fig. 5.13 Input error norm profiles of the improved MS-LCS for three noise levels: $\sigma = 0.05, 0.1$, and 0.2

outperforms both the ideal and improved MS-LCSs due to its ability to minimize the trace of input error covariance matrix through learning gain matrix derivation. However, it should be acknowledged that this particular scheme necessitates complete system information and substantial computational resources, which are deemed unnecessary for practical implementation.

To demonstrate the idea that the generation process of $\|\overline{\boldsymbol{u}}_k\|$ can simulate the generation process of $\mathbb{E}[\|\Delta\boldsymbol{u}_k\|^2]$, Figs. 5.10 and 5.11 show their value profiles and related descending speeds, respectively. Figure 5.10 provides the value profiles of $\mathbb{E}[\|\Delta\boldsymbol{u}_k\|^2]$ corresponding to $\mathbb{E}[\|\Delta\boldsymbol{u}_k^l\|^2]$ in the ideal MS-LCS, and $\|\overline{\boldsymbol{u}}_k\|$ corresponding to $\|\overline{\boldsymbol{u}}_k^l\|$ in the improved MS-LCS. Figure 5.11 provides the value profiles of s_k and $\hat{s}_k$, defined by $s_k = \mathbb{E}[\|\Delta\boldsymbol{u}_k\|^2] - \mathbb{E}[\|\Delta\boldsymbol{u}_{k+1}\|^2]$ and $\hat{s}_k = \|\overline{\boldsymbol{u}}_k\| - \|\overline{\boldsymbol{u}}_{k+1}\|$, respectively. Figure 5.11 shows the inherent principle of the proposed schemes that the generation processes for $\mathbb{E}[\|\Delta\boldsymbol{u}_k\|^2]$ and $\|\overline{\boldsymbol{u}}_k\|$ are similar. By comparing the two figures, the value profiles in Fig. 5.11 are closer to each other than those in Fig. 5.10,

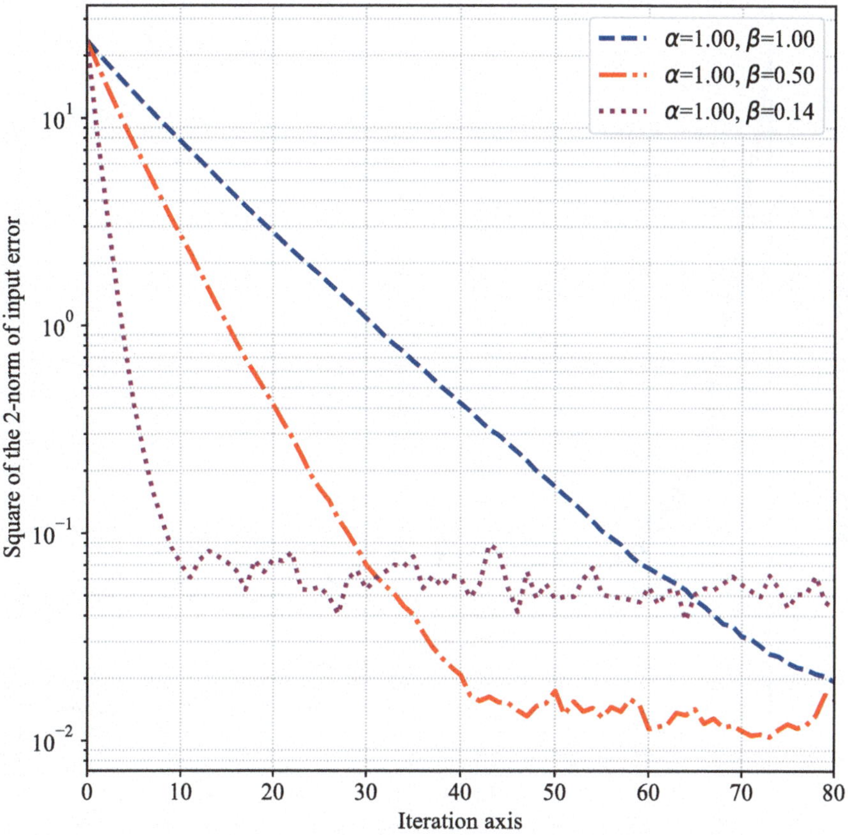

Fig. 5.14 The effect of different β with a fixed α for the improved MS-LCS

indicating that the descending indices of $\|\overline{\boldsymbol{u}}_k\|^2$ and $\mathbb{E}[\|\Delta\boldsymbol{u}_k\|^2]$ are similar although their value profiles differ. Thus, the event trigger (5.124) for $\|\overline{\boldsymbol{u}}_k\|^2$ can satisfactorily improve the learning performance.

The tracking performance of the improved MS-LCS is demonstrated in Fig. 5.12, where the actual outputs at the 40th and 80th iterations are compared with the desired reference. In Fig. 5.12, the top and bottom subplots correspond to the first and second dimensions of the output, respectively. The actual output successfully tracks the desired reference after a few iterations, with fluctuations primarily attributed to random noise.

The norm profiles of the input error for the improved MS-LCS under three different noise levels are illustrated in Fig. 5.13. Specifically, we conducted simulations with $\sigma = 0.05, 0.1$, and 0.2 to examine the proposed scheme's performance. Remarkably, our method successfully reduces the input error to a level lower than each corresponding noise level in all scenarios, thereby demonstrating its remarkable effectiveness and superiority.

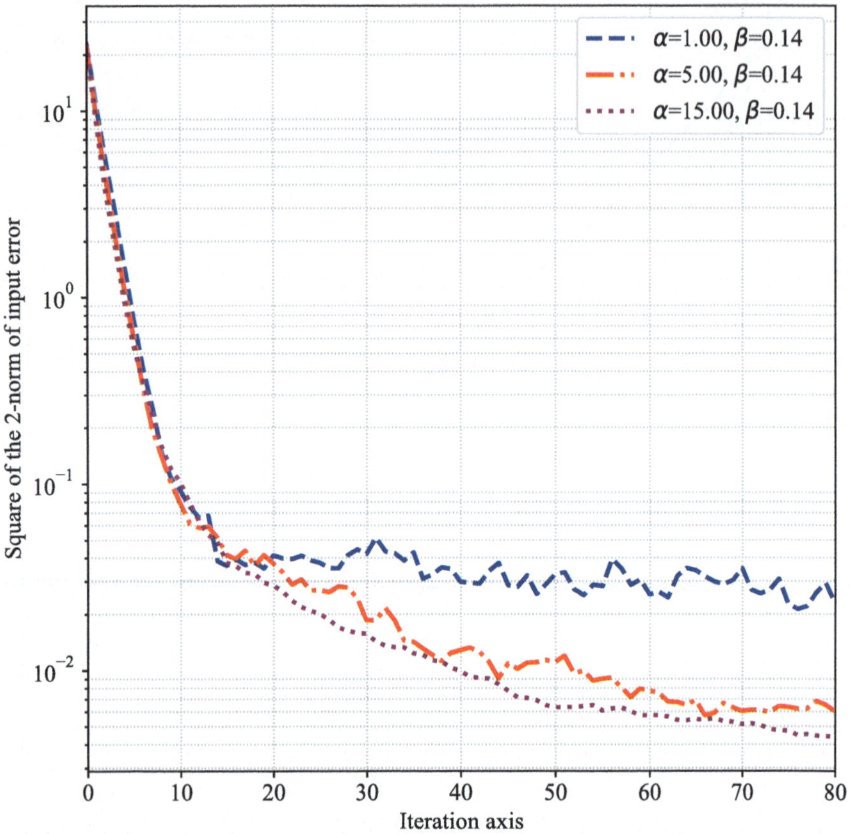

Fig. 5.15 The effect of different α with a fixed β for the improved MS-LCS

Lastly, we demonstrate the superiority of the improved MS-LCS over the practical MS-LCS, elucidating the necessity of incorporating stretching parameters α and β. Figure 5.14 illustrates the impact of β while setting $\alpha = 1$ and varying β as 1, 0.5, and 0.14. The convergence speed at the early iterations significantly accelerates as β decreases due to the larger range for initial gain selection offered by a smaller β. Figure 5.15 demonstrates the influence of α when setting $\beta = 0.14$ and $\alpha = 1$, 5, and 10. A suitable choice of α can effectively enhance transient performance by bringing the value of $\hat{s}_k$ considerably closer to s_k, thereby improving event trigger effectiveness.

Chapter 6
Optimal Gain Design

6.1 Optimal Gain Sequence Design in SILC

In this section, we elaborate the fundamental idea of the optimal gain design in SILC. To this end, we recall the derivations in Sect. 1.4. We first revisit the lifted linear systems with additional noise:

$$\boldsymbol{y}_k = \mathbf{G}\boldsymbol{u}_k + \boldsymbol{d} + \boldsymbol{w}_k, \tag{6.1}$$

where $\boldsymbol{w}_k$ denotes the random noise satisfying $\mathbb{E}\boldsymbol{w}_k = 0$ and $W \triangleq \sup_k \mathbb{E}\|\boldsymbol{w}_k\|^2 < \infty$.

Accordingly, the desired tracking reference $\boldsymbol{y}_d$ satisfies the following equation:

$$\boldsymbol{y}_d = \mathbf{G}\boldsymbol{u}_d + \boldsymbol{d}. \tag{6.2}$$

Then, we obtain the tracking error as follows:

$$\boldsymbol{e}_k = \boldsymbol{y}_d - \boldsymbol{y}_k = \mathbf{G}(\boldsymbol{u}_d - \boldsymbol{u}_k) - \boldsymbol{w}_k. \tag{6.3}$$

The traditional P-type update rule is revised by replacing the constant learning gain with a decreasing gain as follows:

$$\boldsymbol{u}_{k+1} = \boldsymbol{u}_k + \rho_k \mathbf{K}\boldsymbol{e}_k, \tag{6.4}$$

where ρ_k is a variable gain.

Subtracting both sides of (1.16) from $\boldsymbol{u}_d$ and substituting (6.3) yields

$$\Delta\boldsymbol{u}_{k+1} = \Delta\boldsymbol{u}_k - \rho_k\mathbf{K}\mathbf{G}\Delta\boldsymbol{u}_k + \rho\mathbf{K}\boldsymbol{w}_k, \tag{6.5}$$

where $\Delta\boldsymbol{u}_k \triangleq \boldsymbol{u}_d - \boldsymbol{u}_k$ denotes the input error.

D. Shen, *Variable Gain Design in Stochastic Iterative Learning Control*,
Intelligent Control and Learning Systems 13,
https://doi.org/10.1007/978-981-97-8281-9_6

Consider the derivation of $\mathbb{E}\|\Delta \boldsymbol{u}_k\|^2$ as follows:

$$\begin{aligned}\mathbb{E}\|\Delta \boldsymbol{u}_{k+1}\|^2 &= \mathbb{E}\|(\mathbf{I}-\rho\mathbf{KG})\Delta \boldsymbol{u}_k\|^2 + \mathbb{E}\|\rho\mathbf{K}\boldsymbol{w}_k\|^2\\ &\leq \lambda_{\max}((\mathbf{I}-\rho_k\mathbf{KG})^T(\mathbf{I}-\rho_k\mathbf{KG}))\mathbb{E}\|\Delta \boldsymbol{u}_k\|^2 + \rho_k^2\lambda_{\max}(\mathbf{K}^T\mathbf{K})W.\end{aligned} \tag{6.6}$$

Note that $\lambda_{\max}((\mathbf{I}-\rho_k\mathbf{KG})^T(\mathbf{I}-\rho_k\mathbf{KG}))$ is a function of ρ_k and there should exist a constant θ such that $\lambda_{\max}((\mathbf{I}-\rho_k\mathbf{KG})^T(\mathbf{I}-\rho_k\mathbf{KG})) \leq 1-\rho_k\theta$, which leads to

$$\mathbb{E}\|\Delta \boldsymbol{u}_{k+1}\|^2 \leq (1-\rho_k\theta)\mathbb{E}\|\Delta \boldsymbol{u}_k\|^2 + \rho_k^2\lambda_{\max}(\mathbf{K}^T\mathbf{K})W. \tag{6.7}$$

Thus, we may view the right-side as a function of ρ_k, that is, $f(\rho_k) \triangleq (1-\rho_k\theta)\mathbb{E}\|\Delta \boldsymbol{u}_k\|^2 + \rho_k^2\lambda_{\max}(\mathbf{K}^T\mathbf{K})W$. Consequently, to achieve a fast convergence, we consider to minimize the function $f(\cdot)$, which leads to an optimal selection of ρ_k. In particular, the optimal gain ρ_k^* is the root of the following derivative:

$$\frac{\mathrm{d}f(\rho_k)}{\mathrm{d}\rho_k} = -\theta\mathbb{E}\|\Delta \boldsymbol{u}_k\|^2 + 2\rho_k\lambda_{\max}(\mathbf{K}^T\mathbf{K})W. \tag{6.8}$$

The above derivations are not rigorous or concrete, but they aid in comprehending the fundamental concept of optimal gain design. Since the index function $f(\rho_k)$ is an upper bound rather than the actual value of $\mathbb{E}\|\Delta \boldsymbol{u}_{k+1}\|^2$, the derived value of ρ_k^* represents a suboptimal selection. However, throughout this chapter, we still refer to it as the optimal gain design because it reflects the optimization idea behind learning gain. Additionally, it is observed that the optimal gain ρ_k^* varies with each iteration and depends on the input energy at each step. Therefore, careful consideration should be given to formulating this specific optimal gain. Two typical design procedures will be elaborated in Sects. 6.2 and 6.3.

6.2 Optimal Gain Sequence in Probabilistic Quantization Problem

6.2.1 *Quantized Iterative Learning Control*

Driven by advancements in communication technology, an increasing number of systems are adopting networked implementations to enhance facilitation, flexibility, and robustness (Zhang et al. 2016, 2017; Mahmoud and Hamdan 2018). A networked system typically comprises a plant and a controller situated at different locations that communicate through long-distance transmission channels. However, this implementation introduces new control challenges such as data dropouts, communication congestion, limited bandwidth, and packet scheduling. To address these issues effectively, employing a quantization mechanism prior to data transmission has emerged as a promising solution (Han et al. 2016; Shao and Shi 2020; Xu et al. 2017). In this mechanism, continuous variable values are discretized into pre-defined sets of discrete values tailored for the system's units. One of the most critical challenges faced by quantized control is accurately reconstructing data from imprecise received signals in a communication system.

Quantized learning control (QLC) has been extensively studied, with a focus on two types of quantizers: logarithmic and uniform quantizers. Previous work introduced the concept of output quantization using a logarithmic quantizer for learning control, where the measured output is first quantized before being sent to the controller for input updating (Bu et al. 2015, 2017). However, due to the inherent quantization error in logarithmic quantizers, convergence of tracking error can only be proven within a bounded zone. To enhance tracking precision, an error quantization mechanism was proposed for deterministic systems in Xu et al. (2017) and extended to stochastic systems in Shen and Xu (2016). In this mechanism, the desired reference is transmitted to the plant prior to operation; subsequently, the tracking error is locally calculated and then quantized using a logarithmic quantizer before transmission. Although this approach achieves zero-error convergence of input error by reducing the tracking error towards zero through infinite possible states around zero in logarithmic quantizers, it is impractical due to limitations in realizing infinite levels of quantization within finite scopes. To overcome this limitation, we employ a common and cost-effective uniform quantizer on the output side incorporating an encoding-decoding mechanism as proposed in Zhang and Shen (2018), ensuring zero-error convergence of our learning control scheme (LCS). This convergence can be achieved even with a finite-level uniform quantizer by appropriately selecting scaling sequences within both encoder and decoder components. The mechanism was extended to the cases where quantization exists on both measurement and actuator sides in Shen and Zhang (2022) and additional data dropouts are coupled with the quantized transmission in the communication channels (Huo and Shen 2020).

Summarizing the literature, we observe several limitations. Firstly, both logarithmic and uniform quantizers only achieve bounded convergence of the input/tracking error without additional means (Bu et al. 2015, 2017). To attain asymptotic zero-error tracking performance, it is necessary to incorporate supplementary mechanisms such as error transmission and encoding-decoding. However, the transmission of desired reference beforehand in the error transmission mechanism is generally impractical due to shared communication channels between the plant and controller (Xu et al. 2017; Shen and Xu 2016). Moreover, the encoding-decoding mechanism requires local computation ability at the plant site which limits its applicability in scenarios with substantial additive channel noise (Zhang and Shen 2018; Shen and Zhang 2022; Huo and Shen 2020). These limitations motivate us to propose a probabilistic quantization mechanism that ensures perfect tracking performance without configuration adjustment or reliance on local computation resources. This aspect constitutes a primary novelty in our proposed schemes.

In this section, we propose an integrated framework that incorporates a probabilistic quantizer to optimize tracking performance in the presence of channel noise.

Specifically, we consider a common control configuration where additive noise is present in the communication channel. The output is initially quantized using a probabilistic quantizer, which randomly generates a quantization level based on the specific output value. This approach differs significantly from deterministic logarithmic and uniform quantizers that produce fixed quantization levels. We investigate two learning control schemes with decreasing and optimal designs for the learning gains. For the scheme with decreasing gain sequence, we establish strict asymptotic zero-error tracking performance by specifying sufficient conditions for the gain sequence and its relation to convergence speed. For the scheme with recursive gain sequence, we propose an index for performance evaluation and recursively acquire an optimal gain sequence. Additionally, we provide a careful estimation of convergence speed for the recursive gain sequence to verify asymptotic zero-error tracking performance. In summary, this section comprehensively analyzes these learning control schemes using a probabilistic quantizer.

6.2.2 Problem Formulation

Consider the following discrete-time linear systems

$$\begin{aligned} x_k(t+1) &= A_t x_k(t) + B_t u_k(t), \\ y_k(t) &= C_t x_k(t), \end{aligned} \tag{6.9}$$

where t denotes the discrete-time index and k denotes the iteration number, $t = 0, \ldots, n$ with n being the iteration length and $k = 1, 2, \ldots$. The variables $x_k(t) \in \mathbf{R}^o$, $u_k(t) \in \mathbf{R}^p$, and $y_k(t) \in \mathbf{R}^q$ are the state, input, and output, respectively. A_t, B_t, and C_t are system matrices with appropriate dimensions.

For the system, we need the following assumptions.

Assumption 6.1 The input/output coupling matrix $C_{t+1}B_t$ is of full-column rank, $\forall t$.

Assumption 6.2 The desired reference $y_r(t)$ is realizable in the sense that there exist initial state $x_r(0)$ and input $u_r(t)$ satisfying that

$$\begin{aligned} x_r(t+1) &= A_t x_r(t) + B_t u_r(t), \\ y_r(t) &= C_t x_r(t). \end{aligned} \tag{6.10}$$

Assumption 6.3 The state initialization is reset to the desired state $x_r(0)$ for all iterations; that is, $x_k(0) = x_r(0)$.

In this section, we investigate the problem of networked learning control through communication channels. The system configuration is depicted in Fig. 6.1, where the output undergoes quantization prior to transmission in order to alleviate the communication burden. The quantization process is denoted by a capital "$\mathcal{Q}$". Additionally,

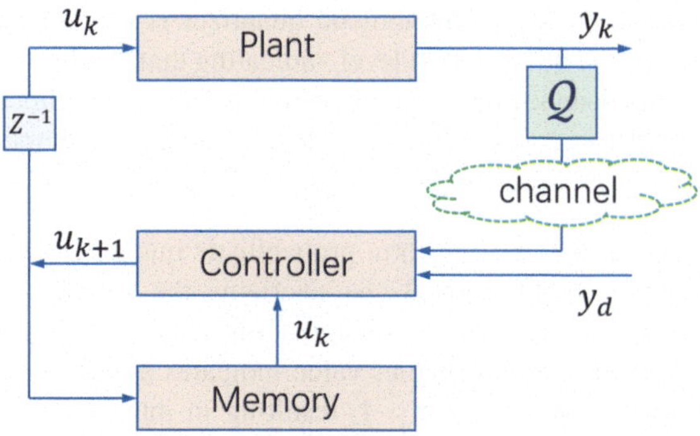

Fig. 6.1 Configuration of networked learning control with a measurement quantizer

additive noise affects the transmitted signal. Throughout this section, $y_k(t)$ represents the actual output while $y_k^{\circ}(t)$ corresponds to its quantized counterpart. The received signal is denoted by $y_k^*(t)$, represented as

$$y_k^*(t) = y_k^{\circ}(t) + \omega_k(t), \tag{6.11}$$

where $\omega_k(t)$ denotes the channel noise. We employ the following assumption on the channel noise.

Assumption 6.4 The channel noise $\omega_k(t)$ is of zero mean and bounded variance, i.e.,

$$\mathbb{E}[\omega_k(t)] = 0, \quad \mathbb{E}[\|\omega_k(t)\|^2] \le \sigma_\omega. \tag{6.12}$$

Remark 6.1 Our work disregards system noise and initialization errors. However, the extension to the general case involving system noise and initialization error is straightforward due to the fact that such noise and error can be effectively captured by the channel noise $\omega_k(t)$, assuming zero mean and bounded variance. Notably, both system noise and initialization error can corrupt the actual output while being quantized; nevertheless, their impact on the learning process remains consistent with that of channel noise. Furthermore, Assumption 6.4 can be extended to encompass colored stationary noise, allowing for seamless implementation of our proposed schemes within an augmented system incorporating a model of this colored noise.

In this section, we present the uniform probabilistic quantizer $\mathcal{Q}$ for the output: $y_k^{\circ}(t) = \mathcal{Q}(y_k(t))$. Let the quantization interval be Δ, which is a reflection of the quantization density. All possible quantized levels are denoted by $j\Delta$, $j \in \mathbf{Z}$. Define $\mathbf{Q} \triangleq \{j\Delta, j \in \mathbf{Z}\}$. For a real number v located in the interval $(j-1)\Delta \le v \le j\Delta$, the output of the quantizer $\mathcal{Q}^{\Delta} : \mathbf{R} \to \mathbf{Q}$ is defined as follows:

$$\mathcal{Q}^{\Delta}(v) = \begin{cases} (j-1)\Delta, & \text{with probability } j - \frac{v}{\Delta} \\ j\Delta, & \text{with probability } \frac{v}{\Delta} - j + 1 \end{cases} \tag{6.13}$$

The output of a probabilistic quantizer is non-deterministic unless the argument equals the quantization level, indicating that it follows a Bernoulli distribution based on the distance between the argument and its neighboring quantization levels. For a vector $\boldsymbol{v} = [v_1, v_2, \ldots, v_m] \in \mathbf{R}^m$, we define the output of the quantizer as $\mathcal{Q}^{\Delta}(\boldsymbol{v}) = [\mathcal{Q}^{\Delta}(v_1), \ldots, \mathcal{Q}^{\Delta}(v_m)]^T$.

Remark 6.2 The uniform probabilistic quantizer $\mathcal{Q}^{\Delta}(\cdot)$ maps the real set $\mathbf{R}$ to the discrete set $\mathbf{Q}$, with Δ characterizing the length of the quantization interval and serving as an index of quantization density. A larger value of Δ indicates lower precision while a smaller value indicates higher precision. The unit quantizer is a special case where $\Delta = 1$, resulting in integer values for transmitted signals. By specifying an expected density of δ, we can generate a quantizer with that density from (6.13), using the relationship $\tau = \Delta/\delta$ and $\mathcal{Q}^{\delta} = \frac{\mathcal{Q}^{\Delta}(\tau v)}{\tau}$. Specifically, we obtain a general probabilistic nonuniform type by defining arbitrary points for quantization levels within intervals such as $[m_j, m_{j+1})$, where $\{m_j\}$ is a sequence of arbitrary points, yielding the general probabilistic quantizer denoted by $\mathcal{Q}^{g}(\cdot)$:

$$\mathcal{Q}^{g}(v) = \begin{cases} m_j, & \text{with probability } \frac{m_{j+1}-v}{m_{j+1}-m_j} \\ m_{j+1}, & \text{with probability } \frac{v-m_j}{m_{j+1}-m_j} \end{cases}$$

Subsequently, we check the property of the quantization. It is shown by Lemma 2 of Aysal et al. (2008) that the probabilistic quantizer (6.13) is equivalent to the dithered quantization scheme $\mathcal{Q}^{d}(\cdot)$:

$$\mathcal{Q}^{d}(v) = \min_{z \in \mathbf{Q}} |z - (v + w)|, \tag{6.14}$$

where w is a uniform random variable in the interval $[-\frac{\Delta}{2}, \frac{\Delta}{2}]$. By employing this expression, it establishes a subtractive dither system (Wannamaker et al. 2000). The subtractive dithering process exhibits statistical independence from its input signal (Schuchman 1964). This independence offers significant convenience for convergence analysis. Furthermore, we present the following statistical result for the quantizer (6.13).

Lemma 6.1 *For an arbitrary value v, the quantized output $\mathcal{Q}^{\Delta}(v)$ satisfies*

$$\mathbb{E}[\mathcal{Q}^{\Delta}(v)] = v, \quad \mathbb{E}[(\mathcal{Q}^{\Delta}(v) - v)^2] \leq \sigma_{\Delta} = \frac{\Delta^2}{4}. \tag{6.15}$$

Proof Since an output of the probabilistic quantizer is a binary-valued random variable, its expectation is given by

$$\mathbb{E}[\mathcal{Q}^{\Delta}(v)] = (j-1)\Delta\left(j - \frac{v}{\Delta}\right) + j\Delta\left(\frac{v}{\Delta} - j + 1\right) = v.$$

Denote the quantization error by $r(v) \triangleq \mathcal{Q}^{\Delta}(v) - v$. We have

$$\begin{aligned}\mathbb{E}[r^2(v)] &= [(j-1)\Delta - v]^2 \left(j - \frac{v}{\Delta}\right) \\ &\quad + (j\Delta - v)^2 \left(\frac{v}{\Delta} - j + 1\right) \\ &= -\left(j\Delta - v + \frac{\Delta}{2}\right)^2 + \frac{\Delta^2}{4}.\end{aligned}$$

The proof is thus completed.

The unbiasedness and finite variance of the quantized output are characterized by Lemma 6.1, which enables us to establish a relatively precise estimation of the tracking performance. In other words, compared to the standard uniform quantizer and logarithm quantizer, the probabilistic quantizer possesses a distinct advantage in providing an unbiased estimate of the original signal.

Notably, the practical implementation of the probabilistic quantizer can adhere to either the original definition or an equivalent dithering process. In the case of the original definition, the probability distribution of the output varies based on the quantization value, resulting in increased computational requirements. Conversely, for the dithering process case, a uniformly distributed random variable w within the interval $[-\Delta/2, \Delta/2]$ can be generated and added to v, denoted as $v' = v + w$. Subsequently, a deterministic approach can be employed to generate the quantized output: $\mathcal{Q}^{\Delta}(v) = \mathcal{Q}^{\det}(v')$, where $\mathcal{Q}^{\det}(\cdot)$ indicates that all values within intervals such as $[j\Delta - \frac{\Delta}{2}, j\Delta + \frac{\Delta}{2})$ are deterministically mapped to $j\Delta$.

The control objective of this section is to design LCSs that can achieve high precision tracking performance while minimizing the communication burden. Specifically, we propose LCSs with decreasing and optimal gains, respectively. The tracking objective is to ensure that $y_r(t) - y_k(t) \to 0$ as $k \to \infty$. To achieve this objective, it is sufficient to guarantee $u_r(t) - u_k(t) \to 0$ in the iteration domain by utilizing (6.9), (6.10), and Assumption 6.3. For the specified LCS, under certain conditions, we demonstrate asymptotic convergence of the input signal towards the desired one, i.e., $u_k(t) \to u_r(t)$.

In this section, we propose an integrated framework comprising three modules to achieve the control objective: a probabilistic quantizer for providing effective information, a learning control scheme for gradually exploiting the inherent operational characteristics, and a decreasing gain sequence for mitigating the influence of channel noise and quantizing errors.

To facilitate subsequent analysis, we employ the lifting formulation in this subsection to transform the state space model. By utilizing this approach, the temporal dynamics are effectively eliminated, enabling a focused examination of the learning process in the iteration domain, which constitutes a fundamental aspect of learning control. However, it is important to emphasize that our proposed LCSs operate on a discrete-time basis and that the lifting formulation serves solely for concise analysis.

Denote the lifting vector of the input and output by $\boldsymbol{u}_k = [u_k^T(0), \ldots, u_k^T(n-1)]^T$ and $\boldsymbol{y}_k = [y_k^T(1), \ldots, y_k^T(n)]^T$. Then, from (6.9), we have

$$\boldsymbol{y}_k = \mathcal{H}\boldsymbol{u}_k + \mathcal{M}x_k(0), \tag{6.16}$$

where

$$\mathcal{H} = \begin{bmatrix} C_1B_0 & 0 & \cdots & 0 \\ C_2A_1B_0 & C_2B_1 & \cdots & 0 \\ \vdots & \vdots & \ddots & \vdots \\ C_nA_{n-1,1}B_0 & C_nA_{n-1,2}B_1 & \cdots & C_nB_{n-1} \end{bmatrix},$$
$$\mathcal{M} = [(C_1A_0)^T, (C_2A_1A_0)^T, \ldots, (C_nA_{n-1,0})^T]^T,$$

with $A_{i,j} \triangleq A_iA_{i-1}\ldots A_j$ for $i \geq j$.

Similarly, we can define $\boldsymbol{u}_r$ and $\boldsymbol{y}_r$ as above by replacing the subscript k with r. Then, we obtain from Assumption 6.2

$$\boldsymbol{y}_r = \mathcal{H}\boldsymbol{u}_r + \mathcal{M}x_r(0). \tag{6.17}$$

Denote the actual tracking error by $e_k(t) \triangleq y_r(t) - y_k(t)$ and its lifting vector by $\boldsymbol{e}_k \triangleq \boldsymbol{y}_r - \boldsymbol{y}_k$. By Assumption 6.3, it is clear that

$$\boldsymbol{e}_k = \mathcal{H}\widetilde{\boldsymbol{u}}_k, \tag{6.18}$$

where $\widetilde{\boldsymbol{u}}_k = \boldsymbol{u}_r - \boldsymbol{u}_k$. The equation (6.18) represents the error-dynamics in the iteration domain, which plays a crucial role in the convergence analysis of learning control. In case of additional uncertainty and/or noise present in the system (6.9), an extra term is introduced on the right-hand side of (6.18), necessitating compensation or elimination.

Due to the probabilistic quantizer and unreliable communication channels (see Fig. 6.1), the controller does not have access to the actual output $y_k(t)$, thereby preventing the utilization of the actual error $e_k(t)$ for driving the learning process. Instead, in accordance with (6.11), only the received signal $y_k^*(t)$ is available.

Denote the virtual tracking error $\epsilon_k(t) \triangleq y_r(t) - y_k^*(t)$. In the lifting formulation, we have

$$\boldsymbol{\epsilon}_k = \boldsymbol{y}_r - \boldsymbol{y}_k^* = \boldsymbol{y}_r - \boldsymbol{y}_k^\circ - \boldsymbol{\omega}_k, \tag{6.19}$$

where $\boldsymbol{\omega}_k = [\omega_k^T(1), \ldots, \omega_k^T(n)]^T$. From the abovementioned analysis, we have $\boldsymbol{y}_k^\circ = \mathcal{Q}^\Delta(\boldsymbol{y}_k)$ and

$$\boldsymbol{\epsilon}_k = \mathcal{H}\widetilde{\boldsymbol{u}}_k - \boldsymbol{r}_k - \boldsymbol{\omega}_k, \tag{6.20}$$

where $\boldsymbol{r}_k = \mathcal{Q}^{\Delta}(\boldsymbol{y}_k) - \boldsymbol{y}_k$. The virtual error dynamics in the iteration domain, as referred to in (6.20), involve random variables that necessitate careful consideration during subsequent design and analysis.

6.2.3 Learning Control Scheme With Decreasing Gain

In this case, the scheme is called LCS-DE, which is formulated as follows:

$$\text{LCS-DE: } u_{k+1}(t) = u_k(t) + \gamma_k L_t \epsilon_k(t+1), \tag{6.21}$$

where L_t is the direction regulation matrix (DRM) and $\{\gamma_k\}$ is a decreasing gain sequence. The decreasing gain γ_k should satisfy the following condition:

$$\gamma_k > 0, \quad \gamma_k \to 0, \quad \sum_{k=1}^{\infty} \gamma_k = \infty. \tag{6.22}$$

The lifting form of (6.21) is given as follows:

$$\boldsymbol{u}_{k+1} = \boldsymbol{u}_k + \gamma_k \mathcal{L} \boldsymbol{\epsilon}_k. \tag{6.23}$$

The main result for LCS-DE is summarized in the following theorem.

Theorem 6.1 *Consider LCS-DE* (6.21) *and the system* (6.9) *under Assumptions 6.1–6.4. If the DRM is designed such that all eigenvalues of $L_t C_{t+1} B_t$ are with positive real parts and the decreasing gain sequence satisfies* (6.22)*, then the input error $\widetilde{\boldsymbol{u}}_k = \boldsymbol{u}_r - \boldsymbol{u}_k$ converges to zero in the mean-square sense as the iteration number increases to infinity, i.e.,*

$$\lim_{k\to\infty} \mathbb{E}[\|\widetilde{\boldsymbol{u}}_k\|^2] = 0. \tag{6.24}$$

Proof Subtracting both sides of (6.23) from the desired input $\boldsymbol{u}_r$ yields that

$$\widetilde{\boldsymbol{u}}_{k+1} = \widetilde{\boldsymbol{u}}_k - \gamma_k \mathcal{L} \boldsymbol{\epsilon}_k. \tag{6.25}$$

Substituting (6.20) into (6.25) leads to that

$$\begin{aligned}\widetilde{\boldsymbol{u}}_{k+1} &= \widetilde{\boldsymbol{u}}_k - \gamma_k \mathcal{L}(\mathcal{H}\widetilde{\boldsymbol{u}}_k - \boldsymbol{r}_k - \boldsymbol{\omega}_k) \\ &= (I - \gamma_k \mathcal{L}\mathcal{H})\widetilde{\boldsymbol{u}}_k + \gamma_k \mathcal{L}(\boldsymbol{r}_k + \boldsymbol{\omega}_k).\end{aligned} \tag{6.26}$$

We introduce a Lyapunov function

$$\mathbf{V}_k = \widetilde{\boldsymbol{u}}_k^T \boldsymbol{\Theta} \widetilde{\boldsymbol{u}}_k, \tag{6.27}$$

where $\boldsymbol{\Theta}$ is a positive-definite matrix such that $(\mathcal{L}\mathcal{H})^T\boldsymbol{\Theta}+\boldsymbol{\Theta}(\mathcal{L}\mathcal{H})=I_{nq}$. Note that $\mathcal{L}$ is a block diagonal matrix and $\mathcal{H}$ is a block lower-triangular matrix; therefore, $\mathcal{L}\mathcal{H}$ is a block lower-triangular matrix. The diagonal blocks of $\mathcal{L}\mathcal{H}$ are $L_tC_{t+1}B_t$ for $t=0,1,\ldots,n-1$. Thus, the set of eigenvalues of $\mathcal{L}\mathcal{H}$ is the union of the sets of eigenvalues of $L_tC_{t+1}B_t$ for $t=0,1,\ldots,n-1$. Because L_t is designed such that all eigenvalues of $L_tC_{t+1}B_t$ are with positive real parts, $\forall t$, the set of eigenvalues of $\mathcal{L}\mathcal{H}$ are all with positive real parts. By the Lyapunov stability theory, the equation $(\mathcal{L}\mathcal{H})^T\boldsymbol{\Theta}+\boldsymbol{\Theta}(\mathcal{L}\mathcal{H})=I_{nq}$ must have a unique solution of a positive-definite matrix $\boldsymbol{\Theta}$. In other words, the Lyapunov function (6.27) is well defined.

Now substituting (6.26) into the Lyapunov function yields that

$$\begin{aligned}\mathbf{V}_{k+1}&=\widetilde{\boldsymbol{u}}_{k+1}^T\boldsymbol{\Theta}\widetilde{\boldsymbol{u}}_{k+1}\\&=\widetilde{\boldsymbol{u}}_k^T(I-\gamma_k\mathcal{L}\mathcal{H})^T\boldsymbol{\Theta}(I-\gamma_k\mathcal{L}\mathcal{H})\widetilde{\boldsymbol{u}}_k\\&\quad+\gamma_k^2(\boldsymbol{r}_k+\boldsymbol{\omega}_k)^T\mathcal{L}^T\boldsymbol{\Theta}\mathcal{L}(\boldsymbol{r}_k+\boldsymbol{\omega}_k)\\&\quad+2\gamma_k\widetilde{\boldsymbol{u}}_k^T(I-\gamma_k\mathcal{L}\mathcal{H})^T\boldsymbol{\Theta}\mathcal{L}(\boldsymbol{r}_k+\boldsymbol{\omega}_k).\end{aligned}\tag{6.28}$$

Note that $\widetilde{\boldsymbol{u}}_k$, $\boldsymbol{r}_k$, and $\boldsymbol{\omega}_k$ are random vectors. To conduct the convergence analysis, we need to apply the conditional expectation with respect to the historical information. To this end, we introduce an increasing σ-algebra $\mathcal{F}_k\triangleq\sigma\{x_r(0),u_1(t),\omega_i(t),r_i(t),1\le i\le k,0\le t\le n\}$. In other words, $\mathcal{F}_k$ indicates the set of all possible cases of the historical iterations up to the kth iteration. According to (6.21), it is noticed that $\boldsymbol{u}_k$ can be generated by all historical information up to the $(k-1)$th iteration; that is, $\boldsymbol{u}_k\in\mathcal{F}_{k-1}$.

Taking conditional expectation to both sides of (6.28) with respect to $\mathcal{F}_{k-1}$, we have

$$\begin{aligned}\mathbb{E}[\mathbf{V}_{k+1}\,|\mathcal{F}_{k-1}]&=\widetilde{\boldsymbol{u}}_k^T(I-\gamma_k\mathcal{L}\mathcal{H})^T\boldsymbol{\Theta}(I-\gamma_k\mathcal{L}\mathcal{H})\widetilde{\boldsymbol{u}}_k\\&\quad+\gamma_k^2\mathbb{E}\left[(\boldsymbol{r}_k+\boldsymbol{\omega}_k)^T\mathcal{L}^T\boldsymbol{\Theta}\mathcal{L}(\boldsymbol{r}_k+\boldsymbol{\omega}_k)\right],\end{aligned}\tag{6.29}$$

where the crossing term (i.e., the last term on the right-hand side of (6.28)) is eliminated because both $\boldsymbol{r}_k$ and $\boldsymbol{\omega}_k$ are independent of $\mathcal{F}_{k-1}$ and they are of zero-mean; that is, $\mathbb{E}[\boldsymbol{r}_k+\boldsymbol{\omega}_k\mid\mathcal{F}_{k-1}]=0$.

We now estimate the terms on the right-hand side of (6.29). First, there is a constant c_1 such that $(\mathcal{L}\mathcal{H})^T\boldsymbol{\Theta}(\mathcal{L}\mathcal{H})\preceq c_1\boldsymbol{\Theta}$. For example, we can let $c_1=\lambda_{\min}((\mathcal{L}\mathcal{H})^T\boldsymbol{\Theta}(\mathcal{L}\mathcal{H}))\lambda_{\max}^{-1}(\boldsymbol{\Theta})$. Consequently, we obtain

$$\begin{aligned}&\widetilde{\boldsymbol{u}}_k^T(I-\gamma_k\mathcal{L}\mathcal{H})^T\boldsymbol{\Theta}(I-\gamma_k\mathcal{L}\mathcal{H})\widetilde{\boldsymbol{u}}_k\\&=\mathbf{V}_k-\gamma_k\widetilde{\boldsymbol{u}}_k^T\widetilde{\boldsymbol{u}}_k+\gamma_k^2\widetilde{\boldsymbol{u}}_k^T(\mathcal{L}\mathcal{H})^T\boldsymbol{\Theta}(\mathcal{L}\mathcal{H})\widetilde{\boldsymbol{u}}_k\\&\le(1-\gamma_k\lambda_{\max}^{-1}(\boldsymbol{\Theta})+\gamma_k^2c_1)\mathbf{V}_k,\end{aligned}\tag{6.30}$$

where the first equality holds by substituting the Lyapunov equation $(\mathcal{L}\mathcal{H})^T\boldsymbol{\Theta}+\boldsymbol{\Theta}(\mathcal{L}\mathcal{H})=I_{nq}$ and the last inequality holds because $\boldsymbol{\Theta}\preceq\lambda_{\max}(\boldsymbol{\Theta})I$.

Note that $\boldsymbol{r}_k$ is independent of $\boldsymbol{\omega}_k$. Both of them are of zero mean. By Assumption 6.4, we have $\mathbb{E}[\|\boldsymbol{\omega}_k\|^2] \leq n\sigma_\omega$. By Lemma 6.1, we have $\mathbb{E}[\|\boldsymbol{r}_k\|^2] \leq nq\sigma_\Delta$. Then,

$$\begin{aligned}
&\mathbb{E}\left[(\boldsymbol{r}_k+\boldsymbol{\omega}_k)^T\mathcal{L}^T\boldsymbol{\Theta}\mathcal{L}(\boldsymbol{r}_k+\boldsymbol{\omega}_k)\right]\\
&=\mathbb{E}[\boldsymbol{r}_k^T\mathcal{L}^T\boldsymbol{\Theta}\mathcal{L}\boldsymbol{r}_k]+\mathbb{E}[\boldsymbol{\omega}_k^T\mathcal{L}^T\boldsymbol{\Theta}\mathcal{L}\boldsymbol{\omega}_k]\\
&\leq \lambda_{\max}(\mathcal{L}^T\boldsymbol{\Theta}\mathcal{L})\{\mathbb{E}[\|\boldsymbol{r}_k\|^2]+\mathbb{E}[\|\boldsymbol{\omega}_k\|^2]\}\\
&\leq c_2 n(\sigma_\omega+q\sigma_\Delta),
\end{aligned} \tag{6.31}$$

where $c_2=\lambda_{\max}(\mathcal{L}^T\boldsymbol{\Theta}\mathcal{L})$.

Then, we have

$$\begin{aligned}
\mathbb{E}[\mathbf{V}_{k+1}\mid\mathcal{F}_{k-1}]\leq{}&(1-\gamma_k\lambda_{\max}^{-1}(\boldsymbol{\Theta})+\gamma_k^2c_1)\mathbf{V}_k\\
&+\gamma_k^2c_2n(\sigma_\omega+q\sigma_\Delta).
\end{aligned} \tag{6.32}$$

Denote $\rho_k=\gamma_k(\lambda_{\max}^{-1}(\boldsymbol{\Theta})-\gamma_kc_1)$ and $\varpi=c_2n(\sigma_\omega+q\sigma_\Delta)$. Taking mathematical expectation to (6.32) yields

$$\mathbb{E}[\mathbf{V}_{k+1}]\leq(1-\rho_k)\mathbb{E}[\mathbf{V}_k]+\gamma_k^2\varpi. \tag{6.33}$$

First, ρ_k is of the same order of magnitude to γ_k. Therefore, after a sufficiently large number, say k_0, we can ensure that $0<\rho_k<1$, $\rho_k\to 0$, and $\gamma_kc_1\leq\frac{1}{2}\lambda_{\max}^{-1}(\boldsymbol{\Theta})$, $\forall k\geq k_0$. Without loss of generality, we assume this property holds for all iterations. Then, $\sum_{k=1}^{\infty}\rho_k\geq\frac{1}{2}\lambda_{\max}^{-1}(\boldsymbol{\Theta})\sum_{k=1}^{\infty}\rho_k=\infty$.

Define $\Phi_{i+1,k}=(1-\rho_i)\Phi_{i,k}$, $\forall i\geq k$, and $\Phi_{k,k}=1$. Because $\sum_{k=1}^{\infty}\rho_k=\infty$, it is evident that $\Phi_{k,i}\xrightarrow[k\to\infty]{}0$, $\forall i$. From (6.33) we obtain

$$\mathbb{E}[\mathbf{V}_{k+1}]\leq\Phi_{k+1,1}\mathbb{E}[\mathbf{V}_1]+\sum_{i=1}^{k}\Phi_{k+1,i+1}\gamma_i^2\varpi. \tag{6.34}$$

It is clear that the first term on the right-hand side of (6.34) tends to zero as iteration number k approaches infinity.

To derive the conclusion, it is sufficient to prove that the last term of (6.34) also converges to zero. To this end, we rewrite it as follows:

$$\sum_{i=1}^{k}\Phi_{k+1,i+1}\gamma_i^2\varpi=\varpi\sum_{i=1}^{k}\Phi_{k+1,i+1}\rho_i\frac{\gamma_i^2}{\rho_i}. \tag{6.35}$$

By direct calculation, it is clear that

$$\sum_{j=i}^{k} \Phi_{k+1,j+1}\rho_j = \sum_{j=i}^{k}\left(\prod_{l=j+1}^{k}(1-\rho_l)\right)\rho_j$$
$$= 1-\prod_{l=i}^{k}(1-\rho_l), \quad (k \geq i). \tag{6.36}$$

Therefore, $\sum_{i=1}^{k} \Phi_{k+1,i+1}\rho_i \xrightarrow[k\to\infty]{} 1$. In addition,

$$\lim_{k\to\infty} \frac{\gamma_k^2}{\rho_k} = \lim_{k\to\infty} \frac{\gamma_k}{\lambda_{\max}^{-1}(\boldsymbol{\Theta}) - \gamma_k c_1} = 0. \tag{6.37}$$

Then, by the Toeplitz Lemma (Li and Hu 2017), we obtain that

$$\sum_{i=1}^{k} \Phi_{k+1,i+1}\rho_i \frac{\gamma_i^2}{\rho_i} \xrightarrow[k\to\infty]{} \lim_{k\to\infty} \frac{\gamma_k^2}{\rho_k} = 0. \tag{6.38}$$

In other words, both terms on the right-hand side of (6.34) converge to zero. Thus, $\lim_{k\to\infty} \mathbb{E}[\mathbf{V}_k] = 0$ and $\lim_{k\to\infty} \mathbb{E}[\|\widetilde{\boldsymbol{u}}_k\|^2] = 0$. The proof is completed.

Theorem 6.1 demonstrates the tracking performance achieved through a decreasing gain sequence. This approach gradually mitigates the impact of quantization error and channel noise, enabling the input sequence to asymptotically converge to a limit. Moreover, it is essential for the decreasing gain sequence to satisfy the condition $\sum_{k=1}^{\infty} \rho_k = \infty$ in order to ensure an adequate learning process during iteration. In other words, the rate at which the gain decreases should not be too rapid. This requirement can be reflected by $\Phi_{k,i}$, as it will not converge to zero if $\sum_{k=1}^{\infty} \rho_k < \infty$.

Here, we analyze the convergence rate of LCS-DE. By referring to (6.34), it is evident that the convergence speed is determined by two terms on the right-hand side. The first term can be approximated as $\Phi_{k+1,1} = \prod_{i=1}^{k}(1-\rho_i) \sim \exp(-\sum_{i=1}^{k}\rho_i) \sim \exp(-\sum_{i=1}^{k}\gamma_i)$. Hence, the convergence rate for this term is explicitly controlled by the divergence rate of $\sum_{i=1}^{k}\rho_i$ or equivalently $\sum_{i=1}^{k}\gamma_i$. Regarding the second term, it should be noted that ρ_k and γ_k are of the same order of magnitudes. Consequently, according to (6.36), the convergence rate depends on the decreasing gain sequence $\{\gamma_k\}$ which limits its maximum value to either $\exp(-\sum_{i=1}^{k}\gamma_i)$ or γ_k. As a special case, if we consider $\gamma_k = a/k$, where $a > 0$, then the convergence rate will not exceed a certain negative power of iteration number depending on system and design parameters. This preliminary analysis highlights that while LCS-DE achieves precise tracking performance, it does so at a slower convergence rate.

Before concluding this section, we present the almost-sure convergence of LCS-DE. To achieve this, we slightly strengthen the condition for the decreasing gain sequence by imposing an additional requirement $\sum_{k=1}^{\infty}\gamma_k^2 < \infty$. This condition ensures that the decreasing gain sequence $\{\gamma_k\}$ does not decrease too slowly. By combining it with the condition $\sum_{k=1}^{\infty} \gamma_k = \infty$, we can determine a feasible range for selecting the decreasing gain.

Proposition 6.1 *Apply the LCS-DE* (6.21) *to the system* (6.9) *under the conditions specified in Theorem 6.1. Then, the input error* $\widetilde{\boldsymbol{u}}_k$ *converges to zero in the almost-sure sense; that is,*

$$\mathbb{P}\left(\lim_{k\to\infty} \|\widetilde{\boldsymbol{u}}_k\|^2 = 0\right) = 1. \tag{6.39}$$

Proof We rewrite (6.32) as follows:

$$\begin{aligned}\mathbb{E}[\mathbf{V}_{k+1} \mid \mathcal{F}_{k-1}] &\le (1 - \gamma_k \lambda_{\max}^{-1}(\boldsymbol{\Theta}) + \gamma_k^2 c_1)\mathbf{V}_k + \gamma_k^2 \varpi \\ &\le \mathbf{V}_k + \gamma_k^2(\varpi + c_1 \mathbf{V}_k).\end{aligned} \tag{6.40}$$

We have proved by Theorem 6.1 that $\mathbb{E}[\mathbf{V}_k]$ converges to zero, thus $\mathbb{E}[\mathbf{V}_k]$ is bounded over the iteration domain. Then, with the strengthened condition of γ_k, we obtain $\sum_{k=1}^{\infty} \gamma_k^2 c_1 \mathbb{E}[\mathbf{V}_k] < \infty$ and $\mathbf{V}_k$ is a nonnegative stochastic process adapted to $\mathcal{G}_k$. By the monotone convergence theorem, it follows that $\sum_{k=1}^{\infty} \gamma_k^2 c_1 \mathbf{V}_k < \infty$ almost surely. By Lemma 2.2, we conclude that $\mathbf{V}_k$ converges almost surely. Again, by Theorem 6.1, it has been proved that $\lim_{k\to\infty} \mathbb{E}[\mathbf{V}_k] = 0$. The limit must be consistent. Therefore, we have (6.39). The proof is completed.

Theorem 6.1 and Proposition 6.1 establish the asymptotic convergence of the input error to zero. Consequently, based on the error dynamics (6.18), it is evident that the tracking error converges to zero in both mean-square and almost-sure senses as the iteration number increases. In other words, precise tracking performance can be achieved by introducing a decreasing gain sequence, which gradually eliminates randomness in the iteration domain. However, it should be noted that the conditions on the decreasing gain sequence [i.e., (6.22)] are considerably general, encompassing a wide range of feasible candidates. Furthermore, this decreasing gain sequence remains independent of the operation process. Nevertheless, practical performance within finite iterations may suffer without considering specific process scenarios. Therefore, an interesting question arises regarding whether an optimal/suboptimal selection of this sequence can be obtained.

6.2.4 Learning Control Scheme With Recursive Gain

In this section, we present a recursive design mechanism for the gain sequence that is optimal or suboptimal based on a specific index. This concept draws inspiration from Yousefian et al. (2012), but our evaluation provides more detailed insights. Unlike LCS-DE in Sect. 6.2.3, where the gain sequence is arbitrarily determined prior to the operation process, our recursive gain sequence formulation is directly linked to tracking performance.

We revisit LCS-DE: $u_{k+1}(t) = u_k(t) + \gamma_k L_t \epsilon_k(t+1)$, and employ the Lyapunov function $\mathbb{E}[\mathbf{V}_k]$ as a performance index with respect to the iteration number k (Yousefian et al. 2012). One may consider this performance index dependent on the matrix $\boldsymbol{\Theta}$. However, $\boldsymbol{\Theta}$ is uniquely determined by the Lyapunov equation $(\mathcal{L}\mathcal{H})^T\boldsymbol{\Theta} + \boldsymbol{\Theta}(\mathcal{L}\mathcal{H}) = I_{nq}$. Thus, the performance index actually depends on $\mathcal{H}$ and $\mathcal{L}$. From (6.33), we have

$$\mathbb{E}[\mathbf{V}_{k+1}] \le (1 - \gamma_k \vartheta)\mathbb{E}[\mathbf{V}_k] + \gamma_k^2 \varpi, \tag{6.41}$$

where $\vartheta = \frac{1}{2}\lambda_{\max}^{-1}(\boldsymbol{\Theta})$, provided that $\gamma_k c_1 \le \frac{1}{2}\lambda_{\max}^{-1}(\boldsymbol{\Theta})$, $\forall k$, i.e., $\gamma_k \le \frac{1}{2c_1}\lambda_{\max}^{-1}(\boldsymbol{\Theta})$.

By (6.41), the performance index for the $(k+1)$th iteration is bounded by the formulation at the right-hand side. In other words, the right-hand side acts as the worst case of the weighted Euclidean norm of the input error. Then, an optimal selection of γ_k can be determined by minimizing this formulation. To this end, we regard the right-hand side as a function of γ_k: $f(\gamma_k) \triangleq (1 - \gamma_k\vartheta)\mathbb{E}[\mathbf{V}_k] + \gamma_k^2\varpi$. Taking the partial derivative of $f(\cdot)$ with respect to γ_k leads to

$$\frac{\partial f(\gamma_k)}{\partial \gamma_k} = -\vartheta \mathbb{E}[\mathbf{V}_k] + 2\gamma_k \varpi. \tag{6.42}$$

Letting this derivative be zero yields the optimal selection

$$\gamma_k^* = \frac{\vartheta}{2\varpi}\mathbb{E}[\mathbf{V}_k]. \tag{6.43}$$

This formulation depends on $\mathbf{V}_k$. Next, we present a recursive formulation of γ_k^*. We consider the worst case as the estimate of the Lyapunov function to derive the recursive gain sequence and we make use of the equality (6.41). Substituting (6.43) into $f(\gamma_k)$, we obtain

$$\begin{aligned} f(\gamma_k^*) &= (1 - \gamma_k^*\vartheta)\mathbb{E}[\mathbf{V}_k] + (\gamma_k^*)^2\varpi \\ &= (1 - \gamma_k^*\vartheta)\frac{2\gamma_k^*\varpi}{\vartheta} + (\gamma_k^*)^2\varpi \\ &= \frac{2\gamma_k^*\varpi}{\vartheta} - (\gamma_k^*)^2\varpi. \end{aligned} \tag{6.44}$$

Then, we have

$$\gamma_{k+1}^* = \frac{\vartheta}{2\varpi}\mathbb{E}[\mathbf{V}_{k+1}] = \frac{\vartheta}{2\varpi} f(\gamma_k^*) = \gamma_k^*\left(1 - \frac{\vartheta}{2}\gamma_k^*\right). \tag{6.45}$$

In short, the optimal gain sequence that minimizes the function $f(\gamma)$ is recursively defined by

$$\gamma_1^* = \frac{\vartheta}{2\varpi}\mathbb{E}[\mathbf{V}_1], \tag{6.46}$$

$$\gamma_k^* = \gamma_{k-1}^*\left(1 - \frac{\vartheta}{2}\gamma_{k-1}^*\right). \tag{6.47}$$

To demonstrate the optimality of the above defined recursive gain sequence, we establish the following proposition, where $\mathbf{L}_k$ corresponds to $\mathbb{E}[\mathbf{V}_k]$.

Proposition 6.2 *Consider the recursive relationship*

$$\mathbf{L}_{k+1} = (1 - \gamma_k\vartheta)\mathbf{L}_k + \gamma_k^2\varpi. \tag{6.48}$$

If the initial value $\mathbf{L}_1$ *satisfies that* $\frac{\vartheta}{2\varpi}\mathbf{L}_1 \le \frac{2}{\vartheta}$, *the recursive gain sequence defined by* (6.46) *and* (6.47) *is optimal in the sense that for any sequence* $\{\gamma_k\}$ *such that* $0 < \gamma_k \le \frac{2}{\vartheta}$, *the outcome of* (6.48) *satisfies* $\mathbf{L}_k \ge \mathbf{L}_k^*$, *where* $\mathbf{L}_k^*$ *is generated by using the optimal sequence* $\{\gamma_k^*\}$.

Proof The proof is carried out by mathematical induction argument. Both cases start from $\mathbf{L}_1$; therefore, the comparison starts from the second iteration.

Initial Step: Consider $k = 2$. We have $\gamma_1^* = \frac{\vartheta}{2\varpi}\mathbf{L}_1$, then

$$\mathbf{L}_2^* = (1 - \gamma_1^*\vartheta)\mathbf{L}_1 + (\gamma_1^*)^2\varpi = \frac{2\varpi\gamma_1^*}{\vartheta} - (\gamma_1^*)^2\varpi.$$

While for an arbitrary gain γ_1, we have

$$\mathbf{L}_2 = (1 - \gamma_1\vartheta)\mathbf{L}_1 + \gamma_1^2\varpi = \frac{2\varpi\gamma_1^*}{\vartheta} - 2\varpi\gamma_1\gamma_1^* + \gamma_1^2\varpi.$$

Then,

$$\mathbf{L}_2 - \mathbf{L}_2^* = \varpi(\gamma_1 - \gamma_1^*)^2 \ge 0.$$

Inductive Step: Assume the conclusion holds for the kth iteration, i.e., $\mathbf{L}_k \ge \mathbf{L}_k^*$. We prove it for the $(k+1)$th iteration. To this end, we have

$$\mathbf{L}_{k+1}^* = \frac{2\varpi}{\vartheta}\gamma_{k+1}^* = \frac{2\varpi}{\vartheta}\gamma_k^*\left(1 - \frac{\vartheta}{2}\gamma_k^*\right),$$

$$\mathbf{L}_{k+1} = (1 - \gamma_k\vartheta)\mathbf{L}_k + \gamma_k^2\varpi.$$

Then,

$$\begin{aligned}\mathbf{L}_{k+1}-\mathbf{L}_{k+1}^{*}&=(1-\gamma_k\vartheta)\mathbf{L}_k+\gamma_k^2\varpi-\frac{2\varpi}{\vartheta}\gamma_k^*\left(1-\frac{\vartheta}{2}\gamma_k^*\right)\\&\geq(1-\gamma_k\vartheta)\mathbf{L}_k^*+\gamma_k^2\varpi-\frac{2\varpi}{\vartheta}\gamma_k^*\left(1-\frac{\vartheta}{2}\gamma_k^*\right)\\&=(1-\gamma_k\vartheta)\frac{2\varpi\gamma_k^*}{\vartheta}+\gamma_k^2\varpi-\frac{2\varpi}{\vartheta}\gamma_k^*\left(1-\frac{\vartheta}{2}\gamma_k^*\right)\\&=\varpi(\gamma_k-\gamma_k^*)^2\geq 0.\end{aligned}$$

By the induction principle, the proof is completed.

Remark 6.3 The gain sequence $\{\gamma_k\}$ in (6.21) is not optimal for a specific process because condition (6.22) is general. In other words, it is independent of the process data and thus incapable of improving the iteration-based transient performance. In contrast, the gain sequence $\{\gamma_k^*\}$, represented by (6.46) and (6.47), serves as an optimal solution in terms of minimizing the index $f(\gamma_k)$. The performance index is minimized along the iteration axis while disregarding potential instability in system dynamics. To address this issue, a classical feedback stabilization strategy can be employed to ensure time-domain stability while incorporating the learning control strategy.

Using the recursive gain sequence $\{\gamma_k^*\}$, we present the LCS-OP as follows:

$$\text{LCS-OP: } u_{k+1}(t)=u_k(t)+\gamma_k^* L_t\epsilon_k(t+1), \tag{6.49}$$

where L_t is the direction regulation matrix.

We proceed to establish the convergence theorem for LCS-OP. To this end, it suffices to demonstrate that the recursive gain sequence defined by (6.46) and (6.47) satisfies condition (6.22). Therefore, we need to provide a meticulous estimation of the rate at which γ_k^* decreases. The ensuing lemma presents a distinct estimate from that provided by Yousefian et al. (2012).

Lemma 6.2 *Consider the following recursion*

$$\gamma_k=\gamma_{k-1}(1-a\gamma_{k-1}),\quad \forall k\geq 1, \tag{6.50}$$

where $a>0$ is a constant and the initial value γ_0 satisfies $0<\gamma_0<\frac{1}{a}$. Then, we have $\gamma_k>0$, $\gamma_k\to 0$, and

$$\gamma_k\sim k^{-1}a^{-1}. \tag{6.51}$$

Proof Note that $0<\gamma_0<\frac{1}{a}$, thus $0<\gamma_1<\frac{1}{a}$. By the induction principle, we have $0<\gamma_k<\frac{1}{a}$, $\forall k$.

From (6.50), it is evident that $\{\gamma_k\}$ is a decreasing sequence. Combining with the fact $\gamma_k > 0$, we have that there is a limitation for the sequence $\{\gamma_k\}$. Denote $\overline{\gamma} = \lim_{k\to\infty} \gamma_k$. Taking limitation to both sides of (6.50) yields $\overline{\gamma} = \overline{\gamma}(1 - a\overline{\gamma})$, which provides the unique solution $\overline{\gamma} = 0$. That is, $\gamma_k \to 0$, or equivalently, $\gamma_k = o(1)$.

To demonstrate the decreasing speed of γ_k, we make a transformation $z_k = a\gamma_k$. Then, we have $z_k \to 0$. Besides, from (6.50) we obtain

$$z_{k+1} = z_k - z_k^2. \tag{6.52}$$

We let $d_k = \frac{1}{z_k}$. It is clear that $d_k \to \infty$. Now, (6.52) becomes

$$\frac{1}{d_{k+1}} = \frac{1}{d_k} - \frac{1}{d_k^2}. \tag{6.53}$$

This further leads to

$$\begin{aligned} d_{k+1} &= d_k \left(1 - \frac{1}{d_k}\right)^{-1} \\ &= d_k \left(1 + \frac{1}{d_k} + O\left(\frac{1}{d_k^2}\right)\right) \\ &= 1 + d_k + O\left(\frac{1}{d_k}\right). \end{aligned} \tag{6.54}$$

Therefore, $d_{k+1} - d_k = 1 + O\left(\frac{1}{d_k}\right)$. For sufficiently large number k_0 and $\forall k \geq k_0$, we have

$$d_{k+1} - d_k > \frac{1}{2}, \quad \text{and} \quad d_k > \frac{1}{4}k. \tag{6.55}$$

Clearly, we have $d_k^{-1} = O(\frac{1}{k})$. Substituting this estimate into (6.54) yields

$$\begin{aligned} d_k &\sim k + O(1) + O(\log k) \\ &= k\left(1 + O\left(\frac{1}{k}\right) + O\left(\frac{\log k}{k}\right)\right). \end{aligned} \tag{6.56}$$

Then, we obtain the estimate for z_k from (6.56) that

$$
\begin{aligned}
z_k = \frac{1}{d_k} &= \frac{1}{k}\left(1 + O\left(\frac{1}{k}\right) + O\left(\frac{\log k}{k}\right)\right)^{-1} \\
&= \frac{1}{k}\left(1 + O\left(\frac{1}{k}\right) + O\left(\frac{\log k}{k}\right)\right) \\
&= \frac{1}{k} + O\left(\frac{\log k}{k^2}\right).
\end{aligned}
\tag{6.57}
$$

In other words, $z_k \sim k^{-1}$. By $z_k = a\gamma_k$, we obtain (6.51). The proof is completed.

By (6.47) and Lemma 6.2, we conclude that

$$
\gamma_k^* \sim \frac{2}{\vartheta k}. \tag{6.58}
$$

Then, it is evident that $\sum_{k=1}^{\infty} \gamma_k^* = \infty$ and $\sum_{k=1}^{\infty} (\gamma_k^*)^2 < \infty$. Thus, the conditions on the gain sequence in Theorem 6.1 and Proposition 6.1 are satisfied. This fact leads to the mean-square and almost-sure convergence of LCS-OP. The results are summarized in the following theorem.

Theorem 6.2 *Consider LCS-OP* (6.49) *and the system* (6.9) *under Assumptions 6.1–6.4. If the DRM is designed such that all eigenvalues of $L_t C_{t+1} B_t$ are with positive real parts and the decreasing gain sequence $\{\gamma_k^*\}$ is defined by* (6.46) *and* (6.47)*, then the input error $\widetilde{\boldsymbol{u}}_k = \boldsymbol{u}_r - \boldsymbol{u}_k$ converges to zero in both mean-square and almost-sure senses as the iteration number increases to infinity, i.e.,*

$$
\lim_{k\to\infty} \mathbb{E}[\|\widetilde{\boldsymbol{u}}_k\|^2] = 0 \quad \textit{and} \quad \mathbb{P}\left(\lim_{k\to\infty} \|\widetilde{\boldsymbol{u}}_k\|^2 = 0\right) = 1. \tag{6.59}
$$

Theorem 6.2 presents the convergence property for LCS-OP, where the optimized decreasing gain sequence is utilized. By referring to (6.58), it can be observed that the selection of the gain sequence, whether optimal or sub-optimal, is inversely proportional to the iteration number. Specifically, a sub-optimal candidate is represented by $\gamma_k = a/k$, where $a > 0$ denotes a constant. As elucidated in Sect. 6.2.3, it should be noted that the convergence speed of (6.49) does not exceed a certain negative power of the iteration number, with this power being dependent on various parameters involved. In essence, the fastest convergence rate achievable for index $\mathbb{E}[\mathbf{V}_k]$ corresponds to an order of $1/k$.

6.2.5 Illustrative Simulations

In this section, we verify the theoretical results by an industrial robot model of the closed-loop joint presented in Zhang et al. (2014), which is given by the following transfer function:

$$G(s) = \frac{948}{s^2 + 42s + 948}.$$

We convert the transfer function to discrete-time state-space represented by the triplet (A, B, C), where

$$A = \begin{bmatrix} 0.6213 & -7.6196 \\ 0.008 & 0.9589 \end{bmatrix}, B = \begin{bmatrix} 0.008 \\ 0 \end{bmatrix},$$
$$C = \begin{bmatrix} 948 & 0 \end{bmatrix}.$$

The iteration length is set to $n = 50$. We set the initial state to $x_k(0) = x_r(0) = 0$ for all k and the initial input profile to $u_0(t) = 0$, $\forall t \in [0, n]$. The channel additive noise is set to $\omega_k(t) \sim N(0, 0.1^2)$. The desired trajectory is given by $y_r(t) = \sin(1.6t/5) + \sin(0.8t/5)$. The direction regulation matrix is $L_t = 1.1$. It can be verified that $L_t C_{t+1} B_t$ are positive real numbers for all time instants. That is, the design conditions in the main theorems are guaranteed. The quantization density is set to $\Delta = 0.5$.

In the following demonstration, simulation results are categorized into four sections. Firstly, we validate the convergence of the proposed schemes and assess their statistical and deterministic performance. Subsequently, we evaluate the robustness of these schemes against system and channel uncertainties. Furthermore, a comparison is made between the proposed schemes and the well-known Kalman filtering-based scheme (KFS) (Saab 2001a), considering both probabilistic quantizer and traditional rounding quantizer approaches. Lastly, we investigate the performance of our proposed schemes by considering three different quantization densities for the probabilistic quantizer.

6.2.5.1 Convergence and Performance of the Proposed Schemes

We first consider the performance under three scenarios of the gain sequence. That is, besides the proposed LCS-DE and LCS-OP, we also consider the scenario where the decreasing gain γ_k is replaced by a constant gain γ. In particular, for the constant gain case, we set $\gamma = 0.07$. For the decreasing gain case, we set $\gamma_k = 0.06/k$ that satisfies condition (6.22). For the recursive gain case, we can calculate that $\vartheta = 5.0495$ and $\varpi = 0.4343$. Then, we have $\gamma_1^* = 0.0828$, while γ_k^* is generated by (6.47). For each scenario, the scheme is run for 60 iterations.

The tracking performance for these three gain scenarios at the final iteration and their corresponding input profiles are illustrated in Fig. 6.2. In both subplots, the dotted, dashed, and dash-dotted lines represent the constant, decreasing, and recursive gain scenarios respectively, while the solid line indicates the desired reference. It can be observed that the tracking performance of the decreasing and recursive gain scenarios exhibits similar behavior, whereas the scenario with a constant gain is marginally inferior to the previous two scenarios. The results depicted in Fig. 6.2 substantiate the efficacy of achieving precise tracking using our proposed schemes.

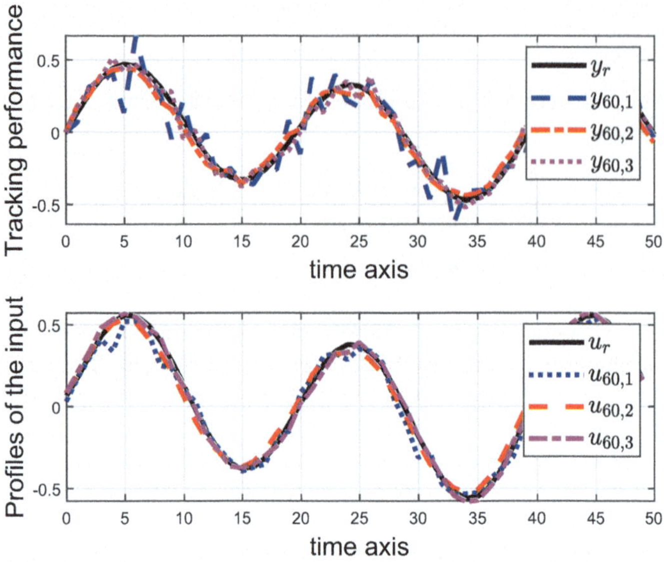

Fig. 6.2 The profiles for three gain scenarios at the 60th iteration and the desired trajectory: the output (above) and the input (bottom).

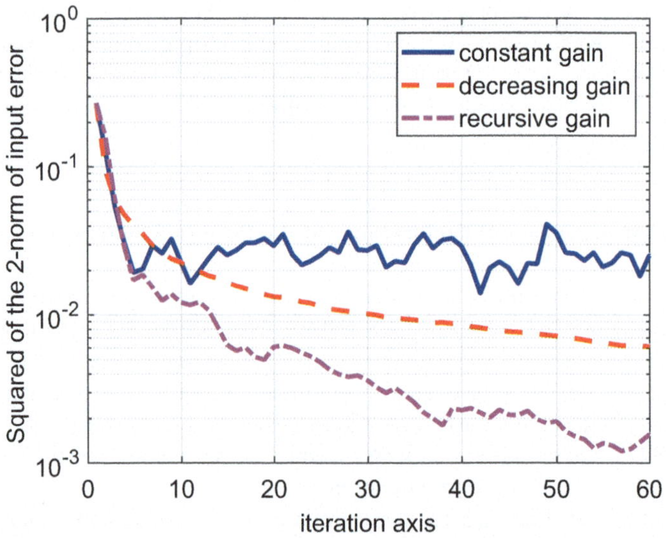

Fig. 6.3 The squared Euclidean norm profiles of the input errors in the iteration domain for three gain scenarios

Table 6.1 Experimental performance of the proposed schemes

	Constat gain	Decreasing gain	Recursive gain
V_1	4.8082×10^{-4}	5.6271×10^{-5}	3.2252×10^{-5}
V_2	0.0311	0.0111	0.0133
M_1	4.7436×10^{-4}	1.4833×10^{-4}	3.1758×10^{-5}
M_2	0.0339	0.0253	0.0158
I_1	0.0418	0.0276	0.0136
I_2	3.1599	2.4386	1.8493

$V_1 = [\text{stdev}(u_r(t) - u_{60}(t))]^2$
$V_2 = \sum_{k=1}^{60}[\text{stdev}(u_r(t) - u_k(t))]^2$
$M_1 = \text{avg}[(u_r(t) - u_{60}(t))^2]$
$M_2 = \sum_{k=1}^{60} \text{avg}[(u_r(t) - u_k(t))^2]$
$I_1 = \max_{t\in[0,50]|u_r(t)-u_{60}(t)|}$
$I_2 = \sum_{k=1}^{60} \max_{t\in[0,50]|u_r(t)-u_k(t)|}$

To understand the difference between the proposed schemes with difference gains, we present the iteration-wise input error profiles in Fig. 6.3, where the dotted, dash-dotted, and solid lines correspond to the constant, decreasing, and recursive gain scenarios, respectively. It can be observed that the dotted profile becomes fluctuating after several iterations. This is due to the fact that the constant gain cannot eliminate the influence of various randomness such as quantization error and additive noise. In contrast, the input error profiles of both decreasing and recursive gain sequences retain a decreasing trend along the iteration axis. Regarding the convergence rate, the scheme with a recursive gain sequence is superior to that with a decreasing gain sequence.

We introduce three measures to quantify the statistical evaluation and deterministic examination of the input errors similarly to Saab (2001a). The measures are the variance, the mean square, and the infinite norm of the input errors for the last iteration and their sum over the 60 iterations. The results are presented in Table 6.1. By examining Table 6.1, the conclusions indicated by Fig. 6.3 are again verified. These observations demonstrate the motivation for introducing the recursive gain.

6.2.5.2 Robustness Against Uncertainties

Unknown uncertainties may exist in the system model and communication channels. Random variables are added to the elements of system matrices to represent model uncertainty, $\varepsilon_{1,k} \sim U(-0.01, 0.01)$, $\varepsilon_{2,k} \sim U(-0.1, 0.1)$, $\varepsilon_{3,k} \sim U(-0.0001, 0.0001)$, and $\varepsilon_{4,k} \sim U(-1, 1)$. Furthermore, the channel noise is described by $\omega_k(t) \sim U(-0.2, 0.2)$ as an alternative to Gaussian distribution. The rest of the parameters are the same as above.

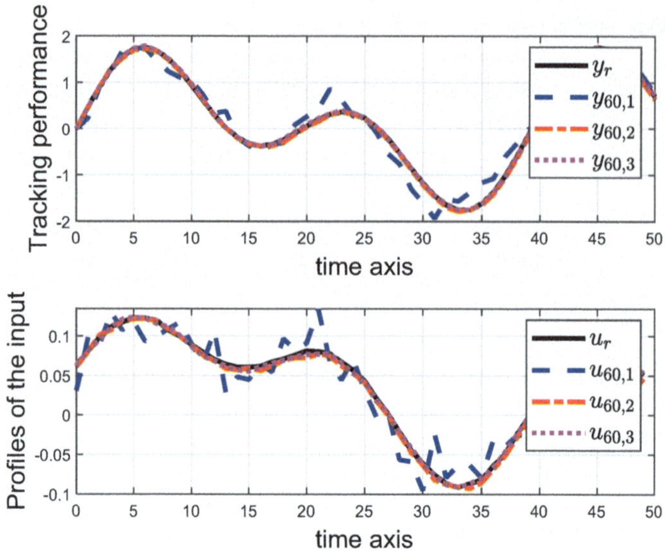

Fig. 6.4 The profiles for three gain scenarios at the 60th iteration in presence of modeling uncertainties and uniformly distributed channel noise: the output (top) and the input (bottom)

$$A = \begin{bmatrix} 0.6213 + \varepsilon_{1,k} & -7.6196 + \varepsilon_{2,k} \\ 0.008 + \varepsilon_{3,k} & 0.9589 + \varepsilon_{1,k} \end{bmatrix}, B = \begin{bmatrix} 0.008 + \varepsilon_{3,k} \\ 0 \end{bmatrix},$$
$$C = \begin{bmatrix} 948 + \varepsilon_{4,k} & 0 \end{bmatrix}.$$

The tracking performance of the proposed schemes in the presence of these uncertainties are given in Fig. 6.4, where the upper and lower plots correspond to the output and input, respectively. The iteration-wise profiles of input errors are illustrated in Fig. 6.5 for the three gain scenarios under examination. It is observed that the profiles in Figs. 6.4 and 6.5 are similar to those of Figs. 6.2 and 6.3.

In addition, we model the channel additive noise $\omega_k(t)$ by Laplace distribution and Cauchy distribution as alternatives to Gaussian distribution. Specifically, we set $\omega_k(t) \sim \text{Laplace}(0, 0.1)$ and $\omega_k(t) \sim \text{Cauchy}(0, 0.01)$ for the simulation. The iteration-wise profiles of the input errors are illustrated in Fig. 6.6 for the three gain scenarios under examination. It can be seen that similar performance is retained under the alternative channel noises. These observations demonstrate that the proposed schemes can be robust against non-Gaussian channel noises.

6.2.5.3 Comparisons with Rounding Quantizer and KFS

To deeply understand the performance of the proposed schemes, we provide comparisons from two perspectives in this subsection. First, we compare the results

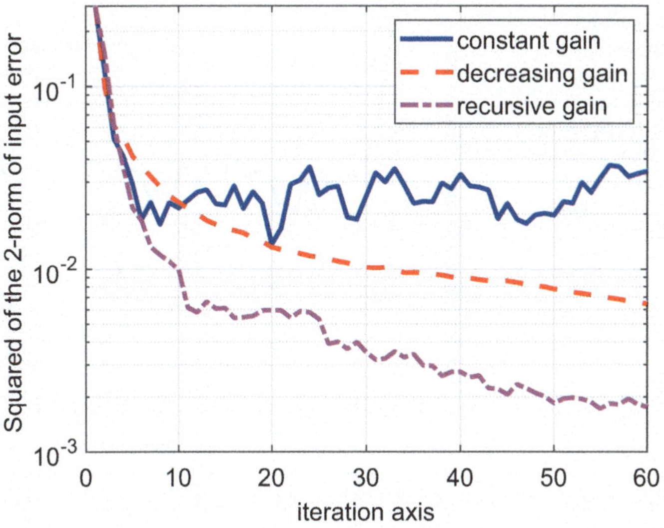

Fig. 6.5 The squared Euclidean norm profiles of the input errors in the iteration domain for three gain scenarios in presence of modeling uncertainties and uniformly distributed channel noise

with the same schemes incorporated with the traditional rounding quantizer to illustrate the superiority of the probabilistic quantizer. A rounding quantizer $\mathcal{Q}_r$ projects $v \in \mathbf{R}$ to its nearest quantization level in $\mathbf{Q} = \{j\Delta, j \in \mathbf{Z}\}$ (Zhu and Chen 2016); that is, $\mathcal{Q}_r(v) = j\Delta$ if $(j - \frac{1}{2})\Delta \leq v < (j + \frac{1}{2})\Delta$. The quantization density is set to $\Delta = 0.4$. We employ the proposed schemes with the following parameters: constant gain $\gamma = 0.09$, decreasing gain $\gamma_k = 0.06/k$, and recursive gain with initial value $\gamma_1^* = 0.0828$. The input error trends versus iterations for these three schemes are depicted in Fig. 6.7. We also provide LCS-OP using a probabilistic quantizer as a benchmark for comparison. Examining Fig. 6.7, we observe that the proposed schemes with a rounding quantizer cannot achieve asymptotically precise tracking performance even for the decreasing and recursive gains. Thus, the introduction of a probabilistic quantizer suffices to achieve fine tracking performance.

Moreover, we compare LCS-OP with KFS provided in Saab (2001a). Both schemes employ the probabilistic quantizer. The KFS in Saab (2001a) recursively generates the direction regulation matrix by minimizing the trace of the covariance matrix of input errors along the time and iteration axes. However, unlike LCS-OP, the KFS in Saab (2001a) requires the knowledge of the system model and statistical model of the random variables in order to optimally generates its learning gains. The input error profile for this scheme is also plotted in Fig. 6.7. It is observed that the performance of both LCS-OP and KFS is similar to each other. This reflects the optimality characteristic of the recursive method presented in Sect. 6.2.4. However, in contrast to KFS, the proposed LCS-OP requires (a) little information of the system model, and (b) less computational efforts. The latter reflects the advantages of our proposed method over the one proposed in Saab (2001a).

Fig. 6.6 The squared Euclidean norm profiles of the input errors in the iteration domain for three gain scenarios in presence of Laplace distributed channel noise (top) and Cauchy distributed channel noise (bottom)

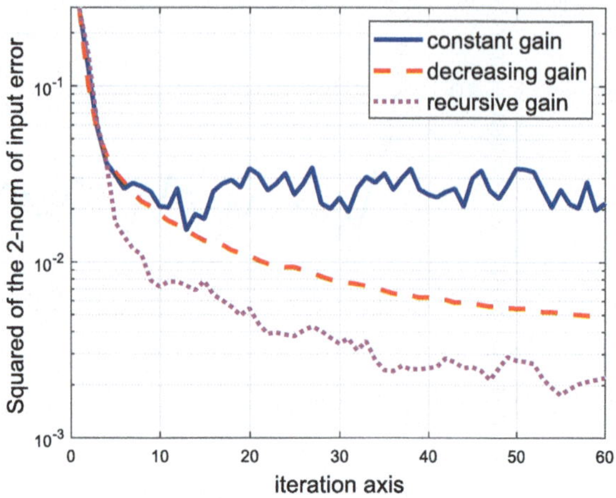

(a)

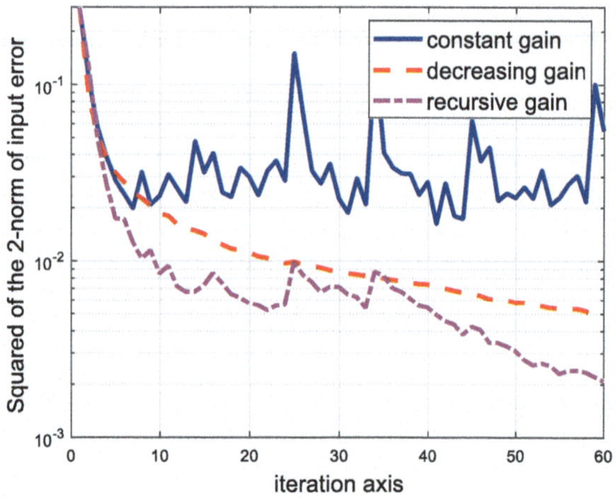

(b)

6.2.5.4 Effect of Quantization Density

In addition, to show the effect of the quantization density of the probabilistic quantizer on the tracking performance, we further consider three specific quantization densities, i.e., $\Delta = 0.3$, $\Delta = 0.5$, and $\Delta = 0.7$, respectively. The results are demonstrated in Fig. 6.8, where the gain selections in the subplots are the same as above. A common characteristic among these subplots is that smaller density would lead to a more precise tracking performance for the specified iteration number. This is consistent with our intuition and verifies the theoretical analysis.

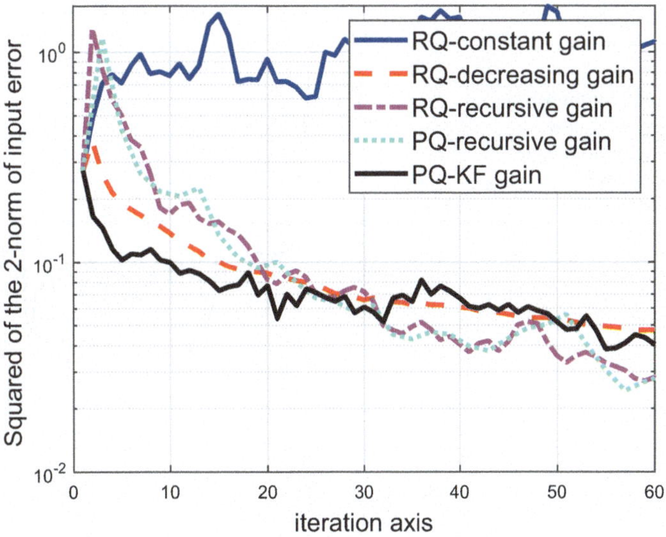

Fig. 6.7 The squared Euclidean norm profiles of the input errors in the iteration domain: RQ and PQ represent the schemes employing the rounding quantizer and the probabilistic quantizer, respectively

6.3 Optimal Gain Sequence in Large-Scale Systems

6.3.1 Decentralized Iterative Learning Control

Large-scale systems (LSSs) have played a significant role in modeling complex systems such as power, chemical engineering, aerospace, transportation, and industrial systems (Zhang et al. 2022; Deng and Yang 2017; Moradvandi et al. 2019; Zhang and Xia 2015; Ren et al. 2022; Wang et al. 2022). An LSS consists of a group of subsystems interconnected by their own states. Therefore, the dynamics of each subsystem are determined not only by the input and the initial state but also by the states of other subsystems. Because the coupling matrices between subsystems are usually unknown, the interactions among subsystems are difficult to access or estimate, which constitutes a major difficulty in controlling the LSSs. Hence, a decentralized design is favorable for LSSs. In a decentralized scheme, each subsystem is controlled based on its information (Zhang et al. 2022; Deng and Yang 2017; Moradvandi et al. 2019; Zhang and Xia 2015). This design elegantly prevents information exchange among subsystems. For example, an optimal estimation framework for LSSs with random transport protocols was successfully established (Ren et al. 2022). An alternative is the distributed scheme (Wang et al. 2022), in which the information of a subsystem and its neighbors is used to improve the control performance, which offers a trade-off between performance enhancement and information exchange burden.

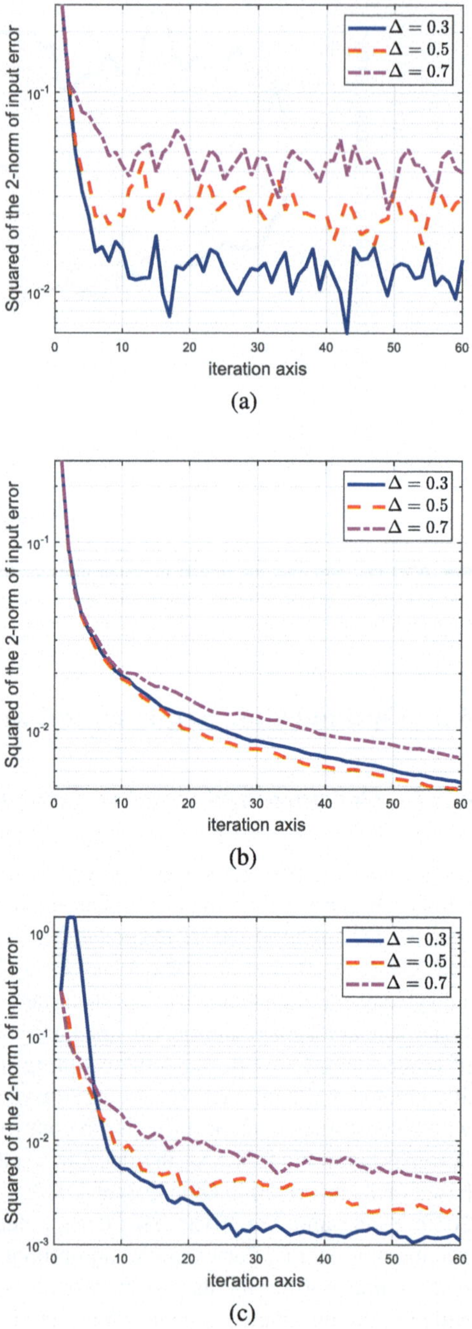

Fig. 6.8 The squared Euclidean norm profiles of the input errors in the iteration domain for three quantization density cases: **a** constant gain, **b** decreasing gain sequence, and **c** recursive gain sequence

Several analyses and design techniques have been proposed to realize high control performance of LSSs in the presence of various uncertainties and nonlinearities. In Karimi and Menhaj (2010); Sui et al. (2014); Tong et al. (2013); Huang and Wang (2014); Ma et al. (2022), approximation approaches, such as neural networks and fuzzy logic systems, were applied to approximate unknown nonlinearities, and backstepping control was applied to ensure stability and satisfactory control performance. In Peng et al. (2017), a decentralized event-triggering mechanism was introduced to solve the networked control problem of LSSs without requiring synchronous measurements. To address an LSS with various unknowns, such as nonlinear functions and dead-zones, and unmeasurable state variables, Sui et al. (2014) proposed a novel decentralized observer-based controller for LSSs, in which the subsystem dynamics were approximated by fuzzy logic systems. However, limited studies have been conducted on the learning control of LSSs because of the inherent characteristic that the interactions among subsystems are complex. Early attempts were made to apply reinforcement learning to certain complex LSSs, such as large industrial processes and traffic signal systems (Li et al. 2020; Wang et al. 2021c). Reinforcement learning control is an effective paradigm for achieving good performance in complex environments (Wang et al. 2021a, b, c).

In this section, we examine repetitive linear time-invariant systems that perform given tasks repeatedly within a finite time interval. Each subsystem is required to track a specified reference trajectory. Repetitive LTI systems are prevalent in industrial applications such as manufacturing processes and collaborative manipulators. For these types of systems, the tracking performance can be gradually improved through ILC as the number of iterations increases. Decentralized learning control for LSSs has been a long-standing issue but with very limited progress. An early attempt was made in Hwang et al. (1993), where a D-type learning scheme was proposed for deterministic continuous-time linear LSSs. The results were extended to systems without identical initialization conditions by introducing an initial state learning algorithm (Wu 2007). Subsequently, the case of non-repetitive reference trajectories was addressed to resolve the hierarchical steady-state optimization problem for linear large-scale industrial processes (Ruan et al. 2008, 2009). However, the aforementioned studies investigated deterministic continuous-time LSSs, where no uncertainty and randomness were involved. To address this issue, in Wang and Chien (2011), nonlinear LSSs subject to unknown uncertainties were studied using an adaptive learning control scheme based on a fuzzy neural learning component and a robust learning component. The presence of random noise in the system dynamics significantly distinguishes the design and analysis of the learning control scheme from existing results due to the coupling effect between interaction and random noise. Consequently, techniques applicable to deterministic or uncertain systems cannot be directly employed to handle randomness. In Shen and Chen (2012), a predefined decreasing gain sequence was introduced to mitigate the impact of random noise, thereby gradually improving tracking performance. This study represented an initial endeavor towards decentralized learning control for stochastic systems. However, if an inappropriate selection of the decreasing gain sequence is made, it can lead to a decrease in convergence speed. Therefore, there remains a need for determining methods that accelerate convergence speed.

In this section, we present an investigation of the learning control problem for LSSs in the presence of random noise in the measured output of each subsystem, rendering techniques for deterministic LSSs inapplicable. In contrast to Shen and Chen (2012), we propose a gain-adaptation mechanism to enhance convergence speed. Specifically, we introduce an index to quantify tracking performance at each iteration and calculate the recursive gain sequence by minimizing this index. We discuss the convergence of learning control schemes using both given and recursive gain sequences. Additionally, we improve the convergence rate practically by designing a suitable direction regulation matrix. Furthermore, we provide a decentralized version of the gain-adaptation mechanism to relax design conditions and specify interconnections among subsystems for novel design of decentralized recursive gain sequences. Moreover, we introduce a state-coupling matrix to describe interactions among subsystems.

6.3.2 Problem Formulation

A large-scale interconnected system consisting of n subsystems is considered, where the dynamics of the ith subsystem are given as

$$\begin{aligned} x_{k,i}(t+1) &= A_i x_{k,i}(t) + B_i u_{k,i}(t) + \sum_{j\neq i} A_{ij} x_{k,j}(t), \\ y_{k,i}(t) &= C_i x_{k,i}(t) + w_{k,i}(t), \end{aligned} \tag{6.60}$$

where t denotes the time index, $t = 0, \ldots, m$, m is the iteration length, k denotes the iteration number, $k = 1, 2, \ldots$, and i denotes the subsystem labels, $i = 1, \ldots, n$. The variables $x_{k,i}(t) \in \mathbf{R}^{o_i}$, $u_{k,i}(t) \in \mathbf{R}^{p_i}$, and $y_{k,i}(t) \in \mathbf{R}^{q_i}$ are the state, input, and output of the ith subsystem, respectively. A_i, B_i, and C_i are the ith system's matrices with appropriate dimensions. The last term in the state equation, that is, $\sum_{j\neq i} A_{ij} x_{k,j}(t)$, reflects the interconnection among these subsystems, where A_{ij} is the connection matrix linking the state of the jth subsystem to the ith subsystem. $w_{k,i}(t)$ denotes measurement noise.

In the system formulation, we consider time-invariant dynamics to maintain consistency in notations throughout this section. The findings of this section can be extended to encompass time-varying scenarios effortlessly. Moreover, for the sake of simplicity in derivations, we have omitted the consideration of state noise. However, if state noise is present in the dynamics, its impact on the output can be expressed as a linear combination of state noise terms originating from interconnected subsystems and can be treated as an integral part of the measurement noise $w_{k,i}(t)$.

We herein consider the decentralized learning tracking problem, where the output of each subsystem is required to track an individual trajectory, whereas only the input and tracking information of the ith subsystem are available for updating its input. The desired reference for the ith subsystem is denoted by $y_{d,i}(t)$. We denote a σ-algebra $\mathfrak{F}_{k,i}^{y} \triangleq \sigma\{u_{0,i}(t), x_{j,i}(0), y_{j,i}(t), 0 \leq t \leq T, 0 \leq j \leq k\}$, which indicates the available information set of the ith subsystem up to the kth iteration. Then, the *admissible control set* for the ith subsystem is defined as

$$U_i \triangleq \{u_{k+1,i}(t) \in \mathfrak{F}_{k,i}^{y}, \sup_k \|u_{k,i}(t)\| < \infty, \forall t, k\}. \tag{6.61}$$

Our control objective is to establish a decentralized learning control algorithm to determine the control sequence $\{u_{k,i}(t), k = 0, 1, \ldots\} \in U_i$, $\forall 1 \leq i \leq n$, such that the following tracking performance index is minimized for the entire large-scale system (6.60):

$$J_t = \sum_{i=1}^{n} J_{t,i} \tag{6.62}$$

where the performance index for the ith subsystem is defined as

$$J_{t,i} = \limsup_{k\to\infty} \frac{1}{k} \sum_{j=1}^{k} \|y_{d,i}(t) - y_{j,i}(t)\|^2, \quad i = 1, 2, \ldots, n. \tag{6.63}$$

To solve this problem, we require the following assumptions for the ith subsystem:

Assumption 6.5 The desired trajectory $y_{d,i}(t)$ is realizable for all subsystems, that is, for a suitable initial state $x_{d,i}(0)$, there exists $u_{d,i}(t)$ such that $\forall 1 \leq i \leq n$,

$$\begin{aligned} x_{d,i}(t+1) &= A_i x_{d,i}(t) + B_i u_{d,i}(t) + \sum_{j\neq i} A_{ij} x_{d,j}(t), \\ y_{d,i}(t) &= C_i x_{d,i}(t), \end{aligned} \tag{6.64}$$

where $x_{d,i}(t)$ is recursively generated by the subsystem dynamics.

The assumption made in this study establishes the feasibility of achieving the desired reference for each subsystem. This assumption serves as a technical tool to facilitate subsequent analyses in this section. If this assumption is found to be invalid, it implies that there exists no input that can simultaneously generate the desired reference for all subsystems. In such cases, one can construct the best achievable reference signals that are closest to the desired reference by minimizing their least square difference. In particular, we can evaluate the performance index (6.62) over the entire time interval, which represents a quadratic function of the input. Consequently, there must exist a minimum value even if it is not possible to precisely track the desired reference for all subsystems simultaneously. Under mild conditions, our proposed scheme remains applicable for achieving this minimum.

Assumption 6.6 For each subsystem, we assume that the input/output coupling matrix $C_i B_i$ is of full column rank.

Assumption 6.7 The initial state for each subsystem is assumed to be identical to the desired state, that is, $x_{k,i}(0) = x_{d,i}(0)$, $\forall k, i$.

Assumption 6.8 For each subsystem i and time instant t, the noise sequences $\{w_{k,i}(t), k = 0, 1, \ldots\}$ are stationary and independent of the iteration number. Moreover, $\mathbb{E}w_{k,i}(t) = 0$, $\sup_k \mathbb{E}\|w_{k,i}(t)\|^2 < \sigma_w$, and $\min_k \mathbb{P}(|z| \le \kappa) > 0$; here, z denotes an arbitrary entry for $w_{k,i}(t)$ and $\kappa > 0$ represents a positive constant.

Before proceeding to the learning control design, we first present a sufficient condition for the control sequence, to realize the control objective; in other words, the tracking performance index (6.62) is minimized for all subsystems.

Theorem 6.3 *Consider large-scale interconnected systems with dynamics of the ith subsystem defined by* (6.60)*, and Assumption 6.5–6.8 hold. We then evaluate the tracking performance index for each subsystem:*

$$J_{t,i} \ge R_{t,i}, \quad i = 1, 2, \ldots, n, \tag{6.65}$$

where

$$R_{t,i} \triangleq \limsup_{k\to\infty} \frac{1}{k} \sum_{j=1}^{k} [w_{j,i}(t)]^T [w_{j,i}(t)].$$

The inequality becomes equal if the admissible control sequences $\{u_{k,i}(t), 0 \le t \le T\}$, $i = 1, \ldots, n$, satisfy $u_{d,i}(t) - u_{k,i}(t) \to 0$ as $k \to \infty$, $\forall i, t$. Here, $u_{d,i}(t)$ is defined in Assumption 6.5. Consequently, $J_t \ge \sum_{i=1}^{n} R_{t,i}$. That is, $\sum_{i=1}^{n} R_{t,i}$ is the minimum of the performance index (6.62).

An admissible control sequence $u_{k,i}(t)$ is called an optimal learning control sequence if it satisfies $u_{d,i}(t) - u_{k,i}(t) \to 0$ as $k \to \infty$.

Proof The state and input errors are denoted as $\Delta x_{k,i} = x_{d,i}(t) - x_{k,i}(t)$ and $\Delta u_{k,i} = u_{d,i}(t) - u_{k,i}(t)$, respectively. Using the system dynamics (6.60) and the reference model (6.64), we obtain

$$\begin{aligned}\Delta x_{k,i}(t+1) &= A_i \Delta x_{k,i}(t) + B_i \Delta u_{k,i}(t) \\ &\quad + \sum_{j \ne i} A_{ij} \Delta x_{k,j}(t).\end{aligned} \tag{6.66}$$

Using Assumption 6.7, the initial state of each subsystem is identical to the desired state; thus, the difference between the output and reference at the initial time instant is actually the measurement noise, that is, $y_{d,i}(0) - y_{k,i}(0) = -w_{k,i}(0)$, $\forall i, k$. This conclusion holds for all subsystems. We prove this theorem using the induction principle with respect to the time label.

Initial Step: We prove the conclusion holds for $J_{1,i}$. From (6.66) and Assumption 6.7, we obtain

$$\begin{aligned}\Delta x_{k,i}(1) &= A_i \Delta x_{k,i}(0) + B_i \Delta u_{k,i}(0) + \sum_{j\neq i} A_{ij} \Delta x_{k,j}(0)\\ &= B_i \Delta u_{k,i}(0).\end{aligned}$$

Therefore, the tracking performance index is given by

$$\begin{aligned}J_{1,i} &= \limsup_{k\to\infty} \frac{1}{k}\sum_{j=1}^{k} \|C_i \Delta x_{j,i}(1) - w_{j,i}(1)\|^2\\ &= \limsup_{k\to\infty} \frac{1}{k}\sum_{j=1}^{k} (\|C_i B_i \Delta u_{j,i}(0)\|^2 + \|w_{j,i}(1)\|^2)\\ &\quad - 2\limsup_{k\to\infty} \frac{1}{k}\sum_{j=1}^{k} (\Delta u_{j,i}(0))^T (C_i B_i)^T w_{j,i}(1).\end{aligned}$$

From the definition of the admissible control set, for the crossing terms, we have

$$\begin{aligned}&\mathbb{E}[(\Delta u_{k,i}(0))^T (C_i B_i)^T w_{k,i}(1)]\\ &= \mathbb{E}\{\mathbb{E}[(\Delta u_{k,i}(0))^T (C_i B_i)^T w_{k,i}(1) \mid \mathfrak{F}^y_{k-1,i}]\}\\ &= 0.\end{aligned}$$

Then, using the law of large numbers, we obtain

$$\limsup_{k\to\infty} \frac{1}{k}\sum_{j=1}^{k} (\Delta u_{j,i}(0))^T (C_i B_i)^T w_{j,i}(1) = 0.$$

Consequently, $J_{1,i} \geq R_{1,i}$; here, $R_{1,i}$ is defined in the theorem, $i = 1, 2, \ldots, n$. If $\Delta u_{k,i}(0) \to 0$ as $k \to \infty$ for all subsystems simultaneously, we have $\Delta x_{k,i}(1) \to 0$ and $J_{1,i} = R_{1,i}$, $i = 1, 2, \ldots, n$. Thus, the conclusion of this theorem is valid.

Inductive Step: We assume that the conclusion holds for $J_{t,i}$ *with* $t = 1, \ldots, s-1$ *and prove its validity for* $J_{s,i}$. From the initial step, we obtained $\Delta x_{k,i}(1) = B_i \Delta u_{k,i}(0)$. Again, by using (6.66),

$$\begin{aligned}\Delta x_{k,i}(2) &= A_i \Delta x_{k,i}(1) + B_i \Delta u_{k,i}(1) + \sum_{j\neq i} A_{ij} \Delta x_{k,j}(1)\\ &= A_i B_i \Delta u_{k,i}(0) + B_i \Delta u_{k,i}(1)\\ &\quad + \sum_{j\neq i} A_{ij} B_j \Delta u_{k,j}(0).\end{aligned}$$

In other words, $\Delta x_{k,i}(2)$ is a linear combination of $\Delta u_{k,j}(0)$, $1 \leq j \leq n$, and $\Delta u_{k,i}(1)$. Similarly, by iterating regression (6.66), it is clear that $\Delta x_{k,i}(s)$ is a linear combination of $\Delta u_{k,j}(t)$, $1 \leq j \leq n$, $t = 0, \ldots, s-1$, and $\Delta u_{k,i}(s)$. The tracking performance index can then be rewritten as

$$\begin{aligned} J_{s,i} &= \limsup_{k\to\infty} \frac{1}{k}\sum_{j=1}^{k} \|C_i \Delta x_{j,i}(s) - w_{j,i}(s)\|^2 \\ &= \limsup_{k\to\infty} \frac{1}{k}\sum_{j=1}^{k} (\|C_i \Delta x_{j,i}(s)\|^2 + \|w_{j,i}(s)\|^2) \\ &\quad - 2\limsup_{k\to\infty} \frac{1}{k}\sum_{j=1}^{k} (\Delta x_{j,i}(s))^T C_i^T w_{j,i}(s), \end{aligned}$$

where $\Delta x_{k,i}(s)$ is a linear combination of the input error at time instants $t = 0, 1, \ldots, s-1$. Similar to the initial step, and by the law of large numbers, the crossing term in the above expansion is equal to zero. Therefore, $J_{s,i} \geq R_{s,i}$, $i = 1, 2, \ldots, n$. Furthermore, if $\Delta u_{k,i}(t) \to 0$ as $k \to \infty$ for all subsystems and time instants before s, we clearly have $\Delta x_{k,i}(t) \to 0$. Therefore, the lower bound of the tracking performance index is achieved, that is, $J_{s,i} = R_{s,i}$, $i = 1, 2, \ldots, n$. Consequently, we obtain $J_t \geq \sum_{i=1}^{n} R_{t,i}$. This completes this proof.

Theorem 6.3 presents the lower bound of the tracking performance index, which is determined by the measurement noise. In other words, if the output can be measured without any noise, it becomes possible to precisely track the desired reference signal, i.e., $J_{t,i} = 0$ or $y_{k,i}(t) \to y_{d,i}(t)$. This proof hints at two key kernels. Conversely, the inequality of $J_{t,i} \geq R_{t,i}$ is guaranteed by the independence between input and measurement noise in each iteration. Furthermore, leveraging induction principle enables us to establish equality between the tracking performance index and its lower bound by demonstrating asymptotic zero-oriented convergence for state error.

The decentralized tracking problem of large-scale interconnected systems is considered in this section. To evaluate the overall tracking performance of the interconnected systems, we integrate all subsystems' tracking performance indices into a single index, denoted as $J_t = \sum_{i=1}^{n} J_{t,i}$. Minimizing this index becomes our specific design objective, which can be achieved by generating an admissible control sequence that ensures $\Delta u_{k,i}(t) \to 0$ for all subsystem labels and time instants. In order to accelerate the diminishing speed of the input error in the presence of measurement noise, a decentralized gain-adaptation mechanism is proposed.

6.3.3 Optimal Gain-Adaptation and Control Scheme

In this section, we present the main findings in five subsections. Firstly, we propose a lifting transformation for the interconnected system to embed time-based dynamics and emphasize iteration-based evolution. Secondly, we introduce the first decentralized learning control scheme with a priori given decreasing sequence of positive real numbers as the learning gain, followed by strict proof of convergence. Thirdly, we incorporate a gain-adaptation mechanism to determine an optimal gain sequence based on practical convergence performance. Additionally, we provide a refined estimation of the decreasing speed for the recursive gain sequence. Fourthly, various issues related to evaluating convergence rate are discussed. Finally, we extend the design of gain adaptation to its decentralized version by revising the metric used for performance evaluation.

6.3.3.1 Lifting Formulation of an Interconnected System

To facilitate the design and analysis of the decentralized learning control scheme described in the subsequent sections, we initially transform the dynamics of each subsystem into a well-established lifting formulation. In this lifting formulation, the relevant variables are organized into a supervector, enabling construction of time-domain dynamics within the provided lifted matrix.

We revisit the subsystem dynamics in (6.60) and denote $A_{ii} \triangleq A_i$. For state $x_{k,i}(1)$, we have

$$x_{k,i}(1) = B_i u_{k,i}(0) + \sum_{j=1}^{n} A_{ij} x_{k,j}(0). \tag{6.67}$$

We proceed to state $x_{k,i}(2)$:

$$x_{k,i}(2) = B_i u_{k,i}(1) + \sum_{j=1}^{n} A_{ij} x_{k,j}(1), \tag{6.68}$$

and substituting (6.67) yields

$$\begin{aligned} x_{k,i}(2) &= B_i u_{k,i}(1) + \sum_{j=1}^{n} A_{ij} \left(B_j u_{k,j}(0) + \sum_{l=1}^{n} A_{jl} x_{k,l}(0) \right) \\ &= B_i u_{k,i}(1) + \sum_{j=1}^{n} A_{ij} B_j u_{k,j}(0) \\ &\quad + \sum_{l=1}^{n} \left(\sum_{j=1}^{n} A_{ij} A_{jl} \right) x_{k,l}(0) \\ &= B_i u_{k,i}(1) + \sum_{j=1}^{n} A_{ij} B_j u_{k,j}(0) + \sum_{j=1}^{n} A_{ij}^{*,1} x_{k,j}(0), \end{aligned} \tag{6.69}$$

where $A_{ij}^{*,2} \triangleq \sum_{l=1}^{n} A_{il} A_{lj}$ denotes the *two-step state-coupling matrix* (SCM) from the jth subsystem to the ith subsystem. For consistent understanding, we can denote $A_{ij}^{*,1} \triangleq A_{ij}$ as the *one-step SCM*.

By iterating the above process for an arbitrary state $x_{k,i}(s+1)$, we obtain

$$
\begin{aligned}
x_{k,i}(s+1) &= B_i u_{k,i}(s) + \sum_{j=1}^{n} A_{ij} x_{k,j}(s) \\
&= B_i u_{k,i}(s) + \sum_{j=1}^{n} A_{ij} \left(B_j u_{k,j}(s-1) + \sum_{l=1}^{n} A_{jl} x_{k,l}(s-1) \right) \\
&= B_i u_{k,i}(s) + \sum_{j=1}^{n} A_{ij} B_j u_{k,j}(s-1) \\
&\quad + \sum_{l=1}^{n} \Big(\sum_{j=1}^{n} A_{ij} A_{jl} \Big) \left(B_l u_{k,l}(s-2) + \sum_{h=1}^{n} A_{lh} x_{k,h}(s-2) \right) \\
&= B_i u_{k,i}(s) + \sum_{j=1}^{n} A_{ij}^{*,1} B_j u_{k,j}(s-1) \\
&\quad + \sum_{j=1}^{n} A_{ij}^{*,2} B_j u_{k,j}(s-2) + \cdots + \sum_{j=1}^{n} A_{ij}^{*,s} B_j u_{k,j}(0) \\
&\quad + \sum_{j=1}^{n} A_{ij}^{*,s+1} x_{k,j}(0),
\end{aligned}
\tag{6.70}
$$

where $A_{ij}^{*,h+1} \triangleq \sum_{1 \le l_1, l_2, \ldots, l_h \le n} A_{il_1} A_{l_1 l_2} \ldots A_{l_h j}$ denotes the $(h+1)$-*multi-step SCM* from the jth subsystem to the ith subsystem, $\forall h$.

The SCM can be recursively defined as follows:

$$
\begin{aligned}
A_{ij}^{*,1} &= A_{ij}, \quad \forall i, j \\
A_{ij}^{*,h} &= \sum_{l=1}^{n} A_{il}^{*,h-1} A_{lj} = \sum_{l=1}^{n} A_{il} A_{lj}^{*,h-1}.
\end{aligned}
$$

Remark 6.4 The effect of other subsystems in the interconnected system is described by introducing a novel concept called SCM. Specifically, the SCM is determined by two factors: the labels assigned to the ordered subsystems and the time instant difference. Taking $A_{ij}^{*,h}$ as an example, the ordered labels indicate that the effect is from the jth subsystem and on the ith subsystem, whereas the superscript h indicates that the effect comes from h time instants before. The calculation of $A_{ij}^{*,h}$ involves n^{h-1} possible paths that can transfer the effect of the jth subsystem to the ith subsystem according to the definition. Specifically, $A_{ii}^{*,h}$ is significantly different from the power of A_i in a traditional state-space system. To illustrate this point, we take $h = 2$ as an example and note that $A_{ii} \triangleq A_i$. In a traditional system, the state transition matrix is described by A_i^2 for $h = 2$. For the SCM defined here, $A_{ii}^{*,2} = A_i^2 + \sum_{l\neq i} A_{il}A_{li}$, where the crossing term $\sum_{l\neq i} A_{il}A_{li}$ represents the output of the coupling effect from the ith subsystem to the other subsystems and then back to itself. The amount of possible transition paths increases exponentially with an increase in the number h.

Now, we lift the input and output with respect to the subsystem label for an arbitrary time instant t:

$$
\begin{aligned}
\boldsymbol{u}_k(t) &= [u_{k,1}^T(t), u_{k,2}^T(t), \ldots, u_{k,n}^T(t)]^T \in \mathbf{R}^{\sum_{i=1}^n p_i},\\
\boldsymbol{y}_k(t) &= [y_{k,1}^T(t), y_{k,2}^T(t), \ldots, y_{k,n}^T(t)]^T \in \mathbf{R}^{\sum_{i=1}^n q_i}.
\end{aligned}
$$

Then, from (6.70) and the output equation in system (6.60), we obtain

$$
\boldsymbol{y}_k(t^+) = \mathcal{B}\boldsymbol{u}_k(t) + \sum_{s=1}^{t} \mathcal{D}_s \boldsymbol{u}_k(t-s) + \mathcal{A}_t \boldsymbol{x}_k^0 + \boldsymbol{w}_k(t^+), \tag{6.71}
$$

where $t^+ \triangleq t+1$, $\boldsymbol{w}_k(t) = [w_{k,1}^T(t), w_{k,2}^T(t), \ldots, w_{k,n}^T(t)]^T$,

$$
\begin{aligned}
\mathcal{B} &= \mathrm{diag}\{C_1B_1, C_2B_2, \ldots, C_nB_n\},\\
\mathcal{D}_s &= \begin{bmatrix} C_1A_{11}^{*,s}B_1 & C_1A_{12}^{*,s}B_2 & \cdots & C_1A_{1n}^{*,s}B_n \\ C_2A_{21}^{*,s}B_1 & C_2A_{22}^{*,s}B_2 & \cdots & C_2A_{2n}^{*,s}B_n \\ \vdots & \vdots & \ddots & \vdots \\ C_nA_{n1}^{*,s}B_1 & C_nA_{n2}^{*,s}B_2 & \cdots & C_nA_{nn}^{*,s}B_n \end{bmatrix},\\
& s = 1, 2, \ldots, t,\\
\mathcal{A}_t &= \begin{bmatrix} C_1A_{11}^{*,t+1} & C_1A_{12}^{*,t+1} & \cdots & C_1A_{1n}^{*,t+1} \\ C_2A_{21}^{*,t+1} & C_2A_{22}^{*,t+1} & \cdots & C_2A_{2n}^{*,t+1} \\ \vdots & \vdots & \ddots & \vdots \\ C_nA_{n1}^{*,t+1} & C_nA_{n2}^{*,t+1} & \cdots & C_nA_{nn}^{*,t+1} \end{bmatrix}.
\end{aligned}
$$

This relationship (6.71) describes the dependence of the output on the multiple inputs of all subsystems involved. We then lift the input, output, and measurement noise with respect to the time label, i.e.,

$$
\begin{aligned}
u_k &= [\boldsymbol{u}_k^T(0), \boldsymbol{u}_k^T(1), \ldots, \boldsymbol{u}_k^T(m-1)]^T,\\
y_k &= [\boldsymbol{y}_k^T(1), \boldsymbol{y}_k^T(2), \ldots, \boldsymbol{y}_k^T(m)]^T,\\
w_k &= [\boldsymbol{w}_k^T(1), \boldsymbol{w}_k^T(2), \ldots, \boldsymbol{w}_k^T(m)]^T.
\end{aligned}
$$

Then, we obtain

$$\mathbf{y}_k = \overline{\mathcal{B}}\mathbf{u}_k + \overline{\mathcal{A}}\mathbf{x}_k^0 + \mathbf{w}_k, \tag{6.72}$$

where

$$\overline{\mathcal{B}} = \begin{bmatrix} \mathcal{B} & 0 & 0 & \cdots & 0 \\ \mathcal{D}_1 & \mathcal{B} & 0 & \cdots & 0 \\ \mathcal{D}_2 & \mathcal{D}_1 & \mathcal{B} & \cdots & 0 \\ \vdots & \vdots & \vdots & \ddots & \vdots \\ \mathcal{D}_{m-1} & \mathcal{D}_{m-2} & \mathcal{D}_{m-3} & \cdots & \mathcal{B} \end{bmatrix},$$
$$\overline{\mathcal{A}} = \left[\mathcal{A}_0^T, \mathcal{A}_1^T, \ldots, \mathcal{A}_{m-1}^T\right]^T,$$
$$\mathbf{x}_k^0 = [x_{k,1}^T(0), x_{k,2}^T(0), \ldots, x_{k,n}^T(0)]^T.$$

Clearly, $\overline{\mathcal{B}}$ is a block lower-triangular matrix with diagonal blocks $C_i B_i, i = 1, \ldots, n$.

Similarly, we can lift the desired dynamics of the reference given in Assumption 6.5. In particular, we obtain

$$\mathbf{y}_d = \overline{\mathcal{B}}\mathbf{u}_d + \overline{\mathcal{A}}\mathbf{x}_d^0, \tag{6.73}$$

where

$$\mathbf{y}_d = [\mathbf{y}_d^T(1), \mathbf{y}_d^T(2), \ldots, \mathbf{y}_d^T(m)]^T,$$
$$\mathbf{y}_d(t) = [y_{d,1}^T(t), y_{d,2}^T(t), \ldots, y_{d,n}^T(t)]^T,$$
$$\mathbf{u}_d = [\mathbf{u}_d^T(0), \mathbf{u}_d^T(1), \ldots, \mathbf{u}_d^T(m-1)]^T,$$
$$\mathbf{u}_d(t) = [u_{d,1}^T(t), u_{d,2}^T(t), \ldots, u_{d,n}^T(t)]^T.$$

6.3.3.2 Decentralized Learning Control Scheme

For large-scale interconnected systems, we first present the following decentralized learning control scheme, which is borrowed from the traditional P-type learning control law (Shen and Chen 2012):

$$u_{k+1,i}(t) = u_{k,i}(t) + \rho_k L_i e_{k,i}(t+1), \tag{6.74}$$

where ρ_k denotes the learning gain, L_i denotes the direction-regulation matrix, and $e_{k,i}(t) \triangleq y_{d,i}(t) - y_{k,i}(t)$ denotes the tracking error of the ith subsystem. The learning gain sequence $\{\rho_k\}$ is introduced here to mitigate the impact of measurement noise after sufficient learning processes, enabling the input sequence to converge towards a limit. This matter will be extensively discussed in the subsequent subsection and further enhanced in subsequent subsections.

Using the lifted formulation in Subsection 6.3.3.1, we obtain

$$\boldsymbol{u}_{k+1}(t) = \boldsymbol{u}_k(t) + \rho_k \mathcal{L} \boldsymbol{e}_k(t+1), \tag{6.75}$$

where $\mathcal{L} = \mathrm{diag}\{L_1, L_2, \ldots, L_n\}$ and $\boldsymbol{e}_k(t) = \boldsymbol{y}_d(t) - \boldsymbol{y}_k(t)$. Further, we obtain

$$\boldsymbol{u}_{k+1} = \boldsymbol{u}_k + \rho_k \overline{\mathcal{L}} \boldsymbol{e}_k, \tag{6.76}$$

where $\overline{\mathcal{L}} = I_m \otimes \mathcal{L}$, $\otimes$ denotes the Kronecker product, and $\boldsymbol{e}_k = \boldsymbol{y}_d - \boldsymbol{y}_k$.

For this decentralized learning control (6.74), we have the following convergence property:

Theorem 6.4 *Consider the large-scale interconnected system* (6.60) *under Assumption 6.5–6.8, and apply the decentralized learning control scheme* (6.74)*. If the direction-regulation matrix L_i satisfies that all eigenvalues of $L_i C_i B_i$ have positive real parts for $i = 1, 2, \ldots, n$ and the learning gain sequence $\{\rho_k\}$ satisfies $\rho_k > 0$, $\rho_k \to 0$, and $\sum_{k=1}^{\infty} \rho_k = \infty$, the generated input sequence $\{\boldsymbol{u}_k\}$ converges to the desired input $\boldsymbol{u}_d$ in the mean-square sense; that is,* $\mathbb{E}[\|\boldsymbol{u}_d - \boldsymbol{u}_k\|^2] \xrightarrow[k\to\infty]{} 0$.

Proof From (6.72), (6.73), and Assumption 6.7, we have

$$\boldsymbol{e}_k = \overline{\mathcal{B}}(\boldsymbol{u}_d - \boldsymbol{u}_k) - \boldsymbol{w}_k. \tag{6.77}$$

Substituting (6.77) into (6.76) leads to

$$\boldsymbol{u}_{k+1} = \boldsymbol{u}_k + \rho_k \overline{\mathcal{L}\mathcal{B}}(\boldsymbol{u}_d - \boldsymbol{u}_k) - \rho_k \boldsymbol{w}_k. \tag{6.78}$$

By subtracting both sides of (6.78) from $\boldsymbol{u}_d$ and denoting $\Delta\boldsymbol{u}_k = \boldsymbol{u}_d - \boldsymbol{u}_k$, we have

$$\Delta\boldsymbol{u}_{k+1} = \Delta\boldsymbol{u}_k - \rho_k \overline{\mathcal{L}\mathcal{B}} \Delta\boldsymbol{u}_k + \rho_k \boldsymbol{w}_k. \tag{6.79}$$

Note that $\overline{\mathcal{B}}$ is a block lower-triangular matrix with diagonal blocks $C_i B_i$, and $\overline{\mathcal{L}}$ is a block diagonal matrix with the diagonal blocks being L_i, $i = 1, \ldots, n$. Thus, $\overline{\mathcal{L}\mathcal{B}}$ is a block lower-triangular matrix in which the diagonal blocks are $L_i C_i B_i$. Thus, the eigenvalues of $\overline{\mathcal{L}\mathcal{B}}$ are combinations of the eigenvalues of $L_i C_i B_i$ for $i = 1, \ldots, n$. Because the eigenvalues of $L_i C_i B_i$ have positive real parts, $\forall i$, all eigenvalues of the large-scale matrix $\overline{\mathcal{L}\mathcal{B}}$ have positive real parts. According to the Lyapunov stability theory (Boyd et al. 1994), for any given positive-definite matrix $\mathcal{S}$, there exists a unique positive-definite matrix $\mathcal{Q}$ such that

$$(\overline{\mathcal{L}\mathcal{B}})^T \mathcal{Q} + \mathcal{Q}(\overline{\mathcal{L}\mathcal{B}}) = \mathcal{S}. \tag{6.80}$$

The following Lyapunov function of the input error for the kth iteration is introduced:

$$\mathbf{V}_k = (\Delta \boldsymbol{u}_k)^T \mathcal{Q}(\Delta \boldsymbol{u}_k). \tag{6.81}$$

Substituting (6.79) into the expression for $\mathbf{V}_{k+1}$, we obtain

$$\begin{aligned}\mathbf{V}_{k+1} &= (\Delta \boldsymbol{u}_{k+1})^T \mathcal{Q}(\Delta \boldsymbol{u}_{k+1})\\ &= \mathbf{V}_k + \rho_k^2 (\Delta \boldsymbol{u}_k)^T (\overline{\mathcal{LB}})^T \mathcal{Q}(\overline{\mathcal{LB}})(\Delta \boldsymbol{u}_k) + \rho_k^2 \boldsymbol{w}_k^T \mathcal{Q} \boldsymbol{w}_k\\ &\quad - \rho_k (\Delta \boldsymbol{u}_k)^T [(\overline{\mathcal{LB}})^T \mathcal{Q} + \mathcal{Q}(\overline{\mathcal{LB}})](\Delta \boldsymbol{u}_k)\\ &\quad + 2\rho_k (\Delta \boldsymbol{u}_k)^T (I - \rho_k \overline{\mathcal{LB}})^T \mathcal{Q} \boldsymbol{w}_k.\end{aligned}$$

We introduce a new σ-algebra $\mathfrak{F}_k \triangleq \sigma\{u_{0,i}(t), x_{j,i}(0), w_{j,i}(t), 0 \le t \le T, 0 \le j \le k, 1 \le i \le n\}$, which indicates the information available for the entire large-scale interconnected system (Chow and Teicher 2003). From (6.74), it is evident that $\boldsymbol{u}_k \in \mathfrak{F}_k$. Taking the conditional mathematical expectation on the last equality with respect to the σ-algebra $\mathfrak{F}_{k-1}$, we analyze all the terms as follows.

First, it is evident that $\mathbb{E}[\mathbf{V}_k \mid \mathfrak{F}_{k-1}] = \mathbf{V}_k$. Similarly, we have

$$\begin{aligned}&\mathbb{E}[(\Delta \boldsymbol{u}_k)^T (\overline{\mathcal{LB}})^T \mathcal{Q}(\overline{\mathcal{LB}})(\Delta \boldsymbol{u}_k) \mid \mathfrak{F}_{k-1}]\\ &= (\Delta \boldsymbol{u}_k)^T (\overline{\mathcal{LB}})^T \mathcal{Q}(\overline{\mathcal{LB}})(\Delta \boldsymbol{u}_k)\\ &\le c_1 \mathbf{V}_k,\end{aligned} \tag{6.82}$$

where $c_1 > 0$ is a constant such that $(\overline{\mathcal{LB}})^T \mathcal{Q}(\overline{\mathcal{LB}}) \le c_1 \mathcal{Q}$.

From Assumption 6.8, $w_{k,i}(t)$ is independent of $\mathfrak{F}_{k-1}$. Thus, we have

$$\begin{aligned}&\mathbb{E}[\boldsymbol{w}_k^T \mathcal{Q} \boldsymbol{w}_k \mid \mathfrak{F}_{k-1}]\\ &\le \lambda_{\max}(\mathcal{Q}) \mathbb{E}[\boldsymbol{w}_k^T \boldsymbol{w}_k]\\ &\le \lambda_{\max}(\mathcal{Q}) nm\sigma_w \triangleq b_w,\end{aligned} \tag{6.83}$$

where $\lambda_{\max}(\cdot)$ denotes the largest eigenvalue of the involved matrix and Assumption 6.8 is applied to the last estimation.

From (6.80), we have

$$\begin{aligned}&\mathbb{E}[(\Delta \boldsymbol{u}_k)^T [(\overline{\mathcal{LB}})^T \mathcal{Q} + \mathcal{Q}(\overline{\mathcal{LB}})](\Delta \boldsymbol{u}_k) \mid \mathfrak{F}_{k-1}]\\ &= (\Delta \boldsymbol{u}_k)^T \mathcal{S}(\Delta \boldsymbol{u}_k)\\ &\ge c_2 \mathbf{V}_k,\end{aligned} \tag{6.84}$$

where $c_2 > 0$ denotes a constant such that $\mathcal{S} \ge c_2 \mathcal{Q}$. The existence of c_2 was guaranteed by the positive-definiteness of $\mathcal{S}$ and $\mathcal{Q}$.

Again, using Assumption 6.8, we obtain

$$\begin{aligned}&\mathbb{E}[(\Delta \boldsymbol{u}_k)^T (I - \rho_k \overline{\mathcal{L}\mathcal{B}})^T \mathcal{Q} \boldsymbol{w}_k \mid \mathfrak{F}_{k-1}]\\&= (\Delta \boldsymbol{u}_k)^T (I - \rho_k \overline{\mathcal{L}\mathcal{B}})^T \mathcal{Q} \mathbb{E}[\boldsymbol{w}_k \mid \mathfrak{F}_{k-1}]\\&= 0.\end{aligned} \tag{6.85}$$

Substituting (6.82)-(6.85) into $\mathbb{E}[\mathbf{V}_{k+1} \mid \mathfrak{F}_{k-1}]$, we obtain

$$\mathbb{E}[\mathbf{V}_{k+1} \mid \mathfrak{F}_{k-1}] \leq (1 - c_2\rho_k + c_1\rho_k^2)\mathbf{V}_k + \rho_k^2 b_w. \tag{6.86}$$

Then, taking the mathematical expectation yields

$$\mathbb{E}[\mathbf{V}_{k+1}] \leq (1 - c_2\rho_k + c_1\rho_k^2)\mathbb{E}[\mathbf{V}_k] + \rho_k^2 b_w. \tag{6.87}$$

The remainder of the proof is completed by applying Lemma 2.1. To this end, we correspond a_k, b_k, and c_k to $\mathbb{E}[\mathbf{V}_k]$, $c_2\rho_k - c_1\rho_k^2$, and $\rho_k^2 b_w$, respectively. Subsequently, we verify the conditions in Lemma 2.1.

Because $\rho_k > 0$ and $\rho_k \to 0$, $0 < c_2\rho_k - c_1\rho_k^2 < 1$ after a sufficient number of iterations, and $c_2\rho_k - c_1\rho_k^2 \to 0$. Moreover, $\sum_{k=1}^{\infty}(c_2\rho_k - c_1\rho_k^2) = \infty$. Additionally, $\frac{\rho_k^2 b_w}{c_2\rho_k - c_1\rho_k^2} = O(\rho_k) \to 0$. In summary, all conditions of Lemma 2.1 are satisfied. Therefore, we have $\mathbb{E}[\mathbf{V}_k] \to 0$, as iteration number k approaches infinity. Note that $\mathcal{Q}$ is a positive-definite matrix. Evidently, $\mathbb{E}[\|\boldsymbol{u}_d - \boldsymbol{u}_k\|^2] \leq \frac{1}{\lambda_{\max}(\mathcal{Q})}\mathbb{E}[\mathbf{V}_k] \to 0$. The proof is completed.

Theorem 6.4 characterizes the convergence property of the decentralized learning control scheme (6.74), where a priori learning gain sequence $\{\rho_k\}$ is selected independently of specific operations. The essential requirement for the learning gain is that it satisfies $\rho_k > 0$, $\rho_k \to 0$, and $\sum_{k=1}^{\infty} \rho_k = \infty$. While Theorem 6.4 provides asymptotic convergence of the proposed learning control scheme, it only describes the limit of the input sequence without revealing its convergence rate. For practical applications, an intuitive question arises as to whether there exists an optimal selection of ρ_k that can improve transient performance of the learning control scheme. This problem will be addressed in the following subsection.

Remark 6.5 The condition for the direction regulation matrix L_i is that all eigenvalues of $L_i C_{i+1} B_i$ have positive real parts for all subsystems. Based on Assumption 6.6, a suitable candidate for L_i is $\tau(C_{i+1}B_i)^T$, where $\tau > 0$ is a tunable parameter. By estimating the coupling matrix $C_{i+1}B_i$, which can be achieved without requiring precise system information, this condition can be ensured to guarantee convergence due to the continuity of the eigenvalue function with respect to each entry of the matrix (Horn and Johnson 1985; Butcher et al. 2008). However, if Assumption 6.5 is invalid and the desired reference cannot be realized by all subsystems simultaneously, this information becomes essential in order to achieve optimal tracking performance in terms of least squares sense (Shen et al. 2021).

6.3.3.3 Recursive Gain Design

Resolving the optimal gain selection problem is challenging due to its dependence on various factors, such as system information and operational conditions. In this subsection, we propose a promising alternative for enhancing convergence performance by selecting an optimal gain sequence based on a specific metric. Therefore, it is essential to establish a metric that effectively describes convergence performance while being easily implementable and computationally efficient. In other words, the metric should provide accessible information and require minimal computational resources.

A natural candidate is the mean-squared input error, $\mathbb{E}[\|\boldsymbol{u}_d - \boldsymbol{u}_k\|^2]$, because it measures the distance between the generated input and the desired input. However, it is difficult to establish a direct regression for this metric owing to the design of the direction regulation matrix L_i. Instead, from (6.87), we observe that $\mathbb{E}[\mathbf{V}_k]$ can serve as an alternative metric, which is straightforward to compute.

In particular, we revisit recursion (6.87) to obtain

$$\mathbb{E}[\mathbf{V}_{k+1}] \leq (1 - \frac{c_2}{2}\rho_k)\mathbb{E}[\mathbf{V}_k] + \rho_k^2 b_w \tag{6.88}$$

under the condition $c_1\rho_k \leq \frac{c_2}{2}$. This condition is valid when ρ_k is sufficiently small in practice.

It is difficult to derive a precise relationship between the current iteration $\mathbb{E}[\mathbf{V}_k]$ and the next iteration $\mathbb{E}[\mathbf{V}_{k+1}]$. However, we can regard the right-hand side of (6.88) as a computation metric because it corresponds to the upper bound of $\mathbb{E}[\mathbf{V}_{k+1}]$. Denote a function of the gain ρ_k: $h(\rho_k) \triangleq (1 - \frac{c_2}{2}\rho_k)\mathbb{E}[\mathbf{V}_k] + \rho_k^2 b_w$. We derive the recursive gain sequence by optimizing the function $h(\rho_k)$:

$$\frac{\partial h(\rho_k)}{\rho_k} = -\frac{c_2}{2}\mathbb{E}[\mathbf{V}_k] + 2\rho_k b_w. \tag{6.89}$$

Letting this derivative equal to zero yields

$$\rho_k^* = \frac{c_2}{4b_w}\mathbb{E}[\mathbf{V}_k]. \tag{6.90}$$

The derived gain is dependent on parameter c_2, performance of the current iteration $\mathbb{E}[\mathbf{V}_k]$, and noise-related term b_w. This dependence reveals the considerations for gain-adaptation in practice.

We now present a recursive computation for ρ_k^*. Substituting (6.90) into function $h(\rho_k)$ yields

$$
\begin{aligned}
h(\rho_k^*) &= (1 - \frac{c_2}{2}\rho_k^*)\mathbb{E}[\mathbf{V}_k] + (\rho_k^*)^2 b_w \\
&= (1 - \frac{c_2}{2}\rho_k^*)\frac{4b_w\rho_k^*}{c_2} + (\rho_k^*)^2 b_w \\
&= \frac{4b_w\rho_k^*}{c_2} - (\rho_k^*)^2 b_w.
\end{aligned}
$$

To derive the recursive formulation, we consider the worst case of $\mathbb{E}[\mathbf{V}_{k+1}]$ as the minimization of the Lyapunov function, that is, $\mathbb{E}[V_{k+1}] = h(\rho_k^*)$. Then, we have

$$
\begin{aligned}
\rho_{k+1}^* &= \frac{c_2}{4b_w}\mathbb{E}[\mathbf{V}_{k+1}] \\
&= \frac{c_2}{4b_w}\left(\frac{4b_w\rho_k^*}{c_2} - (\rho_k^*)^2 b_w\right) \\
&= \rho_k^*\left(1 - \frac{c_2}{4}\rho_k^*\right).
\end{aligned}
\tag{6.91}
$$

The initial value of the gain sequence is $\rho_1^* = (c_2\mathbb{E}[\mathbf{V}_1]/4b_w)$.

To determine the optimality of the recursive gain, we present the following position by checking the special case of (6.88), wherein equality holds, that is, $\mathbf{L}_k = \mathbb{E}[\mathbf{V}_k]$.

Proposition 6.3 *Consider the following recursion*

$$
\boldsymbol{L}_{k+1} = (1 - \frac{c_2}{2}\rho_k)\boldsymbol{L}_k + \rho_k^2 b_w. \tag{6.92}
$$

If the initial value $\mathbf{L}_1$ *satisfies* $\frac{c_2\mathbb{E}[\mathbf{V}_1]}{4b_w} \leq \frac{4}{c_2}$, *then, for any gain sequence* ρ_k, *such that* $0 < \rho_k \leq (4/c_2)$, *the generated output* $\mathbf{L}_k$ *satisfies* $\mathbf{L}_k \geq \mathbf{L}_k^*$, *where* $\mathbf{L}_k^*$ *is the corresponding value of the recursive gain sequence* $\{\rho_k^*\}$ *defined by* (6.91).

Proof This proof applies the principle of mathematical induction. Regardless of the gain sequence applied, the recursion starts from $\mathbf{L}_1$, that is, $\mathbf{L}_1 = \mathbf{L}_1^*$. Thus, the initial step of mathematical induction starts from $k = 2$. For this case, we have

$$
\mathbf{L}_2^* = h(\rho_1^*) = \frac{4b_w\rho_1^*}{c_2} - (\rho_1^*)^2 b_w.
$$

For arbitrary gain ρ_1, we have

$$
\begin{aligned}
\mathbf{L}_2 &= (1 - \frac{c_2}{2}\rho_1)\mathbf{L}_1 + \rho_1^2 b_w \\
&= (1 - \frac{c_2}{2}\rho_1)\mathbf{L}_1^* + \rho_1^2 b_w \\
&= \frac{4b_w\rho_1^*}{c_2} - 2b_w\rho_1^*\rho_1 + \rho_1^2 b_w.
\end{aligned}
$$

where $\rho_1^* = \frac{c_2}{4b_w}\mathbf{L}_1^*$. Then, we have

$$\mathbf{L}_2 - \mathbf{L}_2^* = b_w(\rho_1 - \rho_1^*)^2 \geq 0.$$

Next, we proceed to the induction step. In other words, assuming that $\mathbf{L}_k \geq \mathbf{L}_k^*$ holds for $k = s$, that is, $\mathbf{L}_s \geq \mathbf{L}_s^*$, we prove its validity for $k = s + 1$. To this end, we have

$$\begin{aligned}
\mathbf{L}_{s+1} &= (1 - \frac{c_2}{2}\rho_s)\mathbf{L}_s + \rho_s^2 b_w, \\
\mathbf{L}_{s+1}^* &= \frac{4b_w}{c_2}\rho_{s+1}^* = \frac{4b_w}{c_2}\rho_s^*\left(1 - \frac{c_2}{4}\rho_s^*\right).
\end{aligned}$$

Then, we obtain

$$\begin{aligned}
&\mathbf{L}_{s+1} - \mathbf{L}_{s+1}^* \\
&= (1 - \frac{c_2}{2}\rho_s)\mathbf{L}_s + \rho_s^2 b_w - \frac{4b_w}{c_2}\rho_s^*\left(1 - \frac{c_2}{4}\rho_s^*\right) \\
&\geq (1 - \frac{c_2}{2}\rho_s)\mathbf{L}_s^* + \rho_s^2 b_w - \frac{4b_w}{c_2}\rho_s^*\left(1 - \frac{c_2}{4}\rho_s^*\right) \\
&= (1 - \frac{c_2}{2}\rho_s)\frac{4b_w\rho_s^*}{c_2} + \rho_s^2 b_w - \frac{4b_w}{c_2}\rho_s^*\left(1 - \frac{c_2}{4}\rho_s^*\right) \\
&= b_w(\rho_s - \rho_s^*)^2 \geq 0.
\end{aligned}$$

The proof is completed using the induction principle.

Proposition 6.3 demonstrates the optimality of the recursive gain ρ_k^* generated by (6.91), as it minimizes the worst case of the Lyapunov function. Moreover, this gain sequence adapts to both practical Lyapunov functions and practical input errors, justifying its designation as the recursive gain. However, it remains unclear whether this gain sequence guarantees asymptotical convergence of the learning control scheme. To establish the effectiveness of this gain sequence, it suffices to demonstrate that $\{\rho_k^*\}$ satisfies the conditions outlined in Theorem 6.4. This is verified by the following lemma:

Lemma 6.3 *Consider the following recursion:*

$$\gamma_k = \gamma_{k-1}(1 - \delta\gamma_{k-1}), \tag{6.93}$$

where $\delta > 0$ is a constant and the initial value γ_0 satisfies $0 < \gamma_0 < (1/\delta)$. Then, we obtain $\gamma_k > 0$, $\gamma_k \to 0$, and $\gamma_k \sim k^{-1}\delta^{-1}$.

Remark 6.6 Theorem 6.4 provides conditions for the convergence of a decreasing gain sequence, while Lemma 6.3 offers a rate estimation for the proposed recursive gain sequence that satisfies the optimality criterion presented in Shen and Chen (2012). By comparing these two results, several observations can be made. Firstly, from Lemma 6.3, it is evident that the recursive gain sequence ρ_k^* satisfies the general conditions outlined in Theorem 6.4, making it a special case of the latter. Secondly, unlike the general gain sequence, which remains fixed throughout iterations, the recursive gain sequence is generated by optimizing performance metrics and adapts to practical convergence requirements. Lastly, it should be noted that the rate estimation provided in Lemma 6.3 is asymptotic and holds true as the number of iterations approaches infinity; thus, it does not imply equivalence between early iteration recursive gains and $k^{-1}\delta^{-1}$.

From Lemma 6.3, we note that the recursive gain ρ_k^* generated by (6.91) is positive, approaching zero, and of the order $(4/kc_2)$. Therefore, $\sum_{k=1}^{\infty}\rho_k^* = \infty$, that is, the conditions in Theorem 6.4 are satisfied. Combining these observations, we have the following convergence theorem for recursive gain.

Theorem 6.5 *Consider the large-scale interconnected system* (6.60) *under Assumption 6.5–6.8, and apply the learning control scheme:*

$$u_{k+1,i}(t) = u_{k,i}(t) + \rho_k^* L_i e_{k,i}(t+1), \tag{6.94}$$

where ρ_k^* *is generated by recursion* (6.91) *with the initial value* $\rho_1^* = \frac{c_2\mathbb{E}[V_1]}{4b_w}$. *If the direction regulation matrix* L_i *satisfies all eigenvalues of* $L_iC_iB_i$ *having positive real parts for* $i = 1, 2, \ldots, n$ *and the recursive gain sequence* $\{\rho_k^*\}$ *is generated by* (6.91), *the generated input sequence* $\{\boldsymbol{u}_k\}$ *converges to the desired input* $\boldsymbol{u}_d$ *in the mean-square sense, that is,* $\mathbb{E}[\|\boldsymbol{u}_d - \boldsymbol{u}_k\|^2] \xrightarrow[k\to\infty]{} 0$.

6.3.3.4 Convergence Rate Evaluation

From Subsections 6.3.3.2 and 6.3.3.3 we notice that the convergence rate of the proposed scheme is reflected by the decreasing gain sequence and constant c_2. In this subsection, we provide some evaluations for practical guidelines.

The asymptotic convergence rate is determined by the decreasing gain sequence $\{\rho_k\}$, where the asymptotic convergence rate indicates that the order of the mean-squared input error vanishes to zero. As a general convergence result, Theorem 6.4 shows that the necessary condition of the gain sequence is to ensure an infinity-sum condition; that is, $\sum_{k=1}^{\infty}\rho_k = \infty$. For the specific selection of ρ_k^*, which corresponds to a certain optimality of the tracking performance, it is observed that the recursive gain sequence is of the order $1/k$, that is, the reciprocal of the iteration number.

In practical scenarios, where an infinite learning process is unattainable, it becomes crucial to assess the transient performance of the proposed scheme within a finite number of learning iterations. This performance is closely linked to parameter c_2. Prior to delving into the discussions regarding c_2, we emphasize that the specific convergence rate solely depends on the proposed schemes, i.e., (6.74) and (6.94), rather than on convergence analysis. In other words, evaluating the convergence of $\mathbb{E}[\mathbf{V}_k]$ reflects its practical applicability.

We summarize the effects of c_2 as follows: In (6.87), c_2 reflects the approximate contraction magnitude of the Lyapunov function $\mathbb{E}[\mathbf{V}_k]$. Therefore, within the allowable range, a larger c_2 results in faster convergence speed. Moreover, to derive inequality (6.88), we introduce an additional condition $c_1\rho_k \leq \frac{c_2}{2}$. A larger value of c_2 can provide a wider range for selecting ρ_k. This observation motivated us to consider whether parameter c_2 can be suitably enlarged.

Parameter c_2 is strongly related to the direction regulation matrix $\mathcal{L}$. First, c_2 is introduced in (6.84) to satisfy $\mathcal{S} \geq c_2\mathcal{Q}$, whereas both $\mathcal{S}$ and $\mathcal{Q}$ are derived from the relationship $(\overline{\mathcal{LB}})^T\mathcal{Q} + \mathcal{Q}(\overline{\mathcal{LB}}) = \mathcal{S}$. Therefore, the solution $\mathcal{Q}$ is actually determined by matrix $\mathcal{S}$. In particular, for a given $\mathcal{S}$, we can deduce that $\mathcal{Q} = \int_0^\infty \exp(-(\overline{\mathcal{LB}})^T t)\mathcal{S}\exp(-(\overline{\mathcal{LB}})t)\mathrm{d}t$. Therefore, if $\mathcal{S}$ is given, $\mathcal{Q}$ is determined by $\mathcal{L}$ and then the parameter c_2 is determined accordingly. In other words, parameter c_2 can be optimized with respect to both $\mathcal{S}$ and $\mathcal{L}$. Therefore, we can solve the following optimization problem:

$$\begin{aligned} \max_{\mathcal{S},\mathcal{L}} \quad & c_2 \\ \text{s.t.} \quad & \mathcal{S} \geq c_2\mathcal{Q}, \\ & (\overline{\mathcal{LB}})^T\mathcal{Q} + \mathcal{Q}(\overline{\mathcal{LB}}) = \mathcal{S}, \\ & \overline{\mathcal{LB}} \text{ are with positive eigenvalues.} \end{aligned} \tag{6.95}$$

However, the optimization problem (6.95) is difficult to solve because of several restrictions. First, it is not a convex optimization problem. Moreover, high-dimensional matrices are involved . Additionally, matrices $\mathcal{S}$ and $\mathcal{L}$ are coupled. Therefore, heuristic methods are suggested to obtain an approximate estimate of parameter c_2.

Furthermore, as previously explained, the actual convergence performance is solely determined by the proposed learning control schemes rather than the analysis techniques employed. Additionally, it is essential for the convergence rate to be independent of $\mathcal{S}$, which serves as an auxiliary matrix introduced to establish the Lyapunov function $\mathbf{V}_k$. Therefore, it is reasonable to confine its values to a specific positive definite matrix; for instance, let $\mathcal{S} = I_p$, where $p = nm\sum_{i=1}^n p_i$ represents the dimension. In this case, the optimization problem (6.95) can be simplified as

$$\begin{aligned} \max_{\mathcal{L}} \quad & c_2 \\ \text{s.t.} \quad & I_p \geq c_2\mathcal{Q}, \\ & \mathcal{Q} = \int_0^\infty \exp\left(-(\overline{\mathcal{LB}})^T t - (\overline{\mathcal{LB}})t\right)\mathrm{d}t, \\ & \overline{\mathcal{LB}} \text{ are with positive eigenvalues.} \end{aligned} \tag{6.96}$$

6.3.3.5 Decentralized Recursive Gain and Learning Control

In Subsection 6.3.3.2, we provide a general learning control scheme (6.74), where the design conditions of ρ_k and $\mathcal{L}$ are given in Theorem 6.4. In Subsection 6.3.3.3, the selection of the decreasing gain sequence ρ_k is investigated according to a simplified index (6.88). Subsequently, in Subsection 6.3.3.4, the selection of the direction-regulation matrix $\mathcal{L}$ is given by maximizing parameter c_2 using heuristic methods.

However, it should be noted that the recursive gain sequence $\{\rho_k^*\}$ is not fully decentralized as it requires the expectation of the Lyapunov function $\mathbb{E}[\mathbf{V}_k]$ for all subsystems. The optimization of c_2 is also performed with complete knowledge of the entire system. Nevertheless, these computations are impractical for real-world applications. Therefore, we propose a practical decentralized gain-adaptation mechanism.

First, the gain sequence $\{\rho_k^*\}$ is recursively generated using (6.91). Therefore, it is unnecessary to access $\mathbb{E}[\mathbf{V}_k]$ for all iterations to determine the recursive gain sequence. Instead, only the initial value ρ_1^* is determined. Moreover, by considering the estimation given by Lemma 6.3, the resulting sequence γ_k is independent of the initial value γ_0. Therefore, the initial value $\mathbb{E}[\mathbf{V}_1]$ can be estimated in practice. In this case, although the derived sequence is not optimal with respect to the index $\mathbf{L}_k$, it is still advantageous for applications.

In addition to $\mathbb{E}[\mathbf{V}_1]$, c_2 is also required to derive the recursive gain sequence. To reduce the computational complexity of the centralized lifting system model, we present a decentralized version that lifts the system dynamics. In particular, we lift the input, output, and noise of the ith subsystem at the kth iteration into supervectors, as follows:

$$\begin{aligned}
\boldsymbol{u}_{k,i} &= [u_{k,i}^T(0), u_{k,i}^T(1), \ldots, u_{k,i}^T(m-1)]^T, \\
\boldsymbol{y}_{k,i} &= [y_{k,i}^T(1), y_{k,i}^T(2), \ldots, y_{k,i}^T(m)]^T, \\
\boldsymbol{w}_{k,i} &= [w_{k,i}^T(1), w_{k,i}^T(2), \ldots, w_{k,i}^T(m)]^T.
\end{aligned}$$

From (6.70) and the output equation in system (6.60), we have

$$\boldsymbol{y}_{k,i} = \mathcal{H}_i \boldsymbol{u}_{k,i} + \sum_{j \neq i} \mathcal{M}_{ij} \boldsymbol{u}_{k,j} + \boldsymbol{w}_{k,i} + \mathcal{N}_i \boldsymbol{x}_k^0, \tag{6.97}$$

where

$$\mathcal{H}_i = \begin{bmatrix} C_i B_i & 0 & 0 & \cdots & 0 \\ C_i A_{ii}^{*,1} B_i & C_i B_i & 0 & \cdots & 0 \\ C_i A_{ii}^{*,2} B_i & C_i A_{ii}^{*,1} B_i & C_i B_i & \cdots & 0 \\ \vdots & \vdots & \vdots & \ddots & \vdots \\ C_i A_{ii}^{*,m-1} B_i & C_i A_{ii}^{*,m-2} B_i & \cdots & \cdots & C_i B_i \end{bmatrix},$$

$$\mathcal{M}_{ij} = \begin{bmatrix} 0 & 0 & 0 & \cdots & 0 \\ C_i A_{ij}^{*,1} B_j & 0 & 0 & \cdots & 0 \\ C_i A_{ij}^{*,2} B_j & C_i A_{ij} B_j & 0 & \cdots & 0 \\ \vdots & \vdots & \vdots & \ddots & \vdots \\ C_i A_{ij}^{*,m-1} B_j & C_i A_{ij}^{*,m-2} B_j & \cdots & \cdots & 0 \end{bmatrix}, j \neq i$$

$$\mathcal{N}_i = \begin{bmatrix} C_i A_{i1} & C_i A_{i2} & C_i A_{i3} & \cdots & C_i A_{in} \\ C_i A_{i1}^{*,2} & C_i A_{i2}^{*,2} & C_i A_{i3}^{*,2} & \cdots & C_i A_{in}^{*,2} \\ C_i A_{i1}^{*,3} & C_i A_{i2}^{*,3} & C_i A_{i3}^{*,3} & \cdots & C_i A_{in}^{*,3} \\ \vdots & \vdots & \vdots & \ddots & \vdots \\ C_i A_{i1}^{*,m} & C_i A_{i2}^{*,m} & C_i A_{i3}^{*,m} & \cdots & C_i A_{in}^{*,m} \end{bmatrix}.$$

From (6.97), the output of the ith subsystem is jointly determined by the inputs and initial states of all subsystems. The effect of the other subsystems is reflected by the SCM $A_{ij}^{*,l}$, which depends not only on the labels of subsystems, that is, i and j, but also on the time instant difference, that is, l.

Remark 6.7 The above lifting formulation differs from that in Subsection 6.3.3.1 because of the opposite orders of time and subsystem labels. In Subsection 6.3.3.1, we first lift the vectors with respect to the subsystem label, then we lift them with respect to the time label. This lifting process can effectively address the interconnection effect between subsystems for convergence analysis. However, we merely lift the vectors with respect to the time label. This formulation can highlight the interconnection effect such that proposing a decentralized gain adaptation mechanism is feasible.

We can define $\boldsymbol{u}_{d,i}$ and $\boldsymbol{y}_{d,i}$ similarly to $\boldsymbol{u}_{k,i}$ and $\boldsymbol{y}_{k,i}$, respectively. Then, we obtain

$$\boldsymbol{y}_{d,i} = \mathcal{H}_i \boldsymbol{u}_{d,i} + \sum_{j \neq i} \mathcal{M}_{ij} \boldsymbol{u}_{d,j} + \mathcal{N}_i \boldsymbol{x}_d^0. \tag{6.98}$$

Using Assumption 6.7, we obtain

$$\begin{aligned} \boldsymbol{e}_{k,i} &\triangleq \boldsymbol{y}_{d,i} - \boldsymbol{y}_{k,i} \\ &= \mathcal{H}_i(\Delta \boldsymbol{u}_{k,i}) + \sum_{j \neq i} \mathcal{M}_{ij}(\Delta \boldsymbol{u}_{k,j}) - \boldsymbol{w}_{k,i}. \end{aligned} \tag{6.99}$$

From (6.74), we have

$$\begin{aligned} \Delta \boldsymbol{u}_{k+1,i} &= \Delta \boldsymbol{u}_{k,i} - \rho_k \mathcal{L}_i \boldsymbol{e}_{k,i} \\ &= \Delta \boldsymbol{u}_{k,i} - \rho_k \mathcal{L}_i \mathcal{H}_i(\Delta \boldsymbol{u}_{k,i}) \\ &\quad - \rho_k \sum_{j \neq i} \mathcal{L}_i \mathcal{M}_{ij}(\Delta \boldsymbol{u}_{k,j}) + \boldsymbol{w}_{k,i}, \end{aligned} \tag{6.100}$$

where $\mathcal{L}_i \triangleq I_m \otimes L_i$.

To derive c_2 locally for a given subsystem, say the ith subsystem, we can define the Lyapunov function for the subsystem as $\mathbf{V}_k^i = (\Delta \boldsymbol{u}_{k,i})^T \mathcal{Q}^i (\Delta \boldsymbol{u}_{k,i})$, where $\mathcal{Q}^i$ is a positive-definite matrix satisfying $(\mathcal{L}_i \mathcal{H}_i)^T \mathcal{Q}^i + \mathcal{Q}^i (\mathcal{L}_i \mathcal{H}_i) = \mathcal{S}^i$ and $\mathcal{S}^i$ is an arbitrary positive-definite matrix. In practice, $\mathcal{S}^i$ can be selected as the identity matrix for simplicity.

From (6.100), it is difficult to derive a decentralized recursion of $\mathbf{V}_{k+1}^i$, as in Subsection 6.3.3.2 because of the coupling terms from the other subsystems; the expansion of $\mathbf{V}_{k+1}^i$ depends not only on the input of the ith subsystem, but also on all of the other subsystems. This is caused by the inherent characteristics of LSSs.

Because the lower-triangular matrices $\mathcal{M}_{ij}$ have zero diagonal entries, the influence of the other subsystems on the ith subsystem is insignificant. To derive a decentralized version of the recursive gain sequence, we omit the coupling terms $\sum_{j \neq i} \mathcal{L}_i \mathcal{M}_{ij} (\Delta \boldsymbol{u}_{k,j})$ from (6.100) to obtain $\Delta \boldsymbol{u}_{k+1,i} = \Delta \boldsymbol{u}_{k,i} - \rho_k \mathcal{L}_i \mathcal{H}_i (\Delta \boldsymbol{u}_{k,i}) + \boldsymbol{w}_{k,i}$. Following the steps in Subsection 6.3.3.2, we have a counterpart to (6.88) for the ith subsystem:

$$\mathbb{E}[\mathbf{V}_{k+1}^i] \leq (1 - \frac{c_2^i}{2} \rho_k^i) \mathbb{E}[\mathbf{V}_k^i] + (\rho_k^i)^2 b_w^i, \tag{6.101}$$

where $b_w^i = \lambda_{\max}(\mathcal{Q}^i) m \sigma_w$ and c_2^i satisfy $\mathcal{S}^i \geq c_2^i \mathcal{Q}^i$. Therefore, as mentioned in Subsection 6.3.3.3, we derive the decentralized recursive gain sequence for the ith subsystem as follows:

$$\rho_{k+1}^{i,*} = \rho_k^{i,*} \left(1 - \frac{c_2^i}{4} \rho_k^{i,*} \right) \tag{6.102}$$

where the initial value is $\rho_1^{i,*} = (c_2^i \mathbb{E}[\mathbf{V}_1^i]/4 b_w^i)$. In practical implementations, we can provide an approximate estimate of the initial value to balance the conflict between precision knowledge and the optimal gain sequence.

For the adaptive algorithm (6.102) and Lemma 6.3, we have the following property:

Proposition 6.4 *Consider the recursive gain sequences in* (6.102). *If the initial values satisfy*

$$\rho_1^{i,*} : \rho_1^{j,*} = c_2^j : c_2^i, \quad \forall i \neq j, \tag{6.103}$$

then, for any iteration number k, we have

$$\rho_k^{i,*} : \rho_k^{j,*} = c_2^j : c_2^i, \quad \forall i \neq j. \tag{6.104}$$

Proof This proposition can be proved using a mathematical induction method. The initial step, that is, $k = 1$, is clearly valid by noting (6.103). Assuming that (6.104) holds for iteration number k, that is, $\rho_k^{i,*} : \rho_k^{j,*} = c_2^j : c_2^i, \forall i \neq j$, we prove its validity for iteration number $k+1$. Considering recursion (6.102) for the ith and jth subsystems, we have

$$\rho_{k+1}^{i,*} = \rho_k^{i,*}\left(1 - \frac{c_2^i}{4}\rho_k^{i,*}\right), \quad \rho_{k+1}^{j,*} = \rho_k^{j,*}\left(1 - \frac{c_2^j}{4}\rho_k^{j,*}\right).$$

Multiplying both sides of the upper equality by c_2^i and the lower equality by c_2^j and applying the inductive assumption, we have

$$\begin{aligned} c_2^i\rho_{k+1}^{i,*} &= c_2^i\rho_k^{i,*}\left(1 - \frac{c_2^i}{4}\rho_k^{i,*}\right) \\ &= c_2^j\rho_k^{j,*}\left(1 - \frac{c_2^j}{4}\rho_k^{j,*}\right) \\ &= c_2^j\rho_{k+1}^{j,*}. \end{aligned} \tag{6.105}$$

The proof is completed.

Using the above decentralized recursive gain, we establish the following decentralized learning control scheme

$$u_{k+1,i}(t) = u_{k,i}(t) + \rho_k^{i,*} L_i e_{k,i}(t+1). \tag{6.106}$$

Comparing (6.106) with (6.74) and (6.94), the major difference lies in that the gain sequences of all subsystems are different in (6.106) but identical in (6.74) and (6.94). This point yields a major modification in the convergence analysis.

The convergence theorem is as follows:

Theorem 6.6 *Consider the large-scale interconnected system* (6.60) *under Assumptions 6.5–6.8, and apply the decentralized learning control scheme* (6.106), *in which the gain sequences are generated by* (6.102) *with initial values satisfying* (6.103). *If the direction-regulation matrix L_i satisfies all eigenvalues of $L_i C_i B_i$ having positive real parts for $i = 1, 2, \ldots, n$, the generated input sequence $\{\boldsymbol{u}_k\}$ converges to the desired input $\boldsymbol{u}_d$ in the mean-square sense, that is,* $\mathbb{E}[\|\boldsymbol{u}_d - \boldsymbol{u}_k\|^2] \xrightarrow[k\to\infty]{} 0$.

Proof The theorem can be proved in a manner similar to that of Theorem 6.4. We present the major differences and omit repeatable steps. To this end, we lift the input of all the time instants and subsystems into a supervector $\mathbf{u}_k$. Then, we have

$$\boldsymbol{u}_{k+1} = \boldsymbol{u}_k + \overline{\mathcal{R}}_k \overline{\mathcal{L}} \boldsymbol{e}_k, \tag{6.107}$$

where $\overline{\mathcal{R}}_k = I_m \otimes \mathcal{R}_k$ and $\mathcal{R}_k = \text{diag}\{\rho_k^{1,*} I_{p_1}, \rho_k^{2,*} I_{p_2}, \ldots, \rho_k^{n,*} I_{p_n}\}$.

By subtracting both sides of (6.107) from $\boldsymbol{u}_d$ and substituting the error dynamics in (6.77), we have

$$\Delta \boldsymbol{u}_{k+1} = \Delta \boldsymbol{u}_k - \overline{\mathcal{R}}_k \overline{\mathcal{L}\mathcal{B}} \Delta \boldsymbol{u}_k + \overline{\mathcal{R}}_k \boldsymbol{w}_k. \tag{6.108}$$

From Proposition 6.4, we note that the gain sequences among the various subsystems retain the same proportion for all iterations. Thus, matrix $\mathcal{R}_k$ can be rewritten as

$$\mathcal{R}_k = \rho_k^{1,*} \underbrace{\operatorname{diag}\left\{ I_{p_1}, \frac{c_2^1}{c_2^2} I_{p_2}, \ldots, \frac{c_2^1}{c_2^n} I_{p_n} \right\}}_{\triangleq \Gamma}. \tag{6.109}$$

Then, (6.108) becomes

$$\Delta \boldsymbol{u}_{k+1} = \Delta \boldsymbol{u}_k - \rho_k^{1,*} \overline{\Gamma \mathcal{L}\mathcal{B}} \Delta \boldsymbol{u}_k + \rho_k^{1,*} \overline{\Gamma} \boldsymbol{w}_k, \tag{6.110}$$

where $\overline{\Gamma} = I_m \otimes \Gamma$.

Note that Γ is a diagonal matrix, with its diagonal entries being positive real numbers. Thus, the large-scale matrix $\overline{\Gamma \mathcal{L}\mathcal{B}}$ is still a block lower-triangular matrix whose eigenvalues have positive real parts. Thus, for any given positive-definite matrix $\mathcal{S}$, there exists a unique positive-definite matrix $\mathcal{Q}'$ such that $(\overline{\Gamma \mathcal{L}\mathcal{B}})^T \mathcal{Q}' + \mathcal{Q}'(\overline{\Gamma \mathcal{L}\mathcal{B}}) = \mathcal{S}$ holds true.

Moreover, from Lemma 6.3, $\rho_k^{1,*}$ satisfies $\rho_k^{1,*} > 0$, $\rho_k^{1,*} \to \infty$, and $\sum_{k=1}^{\infty} \rho_k^{1,*} = \infty$. Therefore, following the proof of Theorem 6.4, we can prove this conclusion. Therefore, the proof is completed.

Theorem 6.6 successfully demonstrated that a completely decentralized learning control scheme with a recursive gain sequence generated according to the local tracking performance can ensure asymptotic convergence of the entire LSSs. For simplicity, the gain sequences for different subsystems are required to obey a certain proportion relationship. However, the proportion condition for the initial values can be practically removed because of the essential order estimation using Lemma 6.3. The details would have significantly increased the length of this section. Thus, we omit the discussion of the general case.

The effectiveness of the decentralized learning control scheme with a gain-adaptation mechanism was systematically analyzed. Various design scenarios were considered to establish a comprehensive framework. However, these analyses do not impact the practical implementation. The structure of (6.74), (6.94), and (6.106) can be implemented similarly to the traditional P-type learning control for each subsystem. Furthermore, recursive computation is employed for gain adaptation with an appropriate initial value. To demonstrate the application of the proposed scheme, a simulation involving multiple connected pendulums is presented in the subsequent subsection.

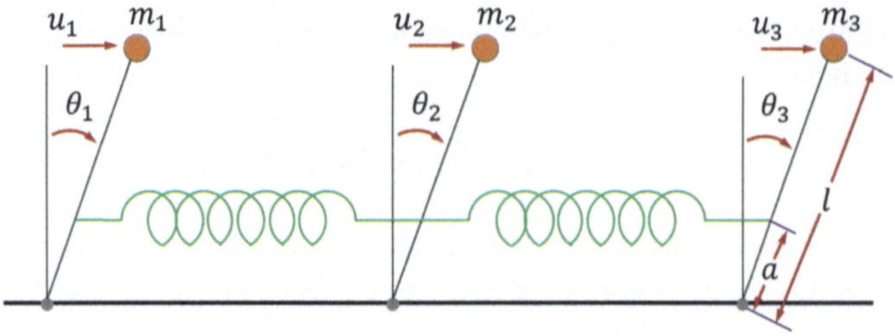

Fig. 6.9 The coupled inverted pendulums

6.3.4 Illustrative Simulations

To compare our results with the existing results, we employed the example of three pendulums coupled with springs (illustrated in Fig. 6.9) as the benchmark, where each pendulum is a single-input-single-output system subject to independent inputs. A simpler case of two identical pendulums was considered in Wu (2007); Fu et al. (2017). The dynamics are formulated as follows:

$$
\begin{aligned}
x_{k,1}(t+1) &= \begin{bmatrix} 0 & 1 \\ \frac{g}{l} - \frac{ha^2}{m_1 l^2} & 0 \end{bmatrix} x_{k,1}(t) + \begin{bmatrix} 0 \\ \frac{1}{m_1 l^2} \end{bmatrix} u_{k,1}(t) + \begin{bmatrix} 0 & 0 \\ \frac{ha^2}{m_2 l^2} & 0 \end{bmatrix} x_{k,2}(t), \\
y_{k,1}(t) &= \begin{bmatrix} 1 & 1 \end{bmatrix} x_{k,1}(t) + w_{k,1}(t), \\
x_{k,2}(t+1) &= \begin{bmatrix} 0 & 1 \\ \frac{g}{l} - \frac{ha^2}{m_2 l^2} & 0 \end{bmatrix} x_{k,2}(t) + \begin{bmatrix} 0 \\ \frac{1}{m_2 l^2} \end{bmatrix} u_{k,1}(t) + \begin{bmatrix} 0 & 0 \\ \frac{ha^2}{m_1 l^2} & 0 \end{bmatrix} x_{k,1}(t) + \begin{bmatrix} 0 & 0 \\ \frac{ha^2}{m_3 l^2} & 0 \end{bmatrix} x_{k,3}(t), \\
y_{k,2}(t) &= \begin{bmatrix} 1 & 1 \end{bmatrix} x_{k,2}(t) + w_{k,2}(t), \\
x_{k,3}(t+1) &= \begin{bmatrix} 0 & 1 \\ \frac{g}{l} - \frac{ha^2}{m_3 l^2} & 0 \end{bmatrix} x_{k,3}(t) + \begin{bmatrix} 0 \\ \frac{1}{m_3 l^2} \end{bmatrix} u_{k,1}(t) + \begin{bmatrix} 0 & 0 \\ \frac{ha^2}{m_2 l^2} & 0 \end{bmatrix} x_{k,2}(t), \\
y_{k,3}(t) &= \begin{bmatrix} 1 & 1 \end{bmatrix} x_{k,3}(t) + w_{k,3}(t),
\end{aligned}
$$

where k and g are the spring and gravity constants, respectively, and the state $x_{k,i}(t)$ is the vector of θ_i and its derivative. The parameters are as follows: $\frac{g}{l} = 1.0$, $\frac{1}{m_i l^2} = 1.0$, $\frac{h}{m_i} = 2.0$, and $\frac{a}{l} = 0.5$ (Wu 2007; Fu et al. 2017). The iteration length is set to $m = 30$. Random noise is subject to a normal distribution with zero mean and variance 0.05.

The tracking references for subsystems 1, 2, and 3 are $y_{d,1}(t) = \sin\left(\frac{2\pi t}{m}\right)$, $y_{d,2}(t) = \cos\left(\frac{2\pi t}{m}\right) - 1$, and $y_{d,3}(t) = 1 - \cos\left(\frac{2\pi t}{m}\right)$. The initial state and input are set to 0. Each algorithm was executed for 150 iterations.

The algorithms used are as follows. The learning gain matrices L_i are set to $L_i = 0.9$, $i = 1, 2, 3$. For the decentralized learning control scheme with decreasing gain (6.76), the gain is set to $\rho_k = 0.36/k$. Without any knowledge of the control system, it is difficult to select the optimal decreasing gain for a specific learning process. For the learning control scheme with the decentralized recursive gain (6.106), the

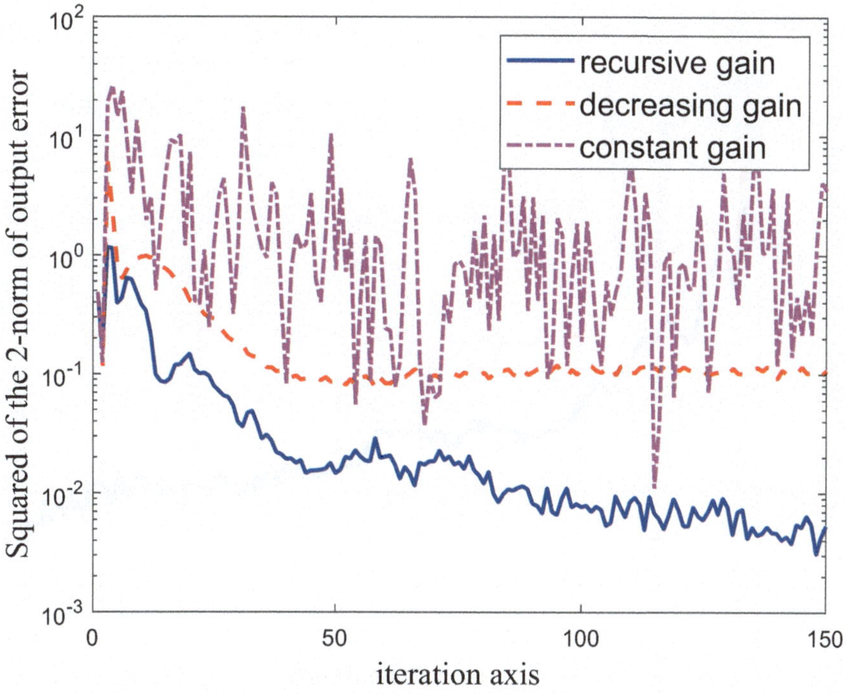

Fig. 6.10 Averaged tracking error profiles along the iteration axis: subsystem 1

recursive gain is generated by the algorithm (6.102), where the parameters are set to $\rho_1^{i,*} = 0.18$, $c_2^i = 3.5$, $i = 1, 3$, and $\rho_1^{2,*} = 0.19$, $c_2^2 = 3.7$. The parameters for the second subsystem are distinct from those of the first and third subsystems owing to their different coupling models. For comparison, we simulated a decentralized learning control scheme with constant gain $\rho = 0.3$. In the figures, we distinguish between the three algorithms by decreasing gain, recursive gain, and constant gain. Notably, both decreasing and constant gains have been widely employed to deal with LSSs in the existing ILC literature.

To observe the improvement in the tracking performance along the iteration axis, we calculated the averaged tracking error for each iteration by $\|\boldsymbol{e}_{k,i}\|^2/m$ for the ith subsystem and kth iteration. For deterministic systems, the maximum tracking error defined by $\max_t \|e_{k,i}(t)\|^2$ is usually selected to measure the tracking performance. However, for stochastic systems with measurement noise, the maximum tracking error may be dominated by the additional noise after a sufficient learning process and thus be unable to reflect the essential tracking performance well. Therefore, the averaged tracking-error smoothing of the influence of additional noise is used to characterize the tracking performance. The averaged tracking error profiles are shown in Figs. 6.10–6.12 for subsystems 1, 2, and 3, respectively. In these figures, the constant gain fails to guarantee an effective improvement in the tracking performance

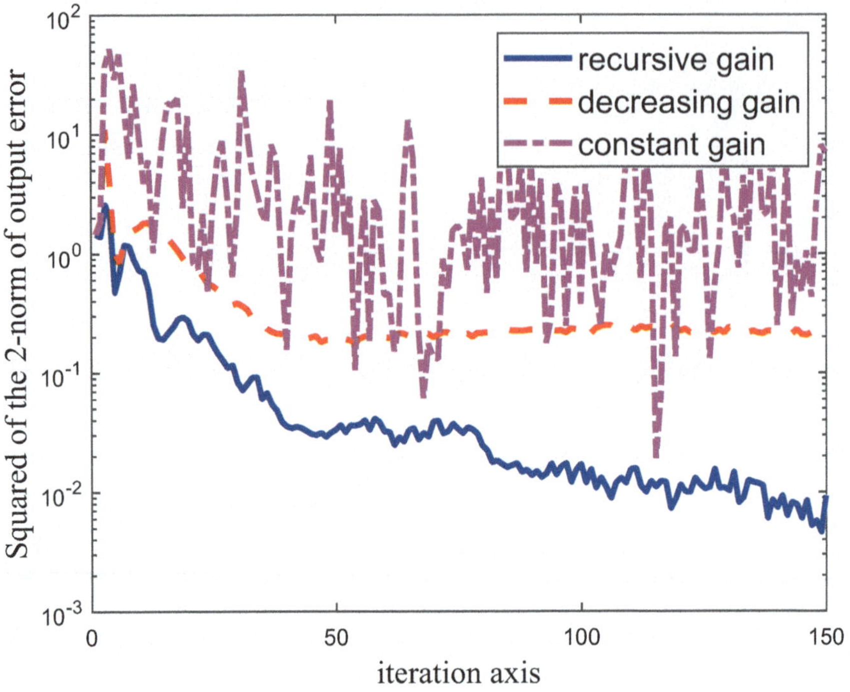

Fig. 6.11 Averaged tracking error profiles along the iteration axis: subsystem 2

because it is affected by random measurement noise. The decreasing gain can realize a continuous improvement of the tracking performance; however, the convergence rate is somewhat slow because it is difficult to provide an optimal gain prior to the learning process. Furthermore, an inappropriate selection of the decreasing sequence may lead to stopping the improvement process. In contrast, the recursive gain generated by the algorithm (6.102) is superior to the other two scenarios. This observation demonstrates the significance of decentralized gain-adaptation.

To intuitively observe the tracking performance, Fig. 6.13 depicts the output profiles of three subsystems at the last iteration for the learning control scheme with the gain generated by (6.102) as well as the desired references. Here, the solid, dashed, and dash-dotted lines present Subsystem 1, 2, and 3, respectively. The lines with and without markers denote the actual outputs and the desired references, respectively. An acceptable tracking performance is achieved at the last iteration, as the output profiles almost overlap the desired reference. Slight deviations of the actual outputs were generated by random measurement noise. This observation demonstrates the learning effectiveness of the proposed scheme in the presence of measurement noise.

To illustrate the effect of the coupling terms between the subsystems, Fig. 6.14 illustrates the averaged error profiles for three subsystems with a recursive gain. It is observed that the decreasing trend of Subsystem 1 is similar to that of Subsystem 3; however, they behave differently from Subsystem 2. This is because Subsystems

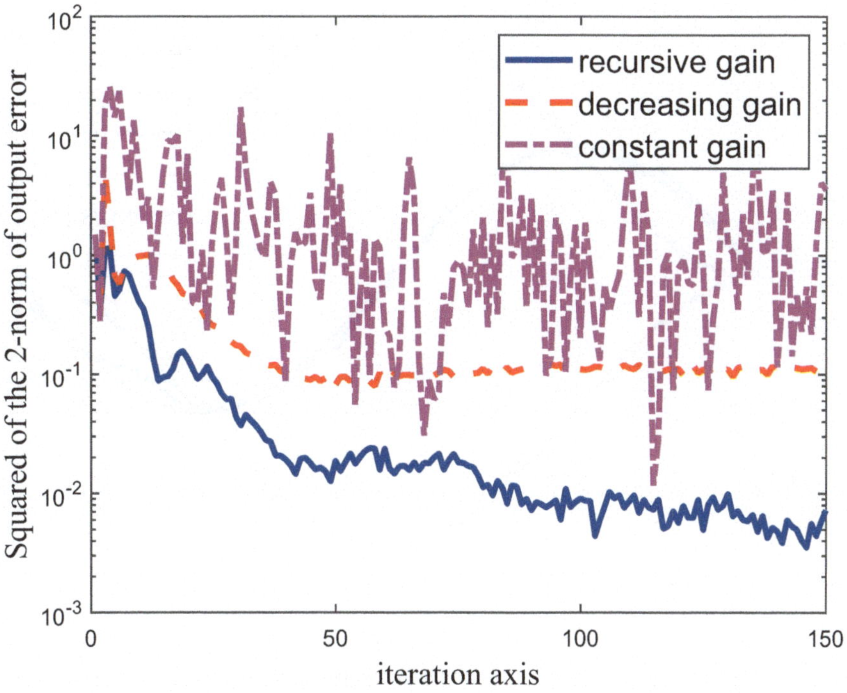

Fig. 6.12 Averaged tracking error profiles along the iteration axis: subsystem 3

1 and 3 have only one spring connection with Subsystem 2, and their coupling effects are similar. In contrast, Subsystem 2 has two coupling terms in its dynamics, which reduces the convergence speed. This observation reflects the primary effect of coupling. Additional explorations are left for future research.

6.4 Optimal Gain Matrix Design in SILC

In addition to optimizing the scalar learning gain, simultaneous optimization of the learning gain and direction regulation matrix can also be pursued. Among various optimization strategies, Saab's Kalman filtering-based approach stands out as the most representative one (Saab 2001a, b, 2003, 2005). This method integrates the learning gain and direction regulation matrix into a single iteration-time-varying matrix, which is recursively calculated by minimizing the trace of the input error's covariance matrix. Hereafter, we provide a concise overview of this strategy.

Consider the following linear stochastic systems

$$\begin{aligned} x(t+1,k) &= A(t)x(t,k) + B(t)u(t,k) + w(t,k), \\ y(t,k) &= C(t)x(t,k) + v(t,k), \end{aligned} \tag{6.111}$$

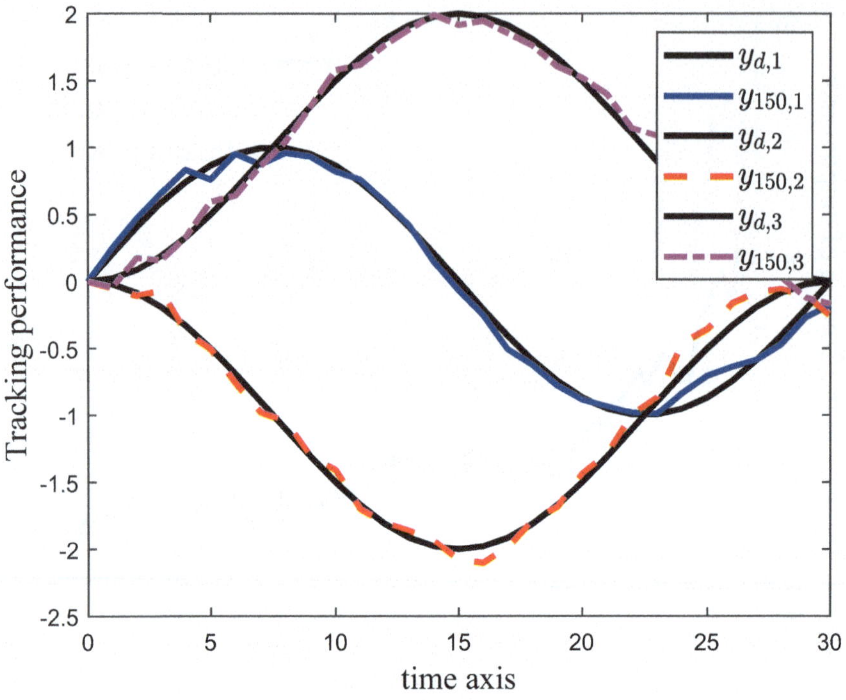

Fig. 6.13 Tracking of the desired references for three subsystems

where $A(t)$, $B(t)$, and $C(t)$ are time-varying system matrices, $w(t,k)$ and $v(t,k)$ are the system noise and measurement noise, respectively.

The update law is given by

$$u(t,k+1) = u(t,k) + K(t,k)[e(t+1,k) - e(t,k)], \tag{6.112}$$

where $K(t,k)$ is the direction regulation matrix.

To analyze the convergence property of the algorithm (6.112), some hypotheses are required. It is first assumed that $y(t,d)$ is realizable. That is, there exist input $u(t,d)$ and initial state $x(0,d) = x_0$ such that

$$\begin{aligned} x(t+1,d) &= A(t)x(t,d) + B(t)u(t,d), \\ y(t,d) &= C(t)x(t,d). \end{aligned} \tag{6.113}$$

Besides, the input/output coupling matrix $C(t+1)B(t)$ is assumed to be full column rank. To keep the notations concise without causing confusion, it is simplified as $C^+B = C(t+1)B(t)$ in the rest of the section. The pre-fix notation Δ denotes the distance between the objective value and the actual value, i.e., $\Delta x(t,k) \triangleq x(t,d) - x(t,k)$, $\Delta u(t,k) = u(t,d) - u(t,k)$.

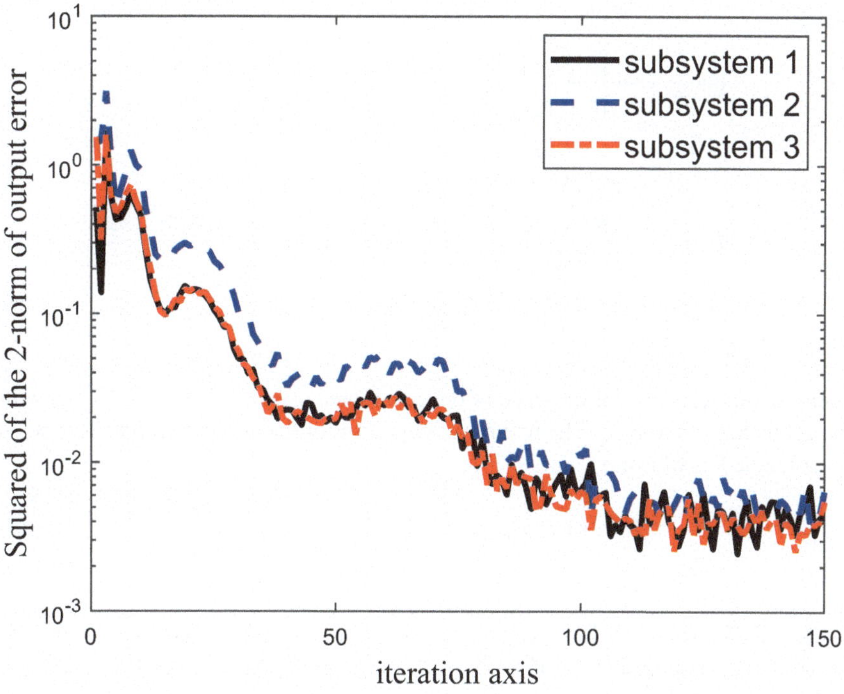

Fig. 6.14 Averaged tracking error profiles for three subsystems with a recursive gain

Both $w(t,k)$ and $v(t,k)$ are assumed as Gaussian white noise with zero mean, where the covariance matrix $Q_t = \mathbb{E}[w(t,k)w(t,k)^T]$ is positive semidefinite while $R_t = \mathbb{E}[v(t,k)v(t,k)^T]$ is positive definite. Moreover, $w(t,k)$ is uncorrelated with $v(s,l)$, $\forall t, s \in [0, N], k, l \in \mathbb{R}$. The initial state is assumed to be a random variable, which is also a Gaussian white noise with zero mean and covariance $P_{x,0} = \mathbb{E}[\Delta x(0,k)\Delta x(0,k)^T]$ being positive semidefinite. Additionally, $\Delta x(0,k)$ is uncorrelated with any noise. Moreover, the initial input satisfies the requirement that the initial input error $\Delta u(t,0)$ is zero-mean white noise with symmetric positive definite covariance $\mathbb{E}[\Delta u(t,0)\Delta u(t,0)^T] = P_{u,0}$.

According to system (6.111) and update law (6.112) with the above assumptions, reference Saab (2001a) obtained the computational algorithm for $K(t,k)$ by minimizing the trace of the input error covariance matrix, and further proved the convergence in the mean square. To show this, the following 2D-Roessor model was first derived as

$$X^+ = \Phi X + \Gamma Z, \tag{6.114}$$

where

$$X^{+} = \begin{bmatrix} \Delta u(t, k+1) \\ \Delta x(t+1, k) \end{bmatrix}, \quad X = \begin{bmatrix} \Delta u(t, k) \\ \Delta x(t, k) \end{bmatrix},$$

$$Z = \begin{bmatrix} w(t, k) \\ v(t+1, k) - v(t, k) \end{bmatrix},$$

$$\Phi = \begin{bmatrix} I - K(t, k)C^{+}B & K(t, k)[C(t) - C(t+1)A(t)] \\ B(t) & A(t) \end{bmatrix},$$

$$\Gamma = \begin{bmatrix} K(t, k)C(t+1) & K(t, k) \\ -I & 0 \end{bmatrix}.$$

Then, set the derivative of the trace of the covariance matrix $P^{+} \triangleq \mathbb{E}(X^{+}X^{+^{T}})$ with respect to $K(t, k)$ to zero. The recursive algorithm for the computation of $K(t, k)$ is thus obtained as follows:

$$K(t, k) = P_{u,k}\Xi^{T}(\Xi P_{u,k}\Xi^{T} + \Lambda_{D,k})^{-1}, \tag{6.115}$$

$$P_{u,k+1} = (I - K(t, k)\Xi)P_{u,k}, \tag{6.116}$$

where $\Xi \triangleq C(t+1)B(t)$, $\Lambda_{D,k} \triangleq (C(t) - C(t+1)A(t))P_{x,k}(C(t) - C(t+1)A(t))^{T} + C(t+1)Q_{t}C(t+1)^{T} + R_{t} + R_{t+1}$, $P_{x,t} = \mathbb{E}[\Delta x(t, k)\Delta x(t, k)^{T}]$, $P_{u,k} = \mathbb{E}[\Delta u(t, k)\Delta u(t, k)^{T}]$.

The following theorem is the major convergence theorem of Saab (2001a).

Theorem 6.7 *For system* (6.111) *and update laws* (6.112), (6.115), *and* (6.116), *if* $C(t+1)B(t)$ *is of full column rank, then* $\forall k, t$, *there exist a consistent norm* $\|\cdot\|$ *such that* $\|I - K(t, h)C(t+1)B(t)\| < 1$. *Consequently,* $\|P_{u,k+1}\| < \|P_{u,k}\|$. *Furthermore,* $P_{u,k} \to 0$, *and* $K(t, k) \to 0$ *uniformly on* $[0, N]$ *as* $k \to \infty$.

Now, we briefly summarize the Kalman filtering-based approach. First, build a 2D-Roessor model with input error $\Delta u(t, k)$ and state error $\Delta x(t, k)$. Next, calculate the derivative of the optimization objective, which is usually the input error covariance $P_{u,k}$, with respect to the direction regulation matrix $K(t, k)$, and set it to zero to obtain the recursive algorithm. Finally, prove the mean square convergence based on the irrelevance of noises and (semi)-positive definiteness of covariance matrices.

Theorem 6.7 has established the framework of a Kalman filtering based SILC approach. However, a lot of information about the system is required to ensure the correct running of algorithms (6.112), (6.115), and (6.116). It is seen that system matrices $A(t)$, $B(t)$, $C(t)$, noise covariances Q_t, R_t, and state error covariance $P_{x,t}$ are all required (see the expression of $\Lambda_{D,k}$).

In Saab (2001b), the term $P_{x,t}$ in (6.115) was removed; then, the algorithms become

$$\widetilde{K}(t, k) = \widetilde{P}_{u,k}\Xi^{T}(\Xi \widetilde{P}_{u,k}\Xi^{T} + \widetilde{\Lambda}_{D})^{-1}, \tag{6.117}$$

$$\widetilde{P}_{u,k+1} = (I - \widetilde{K}(t, k)\Xi)\widetilde{P}_{u,k}, \tag{6.118}$$

where $\widetilde{\Lambda}_D = C(t+1)Q_tC(t+1)^T + R_t + R_{t+1}$. In Saab (2001b), it was proven that the direction regulation matrix $\widetilde{K}(t,h)$ given by (6.117) and (6.118) for update law (6.112) still worked. That is, a theorem similar to Theorem 6.7 holds if $C(t+1)B(t)$ is of full column rank.

It should be especially noted that $\widetilde{P}_{u,k}$ is constructed only for the recursion of the algorithms, and not the real input error covariance. We denote the real input error covariance based on algorithms (6.112), (6.117), and (6.118) as $\overline{P}_{u,k}$. An interesting result is that the convergence of $\widetilde{P}_{u,k}$ and $\overline{P}_{u,k}$ is equivalent, i.e., $\widetilde{P}_{u,k} \to 0 \Leftrightarrow \overline{P}_{u,k} \to 0$. Besides, the rate of convergence is inversely proportional to the iteration number k (Saab 2001b).

So far, it is believed that the removal of the state error covariance does not affect the convergence. However, the input of the modified algorithms is not optimal; thus, (6.115) and (6.116) are called optimal algorithms, whereas (6.117) and (6.118) are called suboptimal algorithms. Note that only D-type algorithms are considered in Saab (2001a, b). Then, it is natural to ask whether similar results hold for P-type algorithms. A positive answer is provided by Saab (2003).

Consider the following P-type update law

$$u(t,k+1) = u(t,k) + K(t,k)e(t+1,k). \tag{6.119}$$

Then, by performing exactly the same steps as in Saab (2001a, b), the recursive computational algorithms are derived. The optimal algorithms are

$$K(t,k) = P_{u,k}\Xi^T(\Xi P_{u,k}\Xi^T + \Lambda_{P,k})^{-1}, \tag{6.120}$$

$$P_{u,k+1} = (I - K(t,k)\Xi)P_{u,k}, \tag{6.121}$$

where $\Lambda_{P,k} = C(t+1)A(t)P_{x,k}(C(t+1)A(t))^T + C(t+1)Q_tC(t+1)^T + R_{t+1}$, while the suboptimal algorithms are

$$\widetilde{K}(t,k) = \widetilde{P}_{u,k}\Xi^T(\Xi \widetilde{P}_{u,k}\Xi^T + \widetilde{\Lambda}_P)^{-1}, \tag{6.122}$$

$$\widetilde{P}_{u,k+1} = (I - \widetilde{K}(t,k)\Xi)\widetilde{P}_{u,k}, \tag{6.123}$$

where $\widetilde{\Lambda}_P = C(t+1)Q_tC(t+1)^T + R_{t+1}$. It has been shown in Saab (2003) that the results of Theorem 6.7 still hold.

Because there are both P-type and D-type algorithms, another natural question is whether there is a relationship between these two kinds of algorithms. As a matter of fact, note that the essential distinction among algorithms (6.115) and (6.116), (6.117) and (6.118), (6.120)and (6.121), and (6.122) and (6.123) is the selection of the bounded positive matrix Λ. This matrix should be $\Lambda_{D,k}$, $\widetilde{\Lambda}_D$, $\Lambda_{P,k}$, or $\widetilde{\Lambda}_P$, respectively. As long as this matrix is bounded, symmetrical, and positive, all of the above algorithms have the same convergence properties when $C(t+1)B(t)$ is of full column rank. In other words, regardless of the algorithm that is selected, the input error covariance matrix will converge to zero, with the rate being inversely proportional to the iteration number.

6.5 Optimal Gain Matrix in Fading Communication

6.5.1 Learning Control over Random Fading Channels

With the rapid advancements in network communication technology, an increasingly large number of systems have embraced a networked control architecture. Networked control architecture refers to the spatial separation of plants and controllers, enabling their communication through wireless or wired networks. This configuration allows operators to work remotely from hazardous and challenging operational environments while enhancing the overall control system's flexibility, stability, and robustness. However, this control structure also introduces potential risks of data corruption during remote data transmission via wireless or wired networks, including issues such as packet loss and data fading.

In this section, we focus on the case of data transmission over fading channels, resulting in imprecise signals modeled by multiplicative randomness (Elia 2005; Xiao et al. 2012; Xu et al. 2017; Dey et al. 2009; Geng et al. 2017). Fading channels refer to the attenuation of transmission signal amplitude caused by channel changes and are commonly observed in wireless communication networks due to reflection, refraction, and diffraction during propagation. This topic has been extensively studied in existing literature using conventional control strategies. For instance, Elia (2005) investigated a remote mean-square stabilization problem for a MIMO system with independent fading channels for each actuator and sensor. They considered state feedback control and provided a mean-square capacity for stabilizing a single-input system as the sum of logarithmic magnitudes of unstable eigenvalues of the plant. Extensions to a MIMO system with multiple fading channels at the input side were explored in Xiao et al. (2012) using linear control strategy. Furthermore, robust stability analysis for multi-topic uncertain systems was presented in Su and Chesi (2017), while input power constraints under fading randomness were studied in Xu et al. (2017) for linear systems.

Due to the detrimental effects of fading channels, received signals often deviate from their original values in a multiplicative manner, which can lead to instability in controlled systems if not corrected prior to usage. This observation highlights the need for addressing the issue of Kalman filtering with faded measurements (Dey et al. 2009; Geng et al. 2017). Existing studies primarily focus on ensuring boundedness of filtered signals and convergence to steady states. In Dey et al. (2009), the authors verified the boundedness of expected error covariance matrix generated by a conventional Kalman filter and proved its convergence to a stable state. Additionally, for scalar measurements with specific fading distributions, upper bounds on expected error covariance were provided in this paper. The problem of Tobit Kalman filtering was investigated in Geng et al. (2017) for discrete-time systems involving both censored and fading measurements. Overall, it is evident from the literature review that research efforts are currently limited in several aspects. Specifically, there have been few reported results on learning tracking control in the presence of fading channels.

This observation motivates us to propose an effective ILC scheme aimed at achieving satisfactory tracking performance.

When considering fading channels, a key challenge lies in effectively handling the randomness introduced by faded signals and potential system noises. To address this issue, selecting an appropriate direction regulation matrix based on specific system performance becomes significantly important. In this context, the conventional fixed direction regulation matrix is inadequate for achieving optimal tracking performance in the presence of diverse types of randomness. For instance, the update law proposed in Shen et al. (2016a), incorporating Arimoto-like and causal invariant gain matrices, ensures almost sure and mean-square convergence even with randomly varying iteration lengths. However, simulation results reveal that the fixed gain matrix fails to converge when additional random noises are present within the system. Numerous studies have addressed the design problem of adapting learning gain matrices for different control topics. In Saab (2001a), a recursive calculation method was proposed to minimize input error covariance with respect to the gain matrix using a Kalman filtering framework - an effective approach for handling various sources of randomness in discrete-time systems. For example, Ahn et al. (2008) successfully applied this framework to stochastic linear systems affected by random data dropouts; experimental results demonstrate that as long as data transmission occurs with a certain probability, the output can converge towards desired trajectories. The aforementioned literature highlights how Kalman filtering theory can enhance tracking performance against stochastic factors - motivating us to employ frameworks from Saab (2001a) and Ahn et al. (2008) for solving learning tracking problems in stochastic linear systems through fading channels.

In this section, we investigate an ILC problem for stochastic linear systems over fading channels, where the fading randomness is characterized by a random variable with known mean and variance. By employing the Kalman filter technique, we derive a recursive direction regulation matrix that is utilized in the learning control algorithm to generate input sequences. We prove that the resulting tracking error converges asymptotically to zero in a mean-square sense along the iteration axis. In contrast to existing research on fading phenomena, our primary contribution lies in proposing a novel approach using Kalman filtering for addressing learning control issues over fading channels. Both output and input fading scenarios are considered, and rigorous analysis as well as illustrative simulations are provided to demonstrate the effectiveness of our proposed scheme.

6.5.2 Problem Formulation

Consider the following stochastic linear time-varying system:

$$\begin{aligned} x_k(t+1) &= A(t)x_k(t) + B(t)u_k(t) + \omega_k(t) \\ y_k(t) &= C(t)x_k(t) + \upsilon_k(t) \end{aligned} \tag{6.124}$$

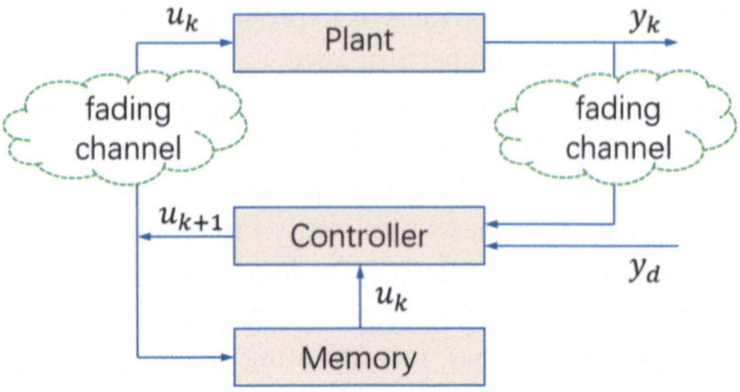

Fig. 6.15 Block diagram of ILC with fading networks

where subscript $k = 1, 2, \ldots$ is the iteration number, and argument t is the time label ($t = 1, 2, \ldots, N$, where N is the iteration length). Variables $x_k(t) \in \mathbf{R}^n$, $u_k(t) \in \mathbf{R}^p$, and $y_k(t) \in \mathbf{R}^q$ are state, input, and output, respectively. System matrices $A(t)$, $B(t)$, and $C(t)$ are with appropriate dimensions with respect to system variables. Variables $\omega_k(t)$ and $\upsilon_k(t)$ are system noise and measurement noise, respectively. Without loss of generality, we assume that the matrix $C(t+1)B(t)$ is of full column rank, implying that the system relative degree is one and the input dimension is not greater than the output dimension. This assumption is commonly satisfied by many practical systems and necessary to a certain extent.

Before further analysis, we need the following assumptions.

Assumption 6.9 The desired trajectory $y_d(t)$ is realizable in the sense that there exist suitable initial state $x_d(0)$ and desired input $u_d(t)$ such that

$$\begin{aligned} x_d(t+1) &= A(t)x_d(t) + B(t)u_d(t) \\ y_d(t) &= C(t)x_d(t). \end{aligned} \tag{6.125}$$

Thus, we can obtain a unique input regarding the desired trajectory as follows:

$$\begin{aligned} u_d(t) =& [(C(t+1)B(t))^T C(t+1)B(t)]^{-1} \\ & (C(t+1)B(t))^T [y_d(t+1) - C(t+1)A(t)x_d(t)]. \end{aligned}$$

Assumption 6.10 The initial input error $u_d(t) - u_0(t)$, $\forall t$, and the initial state error $x_d(0) - x_k(0)$, $\forall k$, are assumed to be zero-mean white noise. Moreover, $\mathbb{E}((u_d(t) - u_0(t))(u_d(t) - u_0(t))^T)$ is a symmetrical positive-definite matrix and $\mathbb{E}((x_d(0) - x_k(0))(x_d(0) - x_k(t))^T)$ is a positive-semidefinite matrix. In addition, $x_d(0) - x_k(0)$ is uncorrelated with $u_d(t) - u_0(t)$, $\omega_k(0)$ and $\upsilon_k(0)$.

Assumption 6.11 The system noise $\omega_k(t)$ and measurement noises $\upsilon_k(t)$ are sequences of independent zero-mean white Gaussian noise such that $\mathbb{E}(\omega_k(t)\omega_k^T(t)) = Q_{11,t}$ is a positive-semidefinite matrix and $\mathbb{E}(\upsilon_k(t)\upsilon_k^T(t)) = Q_{22,t}$ is a positive-definite matrix for all k. Moreover, $\omega_k(t)$ and $\upsilon_k(t)$ are uncorrelated.

The networked system setup is depicted in Fig. 6.15, where the plant and controller establish communication through fading networks. In this implementation, signal transmission can be affected by fading channels. The fading effect in this paper is characterized by a random variable known as the fading gain, which multiplies the original signals. It is assumed that the channels experience stationary fading, and thus statistical information about this fading gain is available and compensated for at the receiver. Specifically, we assume that the random fading gain follows an independent and identically distributed continuous distribution function $f(\theta)$ with respect to k and t, such that $P(\theta_k(t) > 0) = 1$. Additionally, we assume that $\theta_k(t)$ possesses the following statistical properties:

$$\mu = \mathbb{E}[\theta_k(t)], \qquad \sigma^2 = \mathbb{E}[(\theta_k(t) - \mu)^2]. \tag{6.126}$$

Remark 6.8 In this section, we assume a priori knowledge of the fading channel statistics, which is a common assumption in numerous research papers. The fading phenomenon primarily depends on network equipment and physical environments, making it reasonable for us to acquire relatively accurate information about the fading gain $\theta_k(t)$. In cases where statistical information is not available beforehand, online identification strategies can be employed to estimate it by transmitting a unit signal at the beginning of each iteration and calculating an empirical estimation. However, due to space limitations, we omit discussing this topic further.

Due to the introduction of multiplicative randomness by fading channels, the received signal becomes inaccurate, thereby affecting the learning control algorithm and system operation. Hence, it is crucial to establish a suitable learning process that can effectively handle such randomness. The main objective is to design an optimal direction regulation matrix with necessary correction for received signals, ensuring asymptotic convergence of the system output towards the desired trajectory as the iteration number increases. However, due to random noises present in system operation and measurements, it is not possible to conclude that $y_k(t) \to y_d(t)$ as $k \to \infty, \forall t$. The best achievable performance lies in achieving convergence of the output towards the desired trajectory when noises are eliminated. This objective can be attained if we ensure $u_k(t) \to u_d(t)$ as $k \to \infty, \forall t$.

In this section, we sequentially examine two fading scenarios: the output-side fading channel and the input-side fading channel. The subsequent sections provide comprehensive design and analysis procedures for each case, respectively. By segregating the fading channels in this manner, we aim to enhance clarity and facilitate a coherent understanding of our main concept. Furthermore, by employing the same approach, it is possible to extend our methodology to encompass situations where both output and input fading are present.

6.5.3 *Output Fading Case*

In this subsection, we present results for the output fading case. In particular, only the channel from plant to controller suffers from random fading, and the channel from controller to plant is assumed to work without fading.

6.5.3.1 Algorithm Design

The actual output is denoted by $y_k(t)$. The received signal by the controller is denoted by $y_k^{\circ}(t)$. Then, according to the fading model, we have

$$y_k^{\circ}(t) = \theta_k(t) y_k(t). \tag{6.127}$$

It can be seen that the received output signal cannot be directly applied in learning algorithms because it has deviated from the true value. To establish an effective learning algorithm, we modify the received signal by multiplying a mean inverse of the fading gain

$$\hat{y}_k(t) = \mu^{-1} \theta_k(t) y_k(t). \tag{6.128}$$

Thus, $\hat{y}_k(t)$ is corrected output signal. The learning update is given by

$$\begin{aligned} u_{k+1}(t) &= u_k(t) + K_k(t)[y_d(t+1) - \hat{y}_k(t+1)] \\ &= u_k(t) + K_k(t)[y_d(t+1) - \mu^{-1}\theta_k(t) y_k(t+1)], \end{aligned} \tag{6.129}$$

where $K_k(t)$ is a learning control gain matrix with dimension $p \times q$. In order to simplify the writing, we denote input and state error vectors by $\Delta u_k(t) = u_d(t) - u_k(t)$ and $\Delta x_k(t) = x_d(t) - x_k(t)$, respectively. For (6.124), (6.125) and (6.129), we can deduce the expressions of input error and state error.

The state error is shown as follows:

$$\Delta x_k(t+1) = A(t)\Delta x_k(t) + B(t)\Delta u_k(t) - \omega_k(t), \tag{6.130}$$

and the input error is given as follows:

$$
\begin{aligned}
&\Delta u_{k+1}(t) \\
&\quad = \Delta u_k(t) - K_k(t)[y_d(t+1) - \mu^{-1}\theta_k(t)y_k(t+1)] \\
&\quad = [I - K_k(t)C(t+1)B(t)]\Delta u_k(t) \\
&\qquad - K_k(t)C(t+1)A(t)\Delta x_k(t) + K_k(t)C(t+1)\omega_k(t) \\
&\qquad + K_k(t)\upsilon_k(t+1) - (1-\mu^{-1}\theta_k(t))K_k(t)C(t+1) \\
&\qquad \times [A(t)x_k(t) + \omega_k(t) + B(t)u_k(t)] \\
&\qquad - (1-\mu^{-1}\theta_k(t))K_k(t)\upsilon_k(t+1).
\end{aligned} \tag{6.131}
$$

Then, we can combine (6.130) and (6.131) into a two-dimensional Roessor model

$$
\begin{aligned}
&\begin{bmatrix} \Delta u_{k+1}(t) \\ \Delta x_k(t+1) \end{bmatrix} \\
&\quad = \begin{bmatrix} I - K_k C^+ B & -K_k C^+ A \\ B & A \end{bmatrix} \begin{bmatrix} \Delta u_k(t) \\ \Delta x_k(t) \end{bmatrix} \\
&\qquad + \begin{bmatrix} -\tilde{\theta}_k(t) K_k C^+ & -\tilde{\theta}_k(t) K_k \\ 0 & 0 \end{bmatrix} \begin{bmatrix} \omega_k(t) \\ \upsilon_k(t+1) \end{bmatrix} \\
&\qquad + \begin{bmatrix} K_k C^+ & K_k \\ -I & 0 \end{bmatrix} \begin{bmatrix} \omega_k(t) \\ \upsilon_k(t+1) \end{bmatrix} \\
&\qquad + \begin{bmatrix} -\tilde{\theta}_k(t) K_k C^+ B & -\tilde{\theta}_k(t) K_k C^+ A \\ 0 & 0 \end{bmatrix} \begin{bmatrix} u_k(t) \\ x_k(t) \end{bmatrix},
\end{aligned} \tag{6.132}
$$

where, for convenience, we define $K_k = K_k(t)$, $C^+ = C(t+1)$, $B = B(t)$, $A = A(t)$, $\Phi_1 = I - K_k C^+ B$, and $\Phi_2 = K_k(-C^+ A)$. In addition, we introduce $\tilde{\theta}_k(t) = 1 - \mu^{-1}\theta_k(t)$. It is obvious that $\tilde{\theta}_k(t)$ is a zero-mean random variable with variance $\tilde{\sigma} = \sigma^2/\mu^2$.

Writing (6.132) in a compact form, we have

$$
X^+ = \Phi X + \tilde{\theta}_k(t)\Omega\hat{X} + \Psi Z + \tilde{\theta}_k(t)\Upsilon Z \tag{6.133}
$$

where

$$
X^+ = \begin{bmatrix} \Delta u_{k+1}(t) \\ \Delta x_k(t+1) \end{bmatrix} \text{ and } X = \begin{bmatrix} \Delta u_k(t) \\ \Delta x_k(t) \end{bmatrix}
$$

are $(n+p) \times 1$ vectors, and

$$
Z = \begin{bmatrix} \omega_k(t) \\ \upsilon_k(t+1) \end{bmatrix} \text{ and } \hat{X} = \begin{bmatrix} u_k(t) \\ x_k(t) \end{bmatrix}
$$

are a $(n+q) \times 1$ vector and a $(n+p) \times 1$ vector, respectively.

$$
\Phi = \begin{bmatrix} I - K_k C^+ B & -K_k C^+ A \\ B & A \end{bmatrix} \text{ and } \Psi = \begin{bmatrix} K_k C^+ & K_k \\ -I & 0 \end{bmatrix}
$$

are a $(n+p)\times(n+p)$ matrix and a $(n+p)\times(n+q)$ matrix, respectively.

$$\Omega = \begin{bmatrix} -K_k C^+ B & -K_k C^+ A \\ 0 & 0 \end{bmatrix} \text{ and } \Upsilon = \begin{bmatrix} -K_k C^+ & -K_k \\ 0 & 0 \end{bmatrix}$$

are a $(n+p)\times(n+p)$ matrix and a $(n+p)\times(n+q)$ matrix, respectively.

It is assumed that initial state errors and input errors at the initial iteration are of zero-mean, then the right-hand side of (6.133) is of zero-mean. It is proper to refer to $H^+ = \mathbb{E}[X^+ X^{+T}]$ as a covariance matrix. Our objective is to find an optimal direction regulation matrix K_k minimizing the trace of this error covariance matrix. The error covariance matrix H^+ is as follows:

$$\begin{aligned} H^+ = {} & \Phi\mathbb{E}[XX^T]\Phi^T + \tilde{\sigma}\Omega\mathbb{E}[\hat{X}\hat{X}^T]\Omega^T \\ & + \Psi\mathbb{E}[ZZ^T]\Psi^T + \tilde{\sigma}\Upsilon\mathbb{E}[ZZ^T]\Upsilon^T. \end{aligned} \tag{6.134}$$

Let

$$H = \mathbb{E}[XX^T] = \begin{bmatrix} H_{11,k} & H_{12,k} \\ H_{12,k}^T & H_{22,,k} \end{bmatrix},$$
$$V = \mathbb{E}[\hat{X}\hat{X}^T] = \begin{bmatrix} V_{11,k} & V_{12,k} \\ V_{12,k}^T & V_{22,k} \end{bmatrix},$$

where

$$\begin{aligned} H_{11,k} &= \mathbb{E}[\Delta u_k(t)\Delta u_k(t)^T], \\ H_{12,k} &= \mathbb{E}[\Delta u_k(t)\Delta x_k(t)^T], \\ V_{11,k} &= \mathbb{E}[u_k(t)u_k(t)^T], \\ V_{12,k} &= \mathbb{E}[u_k(t)x_k(t)^T], \end{aligned}$$

and

$$\begin{aligned} H_{22,k} &= \mathbb{E}[\Delta x_k(t)\Delta x_k(t)^T], \\ V_{22,k} &= \mathbb{E}[x_k(t)x_k(t)^T]. \end{aligned}$$

As a consequence, H and V are symmetric positive-semidefinite matrices. Moreover, we have

$$Q = \mathbb{E}[ZZ^T] = \begin{bmatrix} Q_{11,t} & 0 \\ 0 & Q_{22,t+1} \end{bmatrix},$$

where the zero sub-matrices $Q_{11,t}$ and $Q_{22,t+1}$ are due to zero cross-correlation between $\omega_k(t)$ and $\upsilon_k(t)$.

We expand the right-hand side of (6.134) to get (6.135)–(6.138) as follows:

$$\Phi\mathbb{E}(XX^T)\Phi^T = \begin{bmatrix}\Phi_1 & \Phi_2\\ B & A\end{bmatrix}\begin{bmatrix}H_{11,k} & H_{12,k}\\ H_{12,k}^T & H_{22,k}\end{bmatrix}\begin{bmatrix}\Phi_1^T & B^T\\ \Phi_2^T & A^T\end{bmatrix}$$
$$= \begin{bmatrix}\begin{pmatrix}\Phi_1 H_{11,k}\Phi_1^T + \Phi_2 H_{12,k}^T\Phi_1^T\\ + \Phi_1 H_{12,k}\Phi_2^T + \Phi_2 H_{22,k}\Phi_2^T\end{pmatrix} & \begin{pmatrix}\Phi_1 H_{11,k}B^T + \Phi_2 H_{12,k}^T B^T\\ + \Phi_1 H_{12,k}A^T + \Phi_2 H_{22,k}A^T\end{pmatrix}\\ \begin{pmatrix}BH_{11,k}\Phi_1^T + AH_{12,k}^T\Phi_1^T\\ + BH_{12,k}\Phi_2^T + AH_{22,k}\Phi_2^T\end{pmatrix} & \begin{pmatrix}BH_{11,k}B^T + AH_{12,k}^T B^T\\ + BH_{12,k}A^T + AH_{22,k}A^T\end{pmatrix}\end{bmatrix}, \tag{6.135}$$

$$\Psi\mathbb{E}(ZZ^T)\Psi^T = \begin{bmatrix}K_kC^+ & K_k\\ -I & 0\end{bmatrix}\begin{bmatrix}Q_{11,t} & 0\\ 0 & Q_{22,t+1}\end{bmatrix}\begin{bmatrix}(K_kC^+)^T & -I\\ K_k^T & 0\end{bmatrix}$$
$$= \begin{bmatrix}\begin{pmatrix}K_kC^+Q_{11,t}(K_kC^+)^T\\ + K_kQ_{22,t+1}K_k^T\end{pmatrix} & -K_kC^+Q_{11,t}\\ -Q_{11,t}(K_kC^+)^T & Q_{11,t}\end{bmatrix}, \tag{6.136}$$

$$\tilde{\sigma}\Upsilon\mathbb{E}(ZZ^T)\Upsilon^T = \tilde{\sigma}\begin{bmatrix}-K_kC^+ & -K_k\\ 0 & 0\end{bmatrix}\begin{bmatrix}Q_{11,t} & 0\\ 0 & Q_{22,t+1}\end{bmatrix}\begin{bmatrix}(-K_kC^+)^T & 0\\ -K_k^T & 0\end{bmatrix}$$
$$= \tilde{\sigma}\begin{bmatrix}(K_kC^+Q_{11,t}(K_kC^+)^T + K_kQ_{22,t+1}K_k^T) & 0\\ 0 & 0\end{bmatrix}, \tag{6.137}$$

$$\tilde{\sigma}\Omega\mathbb{E}(\tilde{X}\tilde{X}^T)\Omega^T = \tilde{\sigma}\begin{bmatrix}\Omega_1 & \Omega_2\\ 0 & 0\end{bmatrix}\begin{bmatrix}V_{11,k} & V_{12,k}\\ V_{12,k}^T & V_{22,k}\end{bmatrix}\begin{bmatrix}\Omega_1^T & 0\\ \Omega_2^T & 0\end{bmatrix}$$
$$= \tilde{\sigma}\begin{bmatrix}\begin{pmatrix}\Omega_1V_{11,k}\Omega_1^T + \Omega_2V_{12,k}^T\Omega_1^T\\ + \Omega_1V_{12,k}\Omega_2^T + \Omega_2V_{22,k}\Omega_2^T\end{pmatrix} & 0\\ 0 & 0\end{bmatrix}. \tag{6.138}$$

Then, we can deduce the trace of H^+ in (6.134) as

$$\begin{aligned}
&\text{trace}\{H^+\}\\
&= \text{trace}\{(\tilde{\sigma}+1)(K_kC^+Q_{11,t}C^{+^T}K_k^T + K_kQ_{22,t+1}K_k^T)\\
&\quad + \tilde{\sigma}[K_k(C^+B)V_{11,k}(C^+B)^TK_k^T\\
&\quad + K_k(C^+A)V_{12,k}^T(C^+B)^TK_k^T\\
&\quad + K_kC^+BV_{12,k}(C^+A)^TK_k^T\\
&\quad + K_kC^+AV_{22,k}(C^+A)^TK_k^T]\\
&\quad + (I - K_kC^+B)H_{11,k}(I - K_kC^+B)^T\\
&\quad - K_k(C^+A)H_{12,k}^T(I - K_kC^+B)^T\\
&\quad + (I - K_kC^+B)H_{12,k}(-C^+A)^T\\
&\quad + K_kC^+AH_{22,k}(C^+A)^TK_k^T + Q_{11,t} + BH_{11,k}B^T\\
&\quad + BH_{12,k}A^T + AH_{12,k}^TB^T + AH_{22,k}A^T\}.
\end{aligned} \tag{6.139}$$

Expanding and rearranging the terms on the right-hand side of this equation, we have

$$
\begin{aligned}
&\text{trace}\{H^+\} \\
&\quad = \text{trace}\{K_k[(\tilde{\sigma}+1)(C^+Q_{11,t}C^{+^T} + Q_{22,t+1}) \\
&\quad + \tilde{\sigma}[C^+BV_{11,k}(C^+B)^T + C^+AV_{12,k}^T(C^+B)^T \\
&\quad + C^+BV_{12,k}(C^+A)^T + C^+AV_{22,k}(C^+A)^T] \\
&\quad + C^+BH_{11,k}(C^+B)^T + C^+AH_{12,k}^T(C^+B)^T \\
&\quad + C^+BH_{12,k}(C^+A)^T + C^+AH_{22,k}(C^+A)^T]K_k^T \\
&\quad + Q_{11,t} + H_{11,k} + K_k[-C^+BH_{11,k} - C^+AH_{12,k}^T] \\
&\quad + [-H_{11,k}(C^+B)^T - H_{12,k}(C^+A)^T]K_k^T \\
&\quad + BH_{11,k}B^T + BH_{12,k}A^T \\
&\quad + AH_{12,k}^T B^T + AH_{22,k}A^T\}.
\end{aligned} \tag{6.140}
$$

Define

$$
D = \left[C^+B \; C^+A\right], \quad H_k^{(1)} = \left[H_{11,k} \; H_{12,k}\right],
$$

and

$$
\begin{aligned}
M_k^{(1)} = & BH_{11,k}B^T + BH_{12,k}A^T + AH_{12,k}^T B^T \\
& + AH_{22,k}A^T + Q_{11,t} + H_{11,k},
\end{aligned}
$$

then (6.140) is transformed as

$$
\begin{aligned}
\text{trace}\{H^+\} = & \text{trace}\{K_k[(\tilde{\sigma}+1)(C^+Q_{11,t}C^{+^T} + C^+Q_{22,t+1}) \\
& + \tilde{\sigma}DVD^T + DHD^T]K_k^T - K_kDH_k^{(1)^T} \\
& - H_k^{(1)}D^TK_k^T + M_k^{(1)}\}.
\end{aligned}
$$

Note that the direction regulation matrix $K_k(t)$ is used to update the input for the next batch. Therefore, the value of $K_k(t)$ is related to the input error at time instant t of the $(k+1)$th iteration and does not affect the state error at time instant $t+1$ of the kth iteration. Then,

$$
\frac{\partial(\text{trace}(H^+))}{\partial K_k} = \frac{\partial(\mathbb{E}[\Delta u_{k+1}(t)\Delta u_{k+1}(t)^T])}{\partial K_k}.
$$

To find K_k such that the trace of input error covariance matrix is minimized, we can set the derivative of trace of H^+ with respect to K_k as follows:

$$
\begin{aligned}
\frac{\partial(\text{trace}(H^+))}{\partial K_k} = & 2K_k[\tilde{\sigma}DVD^T + DHD^T + (\tilde{\sigma}+1) \\
& (C^+Q_{11,t}C^{+^T} + Q_{22,t+1})] - 2H_k^{(1)}D^T.
\end{aligned}
$$

Note that H, V, and $Q_{11,t}$ are positive-semidefinite matrices, and $Q_{22,t+1}$ is a positive-definite matrix. Then, we obtain that $\tilde{\sigma}DVD^T + DHD^T + (\tilde{\sigma}+1)(C^+ Q_{11,t}C^{+^T} + Q_{22,t+1})$ is nonsingular. Let $\frac{\partial(\text{trace}(H^+))}{\partial K_k} = 0$ yielding an optimal direction regulation matrix K_k

$$
\begin{aligned}
K_k =& H_k^{(1)} D^T [\tilde{\sigma} DVD^T + DHD^T \\
&+ (\tilde{\sigma}+1)(C^+ Q_{11,t} C^{+^T} + Q_{22,t+1})]^{-1}.
\end{aligned} \tag{6.141}
$$

6.5.3.2 Convergence Analysis

From (6.129) and (6.130), we have

$$
\begin{aligned}
\Delta x_k(t) =& \left[\prod_{j=0}^{t-1} A^T(j)\right]^T \Delta x_k(0) + \sum_{m=0}^{t-1}\left[\prod_{n=m}^{t-2} A^T(n+1)\right]^T \\
&\times [B(m)\Delta u_k(m) - \omega_k(m)],
\end{aligned} \tag{6.142}
$$

and

$$
\begin{aligned}
\Delta u_k(t) =& \left[\prod_{j=0}^{k-1} \Gamma_{1,j}^T\right]^T \Delta u_0(t) + \sum_{m=0}^{k-1}\left[\prod_{n=m}^{k-2} \Gamma_{1,n+1}^T\right]^T \\
&\times \left[\Gamma_{2,m}\Delta x_m(t) - g_{t,m}\right],
\end{aligned} \tag{6.143}
$$

where

$$
\begin{aligned}
\Gamma_{1,k} &= I - K_k(t)C(t+1)B(t), \\
\Gamma_{2,k} &= -K_k(t)C(t+1)A(t),
\end{aligned}
$$

and

$$
\begin{aligned}
g_{t,k} =& \tilde{\theta}_k(t)K_k\upsilon_k(t+1) - K_k(t)C^+\omega_k(t) - K_k(t)\upsilon_k(t+1) \\
&+ \tilde{\theta}_k(t)K_kC^+[Ax_k(t) + Bu_k(t) + \omega_k(t)].
\end{aligned}
$$

Noting all terms on the right-hand side of both (6.142) and (6.143), we can conclude that $\Delta x_k(0)$, $\Delta u_0(t)$, $\omega_k(0 \le m \le t-1)$, and $g_{t,0\le m\le k-1}$ are all uncorrelated. In addition, these terms are also uncorrelated with $\Delta u_k(0 \le m \le t-1)$ and $\Delta x_{0\le m\le k-1}(t)$. Because of $\Delta u_k(0 \le m \le t-1)$ and $\Delta x_{0\le m\le k-1}(t)$ cannot be represented by each other, thus they are uncorrelated. Consequently, we get $H_{12,k} = 0$.

Next, it is noticed that the covariance between $x_k(t)$ and $u_k(t)$ is the same as the covariance between $\Delta x_k(t)$ and $\Delta u_k(t)$. Since the last two terms $\Delta u_k(0 \le m \le t-$

1) and $\Delta x_{0\le m\le k-1}(t)$ are uncorrelated, hence $u_k(0 \le m \le t-1)$ and $x_{0\le m\le k-1}(t)$ are uncorrelated. Therefore, $V_{12,k} = 0$ and $V_{12,k}^T = 0$.

Then, K_k is reduced to

$$\begin{aligned} K_k =& H_{11,k}(C^+B)^T[\tilde{\sigma}((C^+B)V_{11,k}(C^+B)^T \\ &+ (C^+A)V_{22,k}(C^+A)^T) + (C^+B)H_{11,k}(C^+B)^T \\ &+ (C^+A)H_{22,k}(C^+A)^T \\ &+ (\tilde{\sigma}+1)(C^+Q_{11,t}C^{+T} + Q_{22,t+1})]^{-1}. \end{aligned} \tag{6.144}$$

From (6.131), we obtain the input error covariance matrix

$$\begin{aligned} H_{11,k+1} &= \mathbb{E}[\Delta u_{k+1}(t)\Delta u_{k+1}^T(t)] \\ &= (I - K_kC^+B)H_{11,k}(I - K_kC^+B)^T \\ &\quad + K_k[C^+AH_{22,k}(C^+A)^T \\ &\quad + (\tilde{\sigma}+1)(C^+Q_{11,t}C^{+T} + Q_{22,t+1}) \\ &\quad + \tilde{\sigma}(C^+BV_{11,k}(C^+B)^T \\ &\quad + C^+AV_{22,k}(C^+A)^T)]K_k^T. \end{aligned} \tag{6.145}$$

Define

$$\begin{aligned} M &= C^+B, \\ S_1 &= C^+AH_{22,k}(C^+A)^T \\ &\quad + (\tilde{\sigma}+1)(C^+Q_{11,t}C^{+T} + Q_{22,t+1}) \\ &\quad + \tilde{\sigma}(C^+BV_{11,k}(C^+B)^T + C^+AV_{22,k}(C^+A)^T). \end{aligned}$$

Expanding (6.145), we have

$$\begin{aligned} H_{11,k+1} =& H_{11,k} - K_kMH_{11,k} + K_k(MH_{11,k}M^T \\ &+ S_1)K_k^T - H_{11,k}M^TK_k^T. \end{aligned} \tag{6.146}$$

Substituting (6.144) into (6.146) yields

$$\begin{aligned} H_{11,k+1} =& H_{11,k} - H_{11,k}M^T(MH_{11,k}M^T + S_1)^{-1}MH_{11,k} \\ &- H_{11,k}M^T(MH_{11,k}M^T + S_1)^{-1}MH_{11,k} \\ &+ H_{11,k}M^T(MH_{11,k}M^T + S_1)^{-1}MH_{11,k} \\ =& (I - K_kM)H_{11,k}. \end{aligned} \tag{6.147}$$

Theorem 6.8 *Consider system* (6.124) *with output fading and assume that Assumptions 6.9–6.11 hold. The learning algorithm, presented by* (6.129), (6.144), *and*

(6.147), *guarantees that the input covariance matrix converges to zero asymptotically in the mean-square sense. That is,* $H_{11,k} \to 0$ *as* $k \to \infty$.

Proof Rewrite (6.144) as follows:

$$K_k = H_{11,k}M^T[MH_{11,k}M^T + S_1]^{-1}. \tag{6.148}$$

Note that $H_{22,k}$, $V_{11,k}$, $V_{22,k}$ and $Q_{11,t}$ are positive-semidefinite matrices and $Q_{22,t+1}$ is positive definite, thus S_1 is positive definite. Combining (6.144) and (6.145), we have

$$H_{11,k+1} = (I - K_kM)H_{11,k}(I - K_kM)^T + K_kS_1K_k^T.$$

If $H_{11,k} > 0$, it can be examined from that $H_{11,k+1} \geq K_kS_1K_k^T > 0$, since $S_1 > 0$ and M is a full–column rank matrix. It means that $H_{11,0} > 0$ leads to $H_{11,k} > 0$. Then, taking inverse of both sides of (6.147) with the well-known matrix inversion method yields to

$$\begin{aligned}
H_{11,k+1}^{-1} &= H_{11,k}^{-1}[I - K_kM]^{-1} \\
&= H_{11,k}^{-1}[I - K_k(K_k - M^{-1})^{-1}] \\
&= H_{11,k}^{-1} - M^T(MH_{11,k}M^T + S_1)^{-1} \\
&\quad \times [H_{11,k}M^T(MH_{11,k}M^T + S_1)^{-1} - M^{-1}]^{-1} \\
&= H_{11,k}^{-1} - \left[H_{11,k} - M^{-1}(MH_{11,k}M^T + S_1)M^{-T}\right] \\
&= H_{11,k}^{-1} + M^TS_1^{-1}M.
\end{aligned}$$

Then, we have

$$H_{11,k+1}^{-1} = H_{11,0}^{-1} + kM^TS_1^{-1}M.$$

From the above analysis, we can readily deduce that $H_{11,k+1} < H_{11,k}$, $\forall k$ and $\lim_{k\to\infty} H_{11,k} = 0$.

Theorem 6.8 demonstrates the asymptotic convergence of the generated input sequence to the desired control value as the iteration number increases in a mean-square sense. This convergence is attributed to the enhanced performance of ILC against various sources of randomness, facilitated by Kalman filtering theory. It can be observed from Theorem 6.8 that K_k relies on the variance of the fading variable. In case this variance becomes zero (indicating no impact of fading on the output), our proposed algorithm aligns with that presented in Saab (2001a). This observation establishes a connection between our study's findings and existing literature.

Corollary 6.1 *Consider system* (6.124) *with output fading and assume that Assumptions 6.9–6.11 hold. In addition, assume that there is no system noise and initial state error, i.e.,* $\omega_k(t) = 0$ *and* $\Delta x_k(0) = 0$. *Then, the learning algorithm, presented by* (6.129), (6.144), *and* (6.147), *guarantees that* $\|H_{22,k}\| \to 0$ *as* $k \to \infty$.

Proof Because none of system noise and initial state error exist in the dynamics, (6.142) is reduced to

$$\Delta x_k(t) = \sum_{m=0}^{t-1} \left[\prod_{n=m}^{t-2} A^T(n+1) \right]^T B(m)\Delta u_k(m).$$

Then, we have

$$\begin{aligned}
&\mathbb{E}[\Delta x_k(t)\Delta x_k(t)^T] \\
&= \left\{ \sum_{i,j=0}^{t-1} \left[\prod_{n=i}^{t-2} A^T(n+1) \right]^T B(i)\mathbb{E}[\Delta u_k(i)\Delta u_k(j)^T] \right. \\
&\quad \left. \times B(j)^T \sum_{j=0}^{t-1} \left[\prod_{n=j}^{t-2} A^T(n+1) \right] \right\}.
\end{aligned}$$

Using the basic inequality, we have

$$\begin{aligned}
&\lim_{k\to\infty} \mathbb{E}[\Delta x_k(t)\Delta x_k(t)^T] \\
&\quad \le t \sum_{m=0}^{t-1} \left\{ \left[\prod_{n=m}^{t-2} A^T(n+1) \right]^T B(m) \right. \\
&\quad \times \lim_{k\to\infty} \mathbb{E}[\Delta u_k(m)\Delta u_k(m)^T] B(m)^T \\
&\quad \left. \times \left[\prod_{n=m}^{t-2} A^T(n+1) \right] \right\}.
\end{aligned}$$

Because $\lim_{k\to\infty} H_{11,k} = 0$, $\forall 0 \le m \le t-1$, $\lim_{k\to\infty} \mathbb{E}[\Delta x_k(t)\Delta x_k(t)^T] = 0$.

Remark 6.9 We note that the proposed learning control algorithm (6.144) requires the state error covariance matrix for computation. This covariance matrix can be obtained according (6.130) as follows:

$$\begin{aligned}
H_{22,k}(t+1) &= \mathbb{E}[\Delta x_k(t+1)\Delta x_k(t+1)^T] \\
&= AH_{22,k}(t)A^T + BH_{11,k}(t)B^T + Q_{11,t}.
\end{aligned} \tag{6.149}$$

6.5.4 Input Fading Case

In this subsection, we address the input fading scenario. Specifically, the channel from the controller to the plant experiences random fading, while assuming no fading in the channel from the plant to the controller. In contrast to the previous section, unstable operation may arise due to incorrect reception of input caused by input fading.

6.5.4.1 Algorithm Design

The generated input by controller is $u_k(t)$. The received signal by plant is denoted by $u_k^\circ(t)$. Then, according to the fading model

$$u_k^\circ(t) = \theta_k(t)u_k(t), \tag{6.150}$$

where $\theta_k(t)$ is the random fading variable. In the rest of this section, we retain most notation same as the previous section to save notation.

Clearly, the received input signal is not suitable for driving the system because of its deviation from the computed one. The mean inverse of the fading variable is multiplied to correct the signal

$$\hat{u}_k(t) = \mu^{-1}\theta_k(t)u_k(t), \tag{6.151}$$

where $\hat{u}_k(t)$ denotes the corrected output signal. Then, the dynamics is given by

$$x_k(t+1) = A(t)x_k(t) + B(t)\mu^{-1}\theta_k(t)u_k(t) + \omega_k(t). \tag{6.152}$$

We apply the following update law

$$u_{k+1}(t) = u_k(t) + K_k(t)[y_d(t+1) - y_k(t+1)]. \tag{6.153}$$

The state error and input error can be expressed as follows:

$$\begin{aligned} \Delta x_k(t+1) = &A(t)\Delta x_k(t) + B(t)\Delta u_k(t) \\ &+ B(t)(1 - \mu^{-1}\theta_k(t))u_k(t) - \omega_k(t), \end{aligned} \tag{6.154}$$

and

$$\begin{aligned} \Delta u_{k+1}(t) &= \Delta u_k(t) - K_k(t)[y_d(t+1) - y_k(t+1)] \\ &= [I - K_k(t)C(t+1)B(t)]\Delta u_k(t) \\ &\quad - K_k(t)C(t+1)A(t)\Delta x_k(t) \\ &\quad + K_k(t)C(t+1)\omega_k(t) + K_k(t)\upsilon_k(t+1) \\ &\quad - (1 - \mu^{-1}\theta_k(t))K_k(t)C(t+1)B(t)u_k(t). \end{aligned} \tag{6.155}$$

We can combine (6.154) and (6.155) into a Roessor model

$$\begin{bmatrix} \Delta u_{k+1}(t) \\ \Delta x_k(t+1) \end{bmatrix} = \begin{bmatrix} I - K_k C^+ B & -K_k C^+ A \\ B & A \end{bmatrix} \begin{bmatrix} \Delta u_k(t) \\ \Delta x_k(t) \end{bmatrix} + \begin{bmatrix} K_k C^+ & K_k \\ -I & 0 \end{bmatrix} \begin{bmatrix} \omega_k(t) \\ \upsilon_k(t+1) \end{bmatrix} + \begin{bmatrix} -\tilde{\theta}_k(t) K_k C^+ B & 0 \\ \tilde{\theta}_k(t) B & 0 \end{bmatrix} \begin{bmatrix} u_k(t) \\ x_k(t) \end{bmatrix}. \tag{6.156}$$

Writing (6.156) in a compact form, we have

$$X^+ = \Phi X + \tilde{\theta}_k(t) \Omega \hat{X} + \Psi Z, \tag{6.157}$$

where

$$\Omega = \begin{bmatrix} -K_k C^+ B & 0 \\ B & 0 \end{bmatrix}$$

is a $(n+p) \times (n+q)$ matrix and the other notation is same as in the previous section.

Because initial state errors and input errors at the initial iterations are of zero-mean, the expectation of X^+ is zero. Denote $P^+ = \mathbb{E}[X^+ X^{+T}]$ as a covariance matrix. Out objective is to find an optimal direction regulation matrix K_k minimizing the trace of P^+. To this end, we expand P^+ as follows:

$$\begin{aligned} P^+ = & \Phi \mathbb{E}[X X^T] \Phi^T + \tilde{\sigma} \Omega \mathbb{E}[\hat{X} \hat{X}^T] \Omega^T \\ & + \Psi \mathbb{E}[Z Z^T] \Psi^T. \end{aligned} \tag{6.158}$$

Then, we have

$$\begin{aligned} \text{trace}\{P^+\} = & \text{trace}\{K_k[(C^+ Q_{11,t} C^{+T} + Q_{22,t+1}) \\ & + \tilde{\sigma}(C^+ B) V_{11,k} (C^+ B)^T + C^+ A H_{22,k} (C^+ A)^T \\ & + C^+ B H_{11,k} (C^+ B)^T + C^+ A H_{12,k}^T (C^+ B)^T \\ & + C^+ B H_{12,k} (C^+ A)^T] K_k^T + \tilde{\sigma} B V_{11,k} B^T \\ & + Q_{11,t} + H_{11,k} + K_k[-C^+ B H_{11,k} - C^+ A H_{12,k}^T] \\ & + [-H_{11,k} (C^+ B)^T - H_{12,k} (C^+ A)^T] K_k^T \\ & + B H_{11,k} B^T + B H_{12,k} A^T \\ & + A H_{12,k}^T B^T + A H_{22,k} A^T\}. \end{aligned} \tag{6.159}$$

Setting

$$D = \begin{bmatrix} C^+ B & C^+ A \end{bmatrix}, \quad H_k^{(2)} = \begin{bmatrix} H_{11,k} & H_{12,k} \end{bmatrix},$$

and

$$\begin{aligned} M_k^{(2)} =& BH_{11,k}B^T + BH_{12,k}A^T + AH_{12,k}^T B^T \\ &+ AH_{22,k}A^T + Q_{11,t} + H_{11,k} + \tilde{\sigma} B V_{11,k} B^T, \end{aligned}$$

the trace of P^+ is reduced to

$$\begin{aligned} \text{trace}\{P^+\} = & \text{trace}\{K_k[(C^+ Q_{11,t} C^{+^T} + Q_{22,t+1}) \\ & + \tilde{\sigma}(C^+ B) V_{11,k} (C^+ B)^T + DHD^T] K_k^T \\ & - K_k D H_k^{(2)^T} - H_k^{(2)} D^T K_k^T + M_k^{(2)}\}. \end{aligned}$$

To find K_k minimizing the trace of P^+, we calculate the derivative of trace of P^+ with respect to K_k as follows:

$$\begin{aligned} \frac{\partial(\text{trace} P^+)}{\partial K_k} =& 2K_k[\tilde{\sigma}(C^+ B) V_{11,k} (C^+ B)^T + DHD^T \\ & + (C^+ Q_{11,t} C^{+^T}) + Q_{22,t+1}] - 2H_k^{(2)} D^T. \end{aligned}$$

Note that H, $V_{11,k}$ and $Q_{11,t}$ are positive-semidefinite matrices and $Q_{22,t+1}$ is positive definite, thus $\tilde{\sigma}(C^+ B) V_{11,k} (C^+ B)^T + DHD^T + (C^+ Q_{11,t} C^{+^T}) + Q_{22,t+1}$ is nonsingular. Letting $\frac{\partial(\text{trace} H^+)}{\partial K_k} = 0$, we can obtain a solution of K_k as follows:

$$\begin{aligned} K_k = & H_k^{(2)} D^T [\tilde{\sigma}(C^+ B) V_{11,k} (C^+ B)^T + DHD^T \\ & + (C^+ Q_{11,t} C^{+^T} + Q_{22,t+1})]^{-1}. \end{aligned} \tag{6.160}$$

6.5.4.2 Convergence Analysis

From (6.153) and (6.154), we have

$$\begin{aligned} \Delta u_k(t) = & \left[\prod_{j=0}^{k-1} \Gamma_{1,j}^T \right]^T \Delta u_0(t) + \sum_{m=0}^{k-1} \left[\prod_{n=m}^{k-2} \Gamma_{1,n+1}^T \right]^T \\ & \times \left[\Gamma_{2,m} \Delta x_m(t) - g_{t,m} \right], \end{aligned} \tag{6.161}$$

and

$$\begin{aligned} \Delta x_k(t) = & \left[\prod_{j=0}^{t-1} A^T(j) \right]^T \Delta x_k(0) + \sum_{m=0}^{t-1} \left[\prod_{n=m}^{t-2} A^T(n+1) \right]^T \\ & \times [B(m) \Delta u_k(m) - f_{m,k}], \end{aligned} \tag{6.162}$$

where

$$\begin{aligned}
\Gamma_{1,k} &= I - K_k(t)C(t+1)B(t), \\
\Gamma_{2,k} &= -K_k(t)C(t+1)A(t), \\
g_{t,k} &= -K_k(t)C^{+}\omega_k(t) - K_k(t)\upsilon_k(t+1) \\
&\quad + \tilde{\theta}_k(t)K_kC^{+}Bu_k(t), \\
f_{t,k} &= B\tilde{\theta}_k(t)u_k(t) - \omega_k(t).
\end{aligned}$$

Considering all terms at the right-hand side of (6.161) and (6.162), we can conclude that $\Delta x_k(0)$, $\Delta u_0(t)$, $f_{0\le m\le t-1,k}$, and $g_{t,0\le m\le k-1}$ are all uncorrelated. In addition, these terms are also uncorrelated with $\Delta u_k(0 \le m \le t-1)$ and $\Delta x_{0\le m\le k-1}(t)$. Because of $\Delta u_k(0 \le m \le t-1)$ and $\Delta x_{0\le m\le k-1}(t)$ cannot be represented by each other, they are uncorrelated. Consequently, we get $H_{12,k} = 0$.

Next, it is noticed that the covariance between $x_k(t)$ and $u_k(t)$ is the same as the covariance between $\Delta x_k(t)$ and $\Delta u_k(t)$. Since the last two terms $\Delta u_k(0 \le m \le t-1)$ and $\Delta x_{0\le m\le k-1}(t)$ are uncorrelated, hence $u_k(0 \le m \le t-1)$ and $x_{0\le m\le k-1}(t)$ are uncorrelated. Therefore, $V_{12,k} = 0$ and $V_{12,k}^T = 0$. Then, K_k is reduced to

$$\begin{aligned}
K_k =& H_{11,k}(C^{+}B)^T[\tilde{\sigma}(C^{+}B)V_{11,k}(C^{+}B)^T \\
&+ (C^{+}B)H_{11,k}(C^{+}B)^T + (C^{+}A)H_{22,k} \\
&\times (C^{+}A)^T + (C^{+}Q_{11,t}C^{+^T} + Q_{22,t+1})]^{-1}.
\end{aligned} \tag{6.163}$$

From (6.155), we obtain the input error covariance matrix

$$\begin{aligned}
H_{11,k+1} &= \mathbb{E}[\Delta u_{k+1}(t)\Delta u_{k+1}^T(t)] \\
&= (I - K_kC^{+}B)H_{11,k}(I - K_kC^{+}B)^T \\
&\quad + K_k[C^{+}AH_{22,k}(C^{+}A)^T \\
&\quad + (C^{+}Q_{11,t}C^{+^T} + Q_{22,t+1}) \\
&\quad + \tilde{\sigma}(C^{+}B)V_{11,k}(C^{+}B)^T]K_k^T.
\end{aligned} \tag{6.164}$$

Define

$$\begin{aligned}
M &= C^{+}B, \\
S_2 &= C^{+}AH_{22,k}(C^{+}A)^T \\
&\quad + (C^{+}Q_{11,t}C^{+^T} + Q_{22,t+1}) \\
&\quad + \tilde{\sigma}(C^{+}B)V_{11,k}(C^{+}B)^T.
\end{aligned}$$

Expanding (6.164), we have

$$H_{11,k+1} = H_{11,k} - K_k M H_{11,k} + K_k (M H_{11,k} M^T + S_2) K_k^T - H_{11,k} M^T K_k^T. \tag{6.165}$$

Substituting (6.163) into (6.165) yields

$$\begin{aligned} H_{11,k+1} &= H_{11,k} - H_{11,k} M^T (M H_{11,k} M^T + S_2)^{-1} M H_{11,k} \\ &\quad - H_{11,k} M^T (M H_{11,k} M^T + S_2)^{-1} M H_{11,k} \\ &\quad + H_{11,k} M^T (M H_{11,k} M^T + S_2)^{-1} M H_{11,k} \\ &= (I - K_k C^+ B) H_{11,k}. \end{aligned} \tag{6.166}$$

Theorem 6.9 *Consider system* (6.124) *with input fading and assume that Assumptions 6.9–6.11 hold. The learning algorithm, presented by* (6.153), (6.163), *and* (6.166), *guarantees that the input covariance matrix converges to zero asymptotically in the mean-square sense. That is,* $H_{11,k} \to 0$ *as* $k \to \infty$.

Proof From the above analysis, we can readily deduce that

$$K_k = H_{11,k} M^T [M H_{11,k} M^T + S_2]^{-1}.$$

It means that $H_{11,0} > 0$ leads to $H_{11,k} > 0$. Then, taking inverse of both sides of (6.166) with the well known matrix inversion method yields to

$$\begin{aligned} H_{11,k+1}^{-1} &= H_{11,k}^{-1} [I - K_k M]^{-1} \\ &= H_{11,k}^{-1} - M^T (M H_{11,k} M^T + S_2)^{-1} \\ &\quad \times [H_{11,k} M^T (M H_{11,k} M^T + S_2)^{-1} - M^{-1}]^{-1} \\ &= H_{11,k}^{-1} + M^T S_1^{-1} M. \end{aligned}$$

Then, we have

$$H_{11,k+1}^{-1} = H_{11,0}^{-1} + k M^T S_2^{-1} M.$$

From the above analysis, we can readily deduce that $H_{11,k+1} < H_{11,k}$, $\forall k$ and $\lim_{k\to\infty} H_{11,k} = 0$.

Corollary 6.2 *Consider system* (6.124) *with output fading and assume that Assumptions 6.9–6.11 hold. In addition, assume that there is no system noise and initial state error, i.e.,* $\omega_k(t) = 0$ *and* $\Delta x_k(0) = 0$. *Then, the learning algorithm, presented by* (6.153), (6.163), *and* (6.166), *guarantees that* $\|H_{22,k}\| \to 0$ *as* $k \to \infty$.

Proof The proof is the same as that of Corollary 6.1.

Remark 6.10 To apply the proposed learning control algorithm, we need the state error covariance matrix. It can be derived from (6.154) as follows:

$$\begin{aligned} H_{22,k}(t+1) =&\mathbb{E}[\Delta x_k(t+1)\Delta x_k(t+1)^T] \\ =&AH_{22,k}(t)A^T + BH_{11,k}(t)B^T \\ &+ Q_{11,t} + \tilde{\sigma} BV_{11,k}(t)B^T. \end{aligned} \tag{6.167}$$

6.5.5 Extension to General Case

In the preceding subsections, we have examined the fading of output and input signals separately to elucidate our main concept and emphasize the distinction between these two scenarios. However, in numerous practical applications, both output and input signals are typically transmitted over the same communication networks, resulting in coexisting fading phenomena. In this section, we briefly expound on the extension to this general case.

The input fading variable and the output fading variable are denoted by $\theta_k^{(1)}(t)$ and $\theta_k^{(2)}(t)$, respectively. Moreover, $\theta_k^{(1)}(t)$ is independent of $\theta_k^{(2)}(t)$. In addition, we assume that $\theta_k^{(*)}(t)$ has the following statistical properties

$$\mu_* = \mathbb{E}[\theta_k^{(*)}(t)], \qquad \sigma_*^2 = \mathbb{E}[(\theta_k^{(*)}(t) - \mu_*)^2],$$

where the subscript "$*$" indicates 1 or 2 according to the input fading case or the output fading case.

Define

$$\tilde{\theta}_k^{(*)}(t) = 1 - \mu_*^{-1}\theta_k^{(*)}(t), \qquad \tilde{\sigma}_*^2 = {\mu_*}^2/\sigma_*^2.$$

Again, we denote $C^+ = C(t+1)$, $B = B(t)$, $A = A(t)$, $K_k = K_k(t)$, the state error $\Delta x_k(t) = x_d(t) - x_k(t)$, and the input error vector $\Delta u_k(t) = u_d(t) - u_k(t)$. According to the system (6.124) and reference model (6.125) , we have that

$$\Delta x_k(t+1) =A\Delta x_k(t) + B\mu^{-1}\theta_k^{(1)}(t)u_k(t) + \omega_k(t), \tag{6.168}$$

where the correction to the input is imposed. Therefore, the input error is given by

$$\begin{aligned} \Delta u_{k+1}(t) &= \Delta u_k(t) - K_k(t)[y_d(t+1) - \mu^{-1}\theta_k^{(2)}(t)y_k(t+1)] \\ &= [I - K_kC^+B]\Delta u_k(t) - K_kC^+A\Delta x_k(t) \\ &\quad + K_kC^+\omega_k(t) + K_k\upsilon_k(t+1) \\ &\quad - \tilde{\theta}_k^{(2)}(t)K_k[C^+Ax_k(t) + C^+\omega_k(t) + \upsilon_k(t+1)] \\ &\quad - K_kC^+B[\tilde{\theta}_k^{(1)}(t) + \tilde{\theta}_k^{(2)}(t)\mu_1^{-1}\theta_k^{(2)}(t)]u_k(t). \end{aligned} \tag{6.169}$$

We can combine (6.168) and (6.169) into a Roessor model

$$\begin{aligned}\begin{bmatrix}\Delta u_{k+1}(t)\\ \Delta x_k(t+1)\end{bmatrix} &= \begin{bmatrix}I-K_kC^+B & -K_kC^+A\\ B & A\end{bmatrix}\begin{bmatrix}\Delta u_k(t)\\ \Delta x_k(t)\end{bmatrix}\\ &\quad+\begin{bmatrix}-\tilde{\theta}_k^{(2)}(t)K_kC^+ & -\tilde{\theta}_k^{(2)}(t)K_k\\ 0 & 0\end{bmatrix}\begin{bmatrix}\omega_k(t)\\ \upsilon_k(t+1)\end{bmatrix}\\ &\quad+\begin{bmatrix}K_kC^+ & K_k\\ -I & 0\end{bmatrix}\begin{bmatrix}\omega_k(t)\\ \upsilon_k(t+1)\end{bmatrix}\\ &\quad+\begin{bmatrix}\star & -\tilde{\theta}_k^{(2)}(t)K_kC^+A\\ \tilde{\theta}_k^{(1)}(t)B & 0\end{bmatrix}\begin{bmatrix}u_k(t)\\ x_k(t)\end{bmatrix},\end{aligned} \tag{6.170}$$

where $\star = -K_kC^+B[\tilde{\theta}_k^{(1)}(t)+\tilde{\theta}_k^{(2)}(t)\mu_1^{-1}\theta_k^{(1)}(t)]$.

Similar to the previous analysis procedures, we find that the learning gain that minimizes the trace of the input error covariance matrix is similar to the one presented by (6.144); particularly, we have

$$\begin{aligned}K_k &= H_{11,k}(C^+B)^T[\tilde{\sigma}_2BV_{11,k}B^T\\ &\quad+\tilde{\sigma}_1(C^+A)V_{22,k}(C^+A)^T+(C^+B)H_{11,k}(C^+B)^T\\ &\quad+(C^+A)H_{22,k}(C^+A)^T+\tilde{\sigma}_1\tilde{\sigma}_2(C^+B)V_{11,k}(C^+B)^T\\ &\quad+(\tilde{\sigma}_2+1)(C^+Q_{11,t}C^{+T}+Q_{22,t+1})]^{-1}.\end{aligned} \tag{6.171}$$

The input error covariance matrix becomes

$$\begin{aligned}H_{11,k+1} &= \mathbb{E}[\Delta u_{k+1}(t)\Delta u_{k+1}^T(t)]\\ &= (I-K_kC^+B)H_{11,k}(I-K_kC^+B)^T\\ &\quad+K_k[C^+AH_{22,k}(C^+A)^T\\ &\quad+(\tilde{\sigma}_2+1)(C^+Q_{11,t}C^{+T}+Q_{22,t+1})\\ &\quad+\tilde{\sigma}_2BV_{11,k}B^T+\tilde{\sigma}_1(C^+A)V_{22,k}(C^+A)^T\\ &\quad+\tilde{\sigma}_1\tilde{\sigma}_2(C^+B)V_{11,k}(C^+B)^T]K_k^T\\ &= (I-K_kC^+B)H_{11,k}.\end{aligned} \tag{6.172}$$

Therefore, we can conclude the following theorem.

Theorem 6.10 *Consider system* (6.124) *with both output and input fading and assume that Assumptions 6.9–6.11 hold. The learning algorithm, presented by* (6.129), (6.171), *and* (6.172), *guarantees that the input covariance matrix converges to zero asymptotically in the mean-square sense. That is,* $H_{11,k} \to 0$ *as* $k \to \infty$.

6.5.6 Illustrative Simulations

Consider a discrete time-variant linear MIMO system, where the system matrices are given as follows:

$$A(t) = \begin{bmatrix} 0.08\sin(0.3t) & -0.1 & 0.03t \\ 0.1 & -0.02t & -0.05\cos(0.3t) \\ 0.1 & 0.1 & 0.1 + 0.08\cos(0.3t) \end{bmatrix},$$

$$B(t) = \begin{bmatrix} 1.5 - 0.4\sin^2(0.4\pi t) & 0 \\ 0.02t & 0.02t \\ 0 & 0.5 + 0.2\sin(0.4\pi t) \end{bmatrix},$$

$$C(t) = \begin{bmatrix} 0.4 + 0.2\sin^2(0.3\pi t) & 0.1 & -0.1 \\ 0 & 0.1 & 0.4 - 0.2\sin(0.3\pi t) \end{bmatrix}.$$

For the initial state, it is assumed that $x_k(0) = x_d(0)$ for all k. The random variable $\theta_k(t) \sim N(0.95, 0.1^2)$. The system noises $\omega_k(t) \sim N(0, 0.1^2)$ and measurement noises $\upsilon_k(t) \sim N(0, 0.1^2)$. Without loss of generality, the initial conditions are set to $H_{11,0}(t) = \text{diag}\{20, 20\}$ and $H_{22,k}(0) = \text{diag}\{0, 0, 0\}$. The operation is repeated for 100 iterations.

The desired output within the time interval $t \in [0, 50]$ is given as follows:

$$y_d(t) = \begin{bmatrix} 1.2\sin(\frac{\pi t}{8}) + 1.1(1 - \cos(\frac{\pi t}{6})) \\ -1.2\sin(\frac{\pi t}{8}) - 1.1(1 - \cos(\frac{\pi t}{6})) \end{bmatrix}.$$

The output tracking performance and the convergence of the input error are demonstrated in Figs. 6.16 and 6.17. In Fig. 6.2 (a) and 6.2 (b) for output fading and input fading, respectively, it can be seen that output trajectories at the 40th iteration almost overlap with the desired trajectory, demonstrating a satisfied tracking performance after several learning iterations. The evolution of averaged 2-norm of input errors (defined as $(\sum_{t=0}^{N-1} \|\Delta u_k(t)\|^2)/N$ for the kth iteration) along the iteration axis is demonstrated in Fig. 6.3 (a) and 6.3 (b) for output fading and input fading, respectively. The profiles in both figures decrease rapidly along the iteration axis, implying that the proposed algorithms have favorite performance in handling stochastic systems over fading channels.

To show the better performance of the proposed algorithms than the conventional P-type learning algorithm, we also simulate the algorithms with fixed learning gain matrices. In particular, the selected gain matrix K for both output fading and input fading selects

$$K = \begin{bmatrix} 0.98 & 0.462 \\ -0.01 & 0.574 \end{bmatrix}.$$

The contraction condition $\|I - KC^+B\| < 1$ is satisfied by these selections. The mean-square input error profile for this algorithm along the iteration axis is demon-

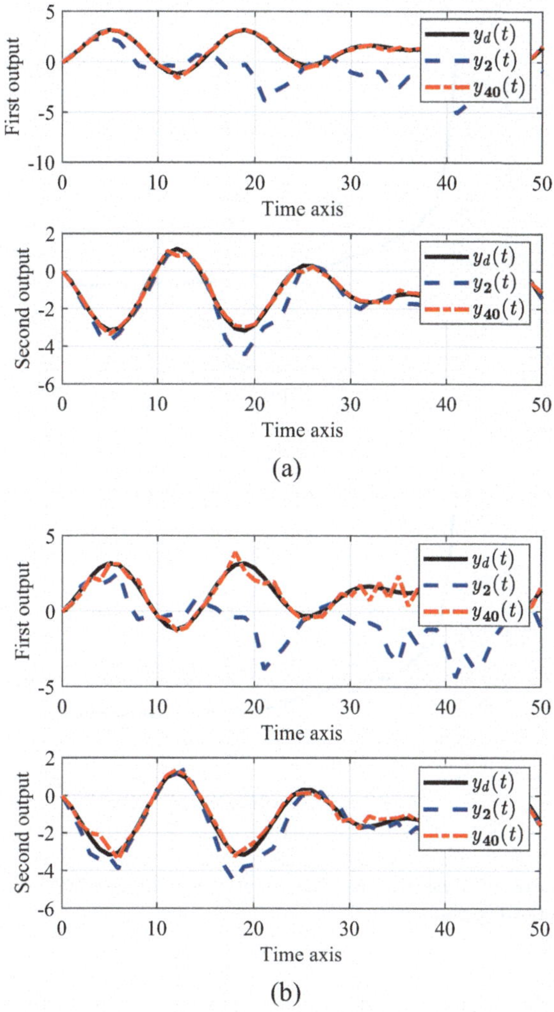

Fig. 6.16 The desired trajectory and tracking profiles at the 2nd and 40th iteration: **a** output fading case; **b** input fading case

strated in Fig. 6.18 (indicated by the dotted lines in both upper and lower figures). It is evident that due to the presence of various random noises, this algorithm with a fixed gain matrix fails to continuously improve the input after several iterations. This observation reveals the distinct advantages of the proposed approach in handling different sources of randomness and achieving competitive tracking performance.

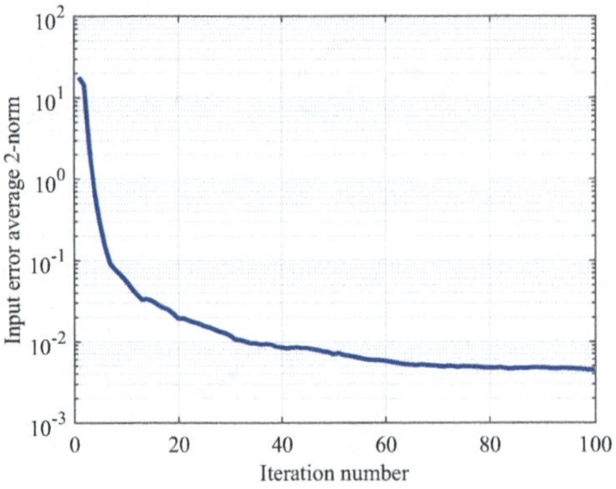

(a)

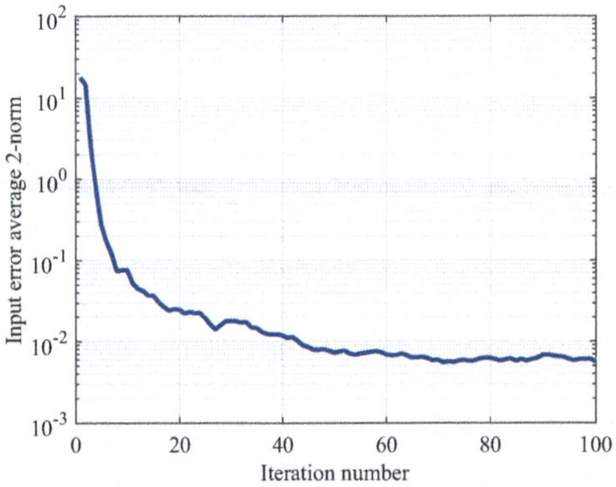

(b)

Fig. 6.17 Evolution of the averaged mean-square input error profile along the iteration axis: **a** output fading case; **b** input fading case

The effect of fading channel is demonstrated by simulating the framework without fading randomness, as a comparison to the algorithm in Saab (2001a). The mean-square input error profile in the iteration domain is shown in Fig. 6.18. It can be observed that the absence of fading leads to a faster convergence speed, which aligns with our intuitive understanding that fading randomness may slow down convergence. Additionally, the output case exhibits a slightly faster convergence rate compared to the input case due to fluctuations caused by fading at the input signal, which deteriorate tracking performance.

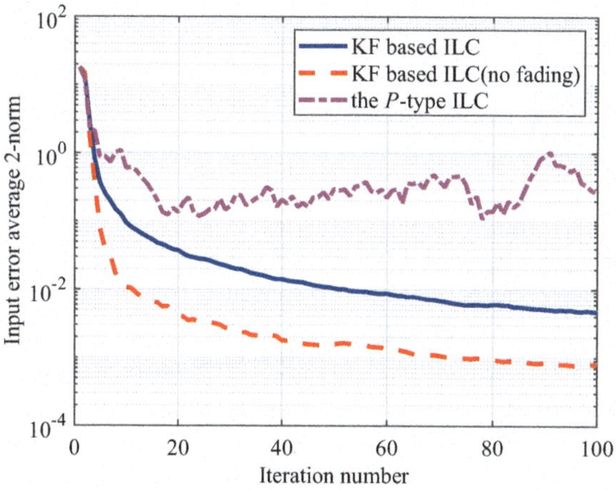

(a)

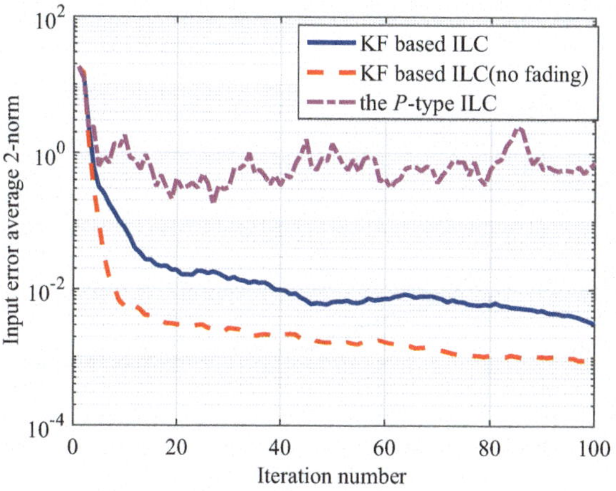

(b)

Fig. 6.18 Evolution of the averaged mean-square input error profile along the iteration axis for different learning control algorithms: **a** output fading case; **b** input fading case

Furthermore, we are interested in investigating the potential impact of different fading distributions on convergence. To address this issue, we examine three fading gain distributions with the following parameters: $\mu = 0.95$ and $\sigma = 0.1$, $\mu = 0.85$ and $\sigma = 0.1$, and $\mu = 0.95$ and $\sigma = 0.05$. The corresponding output and input fading cases are illustrated in Fig. 6.19a, b, respectively. It is evident that altering the mean value of the fading gain does not significantly affect convergence performance;

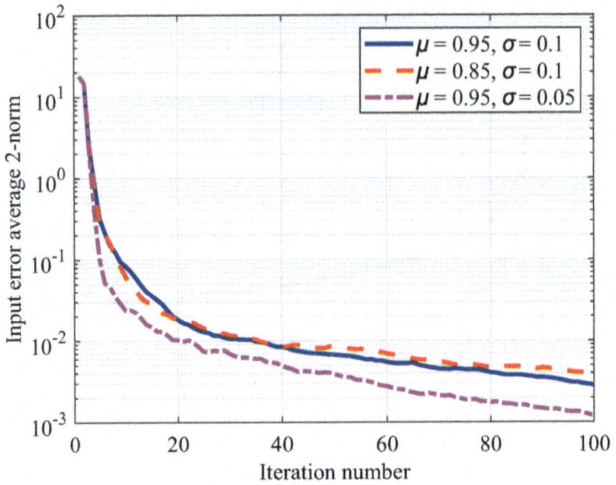

(a)

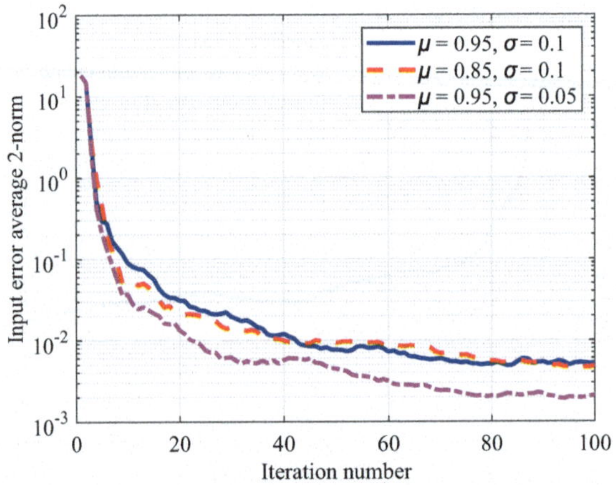

(b)

Fig. 6.19 Averaged input error profiles for different fading parameters: **a** output fading case; **b** input fading case

however, modifying the variance can substantially influence both convergence speed and tracking precision. Notably, a smaller variance of fading gains leads to superior algorithmic performance in our experiments, which aligns with intuitive expectations.

References

Ahn HS, Chen Y (2009) Iterative learning control for multi-agent formation. In: 2009 ICCAS-SICE, pp 3111–3116

Ahn HS, Moore KL, Chen Y (2007) Iterative learning control: robustness and monotonic convergence for interval systems. Springer (2007)

Ahn HS, Moore KL, Chen Y (2008) Discrete-time intermittent iterative learning controller with independent data dropouts. IFAC Proc 41(2):12442–12447

Arimoto S, Kawamura S, Miyazaki F (1984) Bettering operation of robots by learning. J Robot Syst 1(2):123–140

Aysal TC, Coates MJ, Rabbat MG (2008) Distributed average consensus with dithered quantization. IEEE Trans Signal Process 56(10):4905–4918

Bejarano G, Rodríguez D, Lemos JM, Vargas M, Ortega MG (2020) MINLP-based hybrid strategy for operating mode selection of TES-backed-up refrigeration systems. Int Jo Robust Nonlinear Control 30(15):6091–6111

Benaim M (1996) A dynamical system approach to stochastic approximations. SIAM J Control Optim 34(2):437–472

Bensidhoum T, Bouakrif F (2020) Adaptive p-type iterative learning radial basis function control for robot manipulators with unknown varying disturbances and unknown input dead zone. Int J Robust Nonlinear Control 30(10):4075–4094

Bernstein DS, Michel AN (1995) A chronological bibliography on saturating actuators. In J Robust Nonlinear Control 5(5):375–380

Bhaskoro PT, Gilani SIUH, Aris MS (2013) Simulation of energy saving potential of a centralized HVAC system in an academic building using adaptive cooling technique. Energy Convers Manage 75:617–628

Bien Z, Xu JX (2012) Iterative learning control: analysis, design. Springer, Integration and applications

Borkar VS (2009) Stochastic approximation: a dynamical systems viewpoint. Springer (2009)

Boyd S, El Ghaoui L, Feron E, Balakrishnan V (1994) Linear matrix inequalities in system and control theory. Soc Ind Appl Math

Bristow D, Tharayil M, Alleyne A (2006) A survey of iterative learning control. IEEE Control Syst Mag 26(3):96–114

Bu X, Wang T, Hou Z, Chi R (2015) Iterative learning control for discrete-time systems with quantised measurements. IET Control Theory Appl 9(9):1455–1460

D. Shen, *Variable Gain Design in Stochastic Iterative Learning Control*, Intelligent Control and Learning Systems 13,
https://doi.org/10.1007/978-981-97-8281-9

Bu X, Hou Z, Cui L, Yang J (2017) Stability analysis of quantized iterative learning control systems using lifting representation. Int J Adaptive Control Signal Process 31(9):1327–1336

Butcher M, Karimi A, Longchamp R (2008) Iterative learning control based on stochastic approximation. IFAC Proc 41(2):1478–1483

Butcher M, Karimi A (2011) Iterative learning control using stochastic approximation theory with application to a mechatronic system. In: Advances in the theory of control, Signals and systems with physical modeling. Springer, pp 49–64

Campi MC, Sugie T, Sakai F (2008) An iterative identification method for linear continuous-time systems. IEEE Trans Autom Control 53(7):1661–1669

Canudas de Wit C, Noel P, Aubin A, Brogliato B (1991) Adaptive friction compensation in robot manipulators: low velocities. Int J Robot Res 10(3):189–199

Cao YY, Lin Z, Chen BM (2003) An output feedback h_{∞} controller design for linear systems subject to sensor nonlinearities. IEEE Trans Circ Syst I: Fundam Theory Appl 50(7):914–921

Cao Z, Lu J, Zhang R, Gao F (2016) Iterative learning Kalman filter for repetitive processes. J Process Control 46:92–104

Cao Z, Zhang R, Yang Y, Lu J, Gao F (2016) Discrete-time robust iterative learning Kalman filtering for repetitive processes. IEEE Tran Autom Control 61(1):270–275

Chau CK, Qin F, Sayed S, Wahab MH, Yang Y (2010) Harnessing battery recovery effect in wireless sensor networks: experiments and analysis. IEEE J Sel Areas Commun 28(7):1222–1232

Chen BM, Lin Z, Shamash Y (2012) Linear systems theory: a structural decomposition approach. Springer (2012)

Chen G (1991) On a trajectory tracking problem for nonlinear control systems. In: Proceedings of the 30th IEEE conference on decision and control, vol 3, pp 2435–2440

Chen HF, Guo L (1991) Identification and stochastic adaptive control, vol 5. Springer (1991)

Chen HF (2003) Stochastic approximation and its applications. Kluwer Academic Publishers, Dordrecht

Chen Y, Wen C, Gong Z, Sun M (1999) An iterative learning controller with initial state learning. IEEE Trans Autom Control 44(2):371–376

Chen Y, Lü J, Han F, Yu X (2011) On the cluster consensus of discrete-time multi-agent systems. Syste Control Lett 60(7):517–523

Cheng L, Hou ZG, Tan M, Lin Y, Zhang W (2010) Neural-network-based adaptive leader-following control for multiagent systems with uncertainties. IEEE Trans Neural Netw 21(8):1351–1358

Cheng X, Jiang H, Shen D, Yu X (2023) A novel adaptive gain strategy for stochastic learning control. IEEE Trans Cybern 53(8):5264–5275

Chi R, Lv Y, Huang B (2020) Distributed iterative learning temperature control for multi-zone HVAC system. J Franklin Inst 357(2):810–831

Chi R, Hui Y, Wang R, Huang B, Hou Z (2022) Discrete-time-distributed adaptive ILC with non-repetitive uncertainties and applications to building HVAC systems. IEEE Trans Syst Man Cybern: Syst 52(8):5068–5080

Chi R, Zhang H, Huang B, Hou Z (2022) Quantitative data-driven adaptive iterative learning control: from trajectory tracking to point-to-point tracking. IEEE Trans Cybern 52(6):4859–4873

Chow YS, Teicher H (2003) Probability theory: independence, Interchangeability, martingales. Springer

de Dear RJ, Akimoto T, Arens EA, Brager G, Candido C, Cheong K, Li B, Nishihara N, Sekhar S, Tanabe S et al (2013) Progress in thermal comfort research over the last twenty years. Indoor Air 23(6):442–461

Deb K (2008) Multi-objective optimization using evolutionary algorithms. Wiley

Delyon B, Juditsky A (1993) Accelerated stochastic approximation. SIAM J Optim 3(4):868–881

Deng C, Yang GH (2017) Decentralized fault-tolerant control for a class of nonlinear large-scale systems with actuator faults. Inf Sci 382–383:334–349

Dey S, Leong AS, Evans JS (2009) Kalman filtering with faded measurements. Automatica 45(10):2223–2233

Dhar NK, Verma NK, Behera L (2020) An online event-triggered near-optimal controller for Nash solution in interconnected system. IEEE Trans Neural Netw Learn Syst 31(12):5534–5548. https://doi.org/10.1109/TNNLS.2020.2969249
Ding H, Wu J (2007) Point-to-point motion control for a high-acceleration positioning table via cascaded learning schemes. IEEE Trans Ind Electron 54(5):2735–2744
Du DZ, Pardalos PM, Wu W (2013) Mathematical theory of optimization, vol 56. Springer
Durrant-Whyte HF (2012) Integration, coordination and control of multi-sensor robot systems, vol 36. Springer
Elia N (2005) Remote stabilization over fading channels. Syst Control Lett 54(3):237–249
Emmerich MT, Deutz AH (2018) A tutorial on multiobjective optimization: fundamentals and evolutionary methods. Natural Comput 17:585–609
Freeman CT (2012) Constrained point-to-point iterative learning control with experimental verification. Control Eng Practice 20(5):489–498
Freeman CT, Tan Y (2013) Iterative learning control with mixed constraints for point-to-point tracking. IEEE Trans Control Syst Technol 21(3):604–616
Freeman CT, Cai Z, Rogers E, Lewin PL (2011) Iterative learning control for multiple point-to-point tracking application. IEEE Trans Control Syst Technolol 19(3):590–600
Freeman CT, Rogers E, Burridge JH, Hughes AM, Meadmore KL (2015) Iterative learning control for electrical stimulation and stroke rehabilitation. Springer
Freeman C, Tan Y (2011) Point-to-point iterative learning control with mixed constraints. In: Proceedings of the 2011 American control conference, pp 3657–3662
Fu Q, Gu P, Wu J (2017) Decentralized iterative learning control for large-scale interconnected linear systems with fixed initial shifts. Int J Control Autom Syst 15(2):1–10
Geng H, Wang Z, Liang Y, Cheng Y, Alsaadi FE (2017) Tobit Kalman filter with fading measurements. Signal Process 140:60–68
Gholamzadehmir M, Del Pero C, Buffa S, Fedrizzi R et al (2020) Adaptive-predictive control strategy for HAVC systems in smart buildings-A review. Sustain Cities Soc 63(102):480
Grip HF, Saberi A, Wang X (2010) Stabilization of multiple-input multiple-output linear systems with saturated outputs. IEEE Trans Autom Control 55(9):2160–2164
Hall P, Heyde CC (1980) Martingale limit theory and its applications. Academic Press
Han Y, Lu W, Chen T (2013) Cluster consensus in discrete-time networks of multiagents with inter-cluster nonidentical inputs. IEEE Trans Neural Netw Learn Syst 24(4):566–578
Han QL, Liu Y, Yang F (2016) Optimal communication network-based H-∞ quantized control with packet dropouts for a class of discrete-time neural networks with distributed time delay. IEEE Trans Neural Netw Learn Syst 27(2):426–434
Horn RA, Johnson CR (1985) Matrix analysis. Cambridge University Press
Hou Z, Xu JX, Yan J (2008) An iterative learning approach for density control of freeway traffic flow via ramp metering. Transp Res Part C: Emerg Technol 16(1):71–97
Hu J, Hong Y (2007) Leader-following coordination of multi-agent systems with coupling time delays. Physica A: Stat Mech Appl 374(2):853–863
Huang YS, Wang ZY (2014) Decentralized adaptive fuzzy control for a class of large-scale MIMO nonlinear systems with strong interconnection and its application to automated highway systems. Inf Sci 274:210–224
Huang S, Tan KK, Lee TH (2002) Necessary and sufficient condition for convergence of iterative learning algorithm. Automatica 38(7):1257–1260
Huo N, Shen D (2020) Encoding-decoding mechanism-based finite-level quantized iterative learning control with random data dropouts. IEEE Trans Autom Sci Eng 17(3):1343–1360
Hwang DH, Kim BK, Bien Z (1993) Decentralized iterative learning control methods for large scale linear dynamic systems. Int J Syst Sci 24(12):2239–2254
Ji H, Hou Z, Zhang R (2016) Adaptive iterative learning control for high-speed trains with unknown speed delays and input saturations. IEEE Trans Autom Sci Eng 13(1):260–273

Johnston AP, Yuksel S (2014) Stochastic stabilization of partially observed and multi-sensor systems driven by unbounded noise under fixed-rate information constraints. IEEE Trans Autom Control 59(3):792–798

Karimi B, Menhaj MB (2010) Non-affine nonlinear adaptive control of decentralized large-scale systems using neural networks. Inf Sci 180(17):3335–3347

Kelman A, Borrelli F (2011) Bilinear model predictive control of a HVAC system using sequential quadratic programming. In: 18th IFAC World congress. Milano, Italy

Kesten H (1958) Accelerated stochastic approximation. Ann Math Stat 41–59

Khaleghi B, Khamis A, Karray FO, Razavi SN (2013) Multisensor data fusion: a review of the state-of-the-art. Inf Fusion 14(1):28–44

Kolarik J, Toftum J, Olesen BW, Jensen KL (2011) Simulation of energy use, human thermal comfort and office work performance in buildings with moderately drifting operative temperatures. Energy Build 43(11):2988–2997

Kong K, Kniep HC, Tomizuka M (2010) Output saturation in electric motor systems: identification and controller design. J Dyn Syst Measure Control 132(5):051,002

Kreisselmeier G (1996) Stabilization of linear systems in the presence of output measurement saturation. Syst Control Lett 29(1):27–30

Li J, Hu ZC (2017) Toeplitz lemma, complete convergence, and complete moment convergence. Commun Stat-Theory Methods 46(4):1731–1743

Li X, Shen D (2017) Two novel iterative learning control schemes for systems with randomly varying trial lengths. Syst Control Lett 107:9–16

Li T, Zhang JF (2010) Consensus conditions of multi-agent systems with time-varying topologies and stochastic communication noises. IEEE Trans Autom Control 55(9):2043–2057

Li X, Zhu Y, Wang J, Han C (2003) Optimal linear estimation fusion-Part I: unified fusion rules. IEEE Trans Inf Theory 49(9):2192–2208

Li X, Xu JX, Huang D (2014) An iterative learning control approach for linear systems with randomly varying trial lengths. IEEE Trans Autom Control 59(7):1954–1960

Li X, Xu JX, Huang D (2015) Iterative learning control for nonlinear dynamic systems with randomly varying trial lengths. Int J Adaptive Control Signal Process 29(11):1341–1353

Li X, Wen J, Bai EW (2016) Developing a whole building cooling energy forecasting model for on-line operation optimization using proactive system identification. Appl Energy 164:69–88

Li L, Liu Y, Li L, Tan J (2019) Kalman-filtering-based iterative feedforward tuning in presence of stochastic noise: with application to a wafer stage. IEEE Trans Ind Inf 15(11):5816–5826. https://doi.org/10.1109/TII.2019.2906331

Li J, Ding J, Chai T, Lewis FL (2020) Nonzero-sum game reinforcement learning for performance optimization in large-scale industrial processes. IEEE Trans Cybern 50(9):4132–4145

Li Y, Liu Y, Tong S (2021) Observer-based neuro-adaptive optimized control of strict-feedback nonlinear systems with state constraints. IEEE Trans Neural Netw Learn Syst 1–15

Lin Z, Hu T (2001) Semi-global stabilization of linear systems subject to output saturation. Syst Control Lett 43(3):211–217

Liu B, Chen T (2008) Consensus in networks of multiagents with cooperation and competition via stochastically switching topologies. IEEE Trans Neural Netw 19(11):1967–1973

Liu H, Xie G, Wang L (2012) Containment of linear multi-agent systems under general interaction topologies. Syst Control Lett 61(4):528–534

Liu YJ, Tang L, Tong S, Chen CP, Li DJ (2014) Reinforcement learning design-based adaptive tracking control with less learning parameters for nonlinear discrete-time mimo systems. IEEE Trans Neural Netw Learn Syst 26(1):165–176

Li W, Xiao Y, Bi S, Du G (2013) Automatic elliptical trajectory planning algorithm for pick and place operation. In: Proceedings of the 2013 international conference on advanced mechatronic systems, pp 36–39

Longman R, Mombaur K (2006) Investigating the use of iterative learning control and repetitive control to implement periodic gaits. In: Fast motions in biomechanics and robotics: optimization and feedback control. Springer, pp 189–218

Loria A (2016) Observers are unnecessary for output-feedback control of Lagrangian systems. IEEE Trans Autom Control 61(4):905–920

Ma H, Zhou Q, Li H, Lu R (2022) Adaptive prescribed performance control of a flexible-joint robotic manipulator with dynamic uncertainties. IEEE Trans Cybern 52(12):12905–12915

Mahmoud MS, Hamdan MM (2018) Fundamental issues in networked control systems. IEEE/CAA J Automatica Sinica 5(5):902–922

Mei J, Lu Z, Hu J, Fan Y (2021) Energy-efficient optimal guaranteed cost intermittent-switch control of a direct expansion air conditioning system. IEEE/CAA J Automatica Sinica 8(11):1852–1866

Meng D, Zhang J (2018) Deterministic convergence for learning control systems over iteration-dependent tracking intervals. IEEE Trans Neural Netw Learn Syst 29(8):3885–3892

Meng D, Jia Y, Du J, Yu F (2013) Tracking algorithms for multiagent systems. IEEE Trans Neural Netw Learn Syst 24(10):1660–1676

Meng D, Jia Y, Du J, Zhang J (2015) High-precision formation control of nonlinear multi-agent systems with switching topologies: A learning approach. Int J Robust Nonlinear Control 25(13):1993–2018

Meyn SP, Tweedie RL (2012) Markov chains and stochastic stability. Springer

Minakais M, Mishra S, Wen JT (2019) Database-driven iterative learning for building temperature control. IEEE Trans Autom Sci Eng 16(4):1896–1906

Mitra A, Das N, Samant RN, Behera L (2016) Control of a 4 DoF Barrett Wam robot—Modeling, control synthesis and experimental validation. In: 2016 IEEE First international conference on control, measurement and instrumentation (CMI), pp 397–402

Moore KL (2012) Iterative learning control for deterministic systems (2012)

Moradvandi A, Shahrokhi M, Malek SA (2019) Adaptive fuzzy decentralized control for a class of mimo large-scale nonlinear state delay systems with unmodeled dynamics subject to unknown input saturation and infinite number of actuator failures. Inf Sci 475:121–141

Ni W, Cheng D (2010) Leader-following consensus of multi-agent systems under fixed and switching topologies. Syst control Lett 59(3–4):209–217

Owens DH (2016) Iterative learning control: an optimization paradigm. Springer

Owens DH, Freeman CT, Van Dinh T (2013) Norm-optimal iterative learning control with intermediate point weighting: theory, algorithms, and experimental evaluation. IEEE Trans Control Syst Technol 21(3):999–1007

Papadopoulos PM, Reppa V, Polycarpou MM, Panayiotou CG (2020) Scalable distributed sensor fault diagnosis for smart buildings. IEEE/CAA J Autom Sinica 7(3):638–655

Park J, Chang PH, Park HS, Lee E (2006) Design of learning input shaping technique for residual vibration suppression in an industrial robot. IEEE/ASME Trans Mechatron 11(1):55–65

Peeters L, Van der Veken J, Hens H, Helsen L, D'haeseleer W (2008) Control of heating systems in residential buildings: current practice. Energy Build 40(8):1446–1455

Peng C, Tian E, Zhang J, Du D (2017) Decentralized event-triggering communication scheme for large-scale systems under network environments. Inf Sci 380:132–144

Peng C, Sun L, Tomizuka M (2017) Distributed and cooperative optimization-based iterative learning control for large-scale building temperature regulation. In: 2017 IEEE international conference on advanced intelligent mechatronics (AIM), pp 1606–1611

Peng C, Zhang W, Tomizuka M (2016) Iterative design of feedback and feedforward controller with input saturation constraint for building temperature control. In: 2016 American control conference (ACC), pp 1241–1246

Perez-Lombard L, Ortiz J, Pout C (2008) A review on buildings energy consumption information. Energy Build 40(3):394–398

Pipeleers G, Moore KL (2014) Unified analysis of iterative learning and repetitive controllers in trial domain. IEEE Trans Autom Control 59(4):953–965

Polyak BT (1987) Introduction to optimization. Optimization Software Inc., Publications Division, New York

Qu G, Shen D, Yu X (2023) Batch-based learning consensus of multiagent systems with faded neighborhood information. IEEE Trans Neural Netw Learn Syst 34(6):2965–2977

Ramamurthy H, Prabhu B, Gadh R, Madni AM (2007) Wireless industrial monitoring and control using a smart sensor platform. IEEE Sens J 7(5):611–618

Ren H, Wang Y, Liu M, Li H (2022) An optimal estimation framework of multi-agent systems with random transport protocol. IEEE Trans Signal Process 70(5):2548–2559

Rousseau D, Rojas Varela J, Chapeau-Blondeau F (2003) Stochastic resonance for nonlinear sensors with saturation. Phys Rev E 67(021):102

Ruan X, Bien ZZ, Park KH (2008) Decentralized iterative learning control to large-scale industrial processes for nonrepetitive trajectory tracking. IEEE Trans Syst Man Cybern-Part A: Syst Humans 38(1):238–252

Ruan X, Wu H, Li N, Wan B (2009) Convergence analysis in sense of Lebesgue-p norm of decentralized non-repetitive iterative learning control for linear large-scale systems. J Syst Sci Complex 22(3):422–434

Saab SS (2001) A discrete-time stochastic learning control algorithm. IEEE Trans Autom Control 46(6):877–887

Saab SS (2001) On a discrete-time stochastic learning control algorithm. IEEE Trans Autom Control 46(8):1333–1336

Saab SS (2003) Stochastic p-type/d-type iterative learning control algorithms. Int J Control 76(2):139–148

Saab SS (2005) Optimal selection of the forgetting matrix into an iterative learning control algorithm. IEEE Trans Autom Control 50(12):2039–2043

Saab SS, Ghanem P (2018) A multivariable stochastic tracking controller for robot manipulators without joint velocities. IEEE Trans Autom Control 63(8):2481–2495

Schuchman L (1964) Dither signals and their effect on quantization noise. IEEE Trans Communi Technol 12(4):162–165

Seel T, Schauer T, Raisch J (2011) Iterative learning control for variable pass length systems. IFAC Proc 44(1):4880–4885

Seel T, Werner C, Raisch J, Schauer T (2016a) Iterative learning control of a drop foot neuroprosthesis-generating physiological foot motion in paretic gait by automatic feedback control. Control Eng Practice 48:87–97

Seel T, Werner C, Schauer T (2016b) The adaptive drop foot stimulator-multivariable learning control of foot pitch and roll motion in paretic gait. Med Eng Phys 38(11):1205–1213

Shao X, Shi Y (2020) Neural adaptive control for MEMS gyroscope with full-state constraints and quantized input. IEEE Trans Ind Inf 16(10):6444–6454

Shen D (2018) Data-driven learning control for stochastic nonlinear systems: multiple communication constraints and limited storage. IEEE Trans Neural Netw Learn Syst 29(6):2429–2440

Shen D (2018) Iterative learning control with passive incomplete information: algorithm design and convergence analysis. Springer

Shen D, Chen HF (2012) Iterative learning control for large scale nonlinear systems with observation noise. Automatica 48(3):577–582

Shen D, Li X (2019) Iterative Learning control for systems with iteration-varying trial lengths: synthesis and analysis. Springer

Shen D, Qu G (2020) Performance enhancement of learning tracking systems over fading channels with multiplicative and additive randomness. IEEE Trans Neural Netw Learn Syst 31(4):1196–1210

Shen D, Wang Y (2014) Survey on stochastic iterative learning control. J Process Control 24(12):64–77

Shen D, Wang Y (2015) Ilc for networked nonlinear systems with unknown control direction through random lossy channel. Syst Control Lett 77:30–39

Shen D, Xu Y (2016) Iterative learning control for discrete-time stochastic systems with quantized information. IEEE/CAA J Autom Sinica 3(1):59–67

Shen D, Xu JX (2017) A novel Markov chain based ILC analysis for linear stochastic systems under general data dropouts environments. IEEE Trans Autom Control 62(11):5850–5857

Shen D, Xu JX (2018) Distributed learning consensus for heterogenous high-order nonlinear multi-agent systems with output constraints. Automatica 97:64–72

Shen D, Xu JX (2019) Adaptive learning control for nonlinear systems with randomly varying iteration lengths. IEEE Trans Neural Netw Learn Systems 30(4):1119–1132

Shen D, Xu JX (2019) Robust learning control for nonlinear systems with nonparametric uncertainties and nonuniform trial lengths. Int J Robust Nonlinear Control 29(5):1302–1324

Shen D, Xu JX (2020) An iterative learning control algorithm with gain adaptation for stochastic systems. IEEE Trans Autom Control 65(3):1280–1287

Shen D, Yu X (2022) Learning tracking over unknown fading channels based on iterative estimation. IEEE Trans Neural Netw Learn Syst 33(1):48–60

Shen D, Yu X (2023) Iterative learning control over random fading channels. CRC Press

Shen D, Zhang C (2022) Zero-error tracking control under unified quantized iterative learning framework via encoding-decoding method. IEEE Trans Cybern 52(4):1979–1991

Shen D, Zhang W, Wang Y, Chien CJ (2016a) On almost sure and mean square convergence of p-type ILC under randomly varying iteration lengths. Automatica 63:359–365

Shen D, Zhang W, Xu JX (2016b) Iterative learning control for discrete nonlinear systems with randomly iteration varying lengths. Syst Control Lett 96:81–87

Shen D, Han J, Wang Y (2017) Stochastic point-to-point iterative learning tracking without prior information on system matrices. IEEE Trans Autom Sci Eng 14(1):376–382

Shen D, Liu C, Wang L, Yu X (2021) Iterative learning tracking for multisensor systems: a weighted optimization approach. IEEE Trans Cybern 51(3):1286–1299

Shen D, Qu G, Yu X (2021) Averaging techniques for balancing learning and tracking abilities over fading channels. IEEE Trans Autom Control 66(6):2636–2651

Shen D, Huo N, Saab SS (2022) A probabilistically quantized learning control framework for networked linear systems. IEEE Tran Neural Netw Learn Syst 33(12):7559–7573

Shi J, He X, Zhou D (2016) Iterative learning control for nonlinear stochastic systems with variable pass length. J Franklin Inst 353(15):4016–4038

Son TD, Ahn HS (2011) Terminal iterative learning control with multiple intermediate pass points. In: Proceedings of the 2011 American control conference. IEEE, pp 3651–3656

Son TD, Ahn HS, Moore KL (2013) Iterative learning control in optimal tracking problems with specified data points. Automatica 49(5):1465–1472

Song E, Zhu Y, Zhou J, You Z (2007) Optimal Kalman filtering fusion with cross-correlated sensor noises. Automatica 43(8):1450–1456

Su L, Chesi G (2017) Robust stability analysis and synthesis for uncertain discrete-time networked control systems over fading channels. IEEE Trans Autom Control 62(4):1966–1971

Su J, Wang J, Xi Y (2004) Incremental learning with balanced update on receptive fields for multi-sensor data fusion. IEEE Trans Syst Man Cybern Part B (Cybernetics) 34(1):659–665

Sui S, Tong S, Li Y (2014) Observer-based adaptive fuzzy decentralized control for stochastic large-scale nonlinear systems with unknown dead-zones. Inf Sci 259:71–86

Sun S, Lin H, Ma J, Li X (2017) Multi-sensor distributed fusion estimation with applications in networked systems: a review paper. Inf Fusion 38:122–134

Tan Y, Xu JX, Norrlöf M, Freeman C (2011) On reference governor in iterative learning control for dynamic systems with input saturation. Automatica 47(11):2412–2419

Thomas B, Soleimani-Mohseni M, Fahlen P (2005) Feed-forward in temperature control of buildings. Energy Build 37(7):755–761

Tong S, Li Y, Jing X (2013) Adaptive fuzzy decentralized dynamics surface control for nonlinear large-scale systems based on high-gain observer. Inf Sci 235:287–307

Tong S, Sun K, Sui S (2018) Observer-based adaptive fuzzy decentralized optimal control design for strict-feedback nonlinear large-scale systems. IEEE Trans Fuzzy Syst 26(2):569–584

Uchiyama M (1978) Formation of high-speed motion pattern of a mechanical arm by trial. Trans So Instrum Control Eng 14(6):706–712

Van de Wijdeven J, Bosgra O (2008) Residual vibration suppression using Hankel iterative learning control. Int J Robust Nonlinear Control: IFAC-Affil J 18(10):1034–1051

Van de Wijdeven J, Donkers T, Bosgra O (2009) Iterative learning control for uncertain systems: robust monotonic convergence analysis. Automatica 45(10):2383–2391

Van Pham T, Nguyen DH, Banjerdpongchai D (2017) Decentralized iterative learning control of building temperature control system. In, (2017) SICE International symposium on control systems (SICE ISCS). Okayama, Japan

Venkat AN, Hiskens IA, Rawlings JB, Wright SJ (2008) Distributed MPC strategies with application to power system automatic generation control. IEEE Trans Control Syst Technol 16(6):1192–1206

Wang YC, Chien CJ (2011) Decentralized adaptive fuzzy neural iterative learning control for non-affine nonlinear interconnected systems. Asian J Control 13(1):94–106

Wang Z, Hu G (2018) Economic MPC of nonlinear systems with nonmonotonic Lyapunov functions and its application to HVAC control. Int J Robust Nonlinear Control 28(6):2513–2527

Wang D, Ye Y, Zhang B (2014) Practical iterative learning control with frequency domain design and sampled data implementation. Springer

Wang N, Gao Y, Zhang X (2021) Data-driven performance-prescribed reinforcement learning control of an unmanned surface vehicle. IEEE Trans Neural Netw Learn Syst 32(12):5456–5467

Wang N, Gao Y, Zhao H, Ahn CK (2021) Reinforcement learning-based optimal tracking control of an unknown unmanned surface vehicle. IEEE Trans Neural Netw Learn Syst 32(7):3034–3045

Wang X, Ke L, Qiao Z, Chai X (2021) Large-scale traffic signal control using a novel multiagent reinforcement learning. IEEE Trans Cybern 51(1):174–187

Wang Y, Xiong J, Ho DWC (2022) Distributed LMMSE estimation for large-scale systems based on local information. IEEE Trans Cybern 52(8):8528–8536

Wang Z, Shen B, Liu X (2024) H_∞ filtering with randomly occurring sensor saturations and missing measurements

Wannamaker RA, Lipshitz SP, Vanderkooy J, Wright JN (2000) A theory of nonsubtractive dither. IEEE Trans Signal Process 48(2):499–516

Wu H (2007) Decentralized iterative learning control for a class of large scale interconnected dynamical systems. J Math Anal Appl 327(1):233–245

Xiao N, Xie L, Qiu L (2012) Feedback stabilization of discrete-time networked systems over fading channels. IEEE Trans Autom Control 57(9):2176–2189

Xu JX (1997) Analysis of iterative learning control for a class of nonlinear discrete-time systems. Automatica 33(10):1905–1907

Xu JX, Jin X (2013) State-constrained iterative learning control for a class of MIMO systems. IEEE Trans Autom Control 58(5):1322–1327

Xu JX, Tan Y (2003) Linear and nonlinear iterative learning control, vol 291. Springer

Xu JX, Yan R (2005) On initial conditions in iterative learning control. IEEE Trans Autom Control 50(9):1349–1354

Xu JX, Tan Y, Lee TH (2004) Iterative learning control design based on composite energy function with input saturation. Automatica 40(8):1371–1377

Xu L, Mo Y, Xie L, Xiao N (2017) Mean square stabilization of linear discrete-time systems over power-constrained fading channels. IEEE Trans Autom Control 62(12):6505–6512

Xu Y, Lu R, Peng H, Xie K, Xue A (2017) Asynchronous dissipative state estimation for stochastic complex networks with quantized jumping coupling and uncertain measurements. IEEE Trans Neural Netw Learn Syst 28(2):268–277

Xu Y, Shen D, Bu X (2017) Zero-error convergence of iterative learning control using quantized error information. IMA J Math Control Inf 34(3):1061–1077

Xu Y, Shen D, Zhang XD (2017) Stochastic point-to-point iterative learning control based on stochastic approximation. Asian J Control 19(5):1748–1755

Yang F, Li Y (2009) Set-membership filtering for systems with sensor saturation. Automatica 45(8):1896–1902

Yang S, Xu JX, Li X, Shen D (2017) Iterative learning control for multi-agent systems coordination. Wiley

Yang Y, Hu G, Spanos CJ (2020) HVAC energy cost optimization for a multizone building via a decentralized approach. IEEE Trans Autom Sci Eng 17(4):1950–1960

Yang Y, Srinivasan S, Hu G, Spanos CJ (2021) Distributed control of multizone HVAC systems considering indoor air quality. IEEE Trans Control Syst Technol 29(6):2586–2597

Yang S, Xu JX (2012) Adaptive iterative learning control for multi-agent systems consensus tracking. In: 2012 IEEE international conference on systems, man, and cybernetics (SMC), pp 2803–2808

Yousefian F, Nedić A, Shanbhag UV (2012) On stochastic gradient and subgradient methods with adaptive steplength sequences. Automatica 48(1):56–67

Yu L, Xie D, Jiang T, Zou Y, Wang K (2018) Distributed real-time HVAC control for cost-efficient commercial buildings under smart grid environment. IEEE Internet of Things J 5(1):44–55

Zhang YD, Fang YC (2011) Learning control for systems with saturated output. Acta Automatica Sinica 37(1):92–98

Zhang J, Meng D (2023) Iterative rectifying methods for nonrepetitive continuous-time learning control systems. IEEE Trans Cybern 53(1):338–351

Zhang C, Shen D (2018) Zero-error convergence of iterative learning control based on uniform quantisation with encoding and decoding mechanism. IET Control Theory Appl 12(14):1907–1915

Zhang T, Xia X (2015) Decentralized adaptive fuzzy output feedback control of stochastic nonlinear large-scale systems with dynamic uncertainties. Inf Sci 315:17–38

Zhang B, Ye Y, Zhou K, Wang D (2014) Case studies of filtering techniques in multirate iterative learning control. Control Eng Practice 26:116–124

Zhang R, Hou Z, Chi R, Ji H (2015) Adaptive iterative learning control for nonlinearly parameterised systems with unknown time-varying delays and input saturations. Int J Control 88(6):1133–1141

Zhang R, Hou Z, Ji H, Yin C (2016) Adaptive iterative learning control for a class of non-linearly parameterised systems with input saturations. Int J Syst Sci 47(5):1084–1094

Zhang XM, Han QL, Yu X (2016) Survey on recent advances in networked control systems. IEEE Trans Ind Inf 12(5):1740–1752

Zhang D, Shi P, Wang QG, Yu L (2017) Analysis and synthesis of networked control systems: a survey of recent advances and challenges. ISA Trans 66:376–392

Zhang L, Yin X, Ning Z, Ye D (2017) Robust filtering for a class of networked nonlinear systems with switching communication channels. IEEE Trans Cybern 47(3):671–682

Zhang X, Ou X, Li Z, Chen X, Su CY (2022) All state constrained decentralized adaptive implicit inversion control for a class of large scale nonlinear hysteretic systems with time-delays. Inf Sci 588:52–66

Zhu S, Chen B (2016) Quantized consensus by the ADMM: probabilistic versus deterministic quantizers. IEEE Trans Signal Process 64(7):1700–1713

Index

D. Shen, *Variable Gain Design in Stochastic Iterative Learning Control*, Intelligent Control and Learning Systems 13,
https://doi.org/10.1007/978-981-97-8281-9

The manufacturer's authorised representative in the EU is Springer Nature Customer Service Centre GmbH, Europaplatz 3, 69115 Heidelberg, Germany. If you have any concerns regarding our products, please contact ProductSafety@springernature.com

Printed and bound by CPI Group (UK) Ltd, Croydon, CR0 4YY

01/10/2026

02225934-0014